T0181170

Advances in Intelligent Systems and Computing

Volume 458

Series editor

Janusz Kacprzyk, Polish Academy of Sciences, Warsaw, Poland
e-mail: kacprzyk@ibspan.waw.pl

About this Series

The series "Advances in Intelligent Systems and Computing" contains publications on theory, applications, and design methods of Intelligent Systems and Intelligent Computing. Virtually all disciplines such as engineering, natural sciences, computer and information science, ICT, economics, business, e-commerce, environment, healthcare, life science are covered. The list of topics spans all the areas of modern intelligent systems and computing.

The publications within "Advances in Intelligent Systems and Computing" are primarily textbooks and proceedings of important conferences, symposia and congresses. They cover significant recent developments in the field, both of a foundational and applicable character. An important characteristic feature of the series is the short publication time and world-wide distribution. This permits a rapid and broad dissemination of research results.

Advisory Board

Chairman

Nikhil R. Pal, Indian Statistical Institute, Kolkata, India
e-mail: nikhil@isical.ac.in

Members

Rafael Bello, Universidad Central "Marta Abreu" de Las Villas, Santa Clara, Cuba
e-mail: rbellop@uclv.edu.cu

Emilio S. Corchado, University of Salamanca, Salamanca, Spain
e-mail: escorchado@usal.es

Hani Hagras, University of Essex, Colchester, UK
e-mail: hani@essex.ac.uk

László T. Kóczy, Széchenyi István University, Győr, Hungary
e-mail: koczy@sze.hu

Vladik Kreinovich, University of Texas at El Paso, El Paso, USA
e-mail: vladik@utep.edu

Chin-Teng Lin, National Chiao Tung University, Hsinchu, Taiwan
e-mail: ctlin@mail.nctu.edu.tw

Jie Lu, University of Technology, Sydney, Australia
e-mail: Jie.Lu@uts.edu.au

Patricia Melin, Tijuana Institute of Technology, Tijuana, Mexico
e-mail: epmelin@hafsamx.org

Nadia Nedjah, State University of Rio de Janeiro, Rio de Janeiro, Brazil
e-mail: nadia@eng.uerj.br

Ngoc Thanh Nguyen, Wroclaw University of Technology, Wroclaw, Poland
e-mail: Ngoc-Thanh.Nguyen@pwr.edu.pl

Jun Wang, The Chinese University of Hong Kong, Shatin, Hong Kong
e-mail: jwang@mae.cuhk.edu.hk

More information about this series at http://www.springer.com/series/11156

Jyotsna Kumar Mandal · Suresh Chandra Satapathy
Manas Kumar Sanyal · Vikrant Bhateja
Editors

Proceedings of the First International Conference on Intelligent Computing and Communication

Editors
Jyotsna Kumar Mandal
Department of CSE
University of Kalyani
Kalyani, West Bengal
India

Suresh Chandra Satapathy
Department of Computer Science &
    Engineering
Anil Neerukonda Institute of Technology
    and Sciences
Vishakapatnam, Andhra Pradesh
India

Manas Kumar Sanyal
J K Mandal, Dept of CSE
University of Kalyani
Kalyani, West Bengal
India

Vikrant Bhateja
Department of ECE
Sri Ramswaroop Memorial College of
    Engineering and Management Lucknow
Lucknow, Uttar Pradesh
India

ISSN 2194-5357   ISSN 2194-5365   (electronic)
Advances in Intelligent Systems and Computing
ISBN 978-981-10-2034-6   ISBN 978-981-10-2035-3   (eBook)
DOI 10.1007/978-981-10-2035-3

Library of Congress Control Number: 2016946617

This Springer imprint is published by Springer Nature
The registered company is Springer Science+Business Media Singapore

Preface

The Faculty of Engineering, Technology and Management, University of Kalyani, India organized the First International Conference on Intelligent Computing and Communication, (ICIC2 2016), during 18–19 February 2016 at University of Kalyani. This is third International Conference organized by the Faculty of Engineering, Technology and Management, Kalyani University during last 3 years. Organizing such a mega event, covering all aspects of intelligent computing and communication in computer science and technology, data science, general science, educational research where scopes were not only limited to computer researchers but also include researchers from other fields such as mathematics, chemistry, biology, biochemistry, engineering statistics, management and all other related areas where technologies play a vital role for the society, is an enormous task.

ICIC2 2016 received a huge response in terms of submission of papers across the globe. ICIC2 received papers from various countries outside India. The Organizing Committee of ICIC2 2016 constituted a strong international program committee for reviewing papers. A double-blind review process has been adopted. Each paper has been reviewed by at least two and at most five reviewers. The decision system adopted by EasyChair has been employed and 86 papers have been selected after a thorough double-blind review process. The committee has also checked the plagiarism through professional software. In the first round of plagiarism check, we categorized papers into three groups. We send the papers directly for double-blind review where the percentage of plagiarism of the articles was less than 10 %. We categorized the papers as second category, where the percentage of plagiarism was less than 35 % and more than 10 % and the authors were asked to rewrite and resubmit the modified papers. For those who have submitted revised version we checked for plagiarism again and sent those papers for a blind review on satisfying plagiarism report less than 10 %. All other papers were rejected directly. Ultimately, 71 papers were included for the presentation in ICIC2 2016.

The proceedings of the conference will be published in one volume in Advances in Intelligent Systems and Computing (ISSN: 2194-5357), Springer, indexed by ISI Proceedings, EI-Compendex, DBLP, SCOPUS, Google Scholar and Springerlink, and will be available at http://www.springer.com/series/11156. We convey our sincere gratitude to the authority of Springer for providing the opportunity to publish the proceedings of ICIC2 2016.

The proceedings of the ICIC2 2016 are a collection of high-quality 70 articles. These articles are distributed over eight sections. These are wireless sensor network; image processing and pattern recognition; biomedical image processing and bioinformatics; device system and modeling; communication networks and services; data analytics and data mining; security and cryptography; and cloud computing. The contributions received in respective sections are 6, 14, 6, 12, 5, 12, 9, and 6 papers.

We convey our esteemed gratitude to the Honorable Vice-Chancellor, Prof. (Dr.) Malayendu Saha for his extreme enthusiasm for hosting ICIC2 2016 at the University of Kalyani. Also we convey our deep sense of gratitude to the Deans, Faculty of Engineering, Technology and Management, Faculty of Science, Faculty of Arts and Commerce and Faculty of Education for their continuous support and association in this big event.

Our sincere gratitude to all keynote address presenters, invited speakers, session chairs, high officials of Computer Society of India for their gracious presence in the campus on the occasion.

The ICIC2 2016 has been co-sponsored by the DST PURSE program of Government of India of Kalyani University, UGC and Computer Society of India Division V. We express our sincere gratitude to UGC New Delhi, India and the authority of DST PURSE program, Government of India under University of Kalyani for their financial support. We express our deep gratitude to the Computer Society of India for being the Knowledge Partner and special thanks to Prof. (Dr.) S.C. Satapathy, Chairman, DIV V, Research and Development for promoting this event as CSI Division V event. We would also like to thank the program committee members for their efforts, and the reviewers for completing a big reviewing task in a short span of time. We would also like to thank the authority of NIT Durgapur, India for their immense support in checking plagiarism of the articles. Moreover, we would like to thank all the authors who submitted papers to ICIC2 2016 and made a high-quality technical program possible. Finally, we acknowledge the support received from the faculty members, scholars of Faculty of Engineering, Technology and Management, officers, staffs and the authority of University of Kalyani.

We hope that the articles will be useful for the researchers who are pursuing research in the field of computer science, information technology and related areas. Practicing technologists would also find this volume to be a good source of reference.

Kalyani, India   Jyotsna Kumar Mandal
Visakhapatnam, India   Suresh Chandra Satapathy
Kalyani, India   Manas Sanyal
Lucknow, India   Vikrant Bhateja
February 2016

Organizing Committee

Chief Patron

Malayendu Saha, the Vice-Chancellor, University of Kalyani, India

Honorary Chair

Bipin V. Mehta, the President, Computer Society of India

Organizing Chair

Manas Kumar Sanyal, the Dean, Faculty of Engineering, Technology and Management, University of Kalyani, India

Organizing Co-chair

Utpal Biswas, Department of Computer Science and Engineering, University of Kalyani, India

Program Chair

Jyotsna Kumar Mandal, Department of Computer Science and Engineering, University of Kalyani, India

Program Co-chair

Anirban Mukhopadhyay, Department of Computer Science And Engineering, University of Kalyani, India
Vikrant Bhateja, Department of Electronics and Communication Engineering, SRMGPC, Lucknow, India
K. Srujan Raju, Department of Computer Science and Engineering, CMR Technical Campus, Hyderabad, India

Publication Chair

Suresh Chandra Satapathi, Department of Computer Science and Engineering, ANITS, AU, India

Publication Co-chair

Siba K. Udgata, School of Computing and Information Sciences, University of Hyderabad, India

Registration Chair

Priya Ranjan Sinha Mahapatra, Department of Computer Science and Engineering, University of Kalyani, India

Registration Co-chair

Thirupathi Chellapalli, Department of Business Administration, University of Kalyani, India

Reception and Publicity Chair

Ishita Lahiri, Department of Business Administration, University of Kalyani, India

Website Management Chair

Anirban Mukhopadhyay, Department of Computer Science and Engineering, University of Kalyani, India

International Advisory Committee

Anirban Basu, Vice-President, CSI
R.K. Vyas, Hon-Treasurer, CSI
H.R. Mohan, Immediate Past President, CSI
Devaprasanna Sinha, RVP- Region-II, CSI
Partha Pratim Sarkar, University of Kalyani, India
Manas Kumar Sanyal, University of Kalyani, India
Rajashekhar Bellamkonda, Dean, School of Management Studies, University of Hyderabad, India
J. Mathews, University of Bristol, UK
Deepak Garg, Thapar University, Patiala, India
Pramod K. Meher, Nanyang Technological University, Singapur
M.N. Bin Ismail, National Defense University, KL, Malaysia
M. Kaykobad, BUET, Bangladesh
Yu Yu, Hefei University, China
K. Srujan Raju, CMR Technical Campus, Hyderabad, India
Anirban Mukhopadhyay, Department of Computer Science and Engineering, University of Kalyani, India
Bijay Baran Pal, Department of Mathematics, University of Kalyani, India
Debabrata Sarddar, Department of Computer Science and Engineering, University of Kalyani, India
Ishita Lahiri, Department of Business Administration, University of Kalyani, India
Jyotsna Kumar Mandal, Department of Computer Science and Engineering, University of Kalyani, India
Kalyani Mali, Department of Computer Science and Engineering, University of Kalyani, India
Manas Kumar Sanyal, Department of Business Administration, University of Kalyani, India
Partha Pratim Sarkar, Department of Engineering and Technological Studies, University of Kalyani, India
Priya Ranjan Sinha Mahapatra, Department of Computer Science and Engineering, University of Kalyani, India
Utpal Biswas, Department of Computer Science and Engineering, University of Kalyani, India
Tuhin Mukherjee, Department of Business Administration, University of Kalyani, India

Angshuman Bagchi, Department of Biochemistry and Biophysics, University of Kalyani, India
Bijay Baran Pal, Department of Mathematics, University of Kalyani, India

Local Program/Publication Committee

Anirban Mukhopadhyay, Department of Computer Science and Engineering, University of Kalyani, India
Bijay Baran Pal, Department of Mathematics, University of Kalyani, India
Jyotsna Kumar Mandal, Department of Computer Science and Engineering, University of Kalyani, India
Manas Kumar Sanyal, Department of Business Administration, University of Kalyani, India
Partha Pratim Sarkar, Department of Engineering and Technological Studies, University of Kalyani, India
Subhash Chandra Sarkar, Department of Commerce, University of Kalyani, India
Suresh Chandra Satapathy, Department of Computer Science and Engineering, University of Kalyani, India
Susanta Biswas, Department of Engineering and Technological Studies, University of Kalyani, India
Susmita Lahiri, Department of Environmental Engineering, University of Kalyani, India
Tuhin Mukherjee, Department of Business Administration, University of Kalyani, India
Udaybhanu Bhattacharyya, Department of Rural Development and Management, University of Kalyani, India
Utpal Biswas, Department of Computer Science and Engineering, University of Kalyani, India

Finance Committee

Anirban Mukhopadhyay, Department of Computer Science and Engineering, University of Kalyani, India
Anjan Kumar Das, the Audit and Account Officer, University of Kalyani, India
Jyotsna Kumar Mandal, Department of Computer Science and Engineering, University of Kalyani, India
Manas Kumar Sanyal, Department of Business Administration, University of Kalyani, India
Tuhin Mukherjee, Department of Business Administration, University of Kalyani, India

Editorial Committee

Jyotsna Kumar Mandal, Department of Computer Science and Engineering, University of Kalyani, India
Suresh Chandra Satapathy, Department of Computer Science and Engineering, ANITS, AU, India
Manas Kumar Sanyal, Department of Business Administration, University of Kalyani, India
Vikrant Bhateja, Department of Electronics and Communication Engineering, SRMGPC, Lucknow India

Website Committee

Anirban Mukhopadhyay, Department of Computer Science and Engineering, University of Kalyani, India

Registration Committee

Arindam Sarkar, Department of Computer Science and Engineering, University of Kalyani, India
Arup Sarkar, Department of Computer Science and Engineering, University of Kalyani, India
Ranjan Kumar Mondal, Department of Computer Science and Engineering, University of Kalyani, India
Madhumita Sengupta, Department of Computer Science and Engineering, University of Kalyani, India
Priya Ranjan Sinha Mahapatra, Department of Computer Science and Engineering, University of Kalyani, India
Somnath Mukhopadhyay, Department of IT and Systems, Calcutta Business School, India
Sujoy Chatterjee, Department of Computer Science and Engineering, University of Kalyani, India
Sushil Kumar Mandal, Department of Environmental Management, University of Kalyani, India
Jyotirmay Mondal, Department of Computer Science and Engineering, University of Kalyani, India
Ankita Bose, Department of Computer Science and Engineering, University of Kalyani, India

Asim Maiti, Department of Computer Science and Engineering, University of Kalyani, India
Arpita Baral, Department of Computer Science and Engineering, University of Kalyani, India
Parthajit Roy, The Burdwan University, India

Contents

Part VIII   Wireless Sensor Network

List of Reviewers

A. Srinivasan, MNMJ Engineering College, India
A. Ghosal, B. C. Roy Engineering College, India
Anirban Mukhopadhyay, University of Kalyani, India
A.K. Bhattacharya, NIT Durgapur, India
A.K. Mukhopadhyay, BCREC, India
A. Chaudhuri, Jadavpur University, India
A. Dasgupta, Tech Mahindra, India
Arup Sarkar, University of Kalyani, India
Abhik Mukherjee, IIEST, India
Amit Kr Mishra, IIT Guwahati, India
Animesh Biswas, University of Kalyani, India
A.K. Ghosh, Future Institute of Technology, India
Ashish K. Mukhopadhyay, BITM Shantiniketan, India
A. Damodaram, JNTU, Hyderabad, India
Arindam Sarkar, University of Kalyani, India
Avijit Kar, Jadavpur University, India
Amitabha Nag, Academy of Technology, India
Arindrajit Roy, Academy of Technology, India
Abhoy Mondal, Burdwan University, India
Asit Baran Bhattacharya, University of Kalyani, India
Ankita Bose, University of Kalyani, India
B. Basu, OPSIS System, India
B.K. Dey, Tripura University, India
Bijay Baran Pal, University of Kalyani, India
Balaram Bhattacharya, Vishva BharatiShnatiniketan, India
Biswapati Jana, Vidyasagar University, India
Chandan Bhar, ISM Dhanbad, India
Chiranjeev Kumar, ISM Dhanbad, India
D. Das, Jadavpur University, India
D. Garg, Thapar University, India
D.D. Sinha, Calcutta University, India

D. Tomar, MNMJEC, India
Debotosh Bhattacharjee, CSE, Jadavpur University, India
Dharmveer Rajppot, JIIT, Noida, India
Durgesh Mishra, SAIT, India
Gautam Sen, Jadavpur University, India
Gautam Saha, NEHU, India
Goutam Sanyal, NIT Durgapur, India
G. Sahoo, BIT Mesra, India
G.N. Singh, ISM, Dhanbad, India
G.R. Sinha, SSG Institue, India
Hari Om, ISM, Dhanbad, India
H.R. Vishwakarma, VIT University, India
Indrajit Bhattacharya, KGEC, India
Jyotsna Kumar Mandal, University of Kalyani, India
Jules Raymond Tapamo, UKZN, South Africa
Jimson Matthew, University of Bristol, UK
J.P. Choudhury, KGEC, India
J. Sil, IIEST, India
J.K. Singh, Jadavpur University, India
Jyotirmay Mondal, University of Kalyani, India
K. Dasgupta, KGEC, India
Karthikeyan Subbiah, CUET, Bangladesh
Katrick Chandra Mondal, Jadavpur University, India
Krishnendu Chakraborty, KGEC, India
Kaushik Das Sharma, Calcutta University, India
Madhumita Sengupta, IIIT Kalyani, India
Malay Bhattacharyya, IIEST, India
Manas Kumar Sanyal, University of Kalyani, India
Mihir Narayan Mohanty, SOA University, India
Monalisa Mondal, University of Kalyani, India
M.K. Bhowmik, Tripura University, India
M.K. Naskar, Jadavpur University, India
M. Kaykobad, BUET, Bangladesh
M. Marjit Singh, NERIST, India
Manas Paul, JIS Kalyani, India
Mita Nasipuri, Jadavpur University, India
Nabendu Chaki, Calcutta University, India
Nikhilesh Barik, KajiNajrul University, India
P. Dutta, Visva Bharati University, India
P. Acharya, Logitech, India
P.P. Sarkar, University of Kalyani, India
P. Jha, Purbanchal University, Nepal
P.K. Jana, ISM Dhanbad, India
Priya Ranjan Sinha Mahapatra, University of Kalyani, India
P.S. Avadhani, Andhra University, India

Parthajit Roy, The Burdwan University, India
Prataya Kuila, KIIT, Bhubaneswar, India
R.K. Samanta, North Bengal University, India
R.K. Jena, IMT, India
Rituparna Chaki, WBUT, India
Rabindranath Bera, SMIT, India
Rajat Pal, Calcutta University, India
Ranjan Kumar Mondal, University of Kalyani, India
S.K. Basu, BHU, India
S.K. Nandi, IIT Guwahati, India
S. Mal, HIT, India
S. Dutta, BCREC, India
S. Bhattacharyya, RCCIT, India
S. Mukhopadhyay, Texas University, USA
Sripati Mukhopadhay, The Burdwan University, India
S. Shakya, TU, Nepal
S. Sarkar, Jadavpur University, India
S. Jha, KEC, Nepal
S. Satapathy, Andhra University, India
S. Changder, NIT DGP, India
S. Bandyopadhyay, ISI, Kolkata, India
S. Mondal, KGEC, India
Somnath Mukhopadhyay, Calcutta Business School, India
Sujoy Chatterjee, University of Kalyani, India
S. Sarkar, NIT Jamshedpur, India
Santanu Das, KGEC, India
Sunil Karforma, BU, India
S. Roy, NITTTR, Kolkata, India
S.H. Mneney, UKZN, South Africa
S.K. Acharya, BCKV, Kalyani, India
Samar Sen Sarma, Calcutta University, India
Swarup Das, NBU, India
S.K. Udgata, UOH, India
Utpal Biswas, University of Kalyani, India
Utpal Roy, Visva Bharati, India
U. Maulik, Jadavpur University, India
Uday Kumar R.Y., NIT Surathkal, India
Udaya Sameer, ANITS, Vishakapatnam, India
Utpal Nandy, Vidyasagar University, India
Uttam Kumar Mondal, Vidyasagar University, India
V. Prithiraj, PEC, India
V.S. Despande, MIT College, India
Vikrant Bhateja, SRMGPC, Lucknow, India
Vipin Tyagi, Jaypee University, India
Zaigham Mahmood, University of Derby, UK

About the Editors

Prof. Jyotsna Kumar Mandal received his M.Sc. in Physics from Jadavpur University in 1986 and M. Tech. in Computer Science from University of Calcutta. Professor Mandal was awarded Ph.D. in Computer Science and Engineering by the Jadavpur University in 2000. Presently, he is working as Professor of Computer Science and Engineering and former Dean, Faculty of Engineering, Technology and Management, Kalyani University, Kalyani, Nadia, West Bengal for two consecutive terms. He started his career as Lecturer at NERIST, Arunachal Pradesh in September 1988. He has teaching and research experience of 28 years. His areas of research include coding theory, data and network security, remote sensing and GIS-based applications, data compression, error correction, visual cryptography, steganography, security in MANET, wireless networks and unified computing. He has supervised 15 Ph.D. degree, 2 submitted (2015–16) and 8 ongoing scholars. He has supervised three M.Phil. and more than 50 M.Tech dissertations. He is a life member of the Computer Society of India since 1992, CRSI since 2009, ACM since 2012, IEEE since 2013 and Fellow member of IETE since 2012, Honorary Chairman of CSI Kolkata Chapter. He has chaired more than 30 sessions in various international conferences and delivered more than 50 expert/invited lectures during last 5 years. He has acted as program chair of many international conferences and edited more than 15 volumes of proceedings from Springer Series, Science Direct, etc. He is a reviewer of various international journals and conferences. He has over 360 articles and 6 books published to his credit. He is one of the editors for Springer AISC Series, FICTA 2014, CSI 2013, IC3T 2015, INDIA 2015, and INDIA 2016. He is also the corresponding editors of CIMTA 2013 (Procedia Technology, Elsevier), INDIA 2015 (AISC Springer) and ICIC2 2016 (AISC Springer).

Dr. Suresh Chandra Satapathy is currently working as Professor and Head, Department of CSE at Anil Neerukonda Institute of Technology and Sciences (ANITS), Andhra Pradesh, India. He obtained his Ph.D. in Computer Science and Engineering from JNTU Hyderabad and M.Tech. in CSE from NIT, Rourkela,

Odisha, India. He has 26 years of teaching experience. His research interests include data mining, machine intelligence and swarm intelligence. He has acted as program chair of many international conferences and edited six volumes of proceedings from Springer LNCS and AISC series. He is currently guiding eight scholars for Ph.D. Dr. Satapathy is also Sr. Member of IEEE, Life Member of CSI, Chairman—Division—V (Education and Research) 2015–2017, Computer Society of India. He is Editorial Board Member of International Journal of Ambient Computing and Intelligence (IJACI)—indexed in SCOPUS; International Journal of Advanced Intelligence Paradigms (IJAIP)—indexed in SCOPUS; International Journal of Rough Sets and Data Analysis (IJRSDA)—indexed in DBLP, ACM Digital Lib; International Journal of Statistics, Optimization and Information Computing (IJSOAC)—indexed in SCOPUS; International Journal of Synthetic Emotions (IJSE)—indexed in DBLP, ACM Digital Lib., corresponding editor for Springer AISC Series: FICTA 2013, FICTA 2014, CSI 2013, CSI 2014, IC3T 2015, INDIA 2016, IC3T 2015, ICT4SD 2015.

Dr. Manas Kumar Sanyal obtained his M.Tech. in Computer Science from Calcutta University in 1989. He worked in industries for about 4 years and joined as Lecturer at BCKV in the year 1994 and obtained his Ph.D. in 2003. Presently, he is working as Dean, Faculty of Engineering, Technology and Management in the University of Kalyani, Kalyani, West Bengal. His current research interests include big data, e-governance, e-procurement, data warehousing and data mining. In addition to presenting his work at numerous international conferences, Dr. Sanyal has also organized and chaired several international and national conferences. He has contributed over 100 research papers in various international journals and proceedings of international conferences.

Dr. Vikrant Bhateja is Associate Professor, Department of Electronics and Communication Engineering, Shri Ramswaroop Memorial Group of Professional Colleges (SRMGPC), Lucknow, India. Presently, he is also the Head (Academics and Quality Control) in the same college. He has a total academic teaching experience of 12 years with more than 100 publications in reputed international conferences, journals and online book chapter contributions. His areas of research include digital image processing, computer vision and medical imaging. His core work in medical imaging includes development of techniques for computer aided analysis and diagnosis of life threatening diseases which are published in IEEE Sensors Journal, IEEE Transactions in Instrumentation and Measurements and Elsevier—Computer Methods and Programs in Biomedicine. Professor Vikrant has been on TPC and chaired various sessions from the above domain in international conferences of IEEE and Springer. He has been the track chair and served in the core technical/editorial teams for international conferences: FICTA 2014, CSI 2014, INDIA 2015 and ICICT 2015 under Springer ASIC Series and INDIA Com 2015, ICACCI 2015 under IEEE. He has been publication chair of Springer INDIA 2016 conference, Vishakhapatnam (India); Proceeding editor for Springer AISC

Series for conferences, ICIC2 2016, West Bengal (India) and ICDECT 2016, Pune (India). He has published 3 books under Springer AISC Series and Springer Studies in Computational Intelligence Series as an editor. He is associate editor in International Journals of Convergence Computing (IJConvC), Image Mining (IJIM) and Rough Sets and Data Analysis (IJRSDA) under Inderscience Publishers and IGI Global publications.

Part I
Biomedical Image Processing and Bioinformatics

Part 1

Altered PPI Due to Mutations in hPER2 and CKI Delta Locus, Causing ASPS

Ananya Ali and Angshuman Bagchi

Abstract Protein-Protein Interactions (PPI) have a huge impact on several bio-logical processes viz. enzyme substrate interaction, cell signaling to name a few. It is also well documented that altering PPI can also lead to onset of diseases. We have studied one circadian rhythm sleep disorder with genetic correlation viz. Familial Advanced Sleep Phase Syndrom (FASPS). In the present work, utilizing structural bioinformatics approach we tried to elucidate the differences in pattern of bindings between two different core "clock genes" products viz. human Period protein (hPER2) and human Casein Kinase I delta (hCKId), both in wild type and disease-causing mutated variants. Molecular mechanics calculations have also been used to describe the interactions of wild type and mutant proteins. The results from this study may be useful for the development of newer drugs in patients having impaired circadian rhythm.

Keywords Circadian rhythm sleep disorders · Familial advanced sleep phase syndrome · hPER2 · hCKId · Protein-protein interactions

1 Introduction

Protein-Protein Interactions (PPIs) have a huge impact on several biological pro-cesses viz. enzyme substrate interaction, cell signaling to name a few [1]. It is also well documented that altering PPI can also lead to onset of diseases. In our earlier

Ananya Ali · Angshuman Bagchi (✉)
Department of Biochemistry and Biophysics, University of Kalyani, Kalyani,
Nadia 741235, West Bengal, India
e-mail: angshuman_bagchi@yahoo.com

Ananya Ali
e-mail: aliananya17@gmail.com

© Springer Science+Business Media Singapore 2017
J.K. Mandal et al. (eds.), *Proceedings of the First International Conference on Intelligent Computing and Communication*, Advances in Intelligent Systems and Computing 458, DOI 10.1007/978-981-10-2035-3_1

work we have shown how altered PPI of AANAT, the penultimate enzyme of melatonin biosynthesis, with 14-3-3 protein can be computationally predicted for Delayed Sleep Phase Syndrome (DSPS) patients [2, 3].

Advanced Sleep Phase Syndrome (ASPS) is characterized by an early sleep onset (approx 7:00 pm) and early awakening in morning (about 4:00 am) of affected individual [4]. Though ASPS is not common like DSPS, this rarity of incidence has been mostly due to underestimated reported incidence as ASPS have contributed less social conflicts than DSPS [4, 5]. Familial Advanced Sleep Phase Syndrome (FASPS), where the syndrome has genetic background, is rarer than other ASPS. These FASPS are caused by mutations in two different core "clock genes" products viz. human Period protein (hPER2, mutation pS662G) and human Casein Kinase I delta (hCKId, mutations pT44A and pH46R) [6–9]. hPER2 is among core circadian oscillator proteins which give negative feedback to their own transcription in a proper circadian rhythm. Phosphorylation of hPER2 is necessary for their nuclear transport and degradation. The CKI family proteins (epsilon and delta isomers) phosphorylate and lead to degradation of PER proteins to prevent early entry of PER protein into nucleus, thus keeping the rhythm in proper pace. Phosphorylation of hPER2 has rhythmic pattern with a peak phosphorylation, at the time of maximum hPER transcription repression. Studies have revealed phosphorylation of hPER2 at 659th Serine is required for nuclear localization with increased stability and phosphorylation at other places will lead to its degradation [9–11]. Serine at 662nd residue is at the CKI binding domain of hPER2. This Serine, when phosphorylated, works as recognition for further downstream phosphorylation of hPER2. Thus, due to mutation pS662G the absence of recognition phosphoserine prevents further phosphorylation and leads to hypophosphorylation of hPER2 [6, 12]. Mutations in Casein Kinase I delta (hCKId) protein are studies to reduce its kinase activity and reported to cause similar pattern of hypophosphorylation of hPER2 and shortens circadian periodicity. Notably, all these three mutations show Mendelian autosomal dominant inheritance pattern [12].

2   Materials and Methods

2.1   Sequence Information Search for Wild Type Proteins

Full length wild type sequence information for both proteins has been collected from UniProt. [13]. For hPER2 Uniprot ID. O15055 and for hCKId Uniprot ID. P48730 have been used.

2.2 Structural Information Searches

RCSB PDB [14] has been used to collect the three dimensional structural information of the proteins. For hCKId PDB ID. 3UZP (X ray diffraction, 1.94 Å) has been used to collect structural data. For hPER2 no structural information was present in RCSB PDB.

2.3 Homology Modeling of hPER2

A PSI-BLAST [15] has been done to find proper template for structural modeling of hPER2. But no satisfactory result has been obtained. Using Phyre2, [16] an automated online protein structure modeling tool, hPER2 model has been built for 601–700 residues. The stereo-chemical qualities of the model has been checked with PROCHECK (Ramachandran plot) and Verify 3D, using SAVES, an online server [17, 18]. Residues, present at loop regions, have been modified, using Mod-Loop, online server, if have been found at disallowed regions of Ramachandran Plot.

2.4 Mutation Generation in Wild Type Structures

Using Discovery Studio 4.0 Client tool, mutations have been incorporated at 662nd residue of hPER2 converting a Serine to Glycine and 44th residue of hCKId converting Threonine to Alanine and the 46th of hCKId converting Histidine to Arginine. Energy minimization has been done using steepest descent and conjugate gradient algorithm with CHARMm force-field, keeping the r.m.s. gradient to 0.001 kcal/mol [19].

2.5 Molecular Docking of hPER2 and hCKId Proteins

Wild type and mutated hPER2 and hCKId have been docked, using Zdock server [20]. The partners were chosen such way that one of them should be of wild-type. Thus, for both wild type partners (a) hPER2wt-hCKIdwt (wt stands for wild type) has been docked and for mutated protein three more docked complex has been generated viz. (b) hPER2mut-hCKIdwt (mut for mutated, here: hPER2 pS662G mutation has been docked with wild type CKId), (c) hPER2wt-hCKIdmut1 (mut1 for hCKId, pT44A mutation has been considered) and (d) hPER2wt-hCKIdmut2 (mut2 for hCKId, pH46R mutation has been considered). Among top ten predicted output for each of above mentioned four types, best docked complex has been selected after PROCHECK and Verify 3D run. Final docked complex was then energy minimized using steepest descent with CHARMm force-field, keeping the r.m.s gradient to 0.001 kcal/mole [19, 21].

2.6 Structural Comparison

Using superimposition, Root Mean Squared Deviation (RMSD) values have been collected for both c-alpha and main chain atoms of the wild type and mutant proteins and docked complexes. Binding site analysis and comparison were done by Discovery Studio Package version 2.5.

2.7 Interaction Energy Calculation

Using Discovery Studio Package version 2.5 Interaction energy has been calculated for docked complexes to compare wild type proteins interaction with mutated and wild type interactions.

3   Results and Discussions

3.1 Structural Comparisons

Both binding site analyses and superimposition analyses have not revealed any drastic change (data not shown).

3.2 Interaction Energy Comparison

Both Table 1 and Fig. 1 are showing that free energy changes due to protein protein interactions between wild type-wild type proteins are less negative than PPI complexes of mutated proteins with their wild type counterparts. Mutations located in hCKId are affecting the PPI complex more than mutation in hPER2. Thus we could expect that in mutated form the proteins are making more stable interacting complexes than wild type one.

Table 1 Interaction energy values (Kcal/Mol) for wild type and mutated proteins

PPI Complexes	Wt-Wt	Wt-Mut	Mut1-Wt	Mut2-Wt
Interaction energy values (kcal/Mol)	−108.598	−119.048	−137.076	−150.293

Fig. 1 Interaction Energy (kcal/Mol) of wild type and mutated proteins. WT-WT, both hCKId and hPER2 is wild type; WT-MUT, hPER2 is mutated (pS662G); MUT1-WT, hCKId is mutated (pT44A); MUT2-WT, hCKId is mutated (pH46R). Showing wild type proteins are forming less stable complex

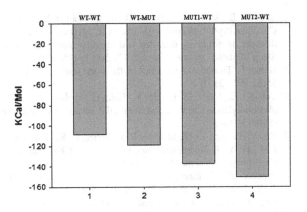

4 Conclusion

Earlier work has shown phospho-switches model describing how several competing phosphorylation sites regulate hPER2 degradation and nuclear localization, thus maintaining proper circadian periodicity and how disrupting this intricate interaction can lead to ASPS [22]. In this work we tried to elucidate the bio-molecular mechanistic details of the interactions of the hPER2 and hCKId proteins and the results are showing similar explanation for ASPS. A more stable interaction between kinase and its substrate might lead to early phosphorylation of multiple sites, thus causing an early degradation before nuclear accumulation, thus shortening the circadian rhythm than normal. The work for the first time provides a rational framework for the analysis of the molecular details of the disease onset.

Acknowledgments Authors highly acknowledge the generous help of bioinformatics infrastructural facility (BIF) Centre (sponsored by DBT), Department of Biochemistry and Biophysics, University of Kalyani for providing the necessary equipments and workstation to carry out the experiments. Ali. A. is also thankful to UGC, India for providing the fellowships.

References

1. Bagchi A, Mort M, Li B, Xin F, Carlise C, Oron T, et al. Analysis of Features from Protein-protein Hetero-complex Structures to Predict Protein Interaction Interfaces Using Machine Learning. Procedia Technol [Internet]. Elsevier B.V.; 2013;10:62–6. Available from: http://linkinghub.elsevier.com/retrieve/pii/S221201731300491X.

2. Bagchi A. Prediction of Protein-Protein Interactions [Internet]. Computational Intelligence and Pattern Analysis in Biological Informatics. 2010. 325–347 p. Available from: http://doi.wiley.com/10.1002/9780470872352.ch15.

3. Ali A, Bhattacharjee S, Bagchi A. Information Systems Design and Intelligent Applications: Structural Analyses of the Mode of Binding Between AANAT Protein with 14-3-3 Protein Involved in Human Melatonin Synthesis. 2015. p. 127–32. Available from: http://link.springer.com/10.1007/978-81-322-2247-7.

4. Sack R. L, Auckley D, Auger R. R, Carskadon M a, Wright K. P, Vitiello M. V, et al. Circadian rhythm sleep disorders: part II, advanced sleep phase disorder, delayed sleep phase disorder, free-running disorder, and irregular sleep-wake rhythm. An American Academy of Sleep Medicine review. Sleep. 2007; 30(11):1484–501.

5. Hamet P, Tremblay J. Genetics of the sleep-wake cycle and its disorders. Metabolism. 2006; 55(SUPPL. 2):10–5.

6. Toh K. L, Jones C. R, He Y, Eide E. J, Hinz W a, Virshup D. M, et al. An hPer2 phosphorylation site mutation in familial advanced sleep phase syndrome. Science. 2001; 291 (5506):1040–3.

7. Xu, Y., Padiath, Q. S., Shapiro, R. E., Jones, C. R., Wu, S. C., Saigoh, N., Saigoh, K., Ptacek, L. J., Fu Y-H. Functional consequences of a CKI-delta mutation causing familial advanced sleep phase syndrome. Nature. 2005;

8. Brennan, K. C., Bates, E. A., Shapiro, R. E., Zyuzin, J., Hallows, W. C., Huang, Y., Lee, H.-Y., Jones, C. R., Fu, Y.-H., Charles, A. C., Ptacek LJ. Casein kinase 1-delta mutations in familial migraine and advanced sleep phase. Sci Transl Med. 2013; Vol. 5(Issue 183).

9. Vanselow K, Vanselow J. T, Westermark P. O, Reischl S, Maier B, Korte T, et al. Differential effects of PER2 phosphorylation: Molecular basis for the human familial advanced sleep phase syndrome (FASPS). Genes Dev. 2006; 20(19):2660–72.

10. Qin, X., Mori, T., Zhang, Y., and Johnson, C. H. (2015). PER2 Differentially Regulates Clock Phosphorylation versus Transcription by Reciprocal Switching of CK1ε Activity. J. Biol. Rhythms 30, 206–216.

11. Kim, J.K., Forger, D. B., Marconi, M., Wood, D., Doran, A., Wager, T., Chang, C., and Walton, K. M. (2013). Modeling and validating chronic pharmacological manipulation of circadian rhythms. CPT Pharmacometrics Syst. Pharmacol. 2, e57.

12. Von Schantz M. Phenotypic effects of genetic variability in human clock genes on circadian and sleep parameters. J Genet. 2008; 87(5):513–9.

13. Leinonen, R., Garcia Diez, F., Binns, D., Fleischmann, W., Lopez, R., & Apweiler R. UniProt archive. Bioinformatics. 2004.

14. Berman H. M. The Protein Data Bank: a historical perspective. Acta Crystallogr A. 2008; 88–95.

15. Camacho, C., Coulouris, G., Avagyan, V., Ma, N., Papadopoulos, J., Bealer, K., & Madden T. L. BLAST+: architecture and applications. BMC Bioinformatics. 2009.

16. Bennett-Lovsey, Herbert, Sternberg & K. No Title. 2008.

17. Lüthy, R., Bowie, J. U., & Eisenberg D. Assessment of protein models with three-dimensional profiles. Nature. 1992;356(6364):83–5.

18. Laskowski, R. A., Rullmannn, J. A., MacArthur, M. W., Kaptein, R., & Thornton J. M. AQUA and PROCHECK-NMR: programs for checking the quality of protein structures solved by NMR. J Biomol NMR. 1996; 477–86.

19. Brooks, B. R., Bruccoleri, R. E., Olafson, B. D., States, D. J., Swaminathan, S., & Karplus M. CHARMM: A program for macromolecular energy, minimization, and dynamics calculations. J Comput Chem. 1983; 4(2):187–217.

20. Pierce, B. G., Wiehe, K., Hwang, H., Kim, B.-H., Vreven, T., & Weng Z. ZDOCK server: interactive docking prediction of protein–protein complexes and symmetric multimers. Bioinformatics. 2014;30(12).

21. Hess, B., Kutzner, C., Van Der Spoel, D., & Lindahl E. GRGMACS 4: Algorithms for highly efficient, load-balanced, and scalable molecular simulation. J Chem Theory Comput. 2008; 4(3):435–47.

22. Min Zhou, Jae Kyoung Kim, Gracie Wee Ling Eng, Daniel B. Forger, and David M. Virshup. A Period2 Phosphoswitch Regulates and Temperature Compensates Circadian Period Molecular Cell 60, 77–88, October 1, 2015 ª2015 Elsevier Inc.

A Comparative Framework of Probabilistic Atlas Segmentation Method for Human Organ's MRI

Sushil Kumar Mahapatra, Sumant Kumar Mohapatra, Sakuntala Mahapatra and Lalit Kanoje

Abstract Recently, different image analysis methods are used for human body parts. But the internal pectoral muscle segmentation of important body parts in a automatic way is widely used. This is also vital for multi modal image registration. Previously, breast MRI image analysis by automatic pectoral muscle segmentation is studied. In this paper, we introduce a comparative framework of probabilistic atlas segmentation method for breast with brain, chest, heart and liver MRI. For breast, brain, heart and liver and chest segmentation, the obtained DSC values are 0.76 ± 0.12, 0.71 ± 0.15, 0.66 ± 0.08, 0.77 ± 0.12 and 0.72 ± 0.13 respectively. The total overlap values for each case are 0.76 ± 0.12, 0.76 ± 0.15, 0.71 ± 0.08, 0.70 ± 0.12 and 0.70 ± 0.13 respectively.

Keywords Multi atlas based segmentation · Breast MRI · Brain MRI · Heart MRI · Liver MRI · Chest MRI

1 Introduction

The MRI of the important body parts is a technique which is used to detect the cancer disease tumor detection of patients. It needs to perform automatic analysis of breast, brain, liver, heart and chest MRI image analysis. Gubern et al. [1] developed

S.K. Mahapatra (✉) · S.K. Mohapatra ·
Sakuntala Mahapatra · Lalit Kanoje
Trident Academy of Technology, Bhubaneswar, Odisha, India
e-mail: mohapatrasushil@gmail.com

S.K. Mohapatra
e-mail: sumsusmeera@gmail.com

Sakuntala Mahapatra
e-mail: mahapatra.sakuntala@gmail.com

Lalit Kanoje
e-mail: lalitkanoje@gmail.com

© Springer Science+Business Media Singapore 2017
J.K. Mandal et al. (eds.), *Proceedings of the First International Conference on Intelligent Computing and Communication*, Advances in Intelligent Systems and Computing 458, DOI 10.1007/978-981-10-2035-3_2

a method to automatically compute breast segmentation in breast MRI. Hence the body breast and air breast surfaces are automatically segmented. By this breast segmentation dice similarity coefficient (DSC) and total overlap values are 0.94 and 0.96 respectively. van der Waal et al. [2] compares different methods for measuring breast density, both visual assessments and automated volumetric density, in a breast cancer screening setting. Van et al. [3] defines about breast body interface by manually in a straight line. The breast value segmentation approach is explained in [4]. Lin et al. [5] presents a fully automatic chest template-based method for cases with different body and breast shapes and different density patterns. Menze et al. [6] introduce a generative probabilistic model that has been designed for tumor lesions to generalize well to stroke images, and the generative discriminative model to be one of the top ranking methods in the BRATS evaluation. Nouranian et al. [7] reduces the segmentation variability and planning time by proposing an efficient learning-based multi-label segmentation algorithm. Alba et al. [8] algorithm for the segmentation of severely abnormal hearts which does not require a priori knowledge of the involved pathology or any specific parameter tuning to be applied to the cardiac image under analysis. Wang et al. [9] used a second derivative information representation by the Hessian matrix to delineate chest wall and air breast boundary.

This paper presents a comparative framework of [1] with brain, chest, heart and liver MRI to automatically segment the above parts of human body. The related work is to compare the technique in [1] with brain, chest, heart and liver segmentation. The above segmentation is verified on 35 MRI cases.

2 Material

To evaluate the segmentation process output result, the data set used which consists of atlases of 35 pre-contrast T1-weighted MR breast, brain, chest, heart and liver MRI scans obtained from different patients. For screening test, the ages of the women are between 25 and 68 years. The above MRI examinations were performed on a 1.5T system (Siemens 1.5T), Magnetom Vision). The clinical imaging parameters is [1] used for whole segmentation process (Table 1).

Table 1 Clinical image parameter

Matrix size	256 × 128 or 256 × 96
Slice wideness	1.33 nm
Slice openness	0.712–1.25 mm
Flip angle	8°, 20°, 25°
Cycle duration	8.1–8.8 ms
Repetition duration	1.8–5.76 ms

3 Probabilistic Atlas Based Segmentation in Breast, Heart, Brain, Liver and Chest

The authors of [1] presented a probabilistic atlas based method by using Bayesian framework. This framework is to provide an accurate probability distribution for the above mentioned muscle area. The authors of this article are trying to utilize this method beyond the breast MRI, also in other important body parts like heart, brain, liver and chest. The whole process is completed in SCB medical college. Cuttack, Odisha, under the supervision of one doctor and one MRI technician. Figures 1, 2, 3, 4 and 5 shows the implementation of the segmentation frame work with Bayesian voxel classification logarithm by the use of probabilistic atlas.

4 Result and Discussion

In this experiment we evaluate the probabilistic segmentation frame works on 35 patients. Each segmented case was not included for the construction of the probabilistic atlas. The quality of the segmentation was measured by the dice similarity coefficient (DSC) and total over lap (Fig. 6).

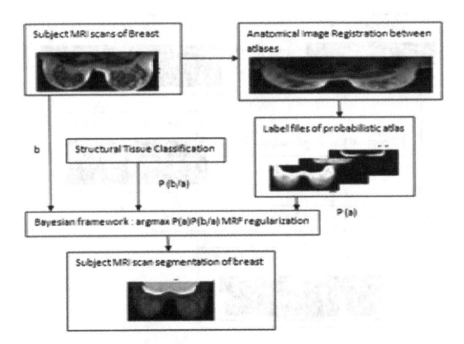

Fig. 1 Probabilistic atlas based segmentation of the pectoral muscle in breast MRI

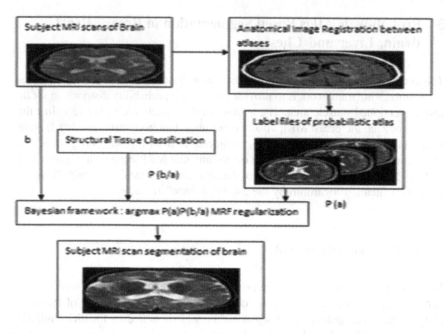

Fig. 2 Probabilistic atlas based segmentation of the pectoral muscle in brain MRI

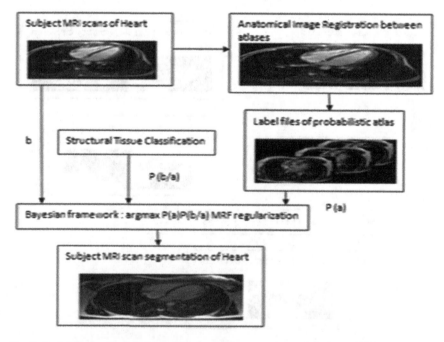

Fig. 3 Probabilistic atlas based segmentation of the pectoral muscle in heart MRI

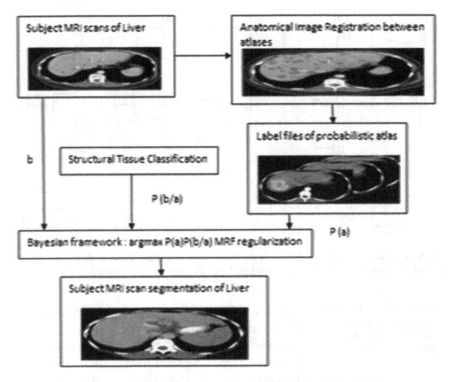

Fig. 4 Probabilistic atlas based segmentation of the pectoral muscle in liver MRI

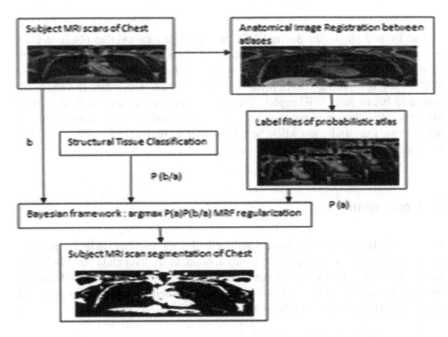

Fig. 5 Probabilistic atlas based segmentation of the pectoral muscle in chest MRI

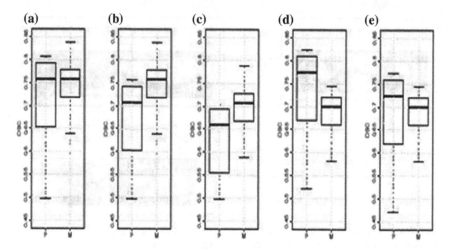

Fig. 6 a–e shows box plot of breast, brain, heart, liver and chest showing their DSC report (P) and total overlap (M) respectively using probabilistic segmentation approach

Table 2 Average DSC and average total overlap obtained from probabilistic atlas segmentation method

MRI of body parts	DSC	Total overlap
Breast MRI	0.76 ± 0.12	0.76 ± 0.12
Brain MRI	0.71 ± 0.15	0.76 ± 0.15
Heart MRI	0.66 ± 0.08	0.71 ± 0.08
Liver MRI	0.77 ± 0.12	0.70 ± 0.12
Chest MRI	0.72 ± 0.13	0.70 ± 0.13

The Table 2 shows the average DSC and average total overlap obtained from probabilistic atlas segmentation methods. The lower DSC value is getting from heart MRI report of 0.66 ± 0.08 as compared to breast MRI 0.76 ± 0.12, chest MRI 0.72 ± 0.13 and the liver MRI DSC 0.77 ± 0.12. The report is very much similar to breast MRI DSC report. The total overlap values exceed 0.70 in each MRI case. Finally, since no previous works performed pectoral segmentation in brain, heart, liver and chest MRI. So this is only a comparative study with [1] to show the DSC and total overlap values.

5 Conclusion

In this work, the probabilistic atlas based methodology has been studied to perform the pectoral muscle segmentation in a breast, brain, heart, liver and chest MRI. This has not been done previously except breast MRI [1]. Fully dedicated probabilistic frameworks have been utilized and tested on 35 different patients. The obtained results are satisfactory with DSC values.

References

1. Albert Gubern-Merida et.al,"Breast Segmentation and Density Estimation in Breast MRI: A Fully Automatic Framework", ieee journal of biomedical and health informatics, Vol. 19, No.1, January 2015, pp. 349–357.
2. Van der Waal D, den Heeten GJ, Pijnappel RM, Schuur KH, Timmers JMH, Verbeek ALM, et al. (2015) Comparing Visually Assessed BI-RADS Breast Density and Automated Volumetric Breast Density Software: A Cross-Sectional Study in a Breast Cancer Screening Setting. PLoS ONE 10(9): e0136667. doi:10.1371/journal.pone.0136667.
3. S. van Engeland, S. Timp and N. Karssemeijer "Finding corresponding regions of interest in mediolateral oblique and craniocaudal mammographic views", Med. Phys., vol. 33, pp. 3203–3212, 2006.
4. K. Nie, J-H. Chen, S. Chan, M-K. I. Chau, H. J. Yu, S. Bahri, T. Tseng, O. Nalcioglu and M-Y. Su "Development of a quantitative method for analysis of breast density based on three-dimensional breast MRI", Med. Phys., vol. 35, pp. 5253–5262, 2008.
5. Muqing Lin et. Al, "Template-based automatic breast segmentation on MRI by excluding the chest region", IEEE Trans. Med Phys, vol. 40, issue 12, Dec 2013, pp 14–18. doi:10.1118/1.4828837.
6. Menze B, Van Leemput K, Riklin Raviv T, Geremia E, Gruber P, Wegener S, Weber MA, Szekely G, Ayache N, Golland P. "A generative probabilistic model and discriminative extensions for brain lesion segmentation—with application to tumor and stroke", IEEE Trans Med Imaging. Vol. 38, Issue. 1, pp. 1–6, Nov 20, 2015.
7. Nouranian S, Ramezani M, Spadinger I, Morris W, Salcudean S, Abolmaesumi P. " Learning-based Multi-label Segmentation of Transrectal Ultrasound Images for Prostate Brachytherapy", Medical Imaging, IEEE Transactions on Volume:05, Issue: 99, Nov 2015, pp. 1–4.
8. Alba X, Pereanez M, Hoogendoorn C, Swift A, Wild J, Frangi A, Lekadir K. "An Algorithm for the Segmentation of Highly Abnormal Hearts using a Generic Statistical Shape Model", IEEE Trans Med Imaging. Volume:PP, Issue: 99, Nov 2015, pp. 1–4.
9. L. Wang, B. Platel, T. Ivanovskaya, M. Harz and H. Hahn "Fully automatic breast segmentation in 3D breast MRI", Proc. IEEE Ninth Int. Symp. Biomed. Imag., pp. 1024–1027, 2012.

Concentration of Acetone Levels in Breath for Monitoring Blood Glucose Level

Sumant Kumar Mohapatra, Sushil Kumar Mahapatra,
Shuvendra Kumar Tripathy and Lalit Kanoje

Abstract Diabetes is a major problem affecting millions of people today and if left unchecked can create enormous implication on the health of the population. Among the various noninvasive methods of detection, breath analysis presents an easier, more accurate and viable method in providing comprehensive clinical care for the disease. This paper examines the concentration of acetone levels in breath for monitoring blood glucose levels and thus predicting diabetes. The analysis uses the support vector mechanism to classify the response to healthy and diabetic samples. For the analysis ten subject samples of acetone levels are taken into consideration and are classified according to three labels which are healthy, type 1 diabetic and type 2 diabetic.

Keywords Acetone level · Blood glucose level · Breath · SVM

1 Introduction

Diabetes can be described as a group of metabolic diseases where the blood glucose level in the body is higher than the normal prescribed parameter. When a person suffers from diabetes, it is seen that their body is either unable to secrete enough insulin or their body is not able to use the insulin produced by the liver. This causes sugar to build up in the blood thus leading to diabetes. There are two major types of

S.K. Mohapatra (✉) · S.K. Mahapatra · S.K. Tripathy · L. Kanoje
Trident Academy of Technology, Bhubaneswar, Odisha, India
e-mail: sumsusmeera@gmail.com

S.K. Mahapatra
e-mail: mohapatrasushil@gmail.com

S.K. Tripathy
e-mail: shuvendra12478@gmail.com

L. Kanoje
e-mail: lalitkanoje@gmail.com

© Springer Science+Business Media Singapore 2017   17
J.K. Mandal et al. (eds.), *Proceedings of the First International Conference on Intelligent Computing and Communication*, Advances in Intelligent Systems and Computing 458, DOI 10.1007/978-981-10-2035-3_3

diabetes which include type 1 and type 2. Type 1 diabetes is the result of the body 's failure to produce enough insulin. While type 2 is a condition in which cells fail to respond to the insulin produced in the body properly. It is seen that the prescribed parameter of blood glucose levels (BGL) in healthy subjects before meals is around 70–80 mg/dL. Sugar less than 100 mg/dL while fasting is considered normal by today's standards. Any BGL higher than normal is considered unhealthy. The most common method of obtaining glucose levels in the body is by drawing blood samples using invasive techniques. A lancet device is used to draw the blood by pricking the figure or the forearm [1]. The droplet of blood obtained is then placed on a disposable strip which consists of a sensing element. A glucometer is then used to calculate the blood glucose levels form these strips. Though this process is highly accurate, it is also painful and inconvenient especially when multiple readings are required to be taken in a day. Thus there is an essential need for a non-invasive technique for monitoring BGL. Investigations show that urine, sweat, saliva, tears and breath contain traces of glucose in them and these traces vary with the levels of glucose in the blood. Therefore these human serums have recently gained recognition as feasible alternatives to using blood for glucose measurement. Extensive research conducted in this area conclude that human breath is a good alternative to monitor and diagnose glucose levels as acetone in the breath has shown a good correlation to BGL. Using breath as a deduction technique allows deduction of blood glucose levels by just exhaling into the monitoring device. It is seen that in the human breath there are number of chemical compounds that relate to different diseases [2, 3]. Traces of acetone in the breath are used for the detection of diabetes. According to the studies it is seen that patients who have diabetes have body cells that are unable to absorb the glucose in blood. In such cases, when the liver breaks down fat for energy, there occurs an abnormal increase in ketone bodies in the patient's blood. Acetone is one of the three kinds of ketone bodies which are volatile, and the body exhales the acetone thus formed. Therefore higher concentration of acetone is found in the exhaled air of a diabetic patient [4]. Initially complicated techniques such as gas chromatography mass spectroscopy, selected ion flow tube mass spectroscopy and cavity ringdown spectroscopy were used to determine the concentrations of acetone in breath. Gas chromatography-mass spectroscopy [5] works on the principal that the difference in the chemical properties between the different molecules in the gaseous mixture and their relative affinity for the stationary phase of the column promotes the separation of the molecules. In the Selected ion flow tube-mass spectroscopy method it was seen that when the neutral analyte molecules of a sample vapor meets the precursor ions they may undergo chemical ionization which depends on their chemical properties, such as their proton affinity or ionization energy [6]. Cavity Ringdown Spectroscopy [7] is an optical spectroscopic technique that measures the absolute extinction by samples that absorb or scatter light. These techniques remain unsuitable for clinical application due to their low portability, complex mechanisms and high cost. The electronic nose or e-nose model [8, 9] overcomes these drawbacks significantly and is emerging as a good alternative suited to clinical applications. Research conducted with e-nose provides substantial data on acetone concentrations in breath [10]. This paper classifies acetone concentration and uses support vector mechanism (SVM) classifier to analyze the data signals. Different acetone concentration levels are classified healthy breath, type 1 diabetic and type 2 diabetic which is then displayed as the output. Initially the acetone levels were classified using SVM classifier but the accuracy levels remained inadequate. To overcome this issue and increase the level of accuracy SVM classification with margin sampling has been adopted.

2 Breath Analysis Procedure

2.1 Proposed Analysis

The proposed analysis system here distinguishes the concentration of acetone levels in the breath as healthy or diabetic. The basic model of the proposed system is described by the following flow chart given in Fig. 1.

Fig. 1 Flow chart of the proposed system

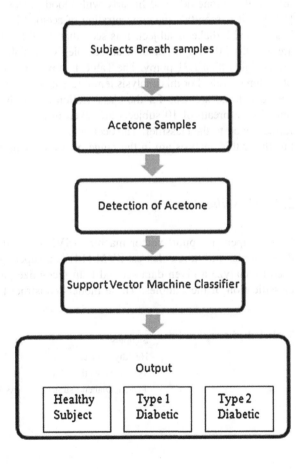

The operation of the system happens in five stages. Initially the subject's breath samples are used for the acetone concentration prediction. The acetone concentration data from various samples are taken individually as input in the next stage. The SVM classification model then classifies the acetone concentration levels based on the thresholds described in the next section. In the final stage the output is displayed under any one of the three classifiers considered for this system. Thus the diabetic and the healthy breath samples are separated in this analysis model.

2.2 Acetone Level Concentration

Human breath consists of many biomarkers which are used for the detection of many diseases. It is explained previously how acetone in the breath is used as the biomarker for the prediction of diabetes. Wang et al. [11] carried out extensive studies and found a linear correlation between the mean concentration of breath acetone and the mean blood glucose levels of each group. It was also observed that the breath acetone declined linearly with blood glucose levels [12]. According to [12, 13], it is seen that the concentration of acetone levels in the breath vary for a healthy and a diabetic subject. It is seen that in all healthy samples acetone levels are less than 0.76 ppmv and in all samples with diabetes acetone levels showed levels higher than 1.71 ppmv. The Table 1 shows the variation of the concentration of acetone levels. For this analysis ten different concentrations of acetone levels are considered and are used for the classification of the data. The acetone concentrations in the breath of 10 subjects are taken in this paper. These data samples are further given to the classifier. The acetone concentration that are detected from the breath for this analysis are in the range of parts per million.

2.3 Classification

In this paper, a support vector machine (SVM) classifier is used to classify the data samples into three classes. SVM [14] is a supervised learning model that is used to analyze a given data set and help recognize patterns that is used for the classification of the samples. These models construct hyper planes that are used

Table 1 Variation of acetone level

Samples	Concentration of acetone(ppm)
Healthy subject	0.22–0.80 ppm
Type 2 Diabetic subject	1.76–3.73 ppm
Type 1 Diabetic subject	As high as 21 ppm

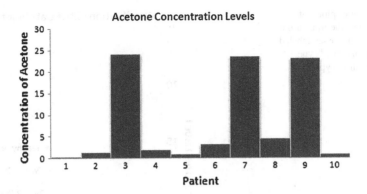

Fig. 2 Distribution of Acetone concentration levels of 10 samples

for the classification and the regression of the data [15]. Here the SVM divides the acetone concentration data which is the input into three different classes. Based on the above mentioned range of acetone concentrations the SVM classifier is used to distinguish them. The acetone levels are classified into their labels which are H for healthy breath, T1 for type 1 diabetes and T2 for type 2 diabetes. From the threshold range mentioned in the previous section it can be seen that there is an order in the levels of acetone concentrations which do not overlap each other and they conform to the guideline of T1 > T2 > H. Here 10 various acetone concentrations are considered for the classification. The input sample vector is taken as these concentrations and the trained samples are the classified data. A set of hyper planes are used here so as to classify the ten samples into the three different classes. The classifier chooses the hyper planes such that the nearest trained data point has the largest distance so that a good segregation of the data points are achieved. The Fig. 2 shows the graphical representation of the acetone concentrations of the subjects considered for this analysis. The bar graph that is plotted clearly shows the variation of the acetone levels of a healthy breath and a diabetic patient. Hence the above thresholds can easily segregate the samples as healthy, type 1 or type 2 diabetes. The scatter plot that distinguishes the acetone levels are represented in the Fig. 3. This figure gives the details of the classification of the acetone levels for a diabetic type 1 and type 2 as well as marks the healthy samples of the ten acetone concentrations which are considered in this paper. It is seen that for the type 2 and type 1 diabetic subjects the levels of acetone in their breath is high compared to the normal breath acetone levels.

Fig. 3 Scatter plot that distinguishes the trained data points into 3 classes labeled _Healthy breath', 'Type 1 Diabetic' and 'Type 2 Diabetic'

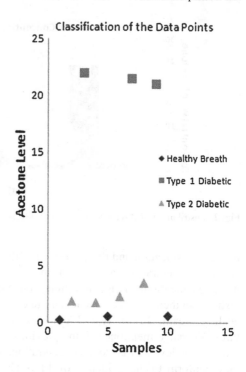

3 Result and Discussion

The classified acetone concentrations of the ten subjects are done using the SVM classifier and the classified acetone level output is tabulated and presented in Table 2 shown below. It is seen that based on the threshold parameters described in the previous section the classification and the labeling of the data has been performed. It is clearly seen that the acetone concentrations in the breath of a diabetic patient (Type 1 and Type 2) is greater than the normal breath standards. When the

Table 2 Variation of acetone level of 10 different patients

Subject	Concentration of acetone (ppm)	Healthy	Type 2	Type 1	Labels
Patient 1	0.23	Yes	No	No	H
Patient 2	1.79	No	Yes	No	T2
Patient 3	24	No	No	Yes	T1
Patient 4	1.87	No	Yes	No	T2
Patient 5	0.78	Yes	No	No	H
Patient 6	3.21	No	Yes	No	T2
Patient 7	23.5	No	No	Yes	T1
Patient 8	4.47	No	Yes	No	T2
Patient 9	23	No	No	Yes	T1
Patient 10	0.23	Yes	No	No	H

ten trained samples were classified, it was found that a higher level of accuracy can be obtained using an SVM classifier with margin sampling. This approach has given the data a clear margin over which they can be segregated. The ten concentrations were labeled into the three categories namely Healthy, Type 1 diabetic and Type 2 diabetic and the samples are placed according to the values in any one of these above mentioned categories.

4 Conclusion and Future Work

This paper investigates the concentration of acetone levels for classification of breath samples in monitoring diabetes. Acetone concentrations were collected and classified for predicting diabetes. The samples were classified using the SVM classifier and were divided into three groups labeled as healthy, type 1 and type 2 diabetes. In this paper an attempt has been made to refine the existing classification approach (healthy and diabetic) further for a more accurate evaluation (healthy, type 1 and type 2 diabetic). The future work will include designing and developing the sensory array for examination of acetone concentrations in real time breath samples.

References

1. Prashanth Makaram, Dawn Owens and Juan Aceros, "Trends in Nanomaterial-Based Non-Invasive Diabetes Sensing Technologies"; Diagnostics 2014.
2. Wolfram Miekisch, Jochen K Schubert, Gabriele F.E Noeldge—Schomburg;—Diagnostic potential of breath analysis—Focus on volatile organic compounds‖; Clinica Chimica Acta 2004.
3. Kim DG Van de Kant, Linda J.T.M van der Sande, Quirijn Jöbsis, Onno C.P van Schayck, Edward Dompeling;—Clinical use of exhaled volatile organic compounds in pulmonary diseases: a systematic review.‖ Respiratory Research 2012.
4. Tassopoulos, C.N.; Barnett, D.; Fraser, T.R;—Breath-acetone and blood-sugar measurements in diabetes‖: Lancet 1969.
5. Chunhui Deng, Jie Zhang, Xiaofeng Yu, Wei Zhang and Xiangmin Zhang; —Determination of acetone in human breath by gas chromatography–mass spectrometry and solid-phase microextraction with on-fiber derivatization‖; Journal of Chromatography 2004.
6. Moorhead, D. Lee, J. G. Chase, A. Moot, K. Ledingham, J. Scotter, R. Allardyce, S. Senthilmohan, and Z. Endre;—Classification Algorithms for SIFT-MS medical diagnosis‖; Proceedings of the 29th Annual International Conference of the IEEE EMBS Cité Internationale, Lyon, France August 23–26, 2007.
7. Chuji Wang, Armstrong Mbi and Mark Shepherd;—A Study on Breath Acetone in Diabetic Patients Using a Cavity Ringdown Breath Analyzer: Exploring Correlations of Breath Acetone With Blood Glucose and Glycohemoglobin A1C‖; IEEE Sensors Journal, Vol. 10, NO. 1, January 2010.
8. P. Wang, Y. Tan, H. Xie, and F. Shen;—A novel method for diabetes diagnosis based on electronic nose‖; Biosensors and Bioelectronics, Vol. 12, No. 9, pp. 1031–1036, 1997.

9. D. Guo, D. Zhang, N. Li, L. Zhang, and J. Yang;—A novel breath analysis system based on electronic olfaction‖; IEEE Transaction on Biomedical Engineering, Vol. 57, No. 11, November 2010.

10. P. Wang, Y. Tan, H. Xie, and F. Shen; "A novel method for diabetes diagnosis based on electronic nose‖; Biosensors and Bioelectronics", Vol. 12, No. 9, pp. 1031–1036, 1997.

11. C. Wang, A. Mbi, and M. Shepherd, "A study on breath acetone in diabetic patients using a cavity ringdown breath analyzer: Exploring correlations of breath acetone with blood glucose and glycohemoglobin a1c", IEEE Sens. J., vol. 10, no. 1, pp. 54–63, Jan. 2010.

12. C. Turner, C. Walton, S. Hoashi, and M. Evans, 'Breath acetone concentration decreases with blood glucose concentration in type I diabetes mellitus patients during hypoglycaemic clamps', ‖ J. Breath Res., vol. 3, no. 4, p. 046004, Dec. 2009.

13. C. Deng, J. Zhang, X. Yu, W. Zhang, and X. Zhang, 'Determination of acetone in human breath by gas chromatography-mass spectrometry and solid-phase microextraction with on-fiber derivatization,' J. Chromatogr. B, vol. 810, no. 2, pp. 269–275, 2004.

14. C.-C. Chang and C.-J. Lin,–LIBSVM: A library for support vector machines,' ACM Trans. Intell. Syst. Technol.', vol. 2, no. 3, 27, pp. 1–27, 2011.

15. C. J. Burges,—A tutorial on support vector machines for pattern recognition,' Data Mining Knowl. Discovery', vol. 2, no. 2, pp. 121–167, 1998.

Computational Molecular Analysis of Human Rhodopsin, Transducin and Arrestin Interactions: An Insight into Signal Transduction for Ophthalmology

Tanushree Mukherjee, Arundhati Banerjee and Sujay Ray

Abstract Retinal G-protein receptor; rhodopsin upon light-activation, gets phosphorylated, experiences conformational shift and interacts with G-protein; transducin. To completely obstruct the signal transduction visual protein; arrestin binds consecutively to disrupt the cationic channels of plasma membrane. Experimented binding assays documents the protein interactions but hitherto computational investigation was undone. This probe aims at the computational study of conformational alterations in rhodopsin upon sequential interactions, accompanied by variations in its surface electrostatic potential and net solvent accessible area. 3D structures of human transducin, arrestin and rhodopsin were analyzed. Residual participation from the optimized and simulated trio-complex (rhodopsin-transducin-arrestin) disclosed that predominantly positively charged amino-acid residues; Arg474, Arg412, Arg229, Arg13, Lys15 and Lys408 from rhodopsin participated with transducin and arrestin forming 9 ionic interactions. Rhodopsin was perceived to interact in a gradual firmer pattern with its partner proteins. This study presents a novel viewpoint into the computational disclosure for participation of concerned visual proteins.

Tanushree Mukherjee and Arundhati Banerjee are equal contributor.

Tanushree Mukherjee · Sujay Ray (✉)
Department of Biotechnology, Bengal College of Engineering & Technology, Durgapur, India
e-mail: raysujay@gmail.com

Tanushree Mukherjee
e-mail: tanu.28.mukherjee@gmail.com

Arundhati Banerjee
Department of Biotechnology, National Institute of Technology, Durgapur, India
e-mail: arundhati.92star@gmail.com

Sujay Ray
Department of Biochemistry and Biophysics, University of Kalyani, Kalyani, Nadia, India

© Springer Science+Business Media Singapore 2017
J.K. Mandal et al. (eds.), *Proceedings of the First International Conference on Intelligent Computing and Communication*, Advances in Intelligent Systems and Computing 458, DOI 10.1007/978-981-10-2035-3_4

25

Keywords Homology modeling · Optimized Rhodopsin-Transducin-Arrestin complex · Protein-protein interactions · Conformational switches · Stability and simulation

Abbreviations

MD   Molecular Dynamics
P.I.C. Protein Interaction Calculator
S1   Rhodopsin before any Interaction
S2   Rhosopsin after Interaction with Transducin
S3   Rhodopsin after Interaction with Arrestin in presence of Transducin

1   Introduction

In darkness or dim light, the rod cells take the active responsibility for the human vision. One such associated vital protein; human rhodopsin is a retinal G-protein receptor that is present in the rod cells and is responsible for the photo transduction process. Light activates the rhodopsin which then undergoes conformational alterations [1, 2]. Due to this conformational shift in rhodopsin, a G-protein; transducin gets instantly triggered to interact with the altered rhodopsin [1, 2]. Binding assay studies from earlier investigations suggests that the α-subunit of transducin now binds to rhodopsin at a definite binding site [1–4] and cyclic guanosine mono-phosphate (cGMP) phosphodiesterase gets activated. This in turn results in the disintegration of the second messenger cGMP [2]. cGMP further inhibits the opening of the cation channels [2]. Finally a complete blockage in signal transduction of ocular physiology is resulted with the participation of arrestin domain with the complex [2]. Therefore the binding of transducin to the light triggered rhodopsin leads to its phosphorylation and further obstruction of photo transduction upon interaction with arrestin. Thus in presence of light, the intracellular stimulus of the retinal rod cells gets modulated to stop the signaling phenomena. Herein, the activation of the retinal cone cell proteins occurs for the vision in light. Though several documentations [5, 6] were yet now stated for the computational and molecular basis of study into the human proteins and their interactions but none indulged with the comparable, in silico and interactive analysis of the three essential human ocular proteins for regulation of signal transduction on exposure to light.

The present computational investigation is concerned with the study of rhodopsin and its conformational switches in response to the binding of its paramount partner proteins; human transducin and subsequently to arrestin. Homology modelling of the G-protein transducin and analysis of the three dimensional structures of rhodopsin and arrestin were performed. After energy optimization and molecular dynamics simulation of the docked trio-protein complex, the residual responsibilities of the individual proteins were estimated. In addition to that, the variation in the conformation of rhodopsin was examined accompanied by the alteration in electrostatic surface potential and net area of solvent accessibility. Therefore, this first residue level, structural and computational study offers with an insight into the detailed molecular phenomena indulged when in presence of light, the rod proteins interact among themselves and switches between the conformations to modulate the signal transduction.

2   Materials and Methods

2.1   Sequence Analysis of Human Transducin

Amino acid sequence for the alpha subunit of Human Transducin having UniProt ID: P11488 was selected from Uniprot KB. The same was validated using NCBI (Accession No. 121032).

2.2   Template Search and Homology Modeling of the Modelled Protein

The protein sequence of interest for modeling human transducin was subjected to SWISS-MODEL [7] for its template exploration first, followed by the homology modeling of the target protein. The template is chosen with the aid of adjustable E-value limit of BLAST [8]. This favors to model proteins with improved stereo-chemical properties utilizing several effective force-fields [7]. The best template, possessing a PDB ID: 3V00 (chain C) was observed to share a sequence identity and query coverage of 92.29 % and 100 % respectively. The template was further analyzed to belong to guanine nucleotide-binding protein G (t) subunit alpha-1 (GNAT1) from *rattus norvegicus*. The homology (comparative) modeled transducin protein was observed to share a root mean square deviation (RMSD) of 0.058 Å when superimposed upon its X-ray template protein.

2.3   Loop Optimization of the Modeled Transducin Protein

ModLoop was utilized for performing efficient loop optimization of the protein to accomplish appropriate conformation of the ψ-φ angles and reduce the inaccuracies in the protein conformation due to distortions in loop regions [9].

2.4 Stereo-Chemical Validation of Transducin Model

The model was validated by VERIFY 3D from the SAVES4 server [10]. The model was also checked on the Ramachandran plot [11] and no residues were observed in the outlier region.

2.5 Structural Analysis of Human Rhodopsin and Arrestin

Crystal structures of Human rhodopsin complex and arrestin were identified from Protein Data Bank [12] and extracted from Discovery Studio Accelyrs 4.1. The rhodopsin protein with PDB ID: 4ZWJ, chain A [13] has 906 residues while the arrestin domain with PDB ID:4R7X, chain A [14] has 185 residues.

2.6 Docking Simulations of the Three Human Proteins Sequentially

Rhodopsin and transducin interactive complex was first produced with the aid of Cluspro 2.0 web server [15]. Out of 10 docked structures, the complex structure having the finest cluster dimension was opted for additional analysis. Rhodopsin, being longer was uploaded as receptor with transducin as ligand. The unstructured residues from both the human proteins were eliminated. Similarly, the so obtained rhodopsin-transducin complex was docked with human arrestin domain to obtain the best trio docked complex structure. Z-DOCK [16] and GRAMM-X [17] helped to obtain the inclusive results for the respective docked complexes at every step.

2.7 Energy Optimization and Refinement Using ModRefiner

Energy minimization and structure refinement of the modeled protein complexes was performed using ModRefiner [18]. Models were refined through high resolution algorithm to mend their unstable geometries and elevate the protein accuracy [18].

2.8 Simulation Studies via Molecular Dynamics for the Trio-Protein Complex

To achieve a steady appropriate protein conformation besides minimizing the net overall energy the assistance of Fragment-Guided Molecular-Dynamics Simulation

(FG-MD) [19]. Knowledge dependant template information as well as physics-based molecular dynamics (MD) simulations was concomitantly utilized to re-organize the pathway for the overall energy in the Molecular Dynamics simulations through simulated annealing [19]. Therefore, the steric clashes were diminished along with the improvement of the torsion angles for the best steady conformation of the complex structure [19].

2.9 Protein–Protein Interaction Calculations

Protein Interaction Calculator (P.I.C.) web server [20] was operated to evaluate and analyze the interaction type involved amongst the proteins in the study. The residual contribution from the rhodopsin-transducin-arrestin complex was therefore analyzed in details. The same was also validated by PyMOL [21] and Accelyrs Discovery Studio 4.1 for consensus results.

2.10 Comparative Analysis of Electrostatic Potential and Net Area of Solvent Accessibility upon the Surface of Rhodopsin

Electrostatic surface potential of the rhodopsin was calculated sequentially using PyMOL [21]. Electrostatic potentials on the surface of rhodopsin in unbound state, followed by rhodopsin in the duo and trio complex respectively were generated in vacuum electrostatics. To compare and analyze comparative estimation for the stronger interaction after interacting with transducin and arrestin, the net solvent accessibility area for human rhodopsin was calculated at the three different stages [22].

2.11 Comparative Analysis of Conformational Switches in Rhodopsin

The conformational shifts in rhodopsin before interaction and after two stages of interaction were estimated and compared with the support of DSSP [23] and Discovery Studio packages from Accelyrs. Earlier investigation suggests that proteins with increased β-sheets and pure α-helical structures accompanied by 3_{10} helices display firmer interaction with steady conformation [23, 24].

3   Results

3.1   Structural Demonstration of Human G-Protein Transducin (α-Subunit)

The homology modeled 3D structure of human transducin (α-subunit), 350 residues long was found to be modeled in parallel to the template protein; GNAT1 possessing PDB ID: 3V00, chain C from *rattus norvegicus*. It comprises 50.3 % residues forming seventeen helices and 12.9 % residues forming four parallel and two anti-parallel β-sheets interspersed with 36.9 % residues forming coils. The entire structure is well depicted in Fig. 1 in its interactive state with rhodopsin and arrestin.

3.2   Analysis of the Residues and Binding Pattern in the Protein Complex

For perceiving the residual participation of the three proteins, nine strong ionic-ionic interactions were perceived for the optimized and simulated trio-complex (Table 1). Several hydrogen bonding interactions were also accomplished.

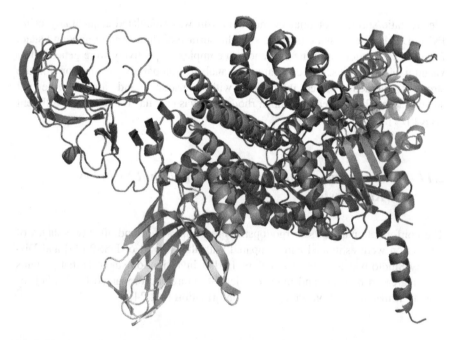

Fig. 1 The interactive view of human rhodopsin (*lime-green*), transducin (with *cyan shaded* helices, *yellow shaded* sheets and *red shaded* coils) and arrestin (*violet shade*)

Table 1 Ionic-Ionic interactions in human Rhodopsin (R)-Transducin (T)-Arrestin (A)

Position	Residue	Protein	Position	Residue	Protein
13	ARG	R	55	ASP	T
15	LYS	R	167	GLU	T
15	LYS	R	55	ASP	T
229	ARG	R	69	GLU	A
408	LYS	R	141	GLU	T
412	ARG	R	232	GLU	T
412	ARG	R	233	ASP	T
474	ARG	R	232	GLU	T
474	ARG	R	235	GLU	T

3.3 Analysis of the Electrostatic Surface Potential and Net Solvent Accessible Surface Area

Furthermore, the variation in the vacuum electrostatic potential calculation from ± 62.322 to ± 52.501 to ± 46.697 (Fig. 2), infers the human rhodopsin protein to get benefitted by the transducin and arrestin interactions for the obstruction of signal transduction. From Fig. 3, the view for the relative analysis of the electrostatic potential values in three different stages of interaction gets illustrated. The blue areas and red areas represent the electrostatically positive and negative regions respectively.

The abrupt decline in the net solvent accessible area value for only human rhodopsin protein from 40330.88 Å^2 before any interaction to 38514.76 and

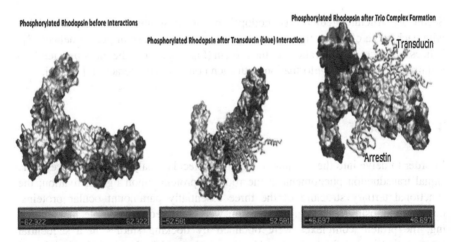

Fig. 2 Comparable analysis of the electrostatic potential upon the protein surface of rhodopsin in three consecutive stages

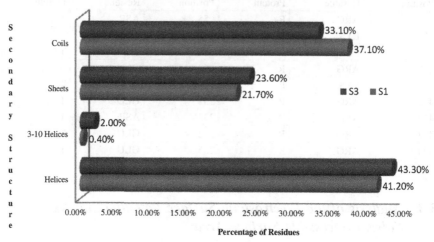

Fig. 3 Comparable study for the conformational alterations in rhodopsin showing percentages of helices, sheets and coils

37730.99 Å^2 post interaction with transducin and arrestin protein respectively, depicts the gradually stronger interaction in the trio complex.

3.4 Comparative Analysis of Conformational Alterations in Rhodopsin

Secondary structure prediction of rhodopsin pre and post interactions ascertained that with each stage of interaction, there exists a notable increase in pure α-helices, 3_{10} helices and sheet concentrations of the protein (Fig. 3). Hence, the rhodopsin tends to acquire a more stable conformation with each consecutive interaction [23, 24].

4 Discussion

In order to delve into the computational and molecular analysis of hindrance in the signal transduction phenomena in the rod cell proteins upon light activation, the functional tertiary structure of the three essentially paramount ocular proteins; human rhodopsin, transducin and arrestin were analyzed. After molecular modeling, the protein complexes were obtained by docking studies in a sequential manner; that is first rhodopsin-transducin complex and further the trio-complex of rhodopsin, transducin and arrestin. The residual contributions for strengthening the final optimized and simulated complex disclosed that only polar positively charged amino-acid residues; arginine from 474, 412, 229, 13 positions and lysine from 15 and 408 positions of rhodopsin interacted strongly with transducin and arrestin. Mainly polar negatively charged; 3 Asp and 5 Glu residues from transducin participated in the strong ionic interaction with rhodopsin while only Glu69 from arrestin was observed to remain indulged in ionic interactions. Several hydrogen bonding interactions were also perceived involving main and side chains of the responsible proteins. Additionally, stability parameters (electrostatic surface potential and net area for solvent accessibility) for alteration in rhodopsin at its three stages (before and after transducin and arrestin interaction) unveiled the retinal G-protein receptor (rhodopsin) to firmly interact with the G-protein (transducin), with an even steadier interaction when arrestin comes into participation with the transducin-rhodopsin complex. Additionally, increase in the helical conformations, 3_{10} helices and β-sheets apprehend rhodopsin to turn conformationally more stable with the gradual sequential interactions.

To concise, this novel in silico study focused upon the basis of the molecular level performance of indispensable retinal proteins upon activation of light and their sequential interaction pattern that gradually but effectively obstructs the signal transduction pathways in ocular physiology. It also explores the alterations in the chief protein; rhodopsin in varied interactive stages. It thus creates a rationale for the residual revelation of modulation in signal transduction in presence of light; when from the cones take active participation for apt vision.

5 Conclusion and Future Scope

This computational and structural insight into the cooperation amongst the three retinal proteins upon photo-activation serves a necessity in computational ophthalmology. The stability and conformational transitions in human retinal rhodopsin (residing in rod cells) upon the chronological interaction with its two partner proteins; transducin and arrestin respectively, discloses the stronger interaction amid the three proteins. Fascinatingly the abundant ionic interactions indulging polar positively charged residues of rhodopsin and negatively charged ones from two partner proteins affirms the optimized and simulated trio-complex to be more interactive. These outcomes altogether apprehends the root core analysis of the modulation of photo transduction phenomena in light. Any mutation or disruption in any of the protein structures might cause fatal retinal degeneracy and associated diseases.

The present investigation thus instigates the future exploration for the mutational analysis in these essential proteins and the process in which the modulation of signaling mechanism gets thus, affected. It would further lead to incite clinical and therapeutic research for a corrective step of the mutations, with supportive drug discovery.

Conflicts of Interest None.

Acknowledgments Authors are grateful for continuous encouragement provided by Dr. Ang-shuman Bagchi, Assistant Professor, Department of Biochemistry and Biophysics, University of Kalyani, India. Authors also deeply acknowledge Department of Biotechnology from National Institute of Technology Durgapur, India and Bengal College of Engineering and Technology, India for their cooperation.

References

1. Artemyev, N.O.: Binding of transducin to light-activated rhodopsin prevents transducin interaction with the rod cGMP phosphodiesterase gamma-subunit. Biochemistry 36(14), 4188–93 (1997)

2. Jason, G. et. al: BenovicSn. Arrestin-Rhodopsin Interaction, Multi-Site Binding Delineated by Peptide Inhibition. The Journal of Biological Chemistry 269(5), 3226–3232 (1994)

3. Lerea, C.L., Somers, D.E., Hurley, J.B., Klock, I.B., Bunt-Milam, A.H.: Identification of specific transducin alpha subunits in retinal rod and cone photoreceptors. Science 234 (4772), 77–80 (1986) doi:10.1126/science.3529395

4. Philips, W.J., Wong, S.C., Cerione R.A.: Rhodopsin/Transducin interactions: Influence of the transducing-βγ subunit complex on the coupling of transducin-α subunit with rhodopsin. Journal of Biological Chemistry 267, 17040–17046 (1992)

5. Simanti, B., Amit, D., Semanti, G., Rakhi, D., Angshuman, B.: Hypoglycosylation of dystroglycan due to T192M mutation: A molecular insight behind the fact. Gene 537, 108–114 (2014)

6. Angshman, B.: Structural characterizations of metal ion binding transcriptional regulator CueR from opportunistic pathogen Pseudomonas aeruginosa to identify its possible involvements in virulence. Appl Biochem Biotechnol (2014) DOI:10.1007/s12010-014-1304-5

7. Biasini, M., et. al.: SWISS-MODEL: modelling protein tertiary and quaternary structure using evolutionary information. Nucleic Acids Research 42(W1), W252-W258 (2014)

8. Altschul, S.F., et. al.: Basic local alignment search tool. Journal of Molecular Biology 25, 403–410 (1990)

9. Fiser, A., Sali, A.: ModLoop: automated modeling of loops in protein structures. Bioinformatics 19 (18), 2500–1 (2003)

10. Eisenberg, D., Luethy, R., Bowie, J.U.: VERIFY3D: assessment of protein models with three-dimensional profiles, Methods Enzymol. 277, 396–404 (1997)

11. Ramachandran, G.N., Sashisekharan, V.: Conformation of polypeptides and proteins, Adv. Protein Chem. 23, 283–438 (1968)

12. Berman, M.H., et. al.: The protein data bank. Nucleic Acids Research 28, 235–242 (2000)

13. Kang, Y., et. al.: Crystal structure of rhodopsin bound to arrestin by femtosecond X-ray laser. Nature 523, 561–567 (2015)

14. Qi, S., O'Hayre, M., Gutkind, J.S., Hurley, J.H.: Insights into ß2-adrenergic receptor binding from structures of the N-terminal lobe of ARRDC3. Protein Sci. 23, 1708–1716 (2014)

15. Comeau, S.R., et al.: ClusPro: An automated docking and discrimination method for the prediction of protein complexes. Bioinformatics. 20, 45–50 (2004)

16. Pierce, B.G. et. al.: ZDOCK Server: Interactive Docking Prediction of Protein-Protein Complexes and Symmetric Multimers. Bioinformatics 30(12), 1771–3 (2014)

17. Tovchigrechko, A., Vakser, I.A.: GRAMM-X public web server for protein-protein docking. Nucleic Acids Research 34, W310-4 (2006)

18. Xu, D., Zhang, Y.: Improving the physical Realism and Structural Accuracy of Protein Models by a Two-step Atomic Level Energy Minimization. Biophysical Journal 101, 2525–2534 (2011)

19. Zhang, J., Liang, Y., Zhang, Y.: Atomic-Level Protein Structure Refinement using Fragment Guided Molecular Dynamics Conformation Sampling. Structure 19, 1784–1795 (2011)
20. Tina, K.G., Bhadra, R., Srinivasan, N.: PIC: Protein Interactions Calculator. Nucleic Acids Research 35 (suppl 2), W473-6 (2007)
21. DeLano, W.L.: The PyMOL Molecular Graphics System. DeLano Scientific, San Carlos, CA, USA (2002)
22. Gerstein, M.: A Resolution Sensitive Procedure for Comparing Protein Surfaces and its Application to the Comparison of Antigen-Combining Sites. Acta Cryst. A(48), 271–276 (1992)
23. Paul, D., Thomas, Ken, Dill, A.: Local and nonlocal interactions in globular proteins and mechanisms of alcohol denaturation. Protein Science 2, 2050–2065 (1993)
24. Toniolo, C., Benedetti, E.: Trends Biochem. Sci 16, 350–353 (1991)

Evaluating the Performance of State of the Art Algorithms for Enhancement of Seismocardiogram Signals

Aditya Sundar and Vivek Pahwa

Abstract Seismocardiography is a new, low cost and non-invasive method for measurement of local vibrations in the sternum due to cardiac activity. Signals recorded using this procedure are termed as seismocardiogram or SCG signals. Analysis of SCG signals provides information about the functionality of the cardiovascular system. Performing an automatic diagnosis using SCG signals involves the use of signal processing, feature extraction and learning machines. However for such methods to yield reliable results, the digitally acquired SCG signals should be accurately denoised and free from artifacts. In this paper, we evaluate the performance of state of the art algorithms in denoising these signals. In our work, clean SCG signals were corrupted with additive white Gaussian noise and the signals were further denoised. Denoising using wavelet transforms, empirical mode decomposition, adaptive filters and morphological techniques has been considered in our work. Standard metrics: mean squared error (MSE), mean absolute error (MAE), signal to noise ratio (SNR), peak signal to noise ratio (PSNR), cross correlation (xcorr) and CPU consumption time have been computed to assess the performance the aforementioned techniques. From our study it is concluded that wavelet thresholding yields the best denoising and is hence the most suitable method for enhancement of real world SCG signals.

Keywords Seismocardiography (SCG) · Signal denoising · Empirical mode decomposition (EMD) · Detrended fluctuation analysis (DFA) · Morphological filtering

Aditya Sundar (✉) · Vivek Pahwa
BITS, Pilani K.K. Birla Goa Campus, Sancoale, India
e-mail: aditsundar@gmail.com

Vivek Pahwa
e-mail: viv.pahwa@gmail.com

© Springer Science+Business Media Singapore 2017
J.K. Mandal et al. (eds.), *Proceedings of the First International Conference on Intelligent Computing and Communication*, Advances in Intelligent Systems and Computing 458, DOI 10.1007/978-981-10-2035-3_5

37

1 Introduction

Seismocardiography involves the non invasive measurement of vibrations in the chest due myocardial movement. SCG is a useful tool in estimation of the mechanical functioning of the cardiovascular system. Analysis of SCG signals shows promise for performing diagnosis of ischemia [1]. SCG is also a simple technique for assessing myocardial contractility [2]. These signals have also been used for developing automatic, real time diagnosis systems. Schlager et al. [3] have proposed a real time system for diagnosis of myocardial ischemia using SCG signals. Building automatic diagnosis systems involves the use of signal processing, feature extraction and machine learning algorithms. The reliability of such systems depends greatly upon the sensitivity of the algorithm to the presence of noise in the signal. Digitally acquired SCG signals contain several artifacts, introduced due to power line interference, improper mounting of the sensor, subject movement during recording and other random environmental artifacts [4]. Thus it is critical that the signal be quickly and accurately denoised so that an appropriate real time diagnosis can be performed. Although SCG denoising is critical due to the aforementioned reasons, an extensive study into the same has not been conducted in previous literature. This was the motivation behind conducting a study into the same. In this paper we evaluate the performance of state of the art algorithms in denoising SCG signals. Denoising using the following methods has been considered in our work:

(1) Wavelet transform: Wavelet soft thresholding, Multivariate wavelet denoising
(2) Empirical mode decomposition: Empirical mode decomposition-detrended fluctuation analysis (EMD-DFA)
(3) Adaptive filters: Normalized least squares filters (NLMS) and Recursive least squares filters (RLS)
(4) Morphological filtering using top hat transform

The gaussian white noise model (WGN) is commonly used in information theory to mimic the effects of random processes. WGN closely resembles the type of noise present in real world biomedical signals [5]. In our method, clean DVP signals (ground truth) were acquired from the Physionet database and corrupted with additive white Gaussian noise of SNR = 10 db. These signals were then denoised and metrics were computed to assess the similarity between the initial, clean and the denoised signals.

Standard metrics: mean squared error (MSE), mean absolute error (MAE), signal to noise ratio (SNR), PSNR (peak signal to noise ratio), cross correlation and CPU consumption time have been computed to assess the performance of the different methods. Figure 1 shows a block diagram of the procedure followed in this paper.

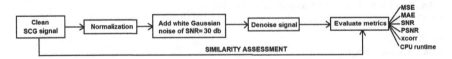

Fig. 1 Block diagram depicting the procedure followed in this paper

2 Methodology

2.1 Database

The signals used in our work have been acquired by IEB research team at Universitat Politecnica de Catalunya [6, 7]. The signals used are available in the Physionet CEBS database. The dataset comprises of simultaneous estimation of ECG, respiratory and seismocardiogram signals [5]. The dataset comprises of the recording of 20 healthy volunteers. Measurements were recorded with the subjects in a still and comfortable position. The measurements were carried out during 3 different intervals: the initial still basal state, state of listening to classical music and the final basal state. The SCG signals were recorded using a triaxial accelerometer. The signals were low-passed at 100 Hz and high-passed at 0.5 Hz. The signals were acquired using Biopac DAQ system. The signals were sampled at frequency of 5 kHz. In our work, we divide have divided the signals into 10 s frames and denoised each of the segments using different techniques. Using this procedure the performance of different methods in denoising a total of 60 signals has been evaluated. Figure 2 shows a sample clean SCG signal.

2.2 Wavelets for Denoising

Since it's advent, wavelet transforms have been used for denoising signals [8]. In this section we investigate the performance of 2 wavelet based denoising methods: wavelet-thresholding and multivariate wavelet denoising (wavelet-PCA) for denoising SCG signals.

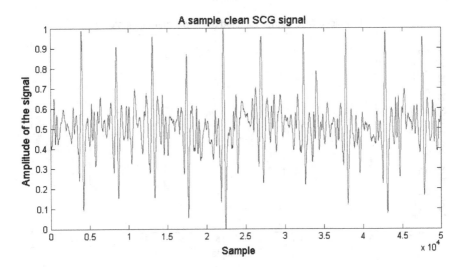

Fig. 2 Sample noisy SCG recording of 10 s

2.2.1 Wavelet-Thresholding for Denoising

Wavelet decomposition coupled with soft thresholding, hard or universal thresholding methods has shown a good performance in denoising biomedical signals and images in previous works [9–11]. In this method the signal is first decomposed into different levels yielding the detail and approximation coefficients. The value of threshold is then computed using certain rules. The wavelet sub-bands are then soft/hard or universally thresholded using this threshold value. The denoised signal is then obtained after reconstructing the thresholded coefficients. A Monte Carlo based approach has been followed in tuning and selection of the optimal denoising parameters. Figure 2 shows the a noisy signal with SNR = 30 db gaussian noise

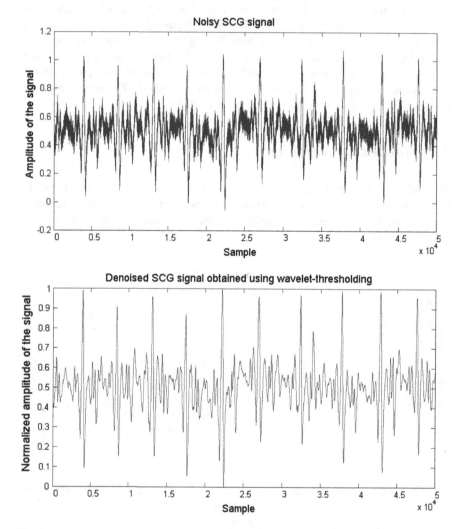

Fig. 3 Denoised signal obtained using wavelet-soft thresholding

added to the signal in Fig. 2, and its denoised version obtained using wavelet-thresholding. The authors have found that biorthogonal 3.3 (bior 3.3) wavelet with 6 levels of decomposition with penalized medium threshold and hard thresholding yields the best denoising. The value of tuning parameter alpha was set equal to 2 and standard deviation of the zero mean Gaussian white noise is set equal to that of the 3rd detail coefficient.

2.2.2 Multivariate Wavelet Denoising (Wavelet-PCA)

Multivariate wavelet denoising is a noise removal algorithm which combines univariate wavelet decomposition with principal component analysis (PCA). This method incorporates univariate wavelet decomposition, in the basis where the assessed noise co-variance matrix is diagonal to non-centered PCA approximations in the wavelet domain. This method was proposed by Aminghafari et al. [12], and since then has been used in several denoising applications. Wavelet-PCA has shown promising results in denoising ECG signals in previous literature and is hence explored for our application [13]. After investigation, it is found that a 5 level wavelet decomposition with a 5th order Coiflet wavelet yields the best denoising results. The Kaiser's rule or the heuristic rule has been used to pick the number of principal components to be retained.

2.3 Denoising Using Empirical Mode Decomposition

Empirical mode decomposition, proposed by Huang is non-linear data driven decomposition tool [14]. EMD works by sifting the signal into detail components called IMFs or intrinsic mode functions and the trend called the residue. In EMD, decomposition occurs iteratively and the sifting processing is terminated using a predefined stopping criterion. In our work, we use the stopping criterion proposed by Rilling et al. [15], where two threshold values are chosen to ensure globally small fluctuations in the mean, while simultaneously taking into account locally large excursions. Threshold values of 0.05 and 0.5 have been chosen as threshold values in our work. The original signal can be retrieved by summing all the decomposed IMFs. Works by Sundar et al. [16, 17] suggest that EMD and similar techniques such as VMD yields a good performance in denoising knee joint VAG signals. Denoising using empirical mode decomposition involves the identification of IMFs that are noisy or that resemble a noise like behavior. Methods such as hurst exponent estimation and detrended fluctuation analysis have been proposed for the same. The hurst exponent however tends to yield spurious scores when dealing with non stationary signals, which is why we adopt the use of empirical mode decomposition-detrended fluctuation analysis (EMD-DFA). Figure 3 shows the noisy and denoised signal obtained using this method. The steps involved in EMD-DFA denoising are listed below:

(1) Decompose the signal into 'N' different Intrinsic mode functions (IMFs) using empirical mode decomposition
(2) Perform detrended fluctuation analysis on each of the IMFs and compute the fractal scaling index value (α)
(3) If the value of $\alpha \leq 0.5$, then discard that IMF
(4) Reconstruct the signal using the leftover IMFs to obtain the denoised signal

2.4 Adaptive Filtering

Adaptive filtering is a method of repeatedly modeling the relationship between filter inputs and outputs subject to certain conditions. The advantage of the adaptive filters over conventional filters is that it self-tunes the filter coefficients for new inputs [18]. In our work, we have considered the NLMS and RLS filters for denoising the SCG signals. For denoising the sample signal shown in Fig. 1, a 30th order FIR filter with a value frequency constraint scalar of 0.01 has been used. The value of NLMS step size is set as 1 and the value NLMS offset was set to 40. The number of taps of the filter was set equal to 32. The NLMS leakage factor is set equal to 1. In denoising using RLS filters, 30th order FIR filter with a value frequency constraint scalar of 0.01 has been used. The value of forgetting factor was set to 0.98. The initial inverse covariance matrix was set equal to a 10 multiplied by an identity matrix of size 31 $\times$ 31.

2.5 Morphological Filtering

Morphological filtering involves a set of non-linear operations, performed on a signal with respect to a small image, termed as the SE or the structuring element [19]. Top hat transform is a morphological filtering method used to extract small details in an image or signal. Works by Bhateja et al. [20] propose that the use of a structuring element such as the one shown in Fig. 4 can be used to estimate the noise in the signal. This estimated noise is then removed from the signal. Figure 5 shows the total morphologically estimated noise in the sample noisy signal in Fig. 1 obtained using top hat transform with the structuring element in Fig. 4.

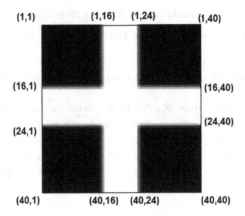

Fig. 4 SE used in top hat transform for estimating the noise

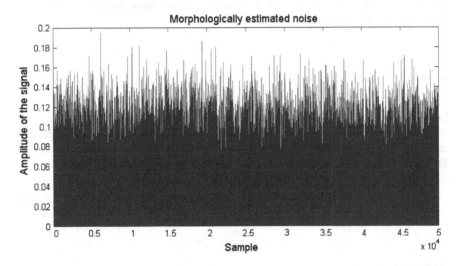

Fig. 5 Total morphologically estimated noise contained in the sample SCG signal in Fig. 1

3 Results Obtained on Denoising

To assess the performance of denoising algorithms metrics: MSE, MAE, SNR, PSNR, xcorr and CPU consumption have been computed. The average value of metrics for obtained on denoising the signals using each methods is calculated and is shown in Table 1. Low values of MSE, MAE and high values of SNR, PSNR indicate accurate denoising. Also the closer the value of xcorr is to 1, the more is the retention in structural integrity after denoising, or better is the enchancement. A small value of CPU consumption time is desired so that a fast, real time diagnosis can be performed.

Table 1 Average value of metrics obtained on denoising 60 SCG signals

Denoising method	MSE	MAE	SNR (db)	PSNR (db)	xcorr	CPU consumption time (s)
Wavelet thresholding	0.00000013	0.000547	57.758393	63.358323	0.999986	0.586163
Wavelet-PCA	0.0000340	0.004646	39.044076	44.644006	0.998976	0.415935
EMD-DFA	0.0000370	0.004830	38.722700	44.322629	0.998897	7.901136
NLMS filter	0.0005660	0.015538	26.873512	32.473442	0.983036	1.035071
RLS filter	0.0005598	0.015496	26.889970	32.50335	0.983099	1.205344
Morphological filters	0.0037760	0.052018	18.645570	24.899094	0.969359	1.418991

From Table 1 it observed that denoising using wavelet thresholding yields the best results in terms of accuracy, as well as a low CPU consumption time. Hence the authors conclude that wavelet thresholding is the most suitable method for denoising real world SCG signals.

4 Conclusion

In this paper we evaluate the performance of state of the art algorithms in enhancement of seismocardiogram signals. Denoising using wavelets, empirical mode decomposition, adaptive filters and morphological filters has been explored. The authors evaluate both the speed as well the accuracy of each method. It is concluded that wavelet thresholding yields the best enhancement in terms of accuracy of enhancement as well as CPU consumption time. We hope this study would be useful to future researches in study and selection of SCG denoising algorithms.

References

1. Korzeniowska K, Bilińska M, Piotrowicz R, "Usefulness of seismocardiography for the diagnosis of ischemia in patients with coronary artery disease", Annals of Noninvasive electrocardiology, Volume 10, Issue 3, Pg. 281–7 (2005), doi:10.1111/j.1542-474X.2005. 00547
2. Mikko Paukkunen, "Seismocardiography: Practical implementation and feasibility", Aalto University publication series, Doctoral dissertations 145/2014, (2014) ISBN 978-952-60-5874-0
3. Kenneth J. Schlager, Bruce H. Boehlen, Stephen H. Gorski, Automated seismic detection of myocardial ischemia and related measurement of cardiac output parameters, US Patent US6024705 A, (2000)
4. Zanetti J.M, Poliac M.O., Crow R.S, Seismocardiography: waveform identification and noise analysis, Proceedings of IEEE Computers in Cardiology 1991, Pg. 49–52 (1991), doi:10.1109/ CIC.1991.169042

5. Goldberger AL, Amaral LAN, Glass L, Hausdorff JM, Ivanov PCh, Mark RG, Mietus JE, Moody GB, Peng C-K, Stanley HE. PhysioBank, PhysioToolkit, and PhysioNet: Components of a New Research Resource for Complex Physiologic Signals. Circulation 101(23):e215-e220 [Circulation Electronic Pages; http://circ.ahajournals.org/cgi/content/full/101/23/e215]; 2000 (June 13).

6. G Gonzalez M.A, Argelagos-Palau A., Fernandez-Chimeno M., Ramos-Castro J, A comparison of heartbeat detectors for the seismocardiogram, Proceedings of IEEE Computing in Cardiology Conference 2013, Pg. 461–464, (2013) ISBN:978-1-4799-0884-4

7. G Gonzale, A Palau. F Chimen, Ramos-Castro J., Differences in QRS Locations due to ECG Lead: Relationship with Breathing", XIII Mediterranean Conference on Medical and Biological Engineering and Computing 2013, IFMBE Proceedings Volume 41, 2014, Pg 962–964 (2013), doi:10.1007/978-3-319-00846-2_238

8. Adriano Z. Zambom and Ronaldo Dias, A Review of Kernel Density Estimation with Applications to Econometrics (2012), arXiv:1212.2812

9. Weyrich N, Warhola G.T, Wavelet shrinkage and generalized cross validation for image denoising, IEEE Transactions on Image Processing, Volume 7, Issue 1, Pg. 82–90.(1998), doi:10.1109/83.650852

10. V.V.K.D.V. Prasad, P. Siddaiah, B. Prabhakara Rao Denoising of Biological Signals Using Different Wavelet Based Methods and their Comparison, Asian Journal of Information Technology, Pg. 146–149, 2008, ISSN:1682-3915

11. E. Castillo, D. P. Morales, A. García, F. Martínez-Martí, L. Parrilla, and A. J. Palma, Noise Suppression in ECG Signals through Efficient One-Step Wavelet Processing Techniques, Journal of Applied Mathematics, Volume 2013, Article ID 763903, http://dx.doi.org/10.1155/2013/763903

12. Mina Aminghafari, Nathalie Cheze, Jean-Michel Poggi, Multivariate denoising using wavelets and principal component analysis, Computational Statistics & Data Analysis, Volume 50, Issue 9, Pg. 2381–2398, (2006), doi:10.1016/j.csda.2004.12.010

13. Sharma L.N, Dandapat S, Mahanta A, Multiscale principal component analysis to denoise multichannel ECG signals, Proc of 5th IEEE Cario International Biomedical Engineering Conference, Pg. 17–20, (2010), doi:10.1109/CIBEC.2010.5716093

14. Flandrin, P, Goncalves, P, Rilling, G, Detrending and denoising with empirical mode decomposition, Proceedings of IEEE 12th European Signal Processing Conference, 2004, pg, 1581–1584, ISBN:978-320-0001-65-7

15. Sundar A, Das C, Pahwa V, Denoising Knee Joint Vibration Signals Using Variational Mode Decomposition, Volume 433 , Advances in Intelligent Systems and Computing pp 719-729, doi:10.1007/978-81-322-2755-7_74

16. Sundar A, Pahwa V, Das C, "A new method for denoising knee joint vibroarthrographic signals", proceedings of 2015 Annual IEEE India Conference (INDICON), pp 1-5, doi:10.1109/INDICON.2015.7443443

17. Sundar A, Pahwa V, Das C, A new method for denoising knee joint Vibration signals, Proceedings of 12th IEEE INDICON (2015)

18. B. Widrow, "Adaptive Filters," Aspects of Network and System Theory, R.E. Kalman and N. DeClaris, eds., Holt, Rinehart and Winston, pp. 563–586, 1971

19. Maragos P, Schafer R.W, Morphological filters–Part I: Their set-theoretic analysis and relations to linear shift-invariant filters, IEEE Transactions on Acoustics, Speech and Signal Processing, Volume 35, Issue 8, Pg. 1153–1169, (1987) doi:10.1109/TASSP.1987.1165259

20. Rishendra Verma, Rini Mehrotra, Vikrant Bhateja, A New Morphological Filtering Algorithm for Pre-Processing of Electrocardiographic Signals, Lecture Notes in Electrical Engineering, Volume 1, Pg. 193–201, (2013), doi:10.1007/978-81-322-0997-3_18

Common Subcluster Mining to Explore Molecular Markers of Lung Cancer

Arnab Sadhu and Balaram Bhattacharyya

Abstract We present a methodology, common subcluster mining, to explore gene expression data for possible biomarkers of lung cancer. Subclusters refer to the peaks formed through superimposition of clusters obtained from expression data of normal samples. Application of the method on the corresponding data sets from diseased samples extracts the genes that undergo high fold changes. The potential candidate genes are examined on the datasets of Stage I through stage IV of the disease. Few genes emerge as indicative molecular markers of lung cancer.

Keywords Common subcluster mining · Lung cancer biomarker · Microarray clustering · Differential gene expression · Molecular marker

1 Introduction

DNA microarray technology measures simultaneous expressions of thousands of genes in both normal and corresponding diseased cells and offers a scope for studying differential patterns. Influence of disease on gene expressional patterns can be a key for exploring possible biomarker of the disease at molecular level. Lung cancer is one of the deadliest types and external symptoms develop after considerable spread. Finding its genetic relation is thus important for early diagnosis. Data from Microarray experiments on cell samples affected with lung cancer is the principal source for the study. Developing appropriate methods for finding aberrations on expression levels of genes while cells are infected by the disease is the basic task of the present study.

Arnab Sadhu · Balaram Bhattacharyya (✉)
Department of Computer and System Sciences, Visva-Bharati University,
Santiniketan 731235, India
e-mail: balaramb@gmail.com; balaram.bhattacharyya@visva-bharati.ac.in

Arnab Sadhu
e-mail: arnabsadhu.rs@visva-bharati.ac.in

© Springer Science+Business Media Singapore 2017   47
J.K. Mandal et al. (eds.), *Proceedings of the First International Conference on Intelligent Computing and Communication*, Advances in Intelligent Systems and Computing 458, DOI 10.1007/978-981-10-2035-3_6

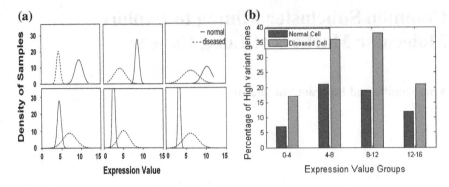

Fig. 1 **a** Distribution of cell samples over expression levels (genes clock-wise from *top left corner*): AGER, SFTPC, SCGB, MMP12, TOX3, COL10A1. **b** Effects of disease on variance of genes

An empirical approach is conducted with computation of probability mass function (pmf) of expressions sliced into small subranges. Nearly 21 % of genes shift corresponding subrange under diseased condition among which some shows significant changes (Fig. 1a). This indicates presence of influence of the disease on functioning of lung cells and thus motivates for detailed study. A study on changes of variance of expressions (Fig. 1b) also suggest the same.

1.1 *Our Contribution*

We propose a method, Common Subcluster Mining (CSM) with the associated algorithm, for discovery of the genes having persistently stable expressional patterns in normal cells. It further mines those genes that deviate from the stable property with sublime changes while cells are infected by the disease. The rest of the paper is organized as follows. Section 2 discusses some earlier works that are related with the present study, Sect. 3 illustrates the proposed method. Section 4 presents the algorithm, its description and complexity, Sect. 5 reports the experimental results, and finally the conclusion in Sect. 6.

2 Related Works

Data mining techniques have wide applications in geneset discovery. Out of those, clustering, being unsupervised, is found to be the most effective in finding target genesets from expression data [1]. Attempts are made to identify genetic relationship of the disease [2] to detect possible attack. Several studies employed variance in expression [3] and the degree of connectivity in gene network [4] for classification of genes.

3 Methods

3.1 Common Subcluster Mining (CSM)

The method aims to capture the sets of genes that express to the same or nearly the same level among all normal samples, i.e., expression behavior is stable under normal condition. On a second phase it attempts to find the individual genes showing significant shift from the respective stable states in the diseased samples. Clustering, being the unsupervised data mining, is chosen to apply on dataset of each normal sample separately. Both crisp and fuzzy clustering techniques are followed to extract the most stable sets of genes (Figs. 2, 3 and 4).

CSM with Crisp Clustering (CSMC) The k-means clustering technique is employed. Seeds are drawn using k-means++ algorithm [5]. Datasets from normal samples are taken in succession for cluster formation. Clusters of close centers are superimposed forming heaps. Heaps of height equal to the number of samples are selected for extracting their central parts. Genes in those parts, irrespective of expression levels, marked as stable set, G_{normal}^{CS}.

The corresponding set in case of diseased samples is denoted by $G_{diseased}^{CS}$. Resulting set of genes (G_ξ) are obtained in order of respective fold change, $\phi(G_i)$:

$$\phi(G_i) = \log_2(\varrho(Diseased_{G_i})/\varrho(Normal_{G_i})) \qquad (1)$$

where ϱ denotes expression.

Fig. 2 Core in normal samples

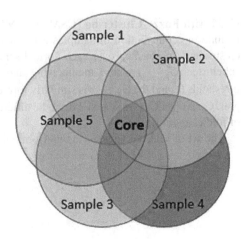

Fig. 3 Distribution of genes into core and non-core part with 8 clusters

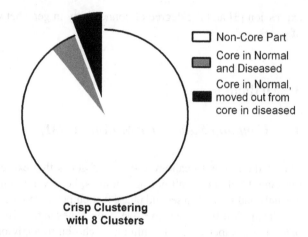

Fig. 4 Normal quatile plot of the t-scores of genes in Table 2

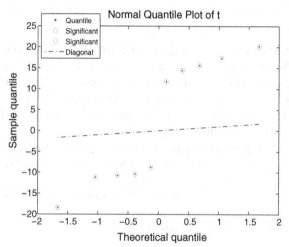

CSM with Fuzzy Clustering (CSMF) CSMF relaxes the crisp boundary so as to include the contribution of samples even moderately distant from the core. FCM clustering is performed over each normal sample. Genes are included in common subcluster (G_{normal}^{CS}) whose membership and sample counts respectively exceeds thresholds θ_1 and θ_2. $G_{diseased}^{CS}$ is similarly formed from the datasets of diseased samples. Genes (set G_n) are selected from the differential set, ($G_{normal}^{CS} \setminus G_{diseased}^{CS}$) in order of fold change (Eq. 1).

The set of selected genes is drawn from union of G_ξ and G_n.

4 Algorithm

Notation used in the algorithm are: G = number of Genes; k = number of clusters; N = number of Samples; F_i = fuzzifier Index (generally selected between 1.5 and 2.5); MV = three dimensional membership value matrix; θ_1 and θ_2 = predefined threshold values;

Complexity of CSM algorithm lies in k-means and fcm clustering. K-means has a complexity of $O(Gki)$ (as number of attribute = 1) and CSMF will have a complexity of $O(Gk^2i)$, as fcm is executed for N samples. So the complexity of the algorithm is $O(Gk^2i)$;

Algorithm CSM

Input: Two sets of GED: **Normal** sets and **Diseased** sets.
Output: Genes which are highly differentially expressed.

1: **procedure** CORE GENESET MINING
2: $\quad G^{CS}_{normal} = CSMC(Normal, G, k); G^{CS}_{diseased} = CSMC(Diseased, G, k);$
3: $\quad G_\xi = HighFoldChange(G^{CS}_{normal} \setminus G^{CS}_{diseased});$
4: $\quad G^{CS}_{normal} = CSMF(Normal, G, k); G^{CS}_{diseased} = CSMF(Diseased, G, k);$
5: $\quad G_\eta = HighFoldChange(G^{CS}_{normal} \setminus G^{CS}_{diseased});$
6: $\quad$ Solution set, $G^{CS} = (G_\xi \cup G_\eta);$
7: **end procedure**
8: **procedure** ($S^{core} = $CSMC(Sample, G, k))
9: $\quad$ select initial centroids m_i, (i=1 to k) by kmeans++ technique;
10: $\quad m_i^{update} = m_i \ \forall m_i(i = 1 \ to \ k);$
11: $\quad$ **for** $p = 1 \ to \ |G|$ **do**
12: $\quad\quad S_i^{(1)} = G_p : ||G_p - m_i||^2 <= ||G_p - m_j||^2 \ \forall \ 1 <= j <= k$
13: $\quad\quad m_i^{update} = \frac{1}{|S_i^{(1)}|} \sum_{G_j \epsilon S_i^{(1)}} G_j$
14: $\quad$ **end for**
15: $\quad$ **while** $m_i \neq m_i^{update} \forall \ m_i \ (i = 1 \ to \ k)$ **do**
16: $\quad\quad$ sort m_i^{update}s in ascending order.
17: $\quad\quad$ **for** j=2 to N **do**
18: $\quad\quad\quad m_i = m_i^{update} \ \forall m_i(i = 1 \ to \ k)$
19: $\quad\quad\quad$ **for** $p = 1 \ to \ |G|$ **do**
20: $\quad\quad\quad\quad S_i^{(j)} = G_p : ||G_p - m_i||^2 <= ||G_p - m_j||^2 \ \forall \ 1 <= j <= k$

21:                $m_i^{update} = \frac{1}{|S_i^{(l)}|} \sum_{G_j \epsilon S_i^{(l)}} G_j$
22:              end for
23:            end for
24:        end while
25:        for $i = $ to k do
26:            $S_i^{core} = S_i^{(1)} \cap S_i^{(2)} \cap S_i^{(3)} \cap .... \cap S_i^{(N)}$
27:        end for
28: end procedure
29: procedure (S^{core} =CSMF(Sample, G, k))
30:      for i=1 to N do
31:          $C(i) = $ Randomly selected initial centroids $\forall m_i(i = 1 \, to \, k)$;
32:          $C, M = fcm(Sample(i), k, F_i)$;
33:          Sort C and corresponding membership values in ascending order;
34:          for $j = 1 \, to \, |G|$ do
35:              for $count = 1 \, to \, k$ do
36:                  MV(j, count, i) = M(j,count);
37:              end for
38:          end for
39:      end for
40:      for count=1 to k do
41:          for i=1 to G do
42:              for j=1 to N do
43:                  if $MV(i, count, j) > \theta_1$ then
44:                      S(i) = S(i) +1;
45:                  end if
46:              end for
48:              if $S(i) > \theta_2$ then
49:                  $S^{core} = S^{core} \, U \, G(i)$;
50:              end if
51:          end for
52:      end for
53: end procedure

5   Results and Discussion

5.1   Dataset

The algorithms CSMC and CSMF are tested on gene expression dataset, GDS3257 [6]. It contains expressions of adenocarcinoma lung and paired normal lung tissues of current, former and never smokers. Four different stages of cancer and corresponding expression levels of 22283 genes is mentioned in the dataset, among them 22215 were test genes and 68 genes were control genes. After removing the genes with missing value entries, 21805 genes remained. A total number of 107 patient samples are collected, 33 among them had both normal and corresponding diseased samples.

Table 1 Numbers of genes in common subcluster

Number of clusters	Size of G_{normal}^{CS} in %		Size of *($G_{normal}^{CS}/G_{diseased}^{CS}$) in %	
	CSMC	CSMF	CSMC	CSMF
4	40	35	15	1
8	11	10.5	6	3
12	3	4	2	4

Table 2 Most affected genes found in both the studies

Possible markers		Fold change	p-value	t-scores	Supported by previous studies
CSMC	CSMF				
AGER	AGER	−0.68	$9.21 \times e^{-31}$	+20.24	[7]
TOX3	TOX3	+0.57	$1.09 \times e^{-13}$	−10.16	[8]
SFTPC	SFTPC	−0.50	$6.51 \times e^{-14}$	+11.79	[9]
CLDN18	CLDN18	−0.50	$5.36 \times e^{-22}$	+17.32	
	SPP1	+0.69	$1.71 \times e^{-28}$	−18.4	[10]
	COL11A1	+0.65	$1.48 \times e^{-13}$	−11.17	[11]
	MMP1	+0.62	$5.26 \times e^{-11}$	−8.72	[12]
	GREM1	+0.57	$1.08 \times e^{-12}$	−10.36	[13]
	CAV1	−0.34	$7.33 \times e^{-20}$	+14.40	[14]
	FHL1	−0.33	$9.63 \times e^{-22}$	+15.66	[15]

5.2 CSMC

The algorithm CSMC is implemented in Matlab and tested on the dataset GDS3257 with number of clusters successively k = 4, 8, 12 and 16. Table 1 shows the results. Genes which changes their domain clusters in diseased samples, are ranked upon their respective fold change (Eq. 1). K-means clustering is applied with 4 clusters, 700 iterations and 3 replications. With the selection criteria $|FoldChange| \geq 0.5$ we found 1 up-regulated and 3 down-regulated genes (Table 2).

5.3 CSMF

The Fuzzy version of CSM returns a few more genes than its crisp counterpart. The CSMF procedure is executed with fuzzifier index = 2.0. θ_1 and θ_2 are set to 0.7 and 83 % (i.e., 30 out of 36 samples) respectively. Like CSMC the numbers of genes in common subcluster is inversely proportional to the number of clusters (Table 1). The result with $k = 8$ is depicted in the Table 2.

Table 3 Stage-wise fold changes of the genes

Gene	Stage 1	Stage 2	Stage 3	Stage 4
AGER	−0.70	−0.62	−0.76	−0.57
TOX3	+0.49	+0.68	+0.64	+0.46
SFTPC	−0.73	−0.63	−0.7	−0.45
CLDN18	−0.52	−0.50	−0.51	−0.38
SPP1	+0.68	+0.67	+0.71	+0.72
COL11A1	+0.55	+0.72	+0.69	+0.74
MMP1	+0.62	+0.58	+0.56	+0.92
GREM1	+0.55	0.64	0.53	+0.64
CAV1	−0.35	−0.40	−0.28	−0.25
FHL1	−0.33	−0.32	−0.35	−0.34

5.4 Possible Markers

Table 2 shows the list of genes obtained from CSMC and CSMF algorithms. While CSMC contains more stringent measures with extraction of only four molecules, CSMF extracts some more due to the contribution from other parts of the clusters leading to a little longer list. Intersection of both the outcome may be considered as possible markers with higher confidence but some of the outcome of only CSMF show high fold change and corroborated by previous studies.

5.5 Sensitivity on Advancement of the Disease

Lung adenocarcinoma is typed into four stages—stage 1 through 4. Table 3 shows increase in fold change for all the genes with advancement of the disease through stages. This is indicative of the level of confidence of the results obtained.

6 Conclusion

The study attempts to capture the genes that possess most stable behavior under normal condition with the aim to trace those which exhibit strong deviation while under diseased condition. The gene set thus obtained are subjected to further studies of their sensitivity on advancement of the disease. The final set thus obtained can be a set of possible molecular markers for lung cancers. The algorithm thus developed can be applied on similar datasets of other diseases also. The method leaves scope for improvement with incorporational biological implications.

References

1. Marcilio CP de Souto, Ivan G Costa, Daniel SA de Araujo, Teresa B Ludermir, and Alexander Schliep. Clustering cancer gene expression data: a comparative study. *BMC bioinformatics*, 9(1):497, 2008.
2. Julie George, Jing Shan Lim, Se Jin Jang, Yupeng Cun, Luka Ozretić, Gu Kong, Frauke Leenders, Xin Lu, Lynnette Fernández-Cuesta, Graziella Bosco, et al. Comprehensive genomic profiles of small cell lung cancer. *Nature*, 524(7563):47–53, 2015.
3. Samuel Manda, Rebecca Walls, and Mark Gilthorpe. A full bayesian hierarchical mixture model for the variance of gene differential expression. *BMC Bioinformatics*, 8(1):124, 2007.
4. Jessica C. Mar, Nicholas A. Matigian, Alan Mackay-Sim, George D. Mellick, Carolyn M. Sue, Peter A. Silburn, John J. McGrath, John Quackenbush, and Christine A. Wells. Variance of gene expression identifies altered network constraints in neurological disease. *PLoS Genet*, 7(8):e1002207, 08 2011.
5. David Arthur and Sergei Vassilvitskii. K-means++: The advantages of careful seeding. In *Proceedings of the Eighteenth Annual ACM-SIAM Symposium on Discrete Algorithms*, SODA '07, pages 1027–1035, Philadelphia, PA, USA, 2007. Society for Industrial and Applied Mathematics.
6. Maria Teresa Landi, Tatiana Dracheva, Melissa Rotunno, Jonine D Figueroa, Huaitian Liu, Abhijit Dasgupta, Felecia E Mann, Junya Fukuoka, Megan Hames, Andrew W Bergen, Sharon E Murphy, Ping Yang, Angela C Pesatori, Dario Consonni, Pier Alberto Bertazzi, Sholom Wacholder, Joanna H Shih, Neil E Caporaso, and Jin Jen. Gene Expression Signature of Cigarette Smoking and Its Role in Lung Adenocarcinoma Development and Survival. *PLoS ONE*, 3(2):e1651, 2008.
7. Junjie Fu, Ravil Khaybullin, Xiao Liang, Madeleine Morin, Amy Xia, Anderson Yeh, and Xin Qi. Discovery of gene regulation pattern in lung cancer by gene expression profiling using human tissues. *Genomics Data*, 3:112–115, 2015.
8. Mathewos Tessema, Christin M Yingling, Marcie J Grimes, Cynthia L Thomas, Yushi Liu, Shuguang Leng, Nancy Joste, and Steven A Belinsky. Differential epigenetic regulation of tox subfamily high mobility group box genes in lung and breast cancers. *PloS one*, 7(4):e34850, 2012.
9. Ruiyun Li, Nevins W. Todd, Qi Qiu, Tao Fan, Richard Y. Zhao, William H. Rodgers, Hong-Bin Fang, Ruth L. Katz, Sanford A. Stass, and Feng Jiang. Genetic deletions in sputum as diagnostic markers for early detection of stage i non-small cell lung cancer. *Clinical Cancer Research*, 13(2):482–487, 2007.
10. Yan Lu, Yijun Yi, Pengyuan Liu, Weidong Wen, Michael James, Daolong Wang, and Ming You. Common human cancer genes discovered by integrated gene-expression analysis. *PLoS ONE*, 2(11):e1149, 11 2007.
11. Kan-kan Wang, Ni Liu, Nikolina Radulovich, Dennis A Wigle, Michael R Johnston, Frances A Shepherd, Mark D Minden, and Ming-Sound Tsao. Novel candidate tumor marker genes for lung adenocarcinoma. *Oncogene*, 21(49):7598–7604, 2002.
12. Wiebke Sauter, Albert Rosenberger, Lars Beckmann, Silke Kropp, Kirstin Mittelstrass, Maria Timofeeva, Gabi Wlke, Angelika Steinwachs, Daniela Scheiner, Eckart Meese, Gerhard Sybrecht, Florian Kronenberg, Hendrik Dienemann, Jenny Chang-Claude, Thomas Illig, Heinz-Erich Wichmann, Heike Bickebller, and Angela Risch. Matrix metalloproteinase 1 (mmp1) is associated with early-onset lung cancer. *Cancer Epidemiology Biomarkers and Prevention*, 17(5):1127–1135, 2008.
13. Michael S Mulvihill, Yong-Won Kwon, Sharon Lee, Li Tai Fang, Helen Choi, Roshni Ray, Hio Chung Kang, Jian-Hua Mao, David Jablons, and Il-Jin Kim. Gremlin is overexpressed in lung adenocarcinoma and increases cell growth and proliferation in normal lung cells. *PloS one*, 7(8):42264, 2012.
14. Noriaki Sunaga, Kuniharu Miyajima, Makoto Suzuki, Mitsuo Sato, Michael A White, Ruben D Ramirez, Jerry W Shay, Adi F Gazdar, and John D Minna. Different roles for caveolin-1 in the

development of non-small cell lung cancer versus small cell lung cancer. *Cancer Research*, 64(12):4277–4285, 2004.

15. Chang Niu, Chaoyang Liang, Juntang Guo, Long Cheng, Hao Zhang, Xi Qin, Qunwei Zhang, Lihua Ding, Bin Yuan, Xiaojie Xu, et al. Downregulation and growth inhibitory role of fhl1 in lung cancer. *International Journal of Cancer*, 130(11):2549–2556, 2012.

Part II
Cloud Computing

Progressing the Security Landscape of Cloud by Incorporating Security Service Level Agreement (Sec-SLA)

Joydeep Choudhury, Indushree Banerjee,
Amitava Nag and Indika Parera

Abstract Gathering personal information of individuals', in return to provide different personalized services, continues to grow. The adaptability and flexibility of cloud that allows mobility of data access and multiple ownerships provide a favorable platform for users and service providers to adapt Cloud services for storing and accessing personal data. However data flow from one level to another service level of cloud may cause data loss or leakage and put the privacy of individuals at risk without them being aware of it. Ensuring privacy of information on Cloud, presents a major challenge to be tackled by future researchers. This paper aims at providing an overall picture of cloud privacy and security at its different level of architecture and discusses the proposed solutions. It will further provide detailed analyses of the various adopted techniques. We will also discuss Security-SLA as a security protection mechanism for cloud users. Further we will try to highlight the areas which can be further researched and make cloud a more secure place to store data.

Keywords Cloud computing · Service level agreement · Security · Privacy · Identity management

Joydeep Choudhury · Indushree Banerjee · Amitava Nag (✉)
Department of Computer Science, Academy of Technology,
Aedconagar, Hooghly 712121, West Bengal, India
e-mail: amitavanag.09@gmail.com

Joydeep Choudhury
e-mail: joydeepchoudhury193@gmail.com

Indushree Banerjee
e-mail: banerjee.indushree@gmail.com

Indika Parera
Department of Computer Science, University of Moratuwa,
Katubedda 10400, Moratuwa, Sri Lanka
e-mail: indika@cse.mrt.ac.lk

© Springer Science+Business Media Singapore 2017
J.K. Mandal et al. (eds.), *Proceedings of the First International Conference on Intelligent Computing and Communication*, Advances in Intelligent Systems and Computing 458, DOI 10.1007/978-981-10-2035-3_7

1 Introduction

Cloud has become an important and growing technology which generates revenue options in both industries as well as in academics, but it is still in the process of evolving and considered an enigma when security and privacy comes into perspective. The term "Cloud" was coined by the telecommunication industry when the providers started to use Virtual Private Network or VPN service data communication [1]. At present cloud services promotes itself as an encouraging technology to deliver infrastructure and resource to its users as pay-as-you-go style and by reducing the cost of IT infrastructure for new and small business [2]. Users in general are provided access to a web based interface for connecting and storing their personal data without being aware of the storage location or resources they are using during the operation of a certain application. Storing their personal data in an unknown location with other consumers can possess several threat especially when unknowingly data could rest on the same resources of a competitor's private information application [3], thus a constant drawback of getting data manhandled and misused becomes a nightmare for many organizations and users concerned about their privacy. Apart from improper access and deletion of data from cloud storage there is a significant privacy issue that makes users rethink about the adaptation of cloud [4] which is losing complete ownership. In spite being clearly predefined in various agreements there is no surety that redundant data have been completely discarded once the partnership or employability of a certain vendor is terminated. Various attempts have been made in past to safeguard privacy of individuals and agencies by utilizing different access control mechanisms and security agreements within cloud, but still it is not clear how provider deals with data and if at all the provider maintains integrity and authentication.

With this paper we try to highlight the major challenges faced by users and organizations acquiring Cloud services in general. Introduction gives an overview of the various topics being covered by the paper. Section 2 provides an in depth discussion of the Cloud architecture being currently implemented by the service providers. Section 3 delves deeper into the limitations and security threats specifically presented by Cloud infrastructures in its different layers. Section 4 will discuss a detailed analysis of the protection methods currently used. Section 5 deals with the Service Level agreement and explains the necessity and utility of SLA with respect to safeguarding users from being victimized. Section 6 introduces the concept of security SLA. Finally the last section provides future direction and concludes the paper.

2 Cloud Architecture

The vital characteristic of cloud that makes it a success are noted as on-demand, pay-as you-go, self-service, ubiquitous network access, which allows geographic area independence, resource pooling and rapid elasticity [5] making it adoptable to small businesses. Privacy and security issues of cloud are inbuilt issues, and in order to completely understand the subatomic reasons for these issues it becomes mandatory to revise the architecture of Cloud. Cloud is divided into four layers in a top down order i.e. application, platform, infrastructure and hardware. These are again grouped together into three layers in accordance to the service oriented business model i.e. Software as a Service (SaaS), Platform as a Service (Paas) and Infrastructure as a Service (Iaas) [6]. Figure 1 illustrates structure of cloud with all layers and an overall utility of each layer. Application layer is the topmost layer of cloud architecture which delivers software that a user need. In business model it is called Software-as-a-Service. It provides networked based access and management of commercially available software from a centralized location to the users [1]. Google App is an example of mostly used SaaS.

Second layer is on the top of infrastructure layer and is mainly responsible for providing all computational resources like programming framework and operating systems. This is called Platform-as-a-Service. Main aim of this level is to reduce the load of direct deployment of application in VMs. Google App engine is an example of PaaS as it provides API support to users for executing storage, database for an

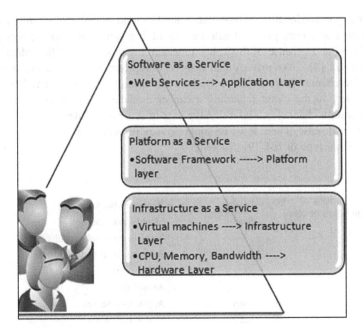

Fig. 1 Cloud structure

application. Infrastructure and Hardware layers are the layers that are responsible managing physical and virtual resources of the cloud. Hardware layer is consists of servers, switches, router and storage units. Data centers are example of hardware layer of cloud. Infrastructure layer is also called virtualization layer as it creates virtual pool of resources. In this level VM technologies, such as Xen, KVM, and VMware etc., are used to partition the physical resources [6].

3 Cloud Security Issues in Different Layers

Cloud is a combination of numerous well established technologies which include grid and distributed computing. Internet is used as a delivery medium to provide services to its users [7]. When the user progress from IaaS to PaaS to SaaS more abstraction of technology is introduced and because of this stored data in cloud are not in direct control of the user and these data are transmitted using Internet makes the user's privacy and security at of higher risk. Whenever a user starts to avail a service, different service level of cloud infrastructure take part in the process and the corresponding privacy mechanisms play a crucial role in the process [2].

3.1 Identity Management

In order to provide location independent data the cloud providers rely on redundant data storage and uses personal information of individual user in the storage to authenticate, and because of this it has to assure the protection of their data inside cloud servers [8]. User privacy can also be compromised while using the communication channel to query the cloud for some information. For example if a user send a query to the cloud regarding a cancer medicine, then an observer of the communication channel can infer that the user or someone related to the user might have cancer disease. There must be some privacy protection mechanism to protect user form this type of risk [9] (Table 1).

Table 1 Security challenges in different layers of cloud

Layers	Issues on that layer
SaaS	Identity management
	Virtualization vulnerability
	Authentication and authorisation
	Data integrity
	Availability
PaaS	Application security
IaaS	Hypervisor attack
	DDoS attack

3.2 Multi Tenancy and VM

Multi tenancy is an essential property of cloud. It helps Cloud Service Providers (CSP) to share resources to different user and make computation very efficient and highly scalable. The main security concern of user is that their personal data might be exposed to the third party as they all are using the same computational space in the cloud. In the virtualized environment where one physical machine hosts multiple users; there is always an associated risk that the other user can monitor its neighbor's activity and lead the other users of the VM accessible to it.

3.3 Attack on Hypervisor

Hypervisor is an application or a computer that creates and run virtual machines to provide the resources to its users. Most common types of attacks are Virtual library checker, encryption attack and migration attack [10]. In migration attack and encryption attack the attacker use the network and virtual machine software vulnerabilities to gain access to the data.

3.4 Availability

Availability means cloud providers must ensure that the service is always available to the authorized user even if a security breach is detected [11]. DDoS attacks makes the services and data unavailable to the user and this makes a real threat for cloud users.

3.5 Lower Layer Issues

The lowest layer of the cloud is hardware layer which is mainly consists of physical machines. A large number of attacks are done in this layer and causes data loss. Most common attack on this layer is Distributed Denial of Service (DDoS) or Denial of Service (DoS) attack. The attacker sends multiple requests to the server in a very short period of time. This technique is called "flooding". With this the server become busy to process unwanted request and thus occupy the bandwidth of the network. Which in due course disrupt service of an authenticate user and prevent access to a service [12]. Cookies poisoning is another type of attack in this physical layer where the attacker modified the cookies into gain access to the cloud.

4   Protection Methods

There are ample amount of work has been done to protect user data from the above attacks, but traditional security issues are still present in cloud environment. In this section we will describe some of solutions of privacy and security issues proposed by researchers. Roy et al. [13] introduced Ariavat, a privacy protection on Map Reduce Systems. This system was built using a combination of mandatory access control and differential privacy technique. The main functionality of the proposed system was to provide end-to-end confidentiality along with integrity and privacy in cloud infrastructure.

IBM, in 2009, introduced a homomorphic encryption scheme to protect data privacy [14]. In this method a user stores it data in an encrypted format in some unknown server, and when user query information from that data set the server then homomorphically computes an encryption of query and send back the cipher text back to the user. In this technique the data and query is fully encrypted and privacy of user is maintained throughout the process.

DDoS attack on cloud is a very common security issues which is described in the last section. To protect cloud form this attack Intrusion Detection Systems (IDS) are developed. Author of [15, 16] explain that intrusion detection is basically a process of monitoring the network flow and analyze every packets to check any attempt of intrusion which violate the integrity, availability or confidentiality of the system. Mohamed et al. proposed collaborative IDS which will work on the IaaS layer of cloud and at the same time prevents the cloud from attack. Modi et al. in [17] proposed another IDS using the Snort and signature algorithm. The proposed framework captured packets from network and compares it with a known attack pattern and if packet give negative result then it allows the packets or else it follows the rules [17].

Authentication of users is done using digital signatures in a combination of SSO (Single Sign-On) and Ldap [11] Shibboleth is now used for web SSO to identify and grant access to the users across or within the organizational boundary. Access control mechanism is largely used in a fully shared system to give permission to the users' to access resources. Data dispersal storage and secure retrieval scheme [18] is one of well discussed approach. The suggested algorithm efficiently reduces some of privacy risk such as server colluding and unauthorized data modification. In the working scenario the system assigns users' data to various domains using some flexible distributed algorithms to maintain the integrity of the data.

5   Service Level Agreement in Cloud

Privacy is still a long-standing topic in Cloud. In cloud privacy, Service level Agreement (SLA) represents an important document which serves as contract between a user and a provider to deliver services. SLAs should cover performance, reliability, and security and privacy of data and on this pre-defined contract the provider is bound to provide the service [19]. Any violation of SLAs will lead to a penalty that can be either monitory or anything other. Various issues that are included in SLA, introduces the following challenge in maintaining privacy:

- Storage—The main concerns are if the stored data is getting mixed with other data from other organization. Another important concern which makes user to think twice before storing confidential data in cloud is that if providers have the right to see the data without notifying the organization.
- Retention—One of the key question that organization needs to know is how long a service provider keeps the data in there server and how the ownership of the data is evaluated.
- Deletion of Data—To provide availability of data all the time cloud providers need to replicate the data. The main concern is, once the data retention period is over, how the user will make sure that all the replicated copies of the data are destroyed.
- Privacy breaches—If a breach occurs how the providers will notify the user and locate who is actually responsible for the breach?

SLA metric is constructed from a few common elements such as name of metric, metric source, duration of sampling, frequency of sampling, scope of testing, target range, weight, reporting process, and penalty/incentive calculation. These elements should be found with every SLA metric used to demonstrate services have been sustained to mutually agreed obligations [20]. Some general metrics are throughput, QoS, bandwidth etc. But SLA can also include some other metrics on the security perspective as it is a contract between user and provider and the users' concern on the data privacy can also be a point which includes security mechanism, security effectiveness as a metrics. The main concern for the user is to know how exactly the CSP deals with their private data. It is quite uncommon for a CSP to specify the security levels for user data associated with their services, hence impeding users from making data security relevant informed decisions. This is known as Quality of Protection (QoP), which includes the capability of a service provider to deliver service according to the security requirement of the user and how well the provider meets the requirements [21, 22].

6 Security SLA

Like SLA there is Sec-SLA (Security Service level agreement), which defines matrices related to security. A Sec-SLA should include [23]:

- Some description of the user required services that the provider is going to provide.
- All the security requirements, along with the monitoring process, which the user and provider are agreed upon before committing

- A detail process of reporting problems, threats or security breach incidents that may arise during the contract period
- Lists of penalties in case any of the party breaks the agreed SLA. This penalty can be either service credit or financial compensation. CSPs may also include restrictions on customer activities and also state some specific moments when the agreement do not apply.
- All the legal and regulatory matters that might happen during the time span which include references to existing legislations and directives that may affect the service as well as the terms under which the SLA will not be valid (Table 2).

In [22] authors had described a life cycle of Sec-SLA which consists of six steps. As security is a very vast portion and it depends upon different user specification so the negotiation between the user and provider is an important part. To create a Sec-SLA there are three steps to follow [18]

- Policy analysis
- Architecture analysis
- Interviews

With the above procedures the provider will evaluate the customer's requirements of web servers, systems and security policy that he will need during the contract period. The authors of [24] introduce the matric on which the negotiation can be done. Services delivered in an "On Demand" condition, require extensive effort in defining security matrices. In traditional SLAs, matrices are mainly QoS, Bandwidth and some portion of security as well. But in Security-SLA the matrices are for example Backup policies, Password management, Secure Network Protocols and Data Transport, data deletion effect etc. [24], which will ensure that the data are stored and also in control of the user. Even after negotiation the users are always worried about the implementation of the agreed security mechanism. To provide user with a privacy management tool the SLA must have been written in a machine readable language. WS-Agreement is a protocol for defining SLAs between

Table 2 User requirement for security in cloud

Level	Layer	Security requirements
Application level	SaaS	• Privacy in multitenant environment • Access control • Software security • Service availability
Virtual level	PaaS and IaaS	• Application security • Virtual cloud protection • Communication security • Management control security • Data security
Physical level	Datacenters	• Hardware security • Network protection • Network resource protection

providers and users [23]. Using this protocol the privacy management tool will help the user to control the storage of data and also modify the security on a move. This tool will help the user to negotiate the security without meeting the provider personally.

7 Future Work and Conclusion

Prospect and existing customers of service providers are demanding confidentiality, integrity, and availability when contracting with vendors for cloud computing [20]. As we are using more and more cloud based services in our daily life, we are giving and storing our information to the vendor side which can be accessed through internet. This paper mainly focuses on the vulnerabilities in cloud with some detail discussion on attacks and challenges faced by cloud technology in the recent times. We have also introduced Sec-SLA and matrices as an approach to help user to secure their personal data in cloud. Security SLA matrices are a medium to gain trust over the user and reduce risk. These matrices must be meaningful and economic cause creating a metrics includes computational cost. Irrelevant metrics can cause and impact on the quality of service by using excessive computational resources [20]. From the above discussion we can see that both the parties are liable to protect their personal information. But CSPs do not include anything in SLA about the security mechanism they are going to provide to their tenants. Apart from this there are few monitoring tools which can be used by the end user to monitor the security measures and at the same time this tool will help them to enforce new security features on their data [25]. Several outstanding issues exist related to cloud security and privacy. Security SLA is still in its early stage. Future research should focus on providing a full view of security that will offer to user and details about the data they stored in the cloud. Along with this there must be a web based framework that will negotiate SLA metrics dynamically and incorporate security according to the user instruction and at the same time give an overview to the vendor as well as user about the performance achieved by the provider. Apart from this there should be another part for further development that can be an extension of the framework which provides an opportunity to view the location of the data stored in the cloud by using meta-data information of the user data. There need to be more specific and detailed research work carried out to make security as a user centric approach, and provide a monitoring and modification tool to help user to choose their security requirement without any human intervention.

References

1. Y. Jadeja and K. Modi, "Cloud computing - concepts, architecture and challenges," in *2012 International Conference on Computing, Electronics and Electrical Technologies (ICCEET)*, 2012, pp. 877–880.
2. G. Zhang, Y. Yang, X. Zhang, C. Liu, and J. Chen, "Key Research Issues for Privacy Protection and Preservation in Cloud Computing," in *2012 Second International Conference on Cloud and Green Computing*, 2012, pp. 47–54.
3. S. Hamouda, "Security and privacy in cloud computing," *2012 Int. Conf. Cloud Comput. Technol. Appl. Manag.*, pp. 241–245, Dec. 2012.
4. S. Surianarayanan and T. Santhanam, "Security issues and control mechanisms in Cloud," *2012 Int. Conf. Cloud Comput. Technol. Appl. Manag.*, pp. 74–76, Dec. 2012.
5. H. Takabi, J. B. D. Joshi, and G.-J. Ahn, "Security and Privacy Challenges in Cloud Computing Environments," *IEEE Secur. Priv. Mag.*, vol. 8, no. 6, pp. 24–31, Nov. 2010.
6. Q. Zhang, L. Cheng, and R. Boutaba, "Cloud computing: state-of-the-art and research challenges," *J. Internet Serv. Appl.*, vol. 1, no. 1, pp. 7–18, Apr. 2010.
7. C. Modi, D. Patel, B. Borisaniya, A. Patel, and M. Rajarajan, "A survey on security issues and solutions at different layers of Cloud computing," *J. Supercomput.*, vol. 63, no. 2, pp. 561–592, Oct. 2012.
8. S. M. Rahaman and M. Farhatullah, "PccP: A model for Preserving cloud computing Privacy," *2012 Int. Conf. Data Sci. Eng.*, pp. 166–170, Jul. 2012.
9. S. De Capitani Di Vimercati, S. Foresti, and P. Samarati, "Managing and accessing data in the cloud: Privacy risks and approaches," *7th Int. Conf. Risks Secur. Internet Syst. Cris. 2012*, 2012.
10. K. Surya, M. Nivedithaa, S. Uma, and C. Valliyammai, "Security issues and challenges in cloud," *2013 Int. Conf. Green Comput. Commun. Conserv. Energy*, pp. 889–893, 2013.
11. D. Zissis and D. Lekkas, "Addressing cloud computing security issues," *Futur. Gener. Comput. Syst.*, vol. 28, no. 3, pp. 583–592, 2012.
12. P. Yadav and S. Sujata, "Security Issues in Cloud Computing Solution of DDOS and Introducing Two-Tier CAPTCHA," *Int. J. Cloud Comput. Serv. Archit.*, vol. 3, no. 3, pp. 25–40, 2013.
13. I. Roy, S. T. V. S. T. V Setty, A. Kilzer, V. Shmatikov, and E. Witchel, "Airavat: Security and privacy for MapReduce," *Proc. 7th USENIX Conf. Networked Syst. Des. Implement.*, pp. 20–20, 2010.
14. D. Chen and H. Zhao, "Data Security and Privacy Protection Issues in Cloud Computing," *2012 Int. Conf. Comput. Sci. Electron. Eng.*, vol. 1, no. 973, pp. 647–651, 2012.
15. M. P. K. Shelke, M. S. Sontakke, and a D. Gawande, "Intrusion Detection System for Cloud Computing," *Int. J. Sci. Technol. Res.*, vol. 1, no. 4, pp. 67–71, 2012.
16. H. Mohamed, L. Adil, T. Saida, and M. Hicham, "A collaborative intrusion detection and Prevention System in Cloud Computing," in *2013 Africon*, 2013, pp. 1–5.
17. C. N. Modi, D. R. Patel, A. Patel, and M. Rajarajan, "Integrating Signature Apriori based Network Intrusion Detection System (NIDS) in Cloud Computing," *Procedia Technol.*, vol. 6, pp. 905–912, 2012.
18. L. Chen and D. B. Hoang, "Novel Data Protection Model in Healthcare Cloud," in *2011 IEEE International Conference on High Performance Computing and Communications*, 2011, pp. 550–555.
19. H. Tianfield, "Security issues in cloud computing," *2012 IEEE Int. Conf. Syst. Man, Cybern.*, pp. 1082–1089, Oct. 2012.
20. M. Hoehl, "Proposal for standard Cloud Computing Security SLAs - Key Metrics for Safeguarding Confidential Data in the Cloud.".
21. R. Schmidt, "Conceptualisation and Lifecycle of Cloud Based Information Systems," *2012 IEEE 16th Int. Enterp. Distrib. Object Comput. Conf. Work.*, pp. 104–113, 2012.

22. K. Bernsmed, M. G. Jaatun, P. H. Meland, and A. Undheim, "Security SLAs for Federated Cloud Services," in *2011 Sixth International Conference on Availability, Reliability and Security*, 2011, pp. 202–209.
23. M. Jaatun, K. Bernsmed, and A. Undheim, "Security SLAs–An Idea Whose Time Has Come?," *Multidiscip. Res. Pract. Inf. Syst.*, pp. 123–130, 2012.
24. S. A. de Chaves, C. B. Westphall, and F. R. Lamin, "SLA Perspective in Security Management for Cloud Computing," in *2010 Sixth International Conference on Networking and Services*, 2010, pp. 212–217.
25. M. Rak, N. Suri, J. Luna, D. Petcu, V. Casola, and U. Villano, "Security as a Service Using an SLA-Based Approach via SPECS," *Requir. Eng. Cloud Comput. (RECC - CloudCom)*, vol. 2, pp. 1–6, 2013.

On Demand IOPS Calculation in Cloud Environment to Ease Linux-Based Application Delivery

Rajesh Bose, Sandip Roy and Debabrata Sarddar

Abstract Today's era of cloud computing and everlasting demands for real-time analysis of the storage data on cloud, it is essential for IT industries to have cognizance about the storage performance. Cloud is elastic computing model where users can hire computing and on demand storage resources from a remote infrastructure and its popularity depends on low cost and on demand availability. Simultaneous execution of huge number of data-intensive applications on the public cloud call for a huge amount of storage in order to access the persistent data leads to degradation of overall system performance. IT personnel have to be assisted with storage performance measurement for prediction of best storage need. Input/Output Operations Per Second (IOPS) calculation helps to determine the amount of I/O's storage to run. This IOPS calculation is incorporated in cloud environment to alleviate Linux based application delivery.

Keywords Big data · Cloud computing · Input/output operations per second (IOPS) · Cloud storage

1 Introduction

The huge amount of data is being produced by the IT industry due to the execution of massive applications on cloud. When an application underperforms IT analysts are looking towards the standard performance benchmark for hard drives. Even though IOPS calculation is a performance measures by which system administrator can clinch about the storage requirement of the current application's bottleneck [1].

Rajesh Bose · Sandip Roy (✉) · Debabrata Sarddar
University of Kalyani, Kalyani, Nadia, West Bengal 741235, India
e-mail: sandiproy86@gmail.com; sandip@klyuniv.ac.in

Rajesh Bose
e-mail: bose.raj00028@gmail.com

Debabrata Sarddar
e-mail: dsarddar1@gmail.com

© Springer Science+Business Media Singapore 2017
J.K. Mandal et al. (eds.), *Proceedings of the First International Conference on Intelligent Computing and Communication*, Advances in Intelligent Systems and Computing 458, DOI 10.1007/978-981-10-2035-3_8

Due to this aforementioned requirement today's researchers have concerned about the system performance and develop a software tool for making alert to the administrator for upgradation of underlying system [2].

In this paper the calculation of IOPS on Cloud System for measuring maximum IOPS of running application within a particular refresh period is discussed, which helps the IT personnel or Organizations to take decision for changing the storage architecture. Here slow application delivery time is inspected to develop an enhanced environment. Our proposed tool can calculate IOPS of applications executing on Linux based system and also alert the system administrator to upgrade the disk system when the storage as the potential bottleneck.

2 Background Study

2.1 Basic Definition of IOPS

The number of input/output operations a storage device can complete within one second is called Input/Output Operations Per Second (IOPS) [3]. The performance characteristics are measured by randomly or sequentially. Depending upon the file size the random and sequential operations are done. When we are concerning large file then sequential operations are done to access of stored operation in contiguous manner otherwise random operations are done to access locations in the storage device in a non-contiguous way.

There are different characteristics for IOPS measures:

- **Sequential Write IOPS**: The average number of sequential write I/O operations that occur per second
- **Sequential Read IOPS**: The average number of sequential read I/O operations that occur per second
- **Random Write IOPS**: The average number of random write I/O operations that occur per second
- **Random Read IOPS**: The average number of random read I/O operations that occur per second
- **Total IOPS**: The total IOPS when performing mixed read and write operations

2.2 Frontend IOPS

Fronted IOPS is the total number of read and write operations per second generated by an application or applications.

2.3 Backend IOPS

Backend IOPS is the total number of read and write operations per second which a storage controller sends to the physical disks. This phenomenon is also known as storage IOPS.

The backend IOPS or storage IOPS is calculated by the formula below:

$Storage\,IOPS =$ Number of RAID Groups $\times$ (((Read Ratio $\times$ Disk Operations/Sec)

$\qquad$ + ((Write Ratio $\times$ Disk Operations/Sec)/ Write Penalty)) $\times$ Quantity of Disk in RAID Group)

$$(1)$$

70 % versus 30 % read/write ratio for 15 K SAS in a single RAID 10, the backend IOPS or storage IOPS is:

$$2 \times (((70\% \times 180) + (30\% \times 180)/2)) \times 8 = 2,448\,\text{IOPS}$$

2.4 RAID Penalty

Write operation can't be completed until both the data and parity info have been written to the disk. If the any of the write operations are failed, waiting for extra time to write the parity info on to disk. This phenomenon is called RAID penalty [3] (Table 1).

2.5 IOPS Calculation

The number of input/output operations is done per second. The formula of IOPS calculation is given below [4] (Table 2):

IOPS per disk $= 1/(((\text{average read seek time} + \text{average write seek time})/2)/1000)$

$\qquad$ + (average rotational latency/1000))

$$(2)$$

The total IOPS of the application is calculated by the following formula [5]:

Total IOPS = Read IOPS + (RAID level based write penalty $\times$ Write IOPS) $\qquad$ (3)

Table 1 Different RAID level penalties

RAID level	Read	Write
RAID 0	1	1
RAID 1 (and 10)	1	2
RAID 5	1	4
RAID 6	1	6

Table 2 Range of IOPS

Disk type	RPM	IOPS range
SATA	5,400	50–75
SATA	7,200	75–100
SAS/FC	10,000	100–125
SSD	N/A	5,000–10,000
SAS/FC	10,000	100–125

To calculate the number of disks is required for frontend IOPS using the following equation [6]:

$$\text{Total number of Disks required} = ((\text{Total Read IOPS} + (\text{Total Write IOPS} \times \text{RAID Penalty}))/\text{Disk Speed IOPS}) \quad (4)$$

2.6 Flow Chart of Our Proposed Model

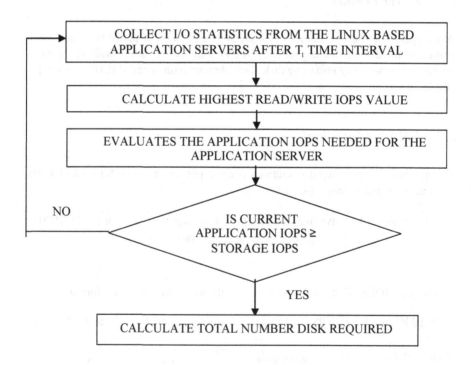

2.7 Discussion of Our Proposed Work

In this manuscript we have proposed a tool which is incorporated in cloud environment that can collect the input output statistics record from the Linux based application servers in different time interval [7]. Thereafter the tool calculates the highest read/write IOPS value. Then tool evaluates the total application IOPS or IOPS needed for these application servers using Eq. (2).

When first time configured the tool, the storage IOPS or backend IOPS value is also provided or calculated by the Eq. (1) and also is compared the application IOPS with backend IOPS. If application IOPS is higher than backend IOPS then the system understand the reason for slow application delivery. To solve the problem our proposed tool now calculates the total number of disks required using Eq. (4).

After that our proposed tool informs the storage administrator the following details: Application IOPS, Backend IOPS and number of disks required to run the application smoothly. From this report an administrator can easily add the required number of disks to improve the performance [7–9].

2.8 Statistical Data Analysis

We have taken these statistical data for our proposed tool which is given below (Table 3).

We have sorted the r/s and w/s values of various device connected with our proposed tool for any certain running application.

device,	r/s,	w/s,	kr/s,	kw/s,	wait,	actv,	svc_t,	%w,	%b
sda5,	1855.8,	685.6,	3623.79,	244.87,	0.1,	0.7	43.1,	6,	5
sda6,	1855.8,	685.6,	2383.79,	236.89,	8.1,	0.9,	31.1,	1,	5
sda7	505.6,	5.8,	288.6,	46.1,	6.2,	1.1,	25.2,	3,	5
sda3	17.8,	43.68,	754.3,	97.9,	1.9,	0.2,	5.1,	0,	2
sda4	17.8,	42.68,	758.8,	94.9,	1.7,	0.5,	7.3,	0,	3
sda1	0.0,	0.0,	0.4,	0.0,	0.1,	0.0,	0.2,	0,	0
sda2	0.0,	0.0,	0.2,	0.0,	0.4,	0.0,	0.1	0,	0

The maximum r/s, w/s and the value of storage IOPS data are stored in file. Application IOPS is computed using r/s and w/s is 3227.199 which is greater than storage IOPS, i.e. 2448. The number of disk(s) required is 2 in this context.

Table 3 Sample outputs

Linux 2.6.18-238.el5 (TestServer) 15/01/2015

avg-cpu:

%user	%nice	%system	%iowait	%steal	%idle
8.57	0.00	1.18	7.66	0.00	82.60

Device	rrqm/s	wrqm/s	r/s	w/s	rsec/s	wsec/s	avgrq-sz	avgqu-sz	await	svctm	%util
sda1	0.00	0.00	0.00	0.00	0.01	0.00	16.01	0.00	1.08	1.07	0.00
sda2	0.00	0.00	0.00	0.00	0.00	0.00	10.47	0.00	1.23	1.22	0.00
sda3	7.14	184.21	17.86	43.68	3818.87	4297.91	131.91	0.07	1.11	0.37	2.30
sda4	7.13	184.21	17.86	43.68	3818.87	4297.91	131.91	0.07	1.11	0.37	2.30
sda5	48.34	3.10	1855.86	685.67	36383.79	236.89	288.11	0.91	7.19	1.23	15.61
sda6	48.34	3.10	1855.86	685.67	36383.79	236.89	288.11	0.91	7.19	1.23	15.61
sda7	58.55	2.48	505.67	5.87	50268.61	146.16	98.57	1.15	2.25	0.48	24.36

3   Conclusion and Future Work

In this context the designed Java package is capable of calculating the application IOPS and number of disk(s) required for any application in Linux environment which helps to improve the overall system performance. As Bigdata application majorly depends on fast disk access, designing such a package for those kind of application in near future is one of the challenging task [3, 10].

References

1. Vasudeva, A., Solving the IO Bottleneck in NextGen DataCenters & Cloud Computing. IMEX Research-Cloud Infrastructure Report (2009–11) 1–20.
2. Vasudeva, A., Solid State Storage: Key to NextGen Enterprise & Cloud Storage. IMEX Research SSD Industry Report (2009–12) 1–40.
3. MEASURING STORAGE PERFORMANCE. White Paper CloudByte (2013) 1–10.
4. Fah, C., What kind of IOPS and throughput do you get from RAID-5/6? – Part 2. (2011).
5. Korenblom, M., Frontend and backend IOPS. (31st October 2011), Retrieved from http://www.storageguru.eu/2011/10/31/frontend-and-backend-iops/.
6. Sudharsan, IOPS, RAID Penalty and Workload Characterization. (25th December 2010), Retrieved from https://sudrsn.wordpress.com/2010/12/25/iops-raid-penalty-and-workload-characterization/.
7. Man pages section 1 M: System Administration Commands. (2006), from http://docs.oracle.com/cd/E19253–01/816-5166/6mbb1kq4q/index.html.
8. Linux and Unix iostat command. (2015), Retrieved from Computer Hope http://www.computerhope.com/unix/iostat.htm.
9. Lowe, S., Calculate IOPS in a Storage array. (12th February 2010), Retrieved from www.techrepublic.com.
10. EMC VNX2 Unified Best Practices for Performance. EMC2 (2015) 1–32.

Intelligent Storage Management Using Cloud Shared Disk Architecture

Subashis Biswas, Nilanjana Roy Chowdhury, Argha Roy
and A.B. Bhattacharya

Abstract In recent years, there is tremendous demand of cutting-edge cloud-based applications in many of the industries. We have proposed in the paper a shared disk cloud database architecture as the basis on which an intelligent data storage management system can be developed for enriching cloud-based web applications. Important features of this proposed architecture are single copied data consistency, dynamic load balancing and high benchmark performance. Based on the software layer, an intelligent data management system for popularizing the concept of SaaS has been pointed out suggesting a cost-effective solution for popularizing the cloud environment.

Keywords Storage area network · Shared disk · Storage architecture · Cloud system · Intelligent data management

1 Introduction

The subject of cloud computing refers to the use of web-based applications and/or server services which are paid for accessing, instead of software or hardware, which are bought and installed locally. Today, organizations of all sizes and bases are

Subashis Biswas
Department of CSE, Netaji Subhash Engineering College, Kolkata, India
e-mail: subashiscse@gmail.com

N.R. Chowdhury · Argha Roy
Department of CSE, Techno India University, Kolkata, India
e-mail: nilanjana.rns@gmail.com

Argha Roy
e-mail: arghacse@gmail.com

A.B. Bhattacharya (✉)
Department of Electronics and Communication Engineering, Techno India University, Kolkata 700091, West Bengal, India
e-mail: bhattacharyaasitbaran@gmail.com

© Springer Science+Business Media Singapore 2017 79
J.K. Mandal et al. (eds.), *Proceedings of the First International Conference on Intelligent Computing and Communication*, Advances in Intelligent Systems and Computing 458, DOI 10.1007/978-981-10-2035-3_9

going to implement this technology. The technology is cost-efficient as it offers unlimited data storage capacity, efficient backup and recovery, besides quick and easy access to any type of information from anywhere and anytime and facilitates fast deployment of web applications [1, 2]. Now-a-days, updated and current database MySQL, DB2, Oracle has robust optimization technology that helps to enable efficiency. In the internet based workload type of environment, sometimes data is hosted on a centralized database as it becomes difficult for an application to partition the data and this is why the scalability of the particular application gets limited. In this paper, we have presented ways to ensure scalability of such applications. In this architecture any number of database nodes can be processed any portion of data. Our approach is based on the data mining technique. Developed algorithm classifies the cloud services and ranks them accordingly [3, 4].

2 Technology Preliminaries

Two types of database management systems are of great significance in present day database market, viz., system of shared nothing and system of data sharing. The first type of the system is comparatively simpler to develop in comparison to the system of data sharing [4]. Data sharing systems are more advantageous than shared nothing systems with respect to the load balancing. First, we consider the Shared Nothing DBMS. The shared disk architecture is shown in Fig. 1 using a block diagram.

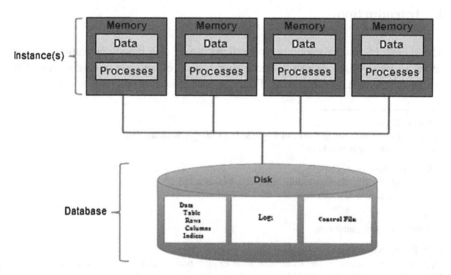

Fig. 1 Block diagram of shared disk architecture

3 Proposed System

In the system every instant should be connected with the processor and the memory [5, 6]. The proposed architecture is shown in Fig. 2. According to our proposed system based on shared disk cloud database architecture, inside a cloud, we have added a new layer which is the software layer. This area has been created in between the memory and the storage. This entire system will function inside a storage area network (SAN). This software layer is connected to the memory and the storage area which consists of all the disks (or specific databases/files). The software layer will maintain the log of data storage and data access including authentication and authorization feature of this software layer. If any particular processor of any other cloud tries to access any specific data of this storage area, then this software layer first checks whether this data access request is authenticated or not. If the data access request is authenticated and authorized, the particular processor is able to access the requested data from the specific database. The location of this requested data will be shown by this software layer. If the data access request is not authenticated then the particular processor will not be able to access the requested data. The data access will be prevented by the software layer. Without the existence of such a software layer, any processor from any other cloud outside the storage area network can access any data it needs from the storage area since there is no question of authorization and authentication.

Outside of SAN area, there will be another opportunity to build low cost cloud environment to popularize the concept. This is because, through this concept, it will

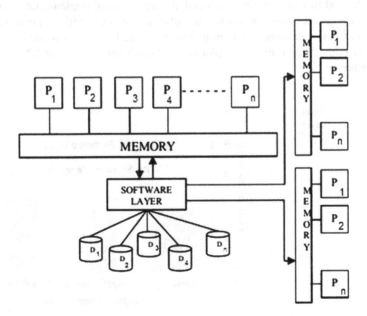

Fig. 2 Proposed architecture

be possible to develop cloud platform without storage. According to this concept, the same storage can be shared by several cloud servers (consisting of memory and processors). Outside the SAN, we need a request for authentication/authorization to secure the transaction. Here, cloud can share common infrastructure which is under the concept of infrastructure as a service (IaaS).

We have deployed the applications over a cloud platform with LAMP facilities. Open source database has been used for testing purpose. The application has been prepared and maintained in a cloud environment and storage has been maintained at other locations through static IP for data processing from various applications. We have tested the application in a real world cloud environment. Proposed layer and performance testing has been done through simulation. According to our simulation, it is giving the same access time. Traditional shared disk architecture are maintaining the same log and it has been processed into the same disk but according to our new layer, unauthentic hit will not be done into the disk. It will reduce to congestion too. According to the simulation graph, it is showing that into the SAN, frequently access database application is structuring data without high latency. SAN Application through software layer is presented in Fig. 3. During the time of outside access/request, it varies over bandwidth to hit the white list table. But, high bandwidth is getting positive response to access the data disk in comparison with SAN application. But it is possible to implement new proposed architecture which is going to reduce hardware cost and cloud can facilitate IaaS for another cloud.

We have developed this software layer in order to prevent unauthorized access of data from this storage area. If this concept is implemented, then all the data will be stored and maintained in only one cloud and several other clouds containing only processors and memory can be developed at very low cost. Implementation of this concept will ensure access of data by any authorized and authenticated processor in any cloud from a common cloud through the software layer. Access latency time is shown in Fig. 4, where there is a plot of bandwidth versus time for SAN request and outside request.

Fig. 3 SAN Application through software layer

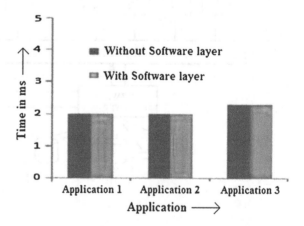

Fig. 4 Access latency time

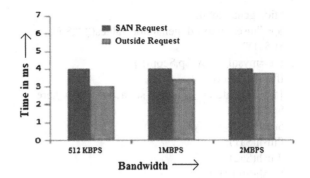

The associated algorithm of SAN operation is shown below:

Algorithm:

SAN_operation
```
If(request)
    {$flag=0;$t=systemtime;
    $s="select * from data_log where file=request and
    active=1 ";
    $rec=mysql_query($s,$con);
    While($row=mysql_query_fetch_array($rec)){
    $a1=row[id]; $a2=row[location];
    $a3=row[signature]; $a4=row[diskid];
    $flag ++;} If($flag==1){
    $p=insert into in_user_log
    values("",$t,$a2,$a4,request,$a3)";
    $re=mysql_query($p,$com);
    If(re){Connect to storage device;}}}
```

Outside_req_operation
```
{If(request) {
    $fg="phigical";
    $mycom=ob_get_contents();
    $ip=p_server['Remote_add'];
    $mac=$_strpos($mycom, $fg);
    $activate=0;
    $s="select * from white_list where whiteip=$ ip and
    mac=$mac";
    $rec=mysql_query($s,$con);
    While ($row=mysql_fetch_arrey($rec))
    {$ a1=row[id];
    a2=row[networkid]; a3=row[accessmode];
    $active=1;} If($activete==1){
    $t=systemtime();
```

```
$tid=genaratetid();
$p="insert into tid_genarator values(" ",$tid,
$t,$a1)";
$re=mysql_arrey($p,$com);}
If(disconnect)
{$p="update tid_genarate set active = 0 where
$tid=$tid";
$re=mysql_query($p,$com);
fflush($a1);
fflush($a2);
fflush($tid);} }
```

4 Analyses

Cloud Computing technology has revolutionized the way in which software can be
used to accomplish storage and retrieval of essential information. Cloud computing
in its form of SaaS has earned worldwide acceptance and popularity on account of
its customer convenience. The cloud offers infinite amount of data storage and this
advantage has enabled it to achieve superiority and preference over traditional
forms of software. The strength of database architecture is proportional to the
scalability and performance of any cloud-based software. In the shared nothing
database architecture the cloud-based system has its own private memory in one or
more disks. According to this system, the clustered processor performs communication by sending message through a network like interconnected computers.
Shared-nothing multiprocessor can scale up to thousand processors since the processors do not interfere among themselves but for each cluster, if separate private
memory and disk exist, the system overall becomes expensive to use. According to
the Shared Disk architecture, the same disk can be used for all the processors or
nodes. Several data tables, logs and control files are maintained on the disk.
According to our new concept, other cloud service providers and individual clients
can share the disk of other cloud platforms but in this way, other cloud service
providers can utilize the same disk. For data access, authentication and authorization checking creates excessive pressure on the disk which can result into data disk
crashing and hacking. Outside the disk, a software layer is proposed. When outside
users of the same SAN access the disk for retrieving data, then data processing will
be done inside the software layer. Log files, table information, metadata and
location will be maintained inside this software layer. If any data request comes
from outside IP without authorization, then it will get rejected at this software layer
and the disk access will be prevented. If any communication needs to take place
inside the SAN network, then there will not be any need for authentication and
authorization. Here, two algorithms are proposed: the SAN operation and the
outside req operation. All the data processing will be accomplished inside the SAN
network without the need of authentication and authorization. In this case, a hit to a

specific disk is possible from only disk location and metadata and this improves the processing time. We have developed and tested our application in a real cloud environment with Xen architecture consisting of 4 hexacore CPUs of 4 GB RAM each on 1 TB storage platform. We have used MySQL database in the communication layer and PhP with Ajax and HTML in the front end for developing our application. We have developed our proposed software layer in this cloud architecture and observed the functioning of our application through this software layer and concluded that there is problem with data access time and that there is uniformity in data layer access time. We have converted various access times into 100 scale with or without the software layer when we find that our application experiences the same data access time. We have also analyzed its corresponding graphical representation. But, time variation was obtained when any data request from outside communicates through a separate algorithm. In this case, the storage is accessed from a web server, from a different IP, through static IP and noted that the time variation completely depends on the bandwidth. In comparison to SAN request, latency time is greater. In case of high bandwidth, the latency time of outside requests can be reduced but it is optimum as per SAN request.

We have used a simulator based on our proposed algorithms. Based on the statement, we have determined conditional code access time [6]. The operation time of outside req operation algorithm is greater than that of SAN operation. By calculating this operation time, we have identified that the latency time is greater. By implementing the concept of Infrastructure-as-a-Service, various clouds will be able to share the storage and it will be possible to develop other low cost cloud environments. Through the software layer, specific data requests are able to hit specific locations. An unauthorized filtration system is maintained to prevent direct hits to the data disk. This reduces power consumption. It may be pointed out that cloud computing related to the cutting-edge cloud-based applications in industries may be regarded as one of the most booming technology and plays a vital role in the possible regulatory changes and thus implementing better applications by using the potential of cloud computing [7].

5 Conclusions

The functioning of shared-disk database architecture depends on inter-nodal messages. These messages alert a node about the status of all the other nodes. In the cluster the number of nodes is found to be proportional to the time taken for a message to reach a particular node from another particular node. As a result, long wait-states are created. This feature affects the scalability of shared-disk databases. On this shared disk architecture, we have worked on a file system. We can bind a flag with this file system so that after getting authentication, it can directly access the data storage. This is because authentication signature will be bound to the file system and several research works will be possible for creating this file system.

Acknowledgments We like to express our sincere thank to Mr. Mayur Goutam, System Engineer, CTRLS and Mr. Avik Chakraborty, System Engineer, CTS for their guidance and help in many parts of the work during implementation Phase.

References

1. Arora, I., Gupta, A.: Cloud Databases: A paradigm shift in databases. International J. of Comp. Sc. 9, 1694-0814 (2012).
2. Melnyk, R. B., Zikopoulos, P. C.: DB2: The Complete Reference. McGraw-Hill, (2001).
3. Microsoft SQL Server, Microsoft, http://www.microsoft.com.
4. Haghighat, M., Zonouz, S., Abdel-Mottaleb, M.: Cloud ID: Trustworthy Cloud-based and Cross-Enterprise Biometric Identification. Exp. Sys. with App. 42, 7905–7916 (2015).
5. Schmidt, E., Rosenberg, J.: How Google Works, Grand Central Publishing, 11 (2014).
6. Franck, C., Song, H., Ferry, N., Fleurey, F.: Evaluating robustness of cloud-based systems. J. of Clou. Comp. 4, doi:10.1186/s13677-015-0043-7 (2015).
7. Ward, J. S., Barker, A.: Observing the clouds: a survey and taxonomy of cloud monitoring. J. of Clou. Comp. 3, 1–30 (2014).

Hybrid Application Partitioning and Process Offloading Method for the Mobile Cloud Computing

Sukhpreet Kaur and Harwinder Singh Sohal

Abstract The application partitioning is the process of the breaking the application processes in the smaller processes for the easy execution and to enable the offloading capabilities of the process. In the proposed model, the process cost evaluation has been calculated in the form of the execution time, from where the threshold is calculated for the offloading decision. At first, the proposed model evaluates the number of instructions followed by the sequencing on the basis of the latter. The proposed model then compute the time cost for every process and make the decision on the basis of the threshold calculating. The experimental results have shown the effectiveness of the proposed model.

Keywords Application partitioning · Early finish time · Instruction set length · Process offloading · Mobile cloud computing

1 Introduction

Cloud computing is the recent technological development in the computing science, which is designed to promote the centralized computing, and is designed to store the user data online on the cloud storage system rather than the desktops, portable devices such as mobiles or tables [1, 2]. The cloud terminology has been evolved from the data centers which are connected together to facilitate the centralized service [3], and it does not let the users to know the actual location of the data center, but the centralized or regional cloud controller (Fig. 1).

The cloud platforms includes the various virtualization combined with the grid computing and distributed computing to provide the cloud platform [4]. The cloud platforms are the computationally rich computing resources equipped with the

Sukhpreet Kaur · H.S. Sohal (✉)
Department of Information Technology, LLRIET, Moga, India
e-mail: harwindersohal23@gmail.com

Sukhpreet Kaur
e-mail: brarsukh91@gmail.com

© Springer Science+Business Media Singapore 2017
J.K. Mandal et al. (eds.), *Proceedings of the First International Conference on Intelligent Computing and Communication*, Advances in Intelligent Systems and Computing 458, DOI 10.1007/978-981-10-2035-3_10

Fig. 1 Mobile application
partitioning model

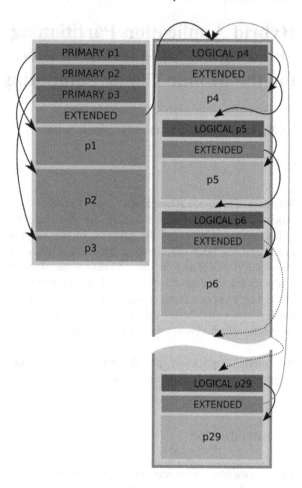

powerful and latest computing devices in order to facilitate the robust computing
applications [5–7]. The cloud platform is the centralized cloud computing resource
built of the number of servers and the computers attached to the cloud computing
through the internet resources [6]. The connected clientele may belong to the
various locations across the globe and owned and operated by the various companies or the individuals.

Cyberinfrastructure: The cyber infrastructure is the environment for the deployment of the robust and flexible applications which supports the large amounts of the
data for storage. The centralized computational environment is the research centric
technology and attracted the focus of the scientists, engineers and the independent
researchers.

Virtualization: The virtualization is the process of using the physical resources in
the abstracted fashion and to avail the use of multiple operating systems on the
single server resource in the cloud computing environment [8]. The virtualization
enables the portable property of the high-level functions which includes the

application interface [9], etc while facilitating the low-level functionalities handled over the lower layer of virtualization using the virtualization manager [10]. The virtualization also enables the aggregation and sharing of the active resources and the processes for the higher computational performance (Fig. 2).

Distributed Computing: can be clearly defined form the its name, as it includes the distributed resources connected using the local network or internet to form of the large computational infrastructure [5, 10].

Mobile cloud computing is the sub-category of the mobile computing which deals with the resource sharing or computational sharing between the mobile and cloud platform to save the computational effort of the mobile device [11] The mobile devices are usually equipped with the low computational resources in comparison with the personal computers or servers and they are also the battery operated devices, which adds the major constraint of the battery lifetime in order to conserve the battery lifetime and to enable the easiness in solving the computationally hard problems from the mobile platforms using the cloud offloading model [12]. The proposed model is the combination of multiple cost calculation methods with directed acyclic graph for the data and process handling between the mobile station and cloud computing platform [2, 3, 8, 10].

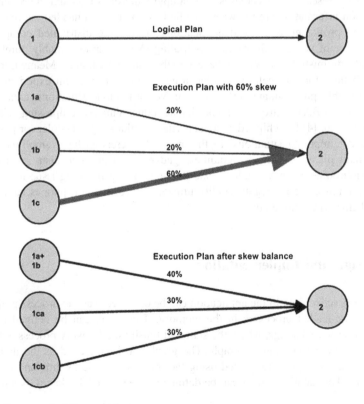

Fig. 2 Process scheduling model

2 Related Work

Feng et al. [1] has built up the Phone2Cloud structure which is utilized for Exploiting calculation offloading for vitality saving money on cell phones in portable distributed computing. Phone2Cloud offloads calculation of an application running on cell phones to the cloud. The goal is to enhance vitality productivity of cell phones and in the meantime, upgrade the application's execution through diminishing its execution time. Thusly, the client's experience can be progressed. The creators have executed the model of Phone2Cloud on Android and Hadoop environment. Two arrangements of analyses, including application tests and situation investigations, are directed to assess the framework. Lian et al. [12] has taken a shot at Energy proficiency on area based applications in portable distributed computing: an overview. Because of the issue's seriousness of battery utilization by the area based administrations in the portable stages, the impressive looks into have concentrated on vitality proficient finding detecting system in the last a couple of years. In this paper, we give a thorough review of late take a shot at low-power outline of LBAs. A diagram of LBAs and distinctive finding detecting innovations utilized today are presented. Strategies for vitality sparing with existing finding advances are researched. Decreases of area upgrading questions and improvements of direction information are likewise specified. Xiao et al. [9] has led the overview on Energy proficient area based applications in versatile distributed computing. With the rise of versatile distributed computing (MCC), an undeniably number of uses and administrations gets to be accessible on cell phones. Meanwhile, the obliged battery force of cell phones has a genuine effect on client experience. As one undeniably predominant sort of uses in versatile cloud situations, area based applications (LBAs) introduce some innate constraints encompassing vitality. Manjinder et al. [4] has chipped away at a vitality enhancing scheduler for portable distributed computing situations. In this paper, the creators have amplified their prior errand planning issue for countless gadgets to a portable distributed computing environment. They ideally tackle the undertaking planning issue for errand task to minimize the aggregate vitality utilization over the cell phones subject to client characterized imperatives.

3 Design and Implementation

The process scheduling in the application can be considered as the process graph of the given or loaded processes in the memory. The graph can be defined as the connected process listings, where the start and finish time of every process is listed in the form of the continuous graph. The graph can be modeled as the dataflow graph of the given processes listed using the process id, process time and process cost entity. The dataflow graph can be defined as the $G = (P, Linkage)$, where P is the array of the components or the processes, and can be defined as the $P = \{1, 2, 3 \ldots N\}$ and Linkage defined the inter-linkage between the processes listed on the output and input components or processes, and can be defined as the Linkage Array $= \{(i, j)|i, j \in P\}$ defines the linkage between the process on the basis of the input and output directions of the processes and (i, j) defines the sequence of the instructions in the process. The process scheduling algorithm defined under this project is a hefty algorithm designed for the purpose of process scheduling. The processes are offloading on the basis of their calculation cost. The process cost evaluation has been defined for the individual process as well as the process trees,

---

Process Scheduling Algorithm.

---

1. Calculate the process cost of each loaded process in the runtime memory

$$\text{Process Weight} = \sum_{i=1}^{N} N * C$$

   Where N is number of instructions, C is cost of process, and i is the index for number of processes.

2. Calculate earliest finish time for every process

   Time Array = Expected Finish Time – Start Time

$$T = \sum_{i=1}^{N} T(1, 2, 3 \ldots N)$$

   Where T is the process cost measured in time, i is the process index, expected finish time is the time on the process scheduling time graph and the start time is the start time of the process on the process scheduling time graph.

3. Sequence the processes according to the early finish time

   Sequential Time Array (Tx) = sort function (Time Array T)

   Where, Tx is the ordered time array and T is the time array computed on the previous step.

4. Then the communication cost to offload (Both uplink and downlink) is computed using the following time equation:

$$CT = \frac{Tx}{Bw} \times 100$$

Where, C_T is the communication cost of one process, Tx is the sequenced process Bw is the available bandwidth.

4   Result Analysis

The results of the proposed model implementation have been obtained in the form of various performance parameters. The proposed model simulation has been designed for the application partitioning and process optimization or offloading in the mobile cloud computing environment. The proposed mode performance evaluation can be performed on the basis of the execution speed, offloading decision accuracy and number of instructions.

The process cost evaluation is requires the many essential parameters for the process time estimation. The process cost calculation requires the number of operands and operators for the purpose of instruction set calculation for every process (Fig. 3). The next step is to sequence the processes on execution graph, which indicates the sequence of the process in the multi-processor environment, where multiple processes can be executed together (Fig. 4).

In Fig. 5, the estimated time cost has been shown in the popular figure of early finish time, which is showing the process time cost calculated in the milliseconds.

Fig. 3 The instruction set
length calculation

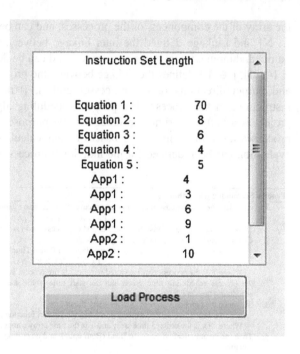

Fig. 4 The sequential order
of the processes in the run
time memory

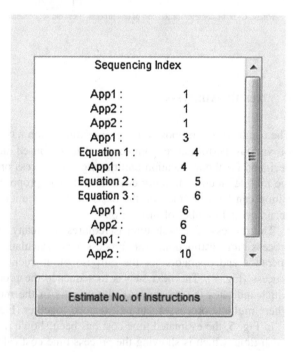

Fig. 5 The estimated time
cost evaluation

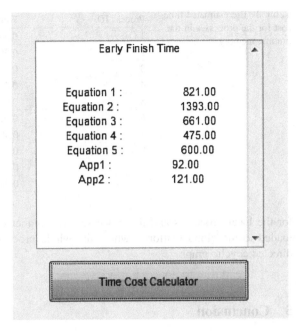

The time estimation is the primary parameter to evaluate the processing of the processes in order to make the offloading decision.

The Table 1 describes the performance of the each process. Total 10 processes have been evaluated for the performance of the proposed model. The results obtained has been collected in the form of process ID, number of instructions and input data size (Table 2).

The proposed model computes the results of the proposed model on the basis of the estimated time cost. The proposed total process size is evaluated on the basis of the number of instructions and size of the input data for the specific process. The proposed model has been defined with the minimum possible process cost evaluation

Table 1 The process cost evaluation

Process ID	No. of instructions	Input data size
1	24	36
2	20	18
3	2	90
4	30	14
5	16	76
6	29	24
7	3	13
8	23	45
9	6	20
10	24	61

Table 2 The estimated time cost for the processes in the runtime memory

Process ID	Estimated time cost
1	0.396667
2	0.394172
3	0.247121
4	0.484109
5	0.993528
6	0.929881
7	0.409774
8	0.348103
9	0.736518
10	0.24128

for the local processes and the processes with higher computational cost are off-loaded to the cloud platform, where the whole process is handled by using the directed acyclic graph.

5   Conclusion

The proposed model has been designed for the application partitioning and process offloading mechanism for the mobile platforms connected under the mobile cloud computing platforms. The proposed model has been designed to minimize the power consumption on the mobile devices by sharing their computational load with the cloud computing platform. The proposed model simulation represents the simulation results have been obtained in the form of the estimated time cost, instruction set and input data size, which gives the accurate the process cost and enables the effective process offloading mechanism. The experimental results have shown the effectiveness of the proposed model.

6   Future Work

In the future the proposed model can be enhanced using the combination of the time cost evaluation with process execution probability value for the dynamic process cost evaluation. The proposed model performance can be also evaluated and compared against the other application partitioning models.

References

1. Xia, Feng, Fangwei Ding, Jie Li, Xiangjie Kong, Laurence T. Yang, and Jianhua Ma (2014). "Phone2Cloud: Exploiting computation offloading for energy saving on smartphones in mobile cloud computing." *Information Systems Frontiers* 16, no. 1 (2014): 95–111.
2. Ma, Xiaoqiang, Yuan Zhao, Lei Zhang, Haiyang Wang, and Limei Peng (2013). "When mobile terminals meet the cloud: computation offloading as the bridge." *Network, IEEE* 27, no. 5 (2013): 28–33.
3. Fadaraliki, David I., and S. Rajendran (2015). "Process offloading from android device to cloud using JADE." In *Circuit, Power and Computing Technologies (ICCPCT), 2015 International Conference on*, pp. 1–5. IEEE, 2015.
4. Nir, Manjinder, Ashraf Matrawy, and Marc St-Hilaire (2014). "An energy optimizing scheduler for mobile cloud computing environments." In *Computer Communications Workshops (INFOCOM WKSHPS), 2014 IEEE Conference on*, pp. 404–409. IEEE, 2014.
5. Flores, Huber, and Satish Srirama . "Adaptive code offloading for mobile cloud applications: Exploiting fuzzy sets and evidence-based learning." In *Proceeding of the fourth ACM workshop on Mobile cloud computing and services*, pp. 9–16. ACM, 2013.
6. Yang, Lei, Jiannong Cao, Yin Yuan, Tao Li, Andy Han, and Alvin Chan (2013). "A framework for partitioning and execution of data stream applications in mobile cloud computing." *ACM SIGMETRICS Performance Evaluation Review* 40, no. 4 (2013): 23–32.
7. Shiraz, Muhammad, Ejaz Ahmed, Abdullah Gani, and Qi Han (2014). "Investigation on runtime partitioning of elastic mobile applications for mobile cloud computing." *The Journal of Supercomputing* 67, no. 1 (2014): 84–103.
8. Durairaj, M., and P. Kannan (2014). "A Novel Approach for Elastic Application Partitioning in Mobile Cloud." In *IEEE-ICAET-4th International Conference on Advances In Engineering & Technology, India*. 2014.
9. Ma, Xiao, Yong Cui, and Ivan Stojmenovic (2012). "Energy efficiency on location based applications in mobile cloud computing: a survey." *Procedia Computer Science* 10 (2012): 577–584.
10. Shiraz, Muhammad, and Abdullah Gani (2014). "A lightweight active service migration framework for computational offloading in mobile cloud computing." *The Journal of Supercomputing* 68, no. 2 (2014): 978–995.
11. Zhang, Weiwen, Yonggang Wen, and H-H. Chen (2014). "Toward transcoding as a service: energy-efficient offloading policy for green mobile cloud." Network, IEEE 28, no. 6 (2014): 67–73.
12. Wang, Lian, Yong Cui, Ivan Stojmenovic, Xiao Ma, and Jian Song (2014). "Energy efficiency on location based applications in mobile cloud computing: a survey." *Computing* 96, no. 7 (2014): 569–585.

Encrypted Data Searching Techniques and Approaches for Cloud Computing: A Survey

Lija Mohan and M. Sudheep Elayidom

Abstract Today, Cloud Computing has paved the way for enormous computing and storage. Cloud servers are third party systems which could be rented on demand basis and paid on usage basis. More and more users are adopting cloud based applications but the only factory that hinders its development is security issue. Users have a fear of trusting a third party system like cloud and they show reluctance to outsource their sensitive information to cloud. Encryption seems to be a direct solution but it limits the computability on data. Hence encryption schemes should be chosen based on the application they need to implement. In this article we study the basic encryption schemes which are widely used in cloud scenario. We compare these schemes in terms of their computational complexity, security, performance etc.

Keywords Cloud · Security · Survey · Encryption schemes · Encrypted data search

1 Introduction

Cloud Computing [1] is synonymous to 'Internet Based Computing' where users could do any type of computations if they have internet connectivity and a web browser to provide the interface. These types of computations are made possible by connecting a lot of virtual resources together and granting access to authorized users. Merits of Cloud computing includes elasticity, reliability, economical computing, pay per usage policy, global accessibility, usability and ease of maintenance.

Lija Mohan (✉) · M. Sudheep Elayidom
Department of Computer Science, School of Engineering, Cochin University
of Science & Technology, Kochi, Kerala, India
e-mail: lija@cusat.ac.in; joinlija@gmail.com

M. Sudheep Elayidom
e-mail: sudheep@cusat.ac.in

© Springer Science+Business Media Singapore 2017   97
J.K. Mandal et al. (eds.), *Proceedings of the First International Conference on Intelligent Computing and Communication*, Advances in Intelligent Systems and Computing 458, DOI 10.1007/978-981-10-2035-3_11

1.1 Challenges Associated with Cloud Computing

In spite all the merits we discussed above, normal users still keep away from cloud fearing that the security and privacy needed by their data will not be met. The cloud model follows a highly dynamic environment where the data will be partitioned and stored in multiple locations. Data will be kept replicated as well. Neither the service providers nor the owners of the data will know about the details of machines where their data will be stored. Providing Fine Grained Access Control [2] is another problem to deal with.

1.2 Motivation for a Solution

To implement solutions for BigData as well as problems involving severe computations, cloud computing is the most appropriate one. But use of public cloud has resulted in a lot of security and privacy issues. Several reports [3–9] reveal the security breaches and data theft took place in real world scenarios.

'Encryption of data' seems to be a first hand solution to ensure secrecy and privacy. But encryption limits the computations that can be performed on data like retrieving a particular file containing a specific keyword or extracting features from an image etc.

There exists a trade-off between security and usability. But the solution here is to apply the security mechanism in a way that it will not limit the functionality as well. This article aims to provide an insight to some encryption schemes that can be applied to cloud system based on the specific applications.

2 Review of Existing Solutions to Enable Encrypted Data Search

Basically we identified six different ways to search on encrypted data, each based on one of the following cryptographic primitives:

2.1 Property Preserving Encryptions (PPE)

PPE schemes [10] encrypt text in such a way that it leaks certain properties of the underlying data.

Different PPE schemes are proposed based on the property that is leaked. The basic one is 'Deterministic Encryption' [11] in which one message always generates same cipher text after encryption. Thus by comparing the cipher text one can determine whether the messages are same. These types of encryptions are hence applicable to problems where similarity is compared.

For e.g. If 'm1' encrypts to 'c1' and 'm2' encrypts to 'c2', then by comparing the value of c1 and c2 we can determine whether m1 is equal to m2.

Order Preserving Encryption (OPE) [12–16], Orthogonality Preserving Encryption etc are some variations of Property Preserving Encryption. Bellare et al. [14] proposed a method where PPE scheme can efficiently applied on securing databases.

Computational Complexity

Search complexity is O(nm), where 'm' is the number of documents, i.e., linear complexity. But data structures like Binary search trees can improve the speed.

Security of PPE

Since encryption on m1 always generate same cipher text security is limited since this can lead to some statistical leakages.

2.2 Functional and Identity-Based Encryption

The concept behind Functional Encryption was first proposed by Sahai and Waters in a conference and later formalized and proved to be practical by Boneh, Sahai, Waters and by O'Neill [17]. Identity Based Encryption, Attribute Encryption, Predicate Encryption etc can be considered as variations of Functional Encryption.

The working of Identity encryption can be explained by a simple real world application: Alice want to send some secret message to Bob. Alice knows that Bob works at Google.

According to Functional Encryption, Google will initialize the security system by generating a pair of master keys (msk, mpk), where one is a secret key and other is public. Google then distributes mpk together with a valid certificate to its authorized employees.

To encrypt a message 'm', Alice will collect Google's Master Public key mpk, and apply the encryption algorithm on 'm' using mpk and Bob's public identity, 'bob@google.com'.

c=E(mpk,'bob@google.com', m).

For Bob, to decrypt the message 'c', Bob generates his secret key using Google's master key and his own id.

sk=KeyGen(msk,'bob@google.com').

Bob recovers the message by applying the decryption algorithm.

m=Dec(sk,c).

The advantage of this method is its simplicity. Without revealing any public key of Bob, Alice can send encrypted messages to him or any person in the organization knowing only the public key of that organization.

In case of attribute based encryption, some attributes approved by Organization will be utilized for encryption. For e.g. Consider a hospital domain. Alice needs to upload a file which can be viewed by a person if he is a 'doctor specialized in oncology with masters degree'. Hence the attributes can be 'doctor', 'MD', 'Oncology' etc.

Computational Complexity

Complexity is O(nm) as the algorithms has to try to decrypt each cipher text in the Encrypted domain. But always m ≪ n, hence the time complexity needed will always be more compared to PPE.

Security

This approach substantially ensures security since neither statistical leakages or brute force attacks exist in the system.

2.3   Fully Homomorphic Encryption

A cryptosystem that supports both addition and multiplication operations on encrypted data is called fully homomorphic encryption (FHE) and is far more powerful. Homomorphic encryption schemes process data in its encrypted form itself. No decryption is needed. Thus these types of applications are best suited for third party computations like cloud computing. Encryption does not reveal any information to external agents.

Homomorphism with respect to addition or multiplication has been made possible since the development of RSA [18] and paillier encryption [20]. They are called partial homomorphic systems. The concept of Fully Homomorphic encryption which made possible additions and multiplications over encrypted data was first proposed by Gentry [19].

Craig Gentry [19] developed lattice based cryptosystem to achieve fully homomorphic property and he was successful in evaluating arbitrary depth circuits. The scheme was also bootstrappable meaning as the circuit grows, the noise rises and ultimately the circuit will get capable of decrypting its own encrypted data i.e., the circuit gains self referential property. Hence in 2012 Gentry along with Vaikundanathan [7] proposed a variation of the original scheme using the property of ideal lattices over integers.

Let us illustrate fully homomorphic symmetric encryption scheme with an example:

Let the shared secret key be an odd number, 101. The domain consist of bits {0, 1}. To encrypt m = 1; Choose a random small prime number r = 5, and large q = 9.

Table 1 Complexity comparison

Dimension	KeyGen	PK size	Re-crypt
512 200,000-bit integers	2.4 s	17 MB	6 s
2048 800,000-bit integers	40 s	70 MB	31 s
8192 3,200,000-bit integers	8 min	285 MB	3 min
32728 13,000,000-bit integers	2 h	2.3 GB	30 min

$$\text{Encryption } (m) = c = m + 2r + pq = 11 + 909 = 920$$

Here cipher text will always be close to a multiple of p.
Therefore, $m \approx$ LSB of distance to nearest multiple of p.

$$\text{Decryption is } m = (c \% p) \% 2 = 11 \% 2 = 1.$$

Computational Complexity

Comparing Fully homomorphic encryption using integers and ideal lattices, the flatter method have exponential complexity which is not at all tolerable. Integer method is assumed to have complexity $\lambda 5$. The table below describes the complexity details (Table 1).

2.4 Oblivious RAM

Oblivious RAM concept was first proposed by Goldreich [18] as a method to implement software protection on third party servers. But at that time it seems to be irrelevant because cloud computing or third party computing were not at all in practice. But now the work has gained so much application context related to cloud storage.

An ORAM scheme basically consist of 3 stages Setup, Read and Write.

- Setup: inputs are

  - security parameter 1K.
  - RAM (memory array) of N items.
  - Outputs: Secret key K and an oblivious memory ORAM.

- Read: A two-party algorithm run between client and server. The client runs the Read function with a secret key K and an index i as input while the server runs the Read Function with an oblivious memory ORAM as input. At the end of the execution, the client receives RAM[i] while the server receives ε, i.e., null.

  Read((K, I),ORAM) = (RAM[i], ε).

Table 2 Critical comparison of searching schemes

Scheme	Search complexity	Search type	Recalculation?
PPE	O(n)	Linear	No
Functional encryption	O(d)	Pre-processed index	No
SSE	O(1)	Pre-processed index	Yes
PEKS	O(n)	Linear	No
Rank ordered	O(d)	Pre-processed index	Yes

Table 3 Summary of major search schemes and their ability to perform certain search options

Scheme	Exact match	Sub match	Case insensitivity	Regex	Proximity	Stemming
Practical technique	Yes	No	No	No	Yes	No
Secure indexes	Yes	Maybe	Maybe	No	No	Maybe
SSE	Yes	Maybe	Maybe	No	No	Maybe
PEKS	Yes	Maybe	Maybe	No	No	Maybe
Rank ordered	No	No	Yes	No	No	Yes

- Write: Two party protocol executed between the client and a server. The client runs the Write function with a key K, an index i and a value v as input and the server runs the Write function with an oblivious memory ORAM as input. At the end of the protocol, the client receives nothing (again denoted as $\mathcal{E}$) and the server receives an updated oblivious memory ORAM' such that the ith location now holds the value v. We represent this as

  Write $((K, i, v), ORAM) = (\mathcal{E}, ORAM')$.

Security of ORAM

ORAM is constructed such that server is unable to derive any information about RAM. Read and Write functions do not leak information about the index and values either.

Computational Complexity

Since FHE has to be implemented in Read and Write phase, ORAM is the slowest of all techniques mentioned above (Tables 2 and 3).

3 Conclusion

Cloud computing is gaining so much interest due to the huge amount of data generated and need for computations to be performed on these data. Security and privacy is the only factor that hinders the usability of cloud. Users of data do not

Table 4 Summary of different encryption schemes

Scheme	Summary
PPE	Fast search, but at the expense of small information leakage
Functional encryption	Easy implementation, Secure but slow search time
FHE	Secure but Application Dependant, We should choose a homomorphic function based on the application context in which it is implemented
ORAM	Most secure solution which hides even the access pattern

trust a third party agent like cloud to store their sensitive data. The solution is encryption. But encryption limits the computability of data. To eliminate such limitations we can choose encryptions that properly match each application. This article surveys the different encryption schemes available in literature and compare them based on factors like security, complexity etc. The table below provides a short summary of all the schemes mentioned (Table 4).

Acknowledgments The authors sincerely thank Department of Science & Technology, India for supporting the research work by granting Inspire Fellowship.

References

1. M. Armbrust, A. Fox, R. Griffith, A. Joseph, R. Katz, A.Konwinski, G. Lee, D. Patterson, A. Rabkin, and M. Zaharia, "A View of Cloud Computing," Comm. ACM, vol. 53, no. 4, pp. 50–58, 2010.
2. Shashank Agrawal, Shweta Agrawal, Saikrishna Badrinarayanan, Abishek Kumarasubramanian, Manoj Prabhakaran, and Amit Sahai. Function Private Functional Encryption and Property Preserving Encryption: New Definitions and Positive Results. Cryptology ePrint Archive, Report 2013/744, 2013.
3. C. Leslie, "NSA Has Massive Database of Americans' Phone Calls," http://usatoday30. usatoday.com/news/washington/2006-05-10/, 2013.
4. R. Curtmola, J.A. Garay, S. Kamara, and R. Ostrovsky, "Searchable Symmetric Encryption: Improved Definitions and Efficient Constructions," Proc. ACM 13th Conf. Computer and Comm. Security (CCS), 2006.
5. C. Wang, N. Cao, J. Li, K. Ren, and W. Lou, "Secure Ranked Keyword Search over Encrypted Cloud Data," Proc. IEEE 30th Int'l Conf. Distributed Computing Systems (ICDCS), 2010.
6. S. Zerr, D. Olmedilla, W. Nejdl, and W. Siberski, "Zerber+r: Top-k Retrieval from a Confidential Index," Proc. 12th Int'l Conf. Extending Database Technology: Advances in Database Technology (EDBT), 2009.
7. M. van Dijk, C. Gentry, S. Halevi, and V. Vaikuntanathan, "Fully Homomorphic Encryption over the Integers," Proc. 29th Ann. Int'l Conf. Theory and Applications of Cryptographic Techniques, H. Gilbert, pp. 24–43, 2010.
8. M. Perc, "Evolution of the Most Common English Words and Phrases over the Centuries," J. Royal Soc. Interface, 2012.
9. O. Regev, "New Lattice-Based Cryptographic Constructions," J. ACM, vol. 51, no. 6, pp. 899–942, 2004.

10. Mihir Bellare, Thomas Ristenpart, Phillip Rogaway, and Till Stegers. Format-preserving encryption. In Michael J. Jacobson Jr., Vincent Rijmen, and Reihaneh Safavi-Naini, editors, Selected Areas in Cryptography, volume 5867 of Lecture Notes in Computer Science, pages 295–312. Springer, 2009.
11. Alexandra Boldyreva, Nathan Chenette, Younho Lee, and Adam O'Neill. Order-preserving symmetric encryption. In Antoine Joux, editor, EUROCRYPT, volume 5479 of Lecture Notes in Computer Science, pages 224–241. Springer, 2009.
12. Alexandra Boldyreva, Nathan Chenette, Younho Lee, and Adam ONeill. Order-preserving symmetric encryption. Cryptology ePrint Archive, Report 2012/624, 2012. http://eprint.iacr.org/.
13. Alexandra Boldyreva, Nathan Chenette, and Adam O'Neill. Order-preserving encryption revisited: Improved security analysis and alternative solutions. In Phillip Rogaway, editor, CRYPTO, volume 6841 of Lecture Notes in Computer Science, pages 578–595. Springer, 2011.
14. Alexandra Boldyreva, Nathan Chenette, and Adam ONeill. Order-preserving encryption revisited: Improved security analysis and alternative solutions. Cryptology ePrint Archive, Report 2012/625, 2012. http://eprint.iacr.org/.
15. Dan Boneh and Xavier Boyen. Efficient selective identity-based encryption without random oracles. J. Cryptology, 24(4):659–693, 2011.
16. Dan Boneh, Xavier Boyen, and Eu-Jin Goh. Hierarchical identity based encryption with constant size ciphertext. In Ronald Cramer, editor, EUROCRYPT, volume 3494 of Lecture Notes in Computer Science, pages 440–456. Springer, 2005.
17. E. Bach and J.O. Shallit. Algorithmic Number Theory. Foundations of computing. MIT Press, 1996.
18. Goldreich, O. "Towards a Theory of Software Protection and simulation by Oblivious RAMs" STOC 87.
19. C. Gentry, "Fully Homomorphic Encryption Using Ideal Lattices," Proc. 41st Ann. ACM Symp. Theory of computing (STOC), pp. 169–178, 2009.
20. Lija Mohan, Sudheep Elayidom, "Fine Grained Access Control and Revocation for secure cloud environment- a polynomial based approach", International Conference on Information and Communication Technologies, to be published in Elsevier Procedia, December 2014.

A Proposed Conceptual Framework to Migrate Microsoft Technology Based Applications to Cloud Environment

Manas Kumar Sanyal, Sudhangsu Das and Sajal Bhadra

Abstract With the evolution of ICT in different business domain, it has become essential to almost all the IT project stakeholders to move applications into cloud for saving IT cost and ensuring sustainability of IT solutions for future. On this fact, it is already proven that cloud implementation has given ample opportunity to reduce overall IT cost of the organization. In cloud implementation, Project stakeholders would have to bear only cost which is project owners ought to pay according to uses of IT resources on pay-per-use cost model. But, in the cloud implementation journey, the most impacting and challenging activity is to do migration of existing application to cloud platform smoothly without interruption of existing application's user's experiences and on-going business activities. In this study, authors' main focus is to propose conceptual framework for assisting to do migration of IT projects in cloud platform. This generic framework will mostly cover End to End processes which require migrating Microsoft based technology applications those are mostly developed in Dot Net and SQL server technology and deployed in Internet Information Server (IIS). The holistic objective of the proposed framework is to facilitate business to provide faster, more reliable, robust and cost effective migration process. This study also has taken in scope to develop very simple proof of concept (POC) to do some sort of validity of the proposed idea.

Keywords Cloud · Migration · Pay-per-use · Cost effective

M.K. Sanyal (✉) · S. Das · S. Bhadra
Department of Business Administration, Kalyani University, Kalyani, India
e-mail: Manas_sanyal@rediffmail.com

S. Das
e-mail: iamsud@gmail.com

S. Bhadra
e-mail: Sajal.bhadra@gmail.com

© Springer Science+Business Media Singapore 2017
J.K. Mandal et al. (eds.), *Proceedings of the First International Conference on Intelligent Computing and Communication*, Advances in Intelligent Systems and Computing 458, DOI 10.1007/978-981-10-2035-3_12

1 Introduction

Nevertheless to mention that cloud implementations are progressing in extreme rapid speed to get cost benefit in case of minimizing IT cost incurred for the organization. Cloud has been provided enormous convenient to users for accessing IT resources and removed the entire individual over burden to manage hardware and software required to run applications [1]. This trend is very significant in case of new initiatives and implementations because business impacts are very minimal as technical and infrastructure architecture of the applications are being designed and formulated by targeting to deploy application in cloud environment. But, this situation is little bit different for the legacy and existing applications because the legacy applications face challenges to get migrated from physical or virtual server to cloud server due to some obvious reason like limitation in application architecture, deployment process, integration dependency, related legacy tools and softwares. The cloud migration process required lot of brainstorming to sketch effective plan from decision to deployment and maintenance. The main risks of cloud migration are to host applications in cloud environment without impacting existing user experience with full functionality. Most of the business takes quite lot of time for doing pre-study and to prepare some sort of proof of concept (POC) before reaching out at final decision whether business should take risk for doing application migration in cloud or not. This pre-study and building POC is the first step for any migration. Generally, Business wants to gather confident about the success of migration from the pre-study. During migration life cycle, there is again required lot of exercise around architecture fine tuning, application migration, deployment process redefining and others relevant tasks. It has been observed that there are number of applications those are getting migrated to cloud environment and still very significant amount of applications are in pipeline to be hosted in cloud environment. In this situation, business stakeholders are very curious to find out some standard process and tools/frameworks which are tested and have capability to do successful migration in any applications overcoming technical and environmental issues. Eventually, this should include all the necessary steps those are mandatory in migration process. So, any standard process and tools/frameworks may help to business to get rid of from the pain require to migrate their application to cloud environment. In this study, an extensive research has been conducted on this high demanding requirement to build some sort of generic framework which may help to business to do successful cloud migration of their applications. These frameworks will cover some suggestions for automation of few activities as per the applicability in different phase of cloud migration. Since, the target of these frameworks to have it generic that's why the scope has been cut down to cover only Microsoft technology based applications those are mainly hosted in Internet Information Server (IIS) and developed in Dot Net, SQL Server, MS ACCESS etc. Further, a tool has been developed based on the proposed frameworks for testing the concept or idea. The result of testing has found very satisfactory which have been included in the simulation and result section of this paper.

2 Literature Review

Software-as-a-Service (SAAS) is a very attractive and successive implementation of cloud initiatives. In cloud migration, it has found that it is often required to do re-engineering of existing applications architecture during the migration for making it compatible as SAAS. Sometimes it requires to do complete re-development of the application which is risky and costly. In a study, Zhao et al. [2] have suggested and proposed an innovative approach to discover and re-use the existing components of the application. Towards achieving the same, Authors have recovered implementation model by reverse engineering, extracted functional model by vertical clustering and identified logical components by horizontal clustering [2].

Cloud migration is not new buzz word at all, the research have been started from last few years as all the legacy system also need to move to cloud for getting cloud benefits. In a comprehensive study, Jamshidi et al. [3] have focused on different existing literature related to cloud migration keeping objective in mind to identify, classify and doing systematic compare among different migration approaches. From the review of 23 selected studies, it has been concluded that cloud migration is still in very early stage and require quite lot of brainstorming to find out steady migration approach, also there is requirement to support automation for different tasks in migration phase [3].

Since Cloud migration have cost and risk both involvements that's why it is required to do proper analysis for taking decision to perform cloud migration of legacy application from on-premises platform. Johnson and Qu [4] have consolidated few analyses on cloud computing risk which has taken consideration of business economics, security and availability. In this study, Authors also have done comparison by evaluating the situation of service requirement situation against two service providers [4].

In cloud environment, hardware resources are being shared based on the application requirement on per demand to utilize resources in best way. So, it is very common scenario that application and data both are not get stored in the same server. In terms of network tropology, application server and data server be located in different geographical location. If network distance in between the application server and data server be longer, in that case it has direct impact on application performance. In the year 2010, Jing Tai Piao and Jun Yan have proposed a conceptual framework to do proper placement of different virtual machines (VM) in cloud environment including migration of VM's to improve overall application performance [5].

Service Oriented Architecture [6, 7] (SOA) is a part of application technical design and decision which facilitates to expose application functionalities as a service among users and others parties irrespective of heterogeneous platform barriers. It has extended reusability and extensibility options of application. Due to huge number of benefits, there is lots of applications those have used SOA very extensively. In a study, Zheng and Wu [8] have proposed capacity planning and infrastructure layout design to migrate services. Here, individual component has been explained in details involved in infrastructure migration in relevant of Services [8].

3   Research Methodology

In this study, extensive researches have been conducted by exploring past literature and websites to gathers background knowledge and to know actual needs of business for cloud migration. So, this is qualitative research on secondary data. Here, Microsoft technology language C# has been considered for designing user interface to make tool very user friendly. Windows PowerShell scripting has been used for automating all the migration related tasks during different phase. Study has been kept limited in scope of Microsoft technology based application to make the tools more generic. The execution environment for the tool will be supported by Internet Information Server (IIS), Dot Net Framework, and Microsoft SQL server/MS Access/MS SQL database.

4   Conceptual Framework to Migrate Applications to Cloud

The proposed conceptual framework has considered seven phases to do End to End (E2E) cloud migration. The scope of these phases is to cover applications those have been developed in Microsoft Technology. The scope have been kept limited in Microsoft Technology because framework has targeted to do some automation using generic script based on some assumption in relation with platform and utility software's.

4.1   Seven Phases Migration Workflow

According to design of the framework, the identified seven phases have to be executed in sequential manner for moderating application migration from On-Premises environment to cloud environment. The required sequence and phases have been shown in Fig. 1 and brief descriptions of different phases have been depicted in below:

Phase-I (Feasibility Study):
Feasibility study is targeted to cover all the potential study like technical, functional, cost, risk, business benefits, and security issue with very serious note.

Fig. 1 7 phase cloud migration workflow's

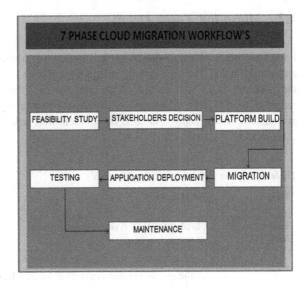

Phase-II (Stakeholders Decision):

This phase is very important in this framework. Here, Business will take final call for **GO** or **NO GO** decision towards doing cloud migration based on the inputs gathered in feasibility study phase.

Phase-III (Platform Build):

In this phase, the necessary softwares, and tools, utilities like IIS, Framework, and SMS/Email server Configuration etc. will be Installed and managed for building cloud platform to install migrated application.

Phase-IV (Migration):

The actual application migration will be conducted in this phase. With application, Database, services and integrated application migration, it will take necessary action to do changes in existing architecture for smooth transformation of application.

Phase-V (Application Deployment):

The focus of this phase is to create application deployment and rollback document for operation team.

Phase-VI (Testing):

This phase will do rigorous testing of application in cloud environment after completion of migration. It will consider functional, regression and performance testing to deliver Error free application to business users.

Phase-VII (Maintenance):

This is last phase of any migration. In this phase, Application will be released for business users. After limited time of Post Go Live (PGL) support, Application operation team will take ownership of the application for continuing maintenance and enhancement.

4.2 *Migration Tasks Detail*

Each individual phase of the proposed framework obviously have multiple major
and minor tasks. The exit criteria of different phases are successful completion of
the different tasks in relevant of the phases. In Table 1, all the different tasks
required to do E2E migration in different phase along with the possibility, if
automation can be accommodated of the identified task have been jotted down and
categorized.

Table 1 7 phase cloud migration task details

Phase	Major tasks	Is automation applicable?
Feasibility study	Cloud type study (Private/public/hybrid)	No
	Data center options	No
	Need and possibility of architectural changes	No
	Challenges of existing application security handling (Authorization/authentication)	No
	Challenges to establish existing integrations of application	No
	Need and possibility to do application performance optimization	No
	Any component require to re-develop from scratch	No
	Scope of automation of existing application features/functionalities	No
	Exploration of reusable component of application	No
	Ball park estimation of overall migration cost	No
	Overall business benefits	No
Stakeholders decision	Cloud type selection	No
	Data center selection	No
	Application architectural change	No
	Budget secure	No
	Project timeline finalization	No
Platform build	Web server (IIS) provisioning	Yes
	Web server configuration as per application demand	Yes
	Framework installation (Dot Net 3.0/3.5/4.0/4.5)	Yes
	Mail/SMS server provisioning	Yes
	Necessary folder creation and permission assignment	Yes
Migration	Architectural evolvement	No
	Application migration	No
	Integration point migration	Yes
	Database migration	Yes
	Services migration	Yes

(continued)

Table 1 (continued)

Phase	Major tasks	Is automation applicable?
Application deployment	Re-writing deployment script	Yes
	Re-writing installation documents	No
	Re-writing Rollback script	Yes
	Re-writing Rollback documents	No
Testing	Functional testing	Yes
	Regression testing	Yes
	Performance testing	Yes
Maintenance	Access matrix preparation for support engineers	No
	Application monitoring system provisioning	Yes
	Functional and technical document preparation and handover to support engineers	No

5 Simulation and Results

In this study, a simple tool has been developed in Microsoft Dot Net Technology with Dot Net Framework 4.0 compatibility to simulate the proposed framework. In this seven phase migration framework, few tasks have been identified and considered to perform through script execution by leveraging the advantage of automation. Snap of simulation tools have been depicted in the below as Figs. 2 and 3.

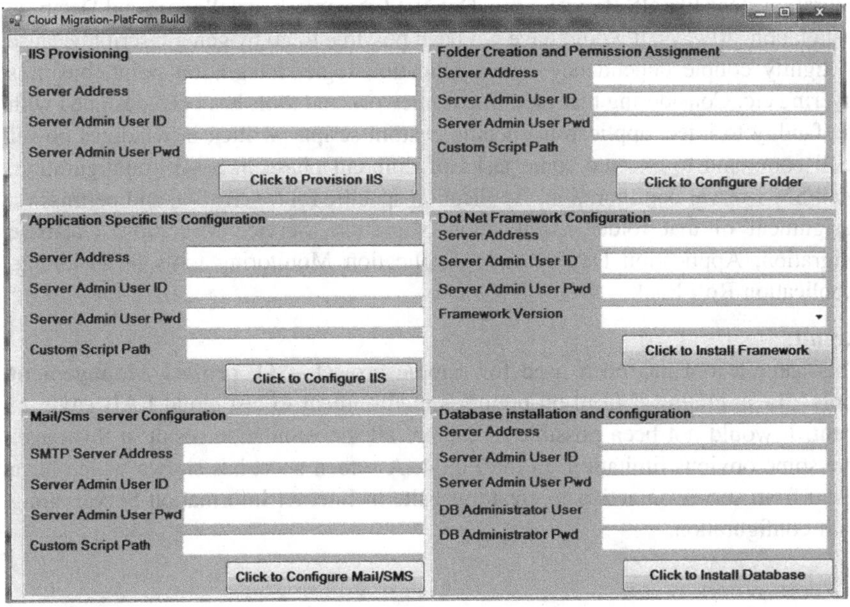

Fig. 2 Screen snap of cloud migration simulation tool for platform build

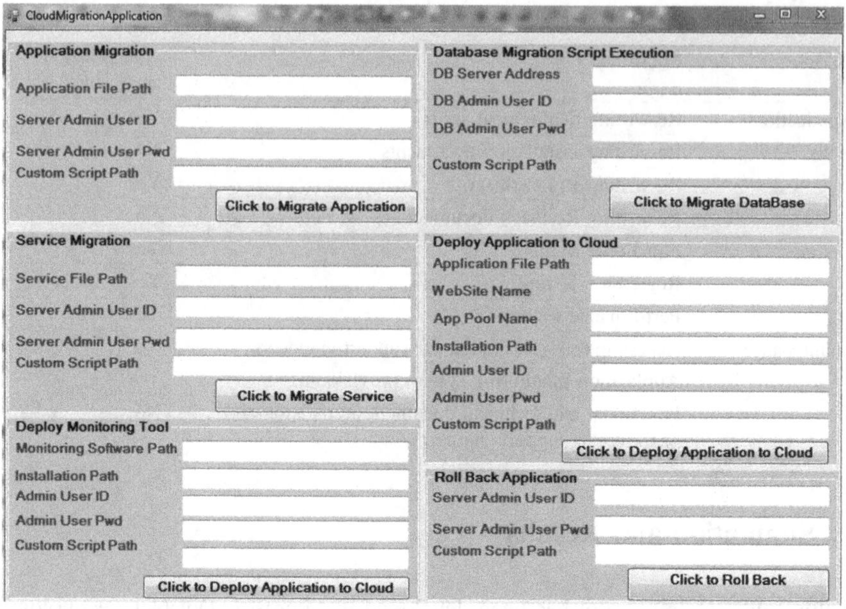

Fig. 3 Screen snap of cloud migration simulation tool for application migration

Upon clicking on the button in tool, it will trigger specific scripts which have been written to accomplish the related tasks. In the framework, generic piece of Dot Net code and power sell script have been incorporated as per convenient to execute few migration steps like IIS Provisioning, Dot Net Framework Installation, and Database installation. Always, it would have not been possible to go by generic script because of tightly couple dependency with application logic, integration point, business layering etc. Considering that fact, The framework and tools has been designed with the facility to inject application specific custom scripts, written in windows power shell command to execute some tasks of different phase like—IIS configuration, Mail/SMS server configuration, Application specific folder creation and permission assignment of that folder, Application Migration, Service Migration, Database Migration, Application Deployment, Application Monitoring tools provisioning, Application Roll Back.

Result:

This sample tool has been used for sample project—**"IT project Management tools"** for migrating it from on-premises environment to test cloud LAB environment. It would not been possible to capture all the simulation result in this paper due some obvious limitation. In the Fig. 4, A screen short has shown which have taken from server on action to create website in Internet Information Server along with configuration.

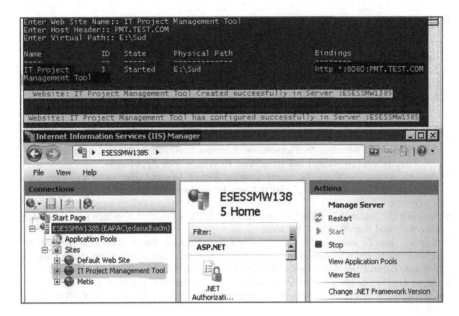

Fig. 4 Screen snap from server

The few simulation test results have been given below:

Platform Build—Generic script:

IIS provision: **Success**
IIS configuration as per application config file: **Success**
Dot Net framework Provision: **Success**

Migration—Application specific script:

Visual studio up gradation—**success**
Data base schema, script up gradation—**success**

Conclusion:

In the recent era, Legacy system cloud implementation is real challenges to the industries due to risk, time and extensive cost. Business never wants to develop new solution for cloud environment by replacing the existing application because of huge cost and time. So, cloud migration of legacy system is the only choice to overcome the problem. In this study, a conceptual framework has been proposed to do application migration from on-premises environment to cloud environment. The framework has been designed with seven phases like **"Feasibility Study"**, **"Stakeholders Decision"**, **"Platform Build"**, **"Migration"**, **"Application Deployment"**, **"Testing"** and **"Maintenance"** for doing E2E migration. The details tasks of each phase also have been explained in this study. A sample tool also has been developed to test the concept. Test result of this tool have found very satisfactory. In future study, it's require more testing and need to consider others parameter for making the concept more generic and may support all type of application migration.

References

1. Brian H. (2008), Cloud computing, Communications ACM, Volume 51, Issue 7, July 2008.
2. Junfeng Z., Jiantao Z., Hongji Y. and Guoping L. (2015), "An Orthogonal Approach to Reusable Component Discovery in Cloud Migration", China Communications.
3. Jamshidi P., Ahmad A., and Pahl C. (2013), "Cloud Migration Research: A Systematic Review", IEEE TRANSACTIONS ON CLOUD COMPUTING, VOL. 1, NO. 2, JULY–DECEMBER 2013.
4. Johnson B., Qu Y. (2012), "A Holistic Model for Making Cloud Migration Decision - A Consideration of Security, Architecture and Business Economics", 2012 10th IEEE international Symposium on Parallel and Distributed Processing with Applications.
5. Piao J.T., Yan, J. (2010), "A Network-aware Virtual Machine Placement and Migration Approach in Cloud Computing", 2010 Ninth International Conference on Grid and Cloud Computing.
6. Atkinson C., Bostan P., Hummel, O. and Stoll D. (2007), A Practical Approach to Web Service Discovery and Retrieval. In ICWS2007.
7. Yen I., Gao T. and Ma H. (2006). A genetic algorithm based QoS analysis tool for reconfigurable service oriented systems. In Advances in Machine Learning Application in Software Engineering, edited by Du Zhang and Jeff Tsai, IDEA Group Publishing, pp. 121–146.
8. Zheng L., Wu, S. (2010), "An Infrastructure for Web Services Migration in Clouds", 2010 International Conference on Computer Application and System Modeling (ICCASM 2010), IEEE.

Part III
Communication Network and Services

An Adaptive Cloud Communication Network Using VSAT with Enhanced Security Implementation

Rajesh Bose, Sudipta Sahana and Debabrata Sarddar

Abstract Cloud computing has opened up a whole new vista wherein it is possible to segregate and compartmentalize the process of constructing infrastructure with the intent of providing certain services to end users from the very business itself. In this paper, we have introduced a communication network that is aimed at augmenting network performance in situations where a company needs to operate a branch through means other than wired or mobile networks. Our proposed model utilizes a mix of VSAT and cable technologies that has been designed to improve efficiency of a cloud data center. The introduction of VSAT connectivity also decreases chances of a network breach as the communication conduit in such is not generally shared. Though, VSAT technology continues to lag behind other forms of communication, its ability to bypass complex routing can be harnessed to improve energy efficiency and network performance while bringing down carbon emission rates.

Keywords Cloud · VSAT · Mobile networks · Energy efficiency · Cloud data center

1 Introduction

In the new paradigm brought about by cloud computing, load balancing to support high performance computing requirements has gained enormous importance. As developers and system administrators have begun migrating web applications from local networks to cloud platforms, so too has the demand risen for optimized

Rajesh Bose · Debabrata Sarddar
University of Kalyani, Kalyani, Nadia 741235, West Bengal, India
e-mail: bose.raj00028@gmail.com

Debabrata Sarddar
e-mail: dsarddar1@gmail.com

Sudipta Sahana (✉)
JIS College of Engineering, Kalyani, Nadia 741235, West Bengal, India
e-mail: ss.jisce@gmail.com

© Springer Science+Business Media Singapore 2017
J.K. Mandal et al. (eds.), *Proceedings of the First International Conference on Intelligent Computing and Communication*, Advances in Intelligent Systems and Computing 458, DOI 10.1007/978-981-10-2035-3_13

117

network speeds to ensure that user-experience is not compromised. The subject of our research are long sessions of connectivity in which applications on client devices can be engaged with web servers for extended lengths of time. An example of such a scenario could be an online shopping service wherein users could be logged onto the web servers of a shopping portal for any length of time. While this does not take impose a load on memory and processing resources, the situation can change very quickly once all the users begin accessing the web application hosted by such shopping portal all at the same time.

In his article titled "Introduction to VSAT Technology" published on the web [1], Mr. Greg Heifner, a pioneer in satellite communications, talks about VSAT as a solution that can be cost-effective in situations where communication networks needs to independently connect sites situated at different geographical locations. Although the article was authored in 2004, it still remains relevant today. VSAT networks offer a flexibility not readily found in other terrestrial services when deployed to offer services covering Internet, local area network, voice or fax communications, video, etc. Installation is quicker, and considerably easier to deploy even in remote locations. However, as the author has pointed out, VSAT communication is not suitable for conducting online gaming sessions where network response times are considered to be a very important factor.

The rest of the paper has been organized in sections which are as follows. We discuss related work and research conducted in the section titled "Related Work". Our approach and introduction to our model has been presented in the section titled "Proposed Work". We elaborate the related algorithm, flowchart and the mode section procedure in this section. In the section titled "Result Analysis" we explain how our model could be utilized to in a cloud environment. Finally, we conclude our paper at the section titled "Conclusion".

2   Related Work

Commercial network offerings which rely on wired TCP/IP networks are susceptible to hijacks and malicious attacks. In the case of VSAT though, such possibilities are largely limited. In their paper titled "VSAT Network Overview" published in IOSR Journal of Electronics and Communication Engineering (IOSR-JECE) e-ISSN: 2278-2834, p-ISSN: 2278-8735. Volume 10, Issue 1, Ver. III (January–February 2015), pp. 18–24, the authors, Hassan et. al. [2] have discussed the components which make up a Very Small Aperture Unit (VSAT). In their work, the authors have explained the significance of the Encryption-Decyrption unit (or the EDU in short) that is responsible for altering the information to be transmitted in a manner such that it is secure. The authors also discuss how VSAT communication can be extended protected against malicious attacks by restricting open access to the Internet. Further, the authors stress on the importance of X.509 certificate authentication protocol to augment security in VSAT transmissions.

In their work titled "A Survey Paper on Security Issues in Satellite Communication Network infrastructure", the authors Shah et. al. [3] have discussed the issue of deploying security techniques to further improve satellite communication. The work of the authors is important considering the fact that VSAT is dependent on satellites to convey information over extremely large distances. The authors discuss the various issues related to security, and data transmission rates that are largely affected by weather patterns. The authors have tried to strike a balance between implementation of security and transmission of data over satellite medium such that the overall rate of delay and packet loss can be lowered as far as possible.

In conducting our research, we found that VSAT has slowly gained in stature among the maritime community. Research suggests that companies with interests in maritime activities for transportation of goods and passengers have benefitted to a large extent by installation VSAT communication equipment on almost all of the vessels that they own. In a report titled "VSAT: Present and Future. A comprehensive survey of maritime VSAT", the authors of the communication agency, Stark Moore Macmillan, which specializes in commercial maritime sector, and iDirect—manufacturers of connectivity platforms and solutions [4], have attempted to conduct an in-depth survey involving experts and professionals in the maritime profession to discuss and understand the use of VSAT technology and how it has been successful in the merchant fleet. The report concludes that although VSAT technology is still to gain a solid footing with ship operators, developments in this field are being watched with interest as the benefits of this technology are considered significantly of being more value than the cost it entails to install and maintain VSAT infrastructure. In our view, this is of paramount interest as nowhere in the world is VSAT of more significance and use as in the open oceans across which the world's global shipping lanes crisscross.

The authors, Hadji et. al. [5] in their research titled "Minimum Cost Maximum Flow Algorithm for Dynamic Resource Allocation in Clouds" proposed an algorithm which the authors call "A minimum cost maximum flow algorithm". The authors, in their course of their work, have identified the need to dynamically balance load and variation of data flow for placement of resources in cloud network. The authors while comparing their method to the Bin-Packing formula, have offered it to choose resources in a manner such that elastic provisioning can be conducted in situations where resources are limited.

3 Proposed Work

In our work, we have directed our research towards facilitating access to cloud infrastructure from remote terrestrial regions where wired connectivity is either unavailable or unreliable to the point where sustained data and information exchange is impossible to conduct. Our model has been tailored to be useful in cases where companies have operations in regions where wired connectivity is not an option. For example, in hilly terrain or in a forested area which do not accord the kind of economic viability necessary to lay cables for wired communication facilities, VSAT assumes a very important role. The immediate economic advantages offered by VSAT in such situations, where other forms of wireless communication may not be very suitable, become, therefore, almost readily apparent. For one, installation of VSAT equipment is considerably cheaper. The bandwidth charges, too, are affordable in comparison to other forms of wired or wireless technologies which would have, otherwise, required considerably larger investment to setup and operate in remote regions.

We have also looked at bolstering security by putting together two forms of network modes, i.e., cable and VSAT. Our model has been designed such that VSAT mode would be selected in cases where security is of paramount interest, and where communication over cable or wired networks to the destination is not possible. To harmonize the transmission over VSAT and wired networks, we have introduced cloud data center manager (CDM) to regulate and control data transmissions. The introduction of a cloud data center manager has an immediate effect of reducing the load on cable or wired networks. In our model, the cloud data center manager (CDM) maintains a 25 % buffer margin to support secure transmissions of sensitive information over VSAT channel. The remaining 75 % VSAT channel capacity, along with the full bandwidth capacity of cable or wired networks, are allotted for support of data transmissions that either do not require secure form of communication, or are too large to be sent over VSAT within a short span of time.

3.1 Algorithm for Data Sending from Cloud Data Center

Step 1   The algorithm behind our proposed model is based on a simple procedure of selection according to priorities of security and size of the data to be transmitted in the form of files

Step 2   On receiving a file for transmission, the cloud data center manager checks the type and details of the contents to be transmitted

Step 3   It then ascertains whether the destination can be reached from source either through single or dual modes

Step 4   In case dual modes are unavailable, then the only mode of data transmission is automatically selected

Step 5   On the other hand, if dual modes exist, the content type is checked

Step 6   If the content to be transferred is found to be of secure type, VSAT mode is selected

Step 7   In case, VSAT mode is unavailable or if the content is found to be larger than the recommended size acceptable for the particular file type for transmission over VSAT, available wired network is selected

Step 8   For the purposes of determining whether a file needs to be transmitted either over VSAT or wired networks, a dedicated cloud data center manager checks the file against a table containing a set of parameters tailored for such selections to be made. This procedure is referred to in our work as the Mode Selection Procedure

Step 9   Finally, the file is transmitted through to the destination on the basis of the selection and availability of the appropriate mode to be so determined by a cloud data center manager.

3.2   Mode Selection Procedure

In order to achieve optimum file transmission, we have incorporated a table of file sizes in our proposed model. Further, we have also used a dedicated cloud data center manager which works to select the appropriate mode of transmission. It is the job of this cloud data center manager to maintain an active index of all wired and VSAT networks available so as to ensure data transmissions can be sped up without compromising security of the data being transmitted en route (Table 1).

Table 1 Mode selection procedure table

Sl. no.	File type	Size	Mode
1	Document or pdf file	Less or equal 5 Mb	VSAT
		More than 5 Mb	Cable
2	Execution file	Less or equal 100 Mb	VSAT
		More than 100 Mb	Cable
3	Object file	Less or equal 5 Mb	VSAT
		More than 5 Mb	Cable
4	Script file	Less or equal 5 Mb	VSAT
		More than 5 Mb	Cable
5	Archive file	Less or equal 25 Mb	VSAT
		More than 25 Mb	Cable
6	Audio file	Any size	Cable
7	Video file	Any size	Cable
8	CAD file	Any size	Cable
9	Database file	Any size	VSAT
10	Others	Any size	Cable

4 Flow Chart

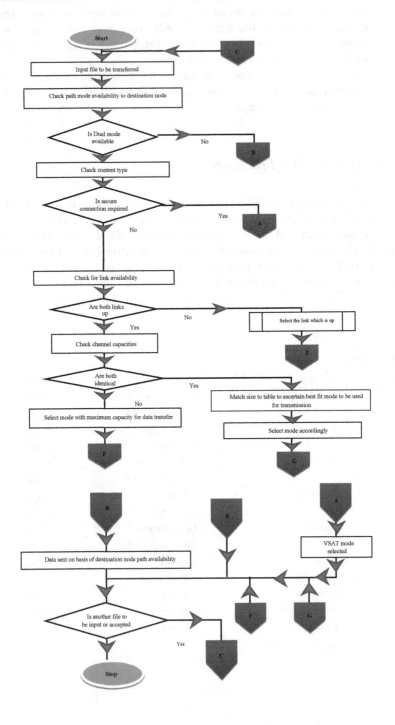

5 Result Analysis

Our research focuses on implementation of using VSAT connectivity as a secure medium of transmission over cloud network. We have shown how VSAT can be used to extend the umbrella of cloud infrastructure in remote areas. We have also introduced our proposed model using which we have postulated how load balancing can be achieved. VSAT installations can significantly assist in reducing communication channel overheads by allowing secure traffic to flow, thereby, bringing down loads on wired networks. As a result, network congestion instances decrease, bandwidth can be properly utilized, security management is enhanced, savings in terms of time indirectly pushes up cost savings, and helps control carbon emissions. In our opinion, our proposed method based on a literature survey conducted by us, has the potential to decrease average times required for routing from source to destination. Our method also introduces secure VSAT transmission which not only eliminates vulnerability, but also eases network bottlenecks and lowers congestion. Figure 1 shows the comparison between proposed and existing method. According to the result analysis our approach takes less average time for routing from source to

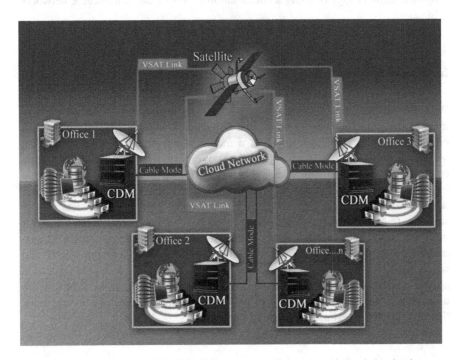

Fig. 1 Proposed diagram of a Cloud Data Center managed data transmission balancing between VSAT and cable modes across cloud networks

Fig. 2 Average routing time versus degree of services

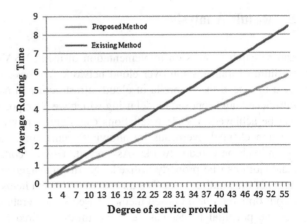

destination due to less congestion and simple VSAT routing using satellite communication. Scope for future exists in the direction of designing more effective pathways for routing secured data in bulk or packets over cloud network using our proposed methodology of using a combination of VSAT and wired/cable networks (Fig. 2).

6 Conclusion

With the advent of cloud computing, computing resources can be leased in a manner that is not only cost-effective, but also efficient in terms of performance and offers robustness. Our paper presents a model which primarily focuses on data security through the introduction of VSAT. We have also introduced cloud data center managers and a mode selection procedure aimed at enabling load balancing in addition to security. VSAT offers an advantage in that it can be deployed anywhere in the world with access to satellite networks. Consequently, data can be securely transmitted without involving complex routing across networks. Our proposed mode selection procedure helps in quickly determining the data transmission mode that needs to be chosen given the availability of both VSAT and wired networks at source and destination points. In our view, our proposed approach strikes a balance between data security and optimum network bandwidth utilization. This, we feel, should offer significant advantages over existing methods that deal in data transmission in cloud networks.

References

1. Greg Heifner in http://www.broadbandproperties.com/2004%20issues/March04-rticles/mar_feature3.pdf
2. Abdalrazig Ibrahim Hassan, Dr. Amin Babiker A/Nabi Mustafa. "VSAT Network Overview" published in IOSR Journal of Electronics and Communication Engineering (IOSR-JECE) e-ISSN: 2278-2834, p-ISSN: 2278-8735. Volume 10, Issue 1, Ver. III (Jan–Feb. 2015), PP 18–24
3. Syed Muhammad Jamil Shah, Ammar Nasir, Hafeez Ahmed "A Survey Paper on Security Issues in Satellite Communication Network infrastructure", in International Journal of Engineering Research and General Science Volume 2, Issue 6, October–November, 2014
4. Stark Moore Macmillan "VSAT: Present and Future. A comprehensive survey of maritime VSAT", Stark Moore http://navcom-solutions.com/images/uploads/VSAT-Present-and-Future-Research-Report.pdf
5. Makhlouf Hadji, D. Zeghlache. "Minimum Cost Maximum Flow Algorithm for Dynamic Resource Allocation in Clouds" in Cloud Computing (CLOUD), 2012 IEEE 5th International Conference on 24–29 June 2012 PP 876–882 ISSN: 2159-6182

Introspecting Effect of Packet Size on End-to-End Network Delay

Uttam Kumar Roy

Abstract A message in computer networks is often divided into frames/packets for various reasons. For a store-and-forward network, a suitable packet-size can drastically reduce delay. In this paper, I have investigated the impact of packet size on delivery time. I have shown that delay is a non-linear function of (i) number of hops the packet traverses, (ii) message size and (ii) the number the message is divided into packets. Since, I can't customize the former two; the last one can suitably be chosen to minimize delay. I found an optimal number of packets that minimizes the delay. Analytical and simulation results show the correctness of the proposed scheme.

Keywords Network · End-to-end delay · Packet size · Hop

1 Introduction

In computer networks, a message is often divided into smaller chunks by various protocol layers. This is done for various reasons such as regulating data flow, error detection, correction and control, link multiplexing, switching etc. However, for a store-and-forward network, the size of packet also has direct impact on end-to-end packet delay hence network performance. By choosing a suitable packet size, it is possible to improve the network performance especially delay.

In this paper, I have investigated the impact of packet size on delivery time with rigorous theoretic derivation. I have shown that delay is a non-linear function of various parameters such as (i) number of hops the packet traverses, (ii) size of the message being broken into and (ii) the number the message is divided into packets. Since, former two are out of our reach; the last one can suitably be chosen to minimize delay. I found an *optimal* number of packets that minimizes the delay. Analytical and simulation results show the correctness of the proposed scheme.

U.K. Roy (✉)
Department of Information Technology, Jadavpur University, Kolkata, India
e-mail: u_roy@it.jusl.ac.in

© Springer Science+Business Media Singapore 2017

127

J.K. Mandal et al. (eds.), *Proceedings of the First International Conference on Intelligent Computing and Communication*, Advances in Intelligent Systems and Computing 458, DOI 10.1007/978-981-10-2035-3_14

2 Related Work

There is relatively scant literature on the impact of number of packets on packet delay. In [1], authors presented design guidelines for WNCS over MANET using the NS2 simulator. In [2, 3], authors estimated the medium access control layer packet delay distribution for IEEE 802.11 considering the differences between busy probability and collision probability. In [4], Anyaegbu and et al. characterized the delay profile of an Ethernet cross-traffic network statically loaded with one of the ITU-T network models and a larger Ethernet inline traffic loaded with uniformly sized packets, showing how the average time interval between consecutive minimum-delayed packets increases with increased network load.

In [5] a TCP control mechanism based on the character of delay distribution and wireless packet loss was proposed. In [6], authors developed an analytic model for congestion in data dissemination protocols and investigate the effect of transmission rate on message delivery latency taking the processing overhead of receiver buffer overflow, which has a significant impact on the results into account. In [7], authors discussed findings from a large-scale study of Internet packet dynamics. They also characterized end-to-end behaviors due to the different directions of the Internet paths, out-of-order delivery and packet corruption.

3 Impact of Packet Size on Delay

Consider a store-and-forward network (Fig. 1) of four nodes A, B, C and D. Node A wants to send a message to D. Since A is not connected to D directly, store-and-forward method is used. In this method node A first forwards the message to B, which in turn forwards the message to C. C does exactly the same as B and the message finally reaches to D.

Figure 2 pictorially describes how everything happens. What is the total time (T_D) taken to deliver this message? Let us find an expression of it.

Consider, following network parameters:

- Number of bits in the message = **m**
- Number of links between source to destination = **k**
- Data rate of each link = **r** bps
- Length of a link = **L** meters
- Propagation speed of a the signal $c = 3 \times 10^8$ m/s
- Propagation delay of each link **d** = **L/c** sec

Fig. 1 A linear store-and-forward network of four nodes. *A* is source, *D* is destination

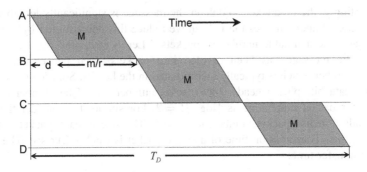

Fig. 2 *A* sends a message to *D* without framing

For simplicity, consider all links are identical and error free. Note that there may be various kind of delay such as nodal processing delay, queuing delay, transmission delay, propagation delay. In this paper, I have investigated the impact of packet size only on the transmission and propagation delay. So, the expression for T_D considering only transmission and propagation delay will then be:

$T_D = \left(3\frac{m}{r} + 3d\right)$ where $\frac{m}{r}$ = message transmission time. In general, if there are k hops between source and destination, the equation becomes:

$$T_D = \left(k\frac{m}{r} + kd\right) \tag{1}$$

Note that the term *kd* represent the total end-to-end propagation delay from A to D. So, this equation is valid even if a network has links of variable length. This delay is directly proportional to the (i) number of hops k, (ii) message transmission time $\frac{m}{r}$ and (iii) propagation delay between source and destination. For a given network configuration, I can't change k, kd. However, This delay can be reduced if I break the message into smaller chunks and deliver them. For example, suppose the message is divided into 3 (three) equal chunks (packets) P_1, P_2 and P_3. The node A can deliver these packets to D using store-and-forward method as shown in Fig. 3.

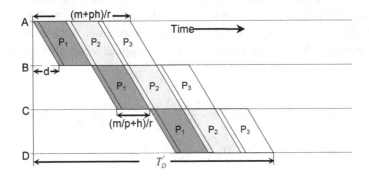

Fig. 3 Message transfer with framing

Interestingly, the time taken to deliver the message is significantly less than the previous case. Important question is "can we reduce this delay further by dividing the message into a suitable number of packets?" Let us examine.

Since the message has been divided into packets, each packet requires a sequence number which is typically incorporated in the header. So, each packet has now m/3 data bits plus h header bits (packet number etc.). The number of bits required for packet numbering is $\lceil \log_2 3 \rceil = 2$. For simplicity, consider that the header only contains packet number, i.e. h = 2. The size of each packet is then: (m/3 + 2) bits. Transmission time of a single packet is (m/3 + 2)/r sec. The total message transfer time:

$$T'_D = 3\left(\frac{\frac{m}{3}+2}{r}\right) + 3d + (3-1)\left(\frac{\frac{m}{3}+2}{r}\right)$$

In general, suppose the message is divided into N number of equal packets P_1, P_2, P_3, ..., P_N. Each packet has now m/N data bits and h = $\log_2 N$ header bits. For simplicity, I omit ceiling function and consider that the header contains only packet number. The size of each packet is then (m/n + $\log_2$N) bits. Transmission time of a single packet is: (m/n + $\log_2$N)/r sec. The total delay for the message transfer is:

$$T'_D = N\left(\frac{\frac{m}{N}+\log_N N}{r}\right) + kd + (k-1)\left(\frac{\frac{m}{N}+\log_N N}{r}\right) \tag{2}$$

This equation is a function of N, k, d, m and r. Since often I have a little control on choosing k, d, m and r, let us find the impact of N on T'_D. Rewriting (2), I get

$$T'_D = \underbrace{\left[\frac{m}{r}+kd\right]}_{X} + \underbrace{\left[\frac{(k-1+N)\log_2 N}{r}\right]}_{Y} + \underbrace{\left[\frac{(k-1)m}{Nr}\right]}_{Z} \tag{3}$$

The expression consists of 3 (three) terms. The term X is free from N; so X has always fixed value for any value of N. Y and Z is respectively monotonically increasing and decreasing function of N (Fig. 4). So, if I plot these three terms and their sum with respect to N, it looks like as shown in the Fig. 4. The resultant function T'_D (=X + Y + Z) first decreases and then increasing with respect to N.

This indicates that T'_D has a minima for some optimal value of N (=N_{opt}) i.e. $\frac{dT'_D}{dN} = 0$

$$\Rightarrow [0] + \left[\frac{1}{r}\left(\log_2 N + 1 + \frac{(k-1)}{N}\right)\right] + \left[-\frac{(k-1)}{r}\frac{m}{N^2}\right] = 0$$

$$\Rightarrow \log_2 N + 1 + (k-1)\left(\frac{1}{N} - \frac{m}{N^2}\right) = 0 \tag{4}$$

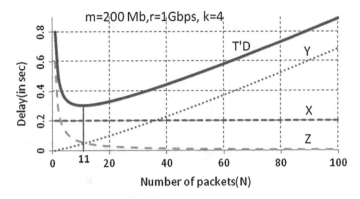

Fig. 4 Message transfer with framing

Since, it is a transcendental equation, I numerically enumerated T'_D (using *fminsearch* function of MATLAB) and found that for m = 200 Mb, r = 1 Gbps, k = 4, the function T'_D has a minima at N_{opt} = 10.675812500000012 ($\approx$11). This means for r = 1 Gbps, k = 4, it is best to divide a 200 Mb message into 11 packets. This will have minimum transmission and propagation delay equal to 0.303004163873043 s. The size of each packet will be $\frac{200}{11}$ + $\lceil\log_2 11\rceil \approx$ 19065022 bits.

This is a very interesting result as it helps us to divide a message into suitable number of packets.

Similarly, I can find out the N_{opt} for other values of m, r and k. A set of N_{opt} is shown in the Table 1 with different hop count and message size.

A series of curves is shown in Fig. 5 with different hop count. It may be noted that as hop count increases, N_{opt} also increases. This means for larger network the message should be divided into more number of packets to have delay minimum.

How exactly the N_{opt} varies with other parameters? To find it I have to find an expression of N_{opt} in terms of other parameters such as k, m etc. Since Eq. 4 is a transcendental equation, let us make a simplified assumption to have an approximate expression of N_{opt}. Let us assume that a packet contains fixed number of h header bits. Then the Eq. 2 reduces to

Table 1 Effect of k and m on N_{opt}

d = 1, m = 200 Mb, r = 1 Mbps																		
k	2	3	4	5	6	7	8	9	10	11	12	13	14	15	16	17	18	19
N_{opt}	7	9	11	12	13	14	15	16	17	18	18	19	20	20	21	21	22	23

d = 1, k = 4 Mb, r = 1 Mbps									
m (in Mb)	100	200	300	400	500	600	700	800	900	1000
N_{opt}	8	11	13	15	16	18	19	20	21	22

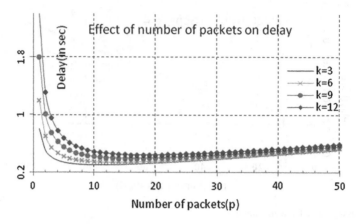

Fig. 5 Impact of number of packets (N) of delay for different hop count

$$T'_D = N\left(\frac{\frac{m}{N}+h}{r}\right) + kd + (k-1)\left(\frac{\frac{m}{N}+h}{r}\right) \tag{5}$$

For minimum delay $\frac{dT'_D}{dN} = 0$

$$\frac{h}{r} - \frac{(k-1)m}{N^2} = 0 \quad \frac{dT'_D}{dN} = 0 \quad \text{i.e. } N_{opt} = \sqrt{\frac{(k-1)m}{h}} \tag{6}$$

This indicates N_{opt} proportional to the square root of k and inversely proportional to the square root of h. This variation is shown in the Figs. 6 and 7 respectively.

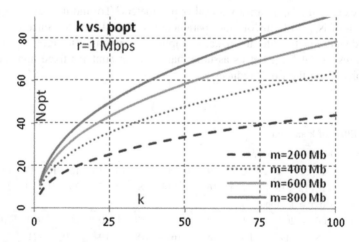

Fig. 6 Optimal number of packets (p_{opt}) versus number of hops (k)

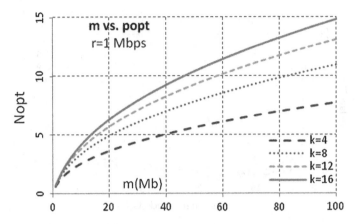

Fig. 7 Optimal number of packets (p_{opt}) versus message size (m)

4 Experimental Result

In this paper, I have assumed all links are error free and have same capacity. In practice, it does not happen and delay equation is not perfect. So test the accuracy of the method, I performed several experiments using NS3 simulator [8, 9].

The experiment was carried out on a network as shown in Fig 8. I haven't used complex network as our intension is to verify how packet size impacts only on end-to-end packet delay due to transmission and propagation. The source is always chosen as A. The destinations chosen are D, G, J and M which are respectively 3, 6, 9, 12 hops distant from A. The network parameters are shown in the Table 2.

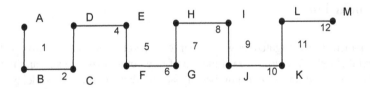

Fig. 8 Network topology used for simulation

Table 2 Various important parameters set in different scenarios

Simulation parameters			
No of nodes	13	Link capacity	1 Gbps
Message size	200 Mb	Number of packets	Varied from 1 to 50
Playground size	740 m × 370 m	Mobility type	NA
Speed	NA	Pause time	NA
Radio characteristics			
Transmitted power		10 mW	
Packet reception power threshold		−100 dBm	

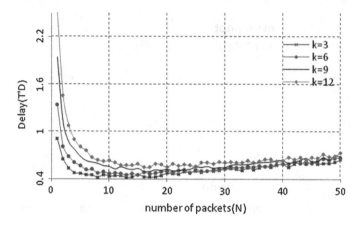

Fig. 9 Experimental result

There are four source-destination pairs: A → D, A → G, A → J and A → M. For each source-destination pair, the 200 Mb message is divided into varying number (1–50) of packets and are delivered using store-and-forward method. Each experiment is carried out 1000 number of times and average result was noted. The result obtained is shown in Fig. 9.

The experimental result almost follows the analytical result (Fig. 5). Note that the delay incurred is slightly more than the theoretical result. This may be due to the processing delay of the simulation software. This proves that the theoretical calculation was correct.

5   Conclusion

In this paper, I investigated how number of packets affects the delivery time. I showed that delay is a non-linear function of hop count, message size and number of packets. I found an *optimal* number of packets that minimizes the delay. Analytical and simulation results showed the correctness of the proposed scheme.

References

1. Hasan, Mohammad Shahidul (et al.): Modeling delay and packet drop in networked control systems using network simulator NS2. International Journal of Automation and Computing 2.2 (2005): 187–194.
2. Alkadeki, Hatm, Xingang Wang, and Michael Odetayo: Estimation of Medium Access Control Layer Packet Delay Distribution for IEEE 802.11. arXiv preprint arXiv:1401.2489(2014).

3. P. Raptis, V. Vitsas, K. Paparrizos, P. Chatzimisios, and A.C. Boucouvalas: Packet Delay Distribution of the IEEE 802.11 Distribution Coordination Function. World of Wireless Mobile and Multimedia Networks June (2005) pp. 299–304.
4. M. Anyaegbu, C.-X. Wang, and W. Berrie: A sample-mode packet delay variation filter for IEEE 1588 synchronization. in Proc. IEEE Int. Conf. ITS Telecommunications, Taipei Taiwan Nov. (2012) pp. 1–6.
5. Yang, Qiuling, Zhigang Jin, and Xiangdang Huang: Research on delay and packet loss control mechanism in wireless mesh networks. Journal of Networks 9.4 (2014): 859–865.
6. Gershinsky, Gidon (et al.): Delay analysis of real-time data dissemination. "Proceedings of the 11th communications and networking simulation symposium. ACM, (2008).
7. Paxson, Vern: End-to-end Internet packet dynamics. ACM SIGCOMM Computer Communication Review. Vol. 27. No. 4. ACM, (1997).
8. EmadAboelela: Computer Networks A System Approach. Edition-3, Networks Simulation Experiment Mannual, University of Massachusetts Dartmouth (2003).
9. ns-3 Documentation, https://www.nsnam.org/doxygen/.

An Ontology Based Context Aware Protocol in Healthcare Services

Anirban Chakrabarty and Sudipta Roy

Abstract Interpretation of medical document requires descriptors to define semantically meaningful relations but due to the ever changing demands in healthcare environment such information sources can be highly dynamic. In these situations the most challenging problem is frequent ontology search keeping with user's interest. To manage this problem efficiently the paper suggests an ontology model using context aware properties of the system to facilitate the search process and allow dynamic ontology modification. The proposed method has been evaluated on Cancer datasets collected from publicly accessible sites and the results confirm its superiority over well known semantic similarity measures.

Keywords Ontology mapping · Context awareness · Search personalization · Healthcare

1 Introduction

Most Healthcare systems contain a large collection of related documents which necessitates a semantic search system to swiftly and accurately identify documents which satisfy user's needs. It has been found that keyword based search engines most often fail to give the expected results relevant to the query context. One of the key factors for personalized information access is the user context [1]. Context can be defined as the user's objective for seeking information [2]. In this paper, context is defined through the notion of ontological profiles which are updated over time to reflect changes in user interests. Ontology consists of a formal conceptualization in a particular field of interest that can be easily visualized or can be considered as a

Anirban Chakrabarty (✉) · Sudipta Roy
Assam Central University, Silchar, India
e-mail: chakrabarty.anirban@redffmail.com

Sudipta Roy
e-mail: Sudipta.it@gmail.com

© Springer Science+Business Media Singapore 2017
J.K. Mandal et al. (eds.), *Proceedings of the First International Conference on Intelligent Computing and Communication*, Advances in Intelligent Systems and Computing 458, DOI 10.1007/978-981-10-2035-3_15

137

description of the elements it contains and when aligned with other ontologies it can help in forming a community of such elements or concepts [3, 4].

The major contributions suggested in this paper is mentioned as follows: (1) It provides an ontology for cancer patients containing both medical and socio economic conditions (2) to search for an optimal treatment plan suited to the patient (3) The ontology can be extended by healthcare experts based on changing treatment options, different disease symptoms and missing information with minimum involvement from external sources. (4) An ontology mapping strategy has been suggested that exploits contextual information of source ontology and supplements it in linguistic and structural similarity to enhance the strength of match with the target ontology.

2 Related Work

Most retrieval systems suffer from keyword barrier phenomenon which refers to the inability of information retrieval systems to convey semantic context of documents [5]. Related work developed a framework for Formal Concept Analysis which extended the Tf-Idf weighting model by introducing ontology dependent concepts [6]. Most search engines do not include user preferences and search context but provide users with a generalized search facility [7]. A related study shows that little work has been done on contextual retrieval to combine search technologies and ontology alignment using context in a single framework to provide the most accurate response for a user's information requirement [8].

A popular method to facilitate information access is through the use of ontology. Researchers have attempted to utilize ontology for improving navigation effectiveness as well as personalized Web search and browsing for generating user profiles [9]. An innovative approach for Ontology based on Concept merging was suggested were first the horizontal technique checks all relationships between concepts at the same level of two ontologies and merges them as defined by WordNet, then the vertical approach completes the merging of concepts at different levels but placed on same branch of the tree [10].

The notion of concept similarity calculation using ontology is that two concepts have a semantic correlation, and there exists a path in class hierarchy diagram. In a pioneering work, Resnik measures the semantic similarity of two words according to the maximum amount of information of their common ancestor node [11]. Leacock and Chodorow proposed a semantic similarity model based on distance which is an easy and natural approach but is heavily dependent on the ontology hierarchy already established [12]. Study of related work shows the use of RIMOM: a dynamic multi strategy ontology alignment framework which uses both linguistic and structural similarity to map source and target ontologies but it is suitable for only for 1:1 mapping [13]. The ontologies can be expressed in various ways. The most common language used to model ontologies is the Web Ontology Language (OWL), which is a form of RDF and is written using a subset of XML [14].

Unfortunately, most of these systems can provide little support in support for knowledge sharing and context reasoning because of their lack of ontology. Recent study focus on the use ontology based personalization system for assisting healthcare industry in patient diagnosis and intervention plans [15].

3 Methodology

In this work ontology has been developed to organize the terms of cancer disease, its symptoms, diagnosis, treatment options, intervention plans, patient condition and other social issues related to disease. On top of this a search algorithm based on context information stored in Ontology has been suggested which extracts all correlated words of the search string and can map from one ontology to another (Figs. 1 and 2).

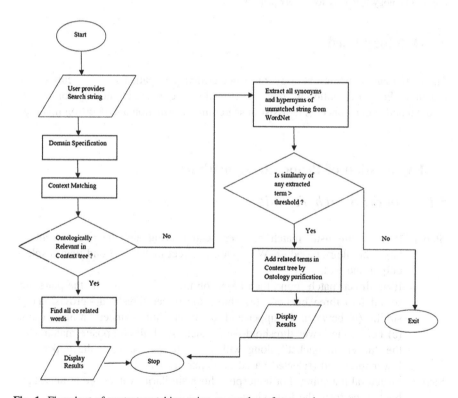

Fig. 1 Flowchart of context matching using an ontology framework

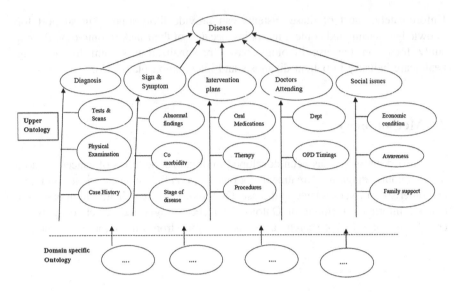

Fig. 2 Ontology hierarchy for cancer patients

4 Ontology Used

There are many existing ontology, but since critically ill patients need accurate and urgent medical intervention we have developed Cancer related Ontology containing medical and social concepts to assist in searching information and decision making.

5 Explanation of Proposed Algorithms

5.1 Context Search Technique

Step 1   Perform linguistic match between search string and the one present in target ontology based on Word Net. If direct match is found then display output and exit.

   If no direct match found then based on the similarity results the pairs are sorted into three buckets: (a) above the upper threshold—provisionally similar (b) between the upper and lower threshold—uncertain bucket and (c) below the lower threshold—no match. Add all composite matches to the "uncertain bucket" along with the concepts between the upper and lower threshold (selected based on experimentation).

Step 2   Structural matching: For concepts whose similarity values are in uncertain bucket, perform the following:

(a) for each pair of concept, compare their parent nodes (when matching is between two ontology) to determine the total parental similarity for updating the original pairs similarity value. If match is found increment the score by 0.1.

(b) for each pair of concepts, compare their grandparent nodes to determine the total similarity for updating the original pairs similarity value. If match is found increment the score by 0.05.

Step 3 Use Word Sense disambiguation on the set of terms in 'Uncertain' and 'Provisionally similar' buckets to select the words actually representing the given context. The WSD method used here takes average of similarity values for each concept searched to show its relatedness to the ontology.

5.2 Ontology Modification

This procedure is called when the context is not found in the ontology tree but the searched context has a comparable relation then the context is dynamically added as a node in the ontology tree in appropriate location. Each ontology node is associated with an Id for comparison.

Pseudo code:

```
Modify (Root,Id,mi,N,M,Left,Right,Par)
Set Ptr = Root, Id= TRUE.
 Apply Algorithm-2 to find the similarity or relevance of searched area.
[If Node found] Set Ptr=Node.
Else if [node was not found but has to be added based on relevance threshold of the
domain set then check immediate ancestor.]
{ Id[NodeO1]->Par=Id [NodeO2]->Par

Add new node.
If{ Id [NodeO1]<Id [NodeO1]->Par
Set Left[Par]=New
Else Set Right[Par]=New.
Set Id of new node.}
Print Updation Successful.
}}
[Else Updation Not Possible. ]
}
```

6 Experiment

6.1 Data Set

The experimental data used in this work has been collected from different cancer related websites which allows open access and made publicly available for research and study purposes [16–21]. The training data set comprising of 2721 documents was used for the representation of the cancer ontology indexing 204 concepts in the hierarchy.

6.2 Experimental Metrics

To perform a comparison of the improvement of proposed method with other established semantic search algorithms used in contemporary literature like k-NN, Resnik similarity, Rocchio based methods [22, 23].

The similarity measure suggested by Resnik, gives the information content (IC) of the Least Common Subsequence (LCS) for two concepts:

$$SimRes = IC(LCS) \tag{6}$$

where IC is defined as:

$$IC(c) = -\log p(c) \tag{7}$$

here $p(c)$ is the probability of encountering a context c in a corpus.

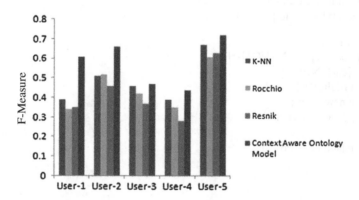

Fig. 3 Comparison of F-measure with other similarity measures for five users

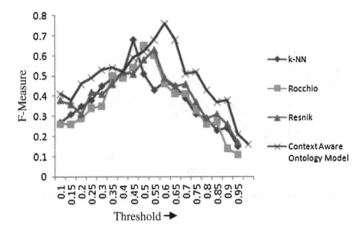

Fig. 4 Comparison of F-measure with varying threshold

The effectiveness of search was measured in terms of Top-n Recall and Top-n Precision [24]. The F-measure defined as $F = 2 * P * R/(P + R)$ is a balanced mean between precision and recall metrics and was used for comparing searches made by five different users.

6.3 Results

See Figs. 3, 4, 5 and 6.

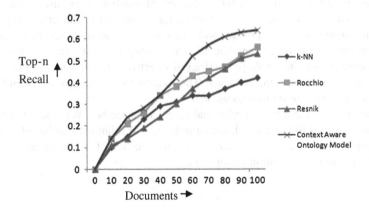

Fig. 5 Variation of average top-n recall in top-n documents using overlap queries

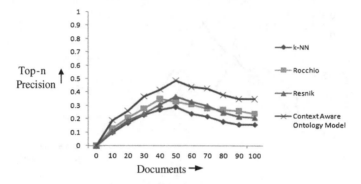

Fig. 6 Variation of average top-n precision in top-n documents using overlap queries

6.4 Discussion of Experimental Results

The comparison of F-measure with other semantic search techniques like k-NN, Rocchio and Resnik methods is depicted in Fig. 3. For all such users there has been improvement in F-measure values in case of our Context Aware Ontology model. Since the value of similarity depends to some extent on the threshold considered in the function so a variation of threshold with F-measure for different similarity techniques have been studied for same set of five users. The highest F-measure value was reached for a threshold of 0.6 in our context aware ontology model as shown in Fig. 4. Figures 5 and 6 gives the comparison of precision and recall values for top-n search results for all similarity techniques as shown above.

7　Conclusion

The work presents a framework for contextual information access using ontologies and demonstrated that the semantic knowledge embedded in cancer ontology can efficiently assist in search process and facilitate dynamic ontology modification. A comparison with other semantic search techniques shows that if contextual information present in ontology is retrieved effectively it can improve the search results based on user's requirements. This search technique can be extended to compare two separate ontologies as well.

It may be noted that the experimental results reported here is based on usage of randomly selected users, a few hundred queries, and a limited number of relevant documents. Future research in this area will consists of much larger scale of experiments and optimization parameters.

References

1. Pei-Min Chen, Fong-Chou Kuo , An information retrieval system based on a user profile, *The Journal of Systems and Software*, vol. 54, pp. 3–8, 2000.
2. Dey, A.K., Salber, D. Abowd, G.D., "A Conceptual Framework and a Toolkit for Supporting the Rapid Prototyping of Context-Aware Applications", *Human-Computer Interaction Journal*, Vol. 16(2–4), pp. 97–166, 2001.
3. Cimiano, P., Ontology Learning and Population from Text: Algorithms, Evaluation and Applications. Springer-Verlag New York, Inc., Secaucus, NJ, USA, 2006.
4. Gargouri, Y., Lefebvre, B., Meunier, J. Ontology Maintenance using Textual Analysis. Proceedings of the Seventh World Multi-Conference on Systemics, Cybernetics and Informatics (SCI). Orlando, USA (2003).
5. M-Y. Chen, H-C. Chu, Y-M. Chen, Developing a semantic enable information retrieval mechanism. *Expert Systems with Applications*, 37:322–340, 2010.
6. Rohana K. Rajapakse, Michael. Denham, Text retrieval with more realistic concept matching and reinforcement learning. *Information Processing and Management*. 42(5), 1260–127, September 2006.
7. Ashraf, J., Khadeer Hussain, O., Khadeer Hussain, F.: A Framework for Measuring Ontology Usage on the Web. In: The Computer Journal. Oxford University Press, 2012.
8. J. Allan, et al. Challenges in information retrieval and language modeling. ACM SIGIR Forum, 37(1):31–47, 2003.
9. Fang Liu, Clement Yu, Personalized Web Search for Improving Retrieval Effectiveness, *IEEE Transactions on Knowledge and Data Engineering*, Vol. 16, No. 1, January 2004.
10. Miyoung Cho, Hanil Kim, and Pankoo Kim. A new method for ontology merging based on concept using wordnet. In Advanced Communication Technology, 2006. ICACT 2006. The 8th International Conference, volume 3, pages 1573/1576, February 2006.
11. Philip Resnik, "Using Information Content to Evaluate Semantic Similarity in a taxonomy", In Proceedings of the *14th International Joint Conference on Artificial Intelligence*. Monte real, Canada, pp. 448–453, 1995.
12. C. Leacock, M. Chodorow. "Combining Local Context and WordNet Similarity for Word Sense Identification", *Computational Linguistics*, vol. 24, no. 1, pp. 147–165, 1998.
13. Juanzi Li, Jie Tang, Yi Li, and Qiong LuoRiMOM: A Dynamic Multistrategy Ontology Alignment Framework, *IEEE Transactions on Knowledge and Data Engineering*, Vol. 21, No. 8, August 2009.
14. Ashraf, J., Khadeer Hussain, O., Khadeer Hussain, F.: A Framework for Measuring Ontology Usage on the Web. In: The Computer Journal. Oxford University Press, 2012.
15. David Riano; Francis Real; Joan Albert Lopez; Fabio Campana; Sara Ercolani; Patrizia Mecocci; Roberta Annicchiarico & Carlo Caltagirone. An Ontology based personalization of Healthcare knowledge to support clinical decisions for chronically ill patients, Journal of Biomedical Informatics, Elsevier, 429–446, 45 (2012).
16. http://www.cancer.org/research/index [Last accessed on 12th October, 2015].
17. http://www.cancercare.org/accessengagementreport [Last accessed on 12th October 2015].
18. http://www.oncolink.org/resources [Last accessed on 7th October, 2015].
19. http://med.stanford.edu/cancer.html [Last accessed on 25th September, 2015].
20. http://www.ncbi.nlm.nih.gov/pubmed [Last accessed on 21st September, 2015].
21. http://www.cdc.gov/cancer/ [Last accessed on 16th September, 2015].
22. Oh-Woog Kwon, Jong Hyeok Lee. Text categorization based on k-nearest neighbor approach for Web site classification, *Information Processing and Management* Volume 39, Issue 1, Pages 25–44, January 2003.

23. Guanyu Gao, Shengxiao Guan. Text categorization based on improved Rocchio algorithm. In Proc. of Systems and Informatics (ICSAI) IEEE International Conference, pp. 2247–2250, 2012.
24. Paulo Cremonesi, Yehuda Koren, Roberto Turrin. Performance of recommender algorithms on top-n recommendation tasks. In Proc. of the fourth ACM conference on Recommender Systems, pp. 39–46, ACM New York, USA, 2010.

Processing ASP.Net Web Services Using Generic Delegation Approach

Vilakshan Saxena, Harshit Santosh and Chittaranjan Pradhan

Abstract For the electronic business applications, web services are usually considered as the design models. Here, our aim is to design an efficient model to deal with both the distributed applications and cooperative applications. In both the cases when it comes to implementation of web services in respective applications, the consumer (developer) has to put an effort to manually provide the reference of the respective web service through a specific set of steps depending upon the target IDE. But what if we have a technique to perform the above mentioned approach in a dynamic and generic fashion without manually adding the web reference for any web service. In this paper, we will represent an efficient approach for interacting with any web service irrespective of its syntax (WSDL) and semantics without adding its web reference. Through this approach the consumer of the web service can access the respective web service dynamically by just mentioning its URL in his/her code and through a little object oriented methodology. Our approach is based on accessing the particular web service by automatic generation of proxy class, delegation, dynamic data type handling through reflections API and producing the desired output in a generic fashion.

Keywords Web services · Reflections API · Proxy class · Web service authentication

Vilakshan Saxena (✉) · Chittaranjan Pradhan
KIIT University, Bhubaneswar, India
e-mail: vilakshankiit@gmail.com

Chittaranjan Pradhan
e-mail: chitaprakash@gmail.com

Harshit Santosh
MIT, Manipal, India
e-mail: harshitsantosh@gmail.com

© Springer Science+Business Media Singapore 2017   147
J.K. Mandal et al. (eds.), *Proceedings of the First International Conference
on Intelligent Computing and Communication*, Advances in Intelligent Systems
and Computing 458, DOI 10.1007/978-981-10-2035-3_16

1   Introduction

Web service is a technique in which one electronic device can communicate to another over a network. It is provided as a software functionality at web node with the services available as utility computing. It describes a methodology of integrating web-based applications using the internet protocol. For data tagging, XML standard is used [1]. To transfer data, SOAP standard is used. Similarly, for the description of the services, WSDL can be used. To get the available services, UDDI standard is used [2].

Using the web service, two software packages exchange data between themselves over the internet. The requesting software system is called as service requester and the processing software system is considered as the service provider [3, 4]. We need a generic software system which is free from any specific programming language. Since XML tags are interpreted by almost all software, it can be used as the data exchange tags in web services. The set of rules used for data communication are defined in Web Services Description Language (WSDL) file [5]. Universal Description, Discovery and Integration (UDDI) is used as the directory of data types and compatible software systems. Once the service provider validates the service requester through WSDL file, Simple Object Access Protocol (SOAP) can be used for data transfer [6, 7].

In the proposed approach, we access the particular web service by automatic generation of proxy class, delegation, dynamic data type handling through Reflections API and producing the desired output in a generic fashion.

2   Related Work

In 2005, John B. Oladosu et al. have done a study on web services; which includes the advantages of the web services over the previous services such as CORBA, COM etc. The study also includes the different components of web service along with the different standards. The application of web services in the e-health domain has also been discussed in the paper [1].

In 2009 C. Boutrous Saab et al. developed a technique on processing of web services by focusing on the two basic features of web service. The first feature deals with the interaction problem provided in the interaction protocol. The second feature deals with the design process of a web service [3].

3   Proxy Class

A client can communicate with the web service using SOAP message. This message encapsulates the input parameters and output parameters in the XML form. A proxy class maps the input and output parameters to the XML elements and sends the message over the network. Using this technique, the proxy class frees the consumer from the communication with web services at SOAP level and permits the consumer to invoke the web services from the development environments supporting web service proxies and SOAP. The proxy class can be added to the development environment/project in Microsoft.NET Framework: (i) using WSDL tool and/or (ii) adding web reference in Microsoft Visual Studio. The proxy classes can also be processed in the similar way in J2EE Framework.

3.1 Using WSDL Tool

The Web Services Description Language (WSDL) tool of. NET Framework SDK allows us to create and use web service proxies. This tool accepts a set of arguments to generate a proxy. When the proxy is generated, the proxy class can be compiled to an assembly file and added as a project item [5, 7].

3.2 Adding Web Reference in Microsoft Visual Studio

When a project needs to consume one or more number of web services, web reference methodology is used. Once the web reference is added to the respective project, then it can be used easily to access corresponding web services over the net using object oriented methodology [8].

4 C# and VB Reflection API

The reflection API is used to create an instance of a type, bind the type to or extract the type from an existing object, and access the methods and its characteristics [9]. The attributes can also be invoked using the reflection. The reflection is used due to these:

- When the attributes of program's metadata needs to be accessed.
- When the types in an assembly needs to be examined and instantiated.
- When new types are built using System.Reflection.Emit class.
- When late binding needs to be performed at run time.

5 NET Web Service Authentication

Web service authentication is performed in order to secure the web service from been hijacked by third party contenders as it travels from source to destination over the network. Web service authentication can be performed using many ways [10]. In .NET framework authentication can be performed using many ways as follows:

- Basic authentication using web service Credentials property.
- Custom Authentication using several security extensions.
- Simple authentication of web services using SOAP and authorization headers which focus on object oriented methodology of creation SOAP header classes and objects and usage of individual member variables for username and password, in which password can be clear text or MD5 hashed.

6 Proposed Algorithm

Here, we propose an algorithm representing an interface for catering of any .NET web services using generic delegation approach. Unlike creating separate classes and objects for each of the web references corresponding with different web services, this interface acts as a delegate for any target web service resulting in creation of one point two way communication for separate web references. Let WS.asmx is the target web service that we want to access with this interface. The steps of the algorithm are as follows:

a. Receive the input parameters regarding the target web service such as

    i. Web Service URL. (www.abc.com/WS.asmx?wsdl).
    ii. Service Name (Can be retrieved from the WSDL file of the web service).
    iii. Method Name (Desired function name to be called).
    iv. Input parameters.
    v. Authentication Class Name.
    vi. Username.
    vii. Password.

    The last three parameters are optional and are required only if there is authentication header associated with the web service with a username and a password.
b. Create an instance of the proxy class and associate it with the provided web service URL as mentioned in (6.a.i) and store the compiled results in 'Compiler Results' class.
c. Using the object mentioned in (6.b), dynamically create instance of the target 'Service Name' as mentioned in (6.a.ii).
d. Once the instance of the target service name is created extract the structure of the method (function) as provided by Method Name parameter as mentioned in (6.a.iii).
e. Using the Reflection API extract the target data type of the Input parameters provided. (As mentioned in (6.a.iv)) as they are boxed into a generic data type. *Note* Here we don't know the data type of the input parameters provided to the interface due to which we have used the Reflection API to detect the data type of the parameters dynamically.

f. After extracting the target data type (As mentioned in (6.e)) extract all the input parameters from provided Input parameters and store it in an array.

g. Search the extracted method structure (As mentioned in (6.d)) among all the methods present in the target web service (As web service can be a collection of many methods) until there is a proper match.

h. Once the target method is found and analyzed process the generic input parameters (As extracted in (6.e)) and store there values in the desired input parameters of the target method to be called.

i. As the data type of the input parameters is determined dynamically, so the provided input parameters are sequentially captured according to the data type in the target method parameters.

j. Once all the parameters are mapped (As mentioned in (6.h)), call the target method.

k. Once the target methods is successfully processed (As mentioned in (6.j)) store the desired output in a generic variable using Reflection API and return it to the calling interface. This output variable will acts as a parent variable which can be correspondingly unboxed into respective data type by the client.

7 Experimental Results

The scope of effectiveness of this approach can be evaluated on the basis of checking this approach in different type of scenarios.

7.1 Authenticated Web Service

Authenticated web services are those web services which accept a valid username and password for connecting a client to its internal resources for various operations. There are various ways to provide an authentication firewall for the web service like:

i. *Windows authentication*:—Authenticity is checked on basis of valid accounts on which clients are logged into using windows user id and password.

ii. *SOAP headers*:—SOAP headers can be defined as a custom way to apply authenticity to a web service as the owner of the web service implements a separate class using SOAP API and SOAP base classes along with the functional classes of the web service for storing and verifying username and password.

iii. *DB driven*:—Owner of the web service stores and processes the valid username and password by storing it in a database table using database level security like encryption and hashing (Fig. 1).

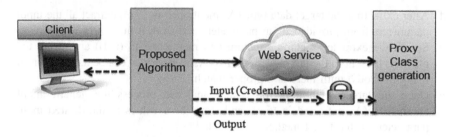

Fig. 1 Authenticated web service

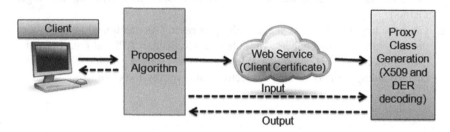

Fig. 2 Web services using client security certificate

Our algorithm handles the authentication of target web service in all the three cases:

i. For windows authentication (7.a.i) our algorithm implements the built in feature of the underlying framework and enforces it to check for valid windows username and password.

ii. For handling SOAP headers authentication (7.a.ii), our algorithm dynamically creates instance of the SOAP header class using 'Compiler Result' support class as mentioned in (6.b) and (6.c). After creating the instance, the algorithm extracts the desired username and password attribute along with their corre-sponding data type using the 'Reflection API' as mentioned in (6.e). It then sets the username (6.a.vi) and password (6.a.vii), through the input parameters passed to the algorithm.

7.2 Web Services with Client Security Certificate

When the applications call the web services, they must be authenticated by the web services. The authentication checking must be performed before the web service authorization. One of the authentication techniques is achieved by using client certificate. The user may receive an "access denied" error message when a client application tries to call a web service. But, when a console application tries to call the same web service, the error message may be omitted. In these cases, the computer system keeps two different certificate stores:

iii. The local machine store, which is used for web applications.
iv. The local user store, which is used for interactive applications.

To enable an application to use a client certificate, you must install the client certificate in the local machine store and the user logged into that machine should also have proper permission to access that certificate (Fig. 2). Our algorithm handles the authentication of the client certificate for accessing web services, (given that the valid client certificate is already installed on client machine) by using a specific validation call back approach in which it checks for the valid client certificate using X509 client certificate base classes and parsing of the DER encoded file. (Key for parsing the DER encoded file is to be exported at the time of installing the client certificate).

8 Catering of the Algorithm

As the effectiveness of the proposed algorithm is observed, here is a brief description about how to cater the algorithm in respective target projects.

a. Add the reference of the algorithm to the target project.
b. Create an object of the respective class associated with the algorithm as shown below:

$$WebServiceAlgorithmobj = new\ WebServiceAlgorithm()$$

c. After creation of the object (8.b), call the algorithm (target function) for processing of the web service.
d. For normal ASP.Net web services, the target function (8.c) can be called as shown below:

$$Return\ Object = obj.TargetFunction$$
$$(Web\ Service\ URL,\ Service\ Name,\ Method\ Name,\ Input\ parameters)$$

where:

(i) First parameter is the URL of the web service.
(ii) Second one is the service name (class of the web service).
(iii) Third one is name of the method to be called to get the output.
(iv) Fourth parameter (8.g) represents collection of input parameter to be used in the method.

e. For authenticated web service, the target function (8.c) can be called as shown below:

Return Object = obj.TargetFunction

(Web Service URL, Service Name, SOAP Authentication

Header Name, Username, Password, Target Function

Name, Input parameters)

where:

All other input parameters are same (8.d) except for three input parameters:—
(SOAP Authentication Header Name) which is the name of the authentication
header associated with the web service and contains the respective username and
password, (Username) is the desired username and (Password) is the desired
password.

Note Above mentioned approach (8.e) highlights authentication using SOAP
header. As far as Windows and DB driven authentication is concerned our
algorithm does not need SOAP header, only username and password is required.

f. As far as web services with client certificate are concerned, the proposed
   algorithm is developed in such a way that it caters these cases implicitly.

g. For populating input parameters for the desired function client can form the
   input structure as follows:

   (i)   Create a class for the input parameter.
   (ii) Declare variables for each input with required data types as needed by the
         web service.
   (iii) Create public caterers for each variable for accessing them.

```
Public Class inputPar
      Data Type Param 1
      Data Type Param 2
      Public Caterer getsetParam1()
            Get
                        Return Param 1
            Set()
                        Param 1= value
            End Set
      End Caterer
      Public Caterer getsetParam2()
            Get
                        Return Param 2
            Set()
                  Param 2= value
                  End Set
      End Caterer
```

   (iv) Assign the variables with the desired values through the public caterers
         defined.

(v) Pass the object of this class as a parameter in the target function (8.c) of the web service.

h. The output of the proposed algorithm returns a parent object which can be correspondingly unboxed into respective data type.

9 Conclusion

In the proposed algorithm, we have presented an approach for processing ASP.Net web services using delegation technique, in which one common delegate will act as medium of communication between clients and any number and type of ASP.Net web services. The proposed algorithm encourages reusability, modularity and generic development. It also reduces time and effort of referencing each web service explicitly and makes target application adaptive to new changes with respect to web services. We can conclude that the proposed algorithm is a generic, robust and effective algorithm for any.Net web service. In future we will try to increase the scope of the algorithm to different additional platforms.

References

1. John B. Oladosu, Funmilola A. Ajala, Olukunle O. Popoola, "On the Use of Web Services Technology in E-Health Applications", Journal of Theoretical and Applied Information Technology, Vol. 12, No. 2, 2010, pp. 94–103.
2. M. Vasko, S. Dustdar, "An Analysis of Web Services Workflow Patterns in Collaxa", Web Services, Lecture Notes in Computer Science, Springer, 2004, pp. 1–14.
3. C. Boutrous Saab, D. Coulibaly, S. Haddad, T. Melliti, P. Moreaux, S. Rampacek, "An Integrated Framework for Web Services Orchestration", International Journal of Web Services Research, Vol. 6, No. 4, 2009.
4. S. Mokarizadeh, P. Kungas, M. Matskin, "Utilizing Web Services Networks for Web Service Innovation, International Conference on Web Services, IEEE, 2014, pp. 646–653.
5. K. Elgazzar, A.E. Hassan, P. Martin, "Clustering WSDL Documents to Bootstrap the Discovery of Web Services", International Conference on Web Services, IEEE, 2010, pp. 147–154.
6. M. Paolucci, T. Kawamura, T. R. Payne, K. Sycara, "Semantic Matching of Web Services Capabilities", International Semantic Web Conference, LNCS, Springer Verlag, 2002, pp. 333–347.
7. McGrawhill Company Inc., "What the heck are Web Services?" Business Week, 2005.
8. A. Alves, A. Arkin, S. Askary, C. Barreto, B. Bloch, F. Curbera, M. Ford, Y. Goland, A. Guizar, N. Kartha, C. K. Liu, R. Khalaf, D. Konig, M. Marin, V. Mehta, S. Thatte, D. V. R. Rijn, P. Yendluri, A. Yiu, "Web Services Business Process Execution Language Version 2.0.", Technical report, OASIS WSBPEL Technical Committee, 2007.
9. C. Peiris, "Creating a .net Web Service", Caulfield, Australia, 2005.
10. Guofeng Yan, Yuxing Peng, Shuhong Chen, Pengfei You, "QoS Evaluation of End to End Services in Virtualized Computing Environments", International Journal of Web Services Research, Vol. 12, No. 1, 2015, pp. 27–44.

Link Based Protection of Light Trail in WDM Mesh Networks

Sampa Rani Bhadra and Utpal Biswas

Abstract An optical bus that connects two nodes of a WDM network and allows multiple communications between those two nodes is called light trail. Failure of a link within a light trail can bring down the communication to halt. In this paper, a link based protection scheme is proposed to solve both single and multiple link failure of light trails in a WDM mesh topology. This scheme is applicable to both static and dynamic light trail.

Keywords Light trail · Survivability · Protection · Restoration · Link based protection · Connection based protection

1 Introduction

An optical bus that connects two nodes of a WDM network and allows multiple communications between those two nodes is called light trail. The first node is called as convener node and the last node is called as end node of the light trail respectively. This is shown in Fig. 1. The concept, architecture, hardware, protocols and set up of light trail are detailed in [1].

The intermediate nodes can also access the light trail. In other words, any two nodes within the light trail can communicate. However, no two nodes can initiate communication at the same time. This is achieved by combining two interesting features of a light trail node [2]. They are 'drop and continue' and passive addition of a signal as shown in Fig. 2. The access unit of a light trail node can partially drop

S.R. Bhadra (✉)
Department of Computer Science, Sree Chaitanya Mahavidyalaya, Habra-Prafulla Nagar,
North 24 Parganas 743268, India
e-mail: sampa.rani.bhadra@gmail.com

Utpal Biswas
Department of Computer Science and Engineering, University of Kalyani, Kalyani 741235,
India
e-mail: utpal01in@yahoo.com

© Springer Science+Business Media Singapore 2017
J.K. Mandal et al. (eds.), *Proceedings of the First International Conference on Intelligent Computing and Communication*, Advances in Intelligent Systems and Computing 458, DOI 10.1007/978-981-10-2035-3_17

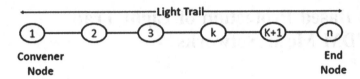

Fig. 1 A light Trail

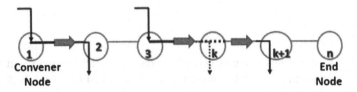

Fig. 2 Drop and continue and passive addition of a signal within a Light Trail

the signal and continue the rest to the next node or completely drop the signal at a node. Also it can insert a signal using local transponder without using any optical switching. This two combined features allow a single light trail (having t number of nodes) to support C(t, 2) source-destination communication without the need of optical switch reconfiguration [3]. All the communication is unidirectional from convener node towards end node.

Another powerful feature of light trail is the ability to expand and contract [3] itself to meet certain kinds of application. A node upstream to the convener node may request communication to a node in a light trail. The communication is allowed by making that upstream node the new convener node. Thus the light trail is expanded in upstream direction. Similarly, a light trail can be expanded in downstream direction when the end node is shifted to any downstream node which was not in the existing light trail. Contraction is realized when the convener node is shifted to any downstream node or the end node is shifted to upstream node.

Based on the two different types of traffic (static or dynamic), the light trail is of two types namely static light trail and dynamic light trail. When satisfying the traffic matrix (the connection requests) which is known beforehand, pre assigned light trails of the network is used for communication. This type of light trail is known as static light trail. When the traffic or the connection request is unknown and arrive dynamically, the light trail is assigned dynamically to satisfy the connection request. This type of light trail is known as dynamic light trail. The main issue after light trail assignment is light trail protection and restoration. Failure of a link or a node or equipment within a light trail halts its communication. Since each of the light trail contains as many as C(t, 2) connections, so failure of a link or a node or an equipment can be catastrophic. The ability of the network to recover from node, link or equipment failure is called network survivability. Survivability techniques are of two types: protection and restoration [4]. Protection is the design of network with spare capacities (wavelength) whereas for restoration the spare capacities are not designed in advance.

Again protection and restoration is of two types: link based and connection based [4]. For the first type, the connections are rerouted against failed components whereas in the second type, a backup trail is assigned to every primary connection.

This paper considers only link based protection against link failure within a light trail. Link failure can be of three types. First type is a single link failure. The second type is not adjacent multiple link failure. The third type is adjacent multiple link failure.

The rest of the paper is organized as follows. Section 2 discusses the related works on protection and restoration of light trail. Section 3 illustrates the problem statement and a brief outline of the solution. Section 4 proposes the heuristic algorithm. Section 5 includes illustration and discussions. Section 6 gives the conclusion.

2 Related Works

The protection and restoration of light trail is a well-researched area. Almost all authors used the concept of primary (working) and backup light trail. The original route used for communication is called the primary trail or working trail and the route(s) used when a primary trail fails is (are) referred to as backup trail(s). Some existing schemes for different network topologies are discussed here.

Dedicated protection scheme (DP) and shared protection scheme (SP) are discussed in [4]. In DP data is sent both on the primary and backup trail simultaneously. The destination uses the signal that has better quality. In SP the backup capacity is shared among multiple primary connections that do not fail together. This backup capacity is reserved and is only used in the case of failure.

For a single link failure two more protection schemes are proposed in [5]. They are Link Based Protection and Connection Based Protection. In connection based scheme two link disjoint light trails, namely primary and backup connection are assigned for every connection request. The primary connection is the working connection and carries signal in case no failure is found. When a failure occurs, the information is propagated to the source through the control channel. The source initiates the communication through the backup light trail. In link based scheme, a backup sub light trail is provided for each link on the light trail. Similar to connection based scheme failure information is sent to the source through the control signal. And a new light trail is established which is basically the remaining part of the failed light trail and the backup sub light trail.

A Ω-shaped protection scheme [6] is proposed for creating a protection light trail after a fiber cut occurs in ring network. When a failure occurs, a new light trail is created in exactly opposite direction to the original light trail whose length is the entire ring circumference minus the failed span, plus the two segments that are remnant of the original light trail. This gives a shape of Ω to the newly created protection light trail. Further some hardware modification is made. All nodes upstream of the fiber cut in span and downstream of the fiber cut are configured in DPC-add mode and DPC-drop mode respectively.

Backup Light-trail Expending Scheme [7] is proposed for WDM mesh network, in which for every primary light trail a link disjoint protection light trail is set up at the time of light trail assignment. When a single link failure is detected during communication, a new light trail is set up using the two working remnant of the primary light trail and the protection light trail.

In survivable light trail network design [8], for each incoming connection request a working as well as backup light trail is found and established to support a single link failure. This scheme is applicable for dynamic light trail routing. If both the working and the backup light trail are found among the candidate light trails, then only connection request is accepted else it is rejected. In other words, a pair of working light trail and backup light trail is assigned to continue communication.

Studying the above works, it is noted that no work is done for multiple link failure. This paper proposes link based protection scheme to solve single link failure, which is extended to multiple link failures. This scheme is applicable to both static and dynamic light trail.

3 Light Trail Protection (LTP)

3.1 Problem Statement

The problem is to provide protection against a single or multiple link failures while routing static as well as dynamic light trail connections within a WDM mesh network.

3.2 Overview of the Solution

The concept of backup link is introduced here. The first step is to assign a backup link between every two adjacent nodes within each newly created [9] light trail. Second step is to detect a link failure in the working light trail. When a link failure is detected, the light trail node drops the failed link, adds the backup link to bypass the affected fibre span and also adds the remnant portion of the working light trail to reach the destination. Thus protection is provided against a single link failure. This scheme is applicable as many times as needed to support non-adjacent and adjacent multiple link failure within a single light trail. It is therefore able to protect the entire light trail and all possible connections within the light trail system for these three cases. Figure 3 is the flow chart of the above discussed solution. To keep the flow chart simple we consider one connection request per light trail.

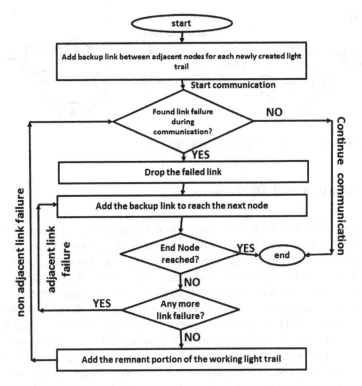

Fig. 3 Simplified flow chart of LTP

4   The Proposed Heuristic

Algorithm LTP
    Step 1: Assign a backup link between every two adjacent nodes within each
    newly created light trail
    For each connection request within the light trail
    Repeat steps 2 to 4 whenever a failure is detected during communication
        Step 2: Drop the failed link
        Repeat step 3 until either a remnant portion of the light trail is found
            or the end node is reached
            Step 3: Add the backup link
      Step 4: Add the remnant portion of the working light trail
    Step 5: End of the algorithm

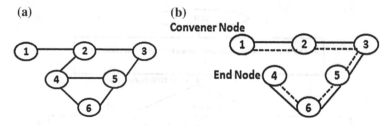

Fig. 4 a Network topology. **b** Light trail with backup links

5 Illustration and Discussions

Three types of the link failure are studied thoroughly and their protection mechanism is illustrated here. Figure 4a shows a WDM mesh topology. Figure 4b shows a newly created light trail with backup links in dotted line. For simplicity the traffic matrix is static and consists two connection requests: (1, 5) and (2, 6). While satisfying the first connection request (1, 5) three cases are considered.

In case 1, a single link failure is found between nodes (2–3). It is shown in Fig. 5a. Algorithm LTP is applied to protect the connection request. Node 1 being the sender initiates the communication. When the optical signal arrives at node 2, the light trail is dropped fully by the light trail access unit [4] and the backup link between node 2 and node 3 is added passively. This enables the signal to propagate to node 3. At node 3 the remnant portion of the light trail is added. The halted signal then reaches the destination. Hence the connection request is satisfied. It is explained pictorially in Fig. 5b. The newly created protection light trail not only protects and satisfies connection request (1–5), it also satisfies the next request (2–6) successfully.

In case 2, a non-adjacent multiple link failure is considered. As shown in Fig. 6a two single link failures are present between nodes (1–2) and (3–5). The necessary steps to protect the connection request (1–5) is shown pictorially in Fig. 6b. When the optical signal tries to propagate to node 2, it detects the first link failure. Applying LTP, the existing light trail is dropped. The backup link between node 1 and node 2 is added passively. This allows the signal to propagate to node 2. Node

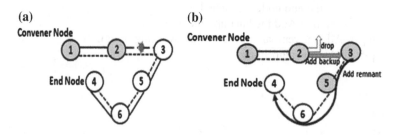

Fig. 5 a Single link failure. **b** Connection request (1–5) protected and satisfied

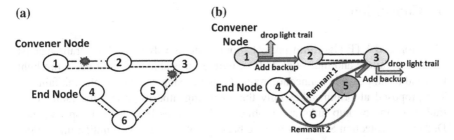

Fig. 6 a Non-adjacent link failure. b Connection request (1–5) protected and satisfied

2 passively adds the remnant portion of the working light trail and the halted communication starts propagating through it. The signal on reaching node 3 is again blocked due to the link failure between (3, 5). The light trail is dropped at node 3 and the backup link is added between node 3 and node 5. Then the signal is propagated to node 5. Since the destination is reached, the communication halts here. However, the destination node is not the end node so the light trail access unit of node 5 passively adds the remnant of the working light trail. Once the light trail is protected, all other requests are satisfied smoothly.

In case 3, adjacent multiple link failure is considered. As shown in Fig. 7a two adjacent single link failures are present between nodes (2–3) and (3–5). The necessary steps to protect the connection request (1–5) is shown pictorially in Fig. 7b. When the optical signal reaches node 2 and tries to propagate to node 3, the same procedure discussed in the case 1 is applied to reach node 3. While adding the remnant portion of the working light trail again a link failure is detected between (3, 5). The light trail access unit of node 3 simply adds the backup link between node 3 and node 5. Then the signal is propagated to node 5. The remnant of the working light trail is added passively. Since node 5 is the destination node the communication stops here. The next connection request (2, 6) is satisfied using the protected light trail.

Since the protection scheme adopted by LTP is dynamic, so it is well suited to protect the light trail for dynamic traffic matrix also.

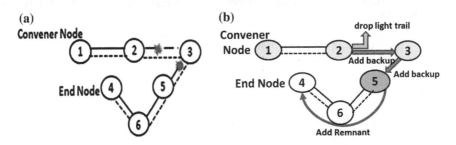

Fig. 7 a Adjacent link failure. b Connection request (1–5) protected and satisfied

6 Conclusion

The proposed LTP algorithm protects both static and dynamic light trails against single link failure and multi-link failure. Further, it is able to protect the entire light trail and its all possible connection requests against single and multiple link failures. The proposed algorithm takes only the switching time to protect the faulty light trail, once the backup links are established. It requires less channel capacity than Dedicated Protection Scheme since the backup links are not used until a link failure occurs. So the proposed algorithm is efficient as it protects multiple link failures in comparison to existing ones which protects only single link failure.

References

1. A. Gumaste and I. Chlamtac, "Light-trails: an optical solution for IP transport", Journal of Optical Networking, May 2004, Vol. 3, Issue 5, pp. 261–281.
2. A. Gumaste and I. Chlamtac, "Light-trails: A Solution for Dynamic Optical Communications", OSA Journal of Optical Networking (JON), May 2004.
3. A. Gumaste and I. Chlamtac, "Light-trails: A Novel Conceptual Framework for Conducting Optical Communications", in 3rd IEEE Workshop on High Performance Switching and Routing, Proceedings of HPSR 2003, Torino Italy June 2003, pp. 251–256.
4. S. Balasubramanian, W. He, A. K. Somani, "Light-trail networks: Design and survivability", in the proceedings of IEEE Conference on Local Computer Networks, 2005, pp. 174–181.
5. W. He, J. Fang, A. K. Somani, "On Survivable Design In Light Trail Optical Networks", in the proceedings of 8th Working Conference on Optical Networks Design and Modelling, Feb, 2004.
6. A. Gumaste and S. Q. Zheng, "Protection and Restoration Scheme for Light-trail WDM Ring Networks", in the proceedings of IEEE Conference on Optical Network Design and Modelling, 2005, pp. 311–320.
7. X. Junwei, W. Hongxiang, J. Yuefeng, "BLE protection scheme for light-trail WDM mesh networks", in the proceedings of Society of Photo-Optical Instrumentation Engineers (SPIE) Conference Series, 2007, Vol. 6784, pp. 67840X–67840X.
8. W. Zhang, G. Xue, J. Tang, K. Thulasiraman, "Dynamic Light Trail Routing and Protection Issues in WDM Optical Networks", in the proceedings of Global Telecommunication Conference, 2005, Globecom'05 IEEE, Vol. 4, pp. 1967–1971.
9. S. R. Bhadra, U. Biswas and M. K. Naskar, "Assignment of Static Light Trail in WDM Networks", in the proceedings of First International Conference on Computational Intelligence: Modeling, Techniques and Applications, CIMTA-2013, Procedia Technology, 2013, Vol. 10, pp. 910–918.

Part IV
Data Analytics and Data Mining

Prediction of Twitter Users' Interest Based on Tweets

Nimita Mangal, Sartaj Kanwar and Rajdeep Niyogi

Abstract In this paper, we try to get the interest of users of certain location based on tweets on certain topics (like entertainment, politics, sports, technology, business, etc.). This study helps us in recommending things to users more accurately. We first analyze the sentiments of tweets and then classify the tweets according to topics. In this way, we are able to get the topic in which users are positively and negatively interested. We have done this experiment on 1500 Indian users and 800 USA users.

Keywords Sentiment analysis · Twitter user · Social media

1 Introduction

Social sites have now become a fast means of communication for spreading news with millions of people in a few seconds. Users share their feelings on Twitter by tweets or on Facebook by posting status. There are 316 million monthly active users on twitter and 500 million tweets are posted per day. Since tweet length is restricted to 140 characters so it is a difficult task to predict the correct interest based on tweets. We can use these tweets for analyzing the interest and get to know the trends going on at any place.

Several works have been done in the field of social networking with emphasis on classification of gender, classification of topic, sentiment analysis of twitter users based on tweets, event detection, community detection, etc. In this paper we make

Nimita Mangal (✉) · Sartaj Kanwar · Rajdeep Niyogi
Department of Computer Science and Engineering, Indian Institute
of Technology Roorkee, Roorkee 247667, India
e-mail: nimitamangal@gmail.com

Sartaj Kanwar
e-mail: kanwarsartaj@gmail.com

Rajdeep Niyogi
e-mail: rajdpfec@iitr.ac.in

© Springer Science+Business Media Singapore 2017   167
J.K. Mandal et al. (eds.), *Proceedings of the First International Conference
on Intelligent Computing and Communication*, Advances in Intelligent Systems
and Computing 458, DOI 10.1007/978-981-10-2035-3_18

an attempt to come up with a method for predicting the interest of users based on sentiments (positive, negative or neutral) and the topic to which tweets are related to get the correct positive or negative interest of users.

In this paper, we are particularly interested in users and their tweets to help them give better recommendation which they need according to their current interest. Tweets are collected using the Twitter4j api in Java. Sentiment analysis has been done using several text files described later which uses word net dictionary and some text file for parsing the word and effectively calculate the impact (either positive or negative) of that word on the sentence. Classification of tweets to which topic it is related has been done using classification of words and tagger file which takes each word of tweet and determines whether it is a noun, verb, or an adjective and relates that word with a topic to which it is closely related with the help of a dictionary.

The paper is organized as follow: Sect. 2 describes the related work. Section 3 describes our method for analyzing tweets. Section 4 describes the results obtained by our method. Section 5 describes the conclusion and future works.

2 Related Work

Different methods have been have proposed for sentiment analysis, finding sentiments in words, sentences, sentiments in topics. Some of these approaches use machine learning, pattern based and natural language processing. In [1] different classifiers in mixed way are used to get better sentiment results. Sentiment analysis of twitter data is studied in [2] and it introduces POS-specific earlier polarity feature and explore the use of tree kernel. Experiments were performed on three models: feature based model uses hundred features only and have same accuracy as that of unigram model that uses ten thousand features. Kernel tree based model first tokenize the tweet into tree by separating punctuation mark, exclamatory mark, negation word and emoticon and prior calculate the polarity of word using word-net dictionary. The unigram model is used as a baseline for the experiments [2].

In [3] two approaches (machine learning and lexical approach) are suggested for sentiment analysis. The machine learning based approach takes text and converts it to a list of words and then takes consecutive pair of words or triplets and calculates some sentiment score based on some code already computed for some set of texts. Using this, new texts are classified into positive, negative or neutral sentiments. In lexical approach, a grammatical structure of language is used and some list of words with sentiment scores and polarities for sentiment score is used. The accuracy of both approaches depends on the training set and the score which is already provided for most of the words.

Many recommender systems provide recommendation using the information based on user profile. [4] suggest a method for user recommendation and the method is based on sentiment volume objectivity. User profiling is done and similarity measure is computed between users (similarity measures based on place, sentiments of tweets). [5] suggest a method for friend recommendation and uses collaborative filtering and graph structure. In [6] semantic users modeling has been done based on twitter posts. [7] suggest a method to predict which political party a twitter user is interested in. [8] analyze the user intentions that are associated at a community level and show how users with similar intentions connect with each other. [9] address the task of user classification in social media using machine learning framework. User profile features such as followers, friends, username, user-location are collected to know about a user. Tweets of user are collected for judging the behavior of user and to classify users of same types. [10] propose two methods for classification of twitter trending topic–one based on textual information and the other based on network structure. In text based model all the hyperlinks are removed from the tweet and then a tokenizer removes stop words and delimited character. Since there is a limitation of 140 characters in tweet, people use acronyms for words and so a vocabulary is used that has full form of these words (e.g. BR is used to represent best regard). The network based approach uses a similarity model to find out the trending topic say X. It searches for five topics that are similar to the topic X and finds out the similarity index.

Most of the above works are related to sentiments, recommenders systems, and trending topic. However these works do not discuss about a user's interest on the topic being discussed by the users. This approach is different from others as we compute the interests of a particular user and the users of a certain location by taking their sentiments (positively or negatively inclined) towards certain topics.

3 Methodology

3.1 Data Collection

Data is collected for the user for which we want to know the interest and behavior by collecting his tweets, number of followers, location, mentions in tweet, retweet count and hashtags. We have collected data about any location by collecting tweets of news channel of that location to get major information about that location and to get trend and interest of users at that location. All these tweets are collected using twitter4j API and all the other information about users is also fetched using this API.

3.2 Sentiment Analysis

Sentiment analysis is done by using senti-strength method. The total positive and negative sentiments are calculated and the two are compared. If the total positive sentiment is larger than 1.5 times the total negative sentiment, the classification is positive, otherwise it is negative. Sentiment analysis uses several text files which contain emoticons, stop words, slang words, frequently used words, interrogations, and English words. These files contain words with some human coded score. First we refine the tweet by removing all hashes, @ and extra spaces to make it more readable. The refined text is then separated into words, emoticons and punctuations. Each word is compared against the word of each file and if a match is found then a score is given to that word according to positive or negative sentiments. Each emoticon is also compared with emoticon lookup table and score is given to each emoticon. Since there is a restriction of 140 characters in tweet, so most users use abbreviations to express feeling using less number of words. Each word is also compared with slang lookup table; if it matches then score is given on that basis. Finally the total score for a text is calculated.

3.3 Classification of Tweet

After sentiment analysis, tweets are classified according to topic to which it is related. Open NLP package is used for classification. We used a tagger file for tagging of sentences. MaxentTagger is a class used for tagging each word in a tweet with its corresponding form whether it is adverb, noun, adjective, etc. Fig. 1 shows the tag given to each word of the tweet.

There are 36 taggers and each word in a tweet belongs to one of these taggers. Each this word tagger pair is compared with ten different categories of topics like entertainment, technology, politics, etc. A method get similarity is called which compares this word with these topics and calculates similarity score if similarity is more than seventy five percent. Figure 2 shows the topic of tweet to which it is related.

Tweet by Shreya Ghoshal: Catching up on movies.. Where else The usual. Mid air! Also, yo! Emirates got wifi onboard but a bit slower than other airlines I think..

```
Problems  @ Javadoc  Declaration  Console ⊠  Call Hierarchy
<terminated> tagtweetssubject [Java Application] C:\Program Files\Java\jre1.8.0_45\bin\javaw.exe (Aug 26, 2015,
Loading default properties from trained tagger taggers/left3words-wsj-0-18.tagger
Reading POS tagger model from taggers/left3words-wsj-0-18.tagger ... done [6.6 sec].
Catching/VBG up/RP on/IN movies/NNS ./. ./. Where/WRB else/RB The/DT usual/JJ ./. Mid
{
   "status": "200 OK",
   "result": [
      {
        "mid": "/m/015w9s",
        "id": "/en/television_movie",
        "name": "Television film",
        "notable": {
          "name": "TV Genre",
          "id": "/tv/tv_genre"
        },
        "lang": "en",
```

Fig. 1 Show how score is given

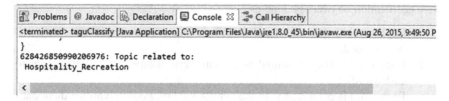

Fig. 2 Topic to which tweet is related

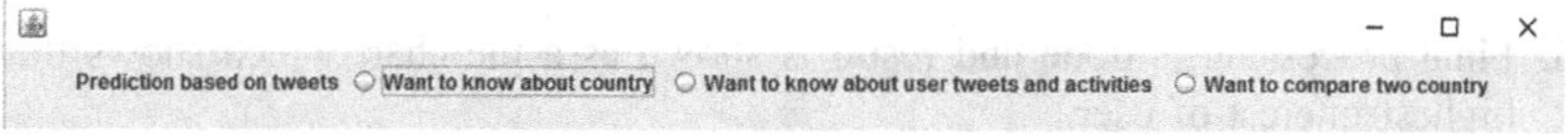

Fig. 3 Home page

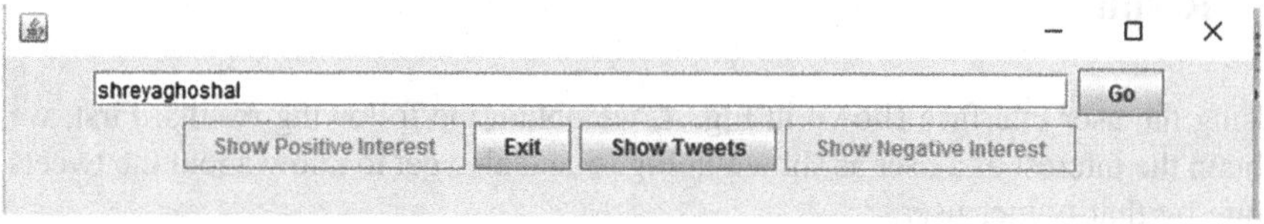

Fig. 4 User interest

3.4 Processing

First the tweets of a user or users of certain place are collected based on our choice as shown in Fig. 3 then sentiment analysis and classification of tweets has been done.

After doing all the process you can click show interest button to get the result as shown in Fig. 4. The results are shown in the form of pie chart.

3.5 Proposed Method

i. Obtain a choice from a user (refer Fig. 3).
ii. User enters a screen name of twitter user about which she wants to know.
iii. A module is run which download current tweets of that twitter user.
iv. Sentiment analysis has been done on the fetched tweet.

    a. Replace all illegal characters (like RT, #, @, etc.) from tweet and a plain text is processed.

    b. Each word is compared with the text files and score is given accordingly.

    c. Final score is computed and tweet is classified as positive, negative or neutral.

v. Classification of tweet is done as:

   a. Replace all illegal characters (like RT, #, @, etc.) from tweet and a plain text is processed.

   b. Each word is then classified as a noun, verb or adjective (assign tagger to each word).

   c. Each word is compared with the similar kind of word in wordnet dictionary and score for category is decided accordingly.

   d. Final score decides to which topic tweet is belonging (like entertainment, politics, etc.).

vi. Final processing is done and result is shown as a pie chart which shows the highest interest of user.

4 Results

Using the user interface shown in Fig. 4, we obtain the following results. First, we obtain the interest of a user as shown in Fig. 5 and also get to know about the tweets done by that twitter user.

Figure 6 shows the sentiments of Indian users and Fig. 7 shows the interest of users of India.

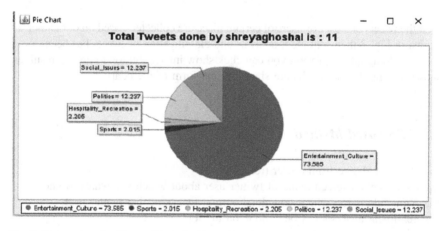

Fig. 5 Tweets done by Shreya Ghoshal with positive interest from 13-08-15 to 21-08-15

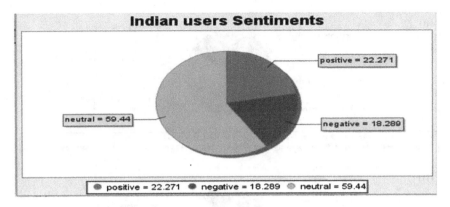

Fig. 6 Indian user's sentiments

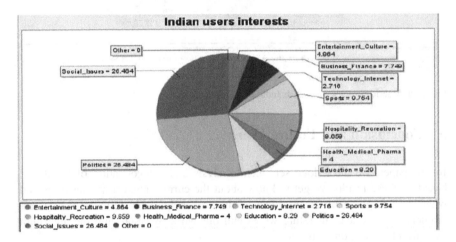

Fig. 7 Indian user interest from 12-08-15 to 21-08-15 tweets

It is useful because it makes you easy to you before launching some product at any location that whether their people are interested or not in such topic. Figure 8a and b gives the comparison results of Indian and US users.

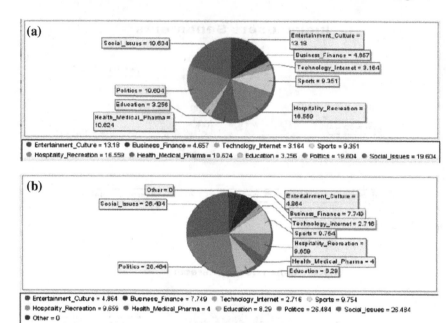

Fig. 8 **a** American user's interest. **b** Indian user's interest

5 Conclusion and Future Work

In this paper, we have done sentiment analysis and categorisation of tweet and based on these results we get to know about the current interest among users and what trend is going on at any place. This work helps in betterment of any recommendation system. In future we extend our work by improving the algorithm used for classification of tweets.

References

1. R. Prabowo1, and, M. Thelwall, "Sentiment Analysis: A Combined Approach," Journal of Informetrics, Vol. 3(2), pp 143–157, 2009.
2. A. Agarwal, B. Xie, I. Vovsha, O. Rambow, and, R. Passonneau, "Sentiment Analysis of Twitter Data," Proceedings of LSM'11 Workshop on Languages in Social Media. Association for Computational Linguistics, pp 30–38, 2011.
3. M. Thelwall, "Heart and Soul: Sentiment Strength detection in the Social Web with SentiStrength," Proceedings of the CyberEmotions, pp 1–14, 2013.
4. D. F. Gurini, F. Gasparetti, A. Micarelli, and, G. Sansonetti, "A Sentiment-Based Approach to Twitter User Recommendation," Proceedings of 5th ACM RecSys workshop on Recommender Systems and the social web, June 2013.

5. V. Agarwal, and, K. K. Bharadwaj, "A collaborative filtering framework for friends recommendation in social networks based on interaction intensity and adaptive user similarity," Journal Social Network Analysis and Mining, Vol. 3, pp 359–379, 2013.
6. F. Abel, Q. Gao, G. Houben, and, K. Tao, "Semantic Enrichment of Twitter Posts for User Profile Construction on the Social Web," Proceedings of the Semantic Web: Research and Applications. Springer Berlin Heidelberg, pp 375–389, 2011.
7. A. Boutet, H. Kim, and, E. Yoneki, "What's in Twitter, I know what parties are popular and who you are supporting now!," Journal Social Network Analysis and Mining, Vol. 3(4), pp 1379–1391, 2013.
8. A. Java, X. Song, T. Finin, and, B. Tseng, "Why we Twitter: Understanding Microblogging Usage and Communities," Proceedings of 9th WebKDD and 1st SNA-KDD Workshop, SanJose, California, USA, pp 56–65, August 2007.
9. M. Pennacchiotti, and A. Popescu, "A Machine Learning Approach to Twitter User Classification," Proceedings of the Fifth ICWSM, pp 281–288, 2011.
10. K. Lee, D. Palsetia, R. Narayanan, Md. Mostofa Ali Patwary, A. Agrawal, and, A. Choudhary, "Twitter Trending Topic Classification," Proceedings of 11th IEEE International Conference on Data Mining Workshops, pp 251–258, December 2011.

Event Detection Over Twitter Social Media

Sartaj Kanwar, Nimita Mangal and Rajdeep Niyogi

Abstract Twitter is a social networking site that allows a large number of users to communicate with each other. Twitter allows users to share their views on different topics ranging from day to day life to what is going in society. Event detection in twitter is the process of detecting popular events using messages generated by the users. Event detection is difficult in twitter as compared to other media because the message known as tweets is only allowed to be less than 140 characters. Moreover the tweets are noisy because there may be personal messages by the user also. The focus of this paper is to find top k popular events from tweets using keywords contained in the tweets. This paper also classified the popular events into different categories.

Keywords Twitter · Event detection · Social media

1 Introduction

Twitter is an electronic medium that has gained popularity in recent time. Twitter allows users to post their messages (known as Tweets) on various events. Lengths of tweets are restricted to be less than 140 characters. Twitter gained popularity as it has more than 100 million users in 2012 who has posted 340 million tweets per day. In 2013 Twitter was among the ten top most visited website and given the title of "SMS of the internet". To post tweets on twitter user has to register himself on the twitter. Users who have not registered themselves on twitter can only read the messages. The relationship between twitter users are of follower and followed type.

Sartaj Kanwar (✉) · Nimita Mangal · Rajdeep Niyogi
Department of Computer Science and Engineering, IIT Roorkee, Roorkee 247667, India
e-mail: kanwarsartaj@gmail.com

Nimita Mangal
e-mail: nimitamangal@gmail.com

Rajdeep Niyogi
e-mail: rajdpfec@iitr.ac.in

© Springer Science+Business Media Singapore 2017   177
J.K. Mandal et al. (eds.), *Proceedings of the First International Conference on Intelligent Computing and Communication*, Advances in Intelligent Systems and Computing 458, DOI 10.1007/978-981-10-2035-3_19

Twitter users have unilateral relationship i.e. when a user A follows user B, it means user A can get all the messages shared by user B but not vice versa. Due to the restriction on length of tweets sometimes it is difficult to put your idea in fixed characters so, users can share link to other objects on internet such as article, blogs, video etc. (termed as artifacts in twitter). As users share what is going around them on twitter and this will be almost instantly. This makes twitter a good source for timely information. Twitter uses the following concept "when you need to know what's going on-in your town or across the globe-get the best of what's happening on twitter".

The goal of this paper is to demonstrate how the popular events can be extracted from tweets posted by the twitter users. Events are the incident that is happening at some place at some point of time. As there may be lots of event going on but we are interested in popular event. Popular event is the event which got more attention as compared to other events. This attention is calculated on the basis of frequency of occurrence of tweets related to that topic. More the tweets generated by the users on particular topic means that topic is more popular to other topics.

In this paper we suggest an algorithm to classify the tweets based on their similarity. This similarity will be governed by the keywords used in the tweets. This approach will use nouns used in tweets as their keyword and merge two tweets into one event when there keyword matches above some threshold. We present a method to calculate the popularity of an event. Popularity of an event is decided by number of tweets are related to that topic. For this purpose we have used count of number of tweets in each and every topic. We present an interface that will ask a user about how many top events he/she is interested in and depending upon the value (k) entered by the user we will display top k events at that point of time. We also discuss an algorithm to examine the type of particular event, for example whether an event is related to politics, entertainment, sports etc. For this we have classified events in different categories depending on the type of keywords used in tweets of that event.

The paper is organized as follow: Sect. 2 describes some related work that has been done till now in this field. Section 3 describes our methodology which we used for different purposes. Section 4 describes the result generated by our method. Section 5 describes the conclusion and future works that can be done in this area.

2 Related Work

In [1] described the event detection algorithm which builds a Key Graph using keywords used in tweets. The edges between two nodes will signify the occurrence of two keywords simultaneously. After that they have used community detection algorithm to cluster different nodes to form the cluster. Inter cluster edges are removed based on betweenness centrality score. In this paper they had used named entities and noun phrases as keywords. Using these keywords they had created the Key Graph. In Key Graph nodes are the keywords and edges between them are formed when two keyword occur simultaneously in a document. In [2] the problem of online social event monitoring over tweet streams for real applications like crisis management and decision making are addressed. They presented location-time constrained topic model for associating location with each event. In [3] presented algorithm for event detection named as New Event Detection (NED). NED is consists of two subtasks i.e. Retrospective and Online NED. In Retrospective NED previously unidentified events are detected and in Online NED new events in text stream are detected.

In [4] authors had used wavelet transformation for event detection. They have used the concept when and how frequency of the signal changes over time. In [5] they posed the problem of identifying events and their user contributed social media documents as a clustering task, where documents have multiple features, associated with domain-specific similarity metrics [6]. They proposed a general online clustering framework, suitable for the social media domain. They developed several techniques for learning a combination of the feature-specific similarity metrics, and used them to indicate social media document similarity in a general clustering framework. They evaluated their proposed clustering framework and the similarity metric learning techniques on two real-world datasets of social media event content.

In [7] considered location with every event as location and event are strongly connected. They considered where, when and what about an event. Firstly preprocessing is performed to remove stop words and irrelevant words. Secondly they performed clustering to automatically group the messages in event. Thirdly they proposed a hotspot event detection method. Fourthly emerging hotspot detection method is performed. Finally visualization model for emerging event is presented.

In [8] presented a platform *TwitInfo* for exploring Tweets regarding to a particular topic. User has to enter the keyword for an event and TwitInfo has provided the message frequency, tweet map, related tweets, popular links and overall sentiment of the event. The *TwitInfo* user interface contained following thing: The user defined name of the event with keywords in the tweet, Timeline interface with y axis containing the volume of the tweet, Geo location along with that event is displayed on the map, Current tweets of selected event are colored red if sentiment of tweet is negative or blue if sentiment of the tweet is positive and Aggregate sentiment of currently selected event using pie chart.

In [9] presented a system known as *TwitterMonitor* detects the real time events in real time. This is done in three steps. In first step bursty keywords are identified i.e. keywords that are occurring at very high rate as compared to others. In second step grouping of bursty keyword is done based on their occurrences. In third and last step additional information about the event is collected.

In [10] built a news processing system for twitter called as *TwitterStand*. They have taken 2000 handpicked seeders for collecting tweets. Seeders are mainly newspaper and television stations because they are supposed to publish news. After that they have separated the junk tweets from news using naïve Bayes classifier. Authors have used online clustering algorithm called leader-follower clustering to cluster the tweets to form events. In [11] proposed a statistical method *MABED* (mention-anomaly-based event detection). They divided the whole process of event detection in three steps. In first step they had detected the events based on mention anomaly. Secondly they have selected words that will best describe each event. After this authors have deleted all the duplicated events or merged the duplicate events. Lastly they have generated a list of top k events.

3   Proposed Algorithm

The basic way to detect the events and news from large amount of data is by using keywords. If two events are similar then up to some extent they contain similar words. This similarity should have some threshold. It means two tweets are only put into one event when they have more same keywords than some threshold. Before going into detail of the algorithm we should understand the architecture and flow of whole algorithm.

To perform the experiments on event detection we have taken 10,000 users. The screen names of all the users are fetched and stored in the database. For collecting this users, we have used initial some users that are supposed to provide news information as personal messages are not useful in event detection. After collecting the seeders we have collected the followers of the seeders. In this way our set of users are built.

After collecting screen names of users we have started storing tweets for those users in our database. For each user we have stored various information such as tweet id, location etc. For keywords we have used nouns present in tweets. Nouns are fetched for every tweet and stored in database.

3.1   Algorithm

Input: Input to our algorithm is tweets along with keywords and hash if provided by the user.

Algorithm: Algorithm is taking tweets one by one and comparing its hash with already formed events. If the hash is same means those two tweets corresponds to the same event. For every tweet we have also stored the number of tweets fall under that event. So put new tweet along with previous tweet in same event and increment the tweet count for that event. Either if hash is not same for new tweet and tweets present in already created events or hash of new tweet entering in system is not known then we have compared the keywords present in the new tweet and tweets present in previous events. If keywords are matched above the threshold for any event then we merge this new tweet with that event. Whenever any tweet is merged in already created event the keyword of that tweet is updated in that event. So in this way the event is learning new keywords about that event on addition of new tweets to that event.

If no previous event has the same kind of keywords, then make a new event and put tweet in that event along with its keywords. Keyword of this tweet is made as keyword of the newly created event. The above steps are done iteratively until all tweets are put into some already created events or new events are created for them. After this we have distinguished events along with count indicating the number of tweets in that event.

Output: For output user is asked that how many popular events he wants, say k. depending upon value of k the algorithm is fetching the top k events depending upon the count of each event that is updated in previous steps.

The interface provides options to users if user wants to have a look at tweets contained in every popular event. Algorithm also provides a pie chart along with every event telling how the tweets under this event are distributed. The distribution is having dimensions such as entertainment, social, games, politics etc. More the percentage of particular type means that event fall under that category.

The whole algorithm is explained below in steps:

S—Set of all tweets along with its keywords and hash. Initially this set is empty
E—Set of all events created at any point of time in algorithm. Initially this set is empty.
p—Threshold for merging tweet into that event.
n_e—New event in the system.
n_t—New tweet in the system.
e—Any event from set E.
max_ev—Event in set E with which maximum keyword of n_t matches.

(i)   Fetch the tweets for usernames from the twitter website.
(ii) Break the tweets in keywords (nouns) and store along with hash of that tweet i.e. S.
(iii) Do steps from (a) to (c) until set S is empty.

    a. Take one new tweet (i.e. n_t) from the set S.
    b. If E is empty create new event n_e and add all keywords of n_t into n_e a. Set tweet count for n_e to 1 as there is only one tweet added to n_e. Also make hash of n_t as the hash of n_e.
    c. If E is not an empty set then one by one check the events in set E i.e. e.

        • If hash of n_t and e matches put n_t into event e and increment the count of event e.
        • If hash of any event e not matches with n_t, then match keywords of every event e in set E with n_t. Select the event (i.e. max_ev) with which maximum keywords are matching.
        • If the similarity of keywords are greater than p, put n_t into max_ev and increment the count of max_ev by one.
        • Else if similarities of keywords are less than p, create new event using n_t.

d. When set S is empty means all tweets are put into some event. Now set E contains all the events from the tweets and along with tweet count.

e. Take input from the user for k (how many top popular events should be shown).

f. Find k events from the set E having maximum tweet count.

4 Results

In results we have shown the snapshots of interface we have created in java. Figure 1 is showing the main frame which will ask the value of k i.e. top k events user is interested in. This frame is also showing top k events. As shown in Fig. 1 that user is interested in top 5 popular events. After clicking popular event button interface has shown top 5 events.

On clicking of tweet button of main frame second frame appears which allow user to select one event among top k events as shown in Fig. 2. As shown below user is interested in 437th event which corresponds to Kasab event.

After user selection it will show the tweets corresponding to that event as shown in Fig. 3.

We also provided user a pie chart corresponding to each event to show how tweets in that event is distributed. Larger the percentage of any class means event fall under that class. As selected event is more falls into category of politics as compared to other classes. Below pie chart is useful when we want to know in which class a particular event will fall (Fig. 4).

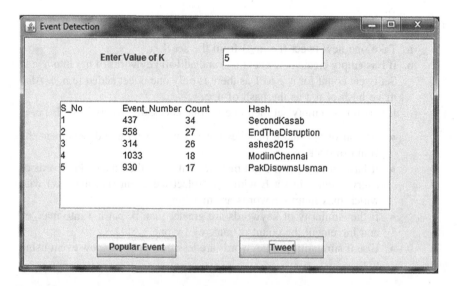

Fig. 1 Top k(k = 5) popular events along with tweet count of each event

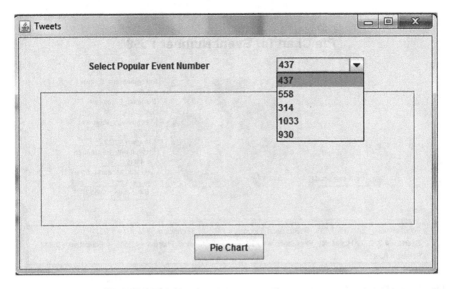

Fig. 2 Drop down to select a particular event

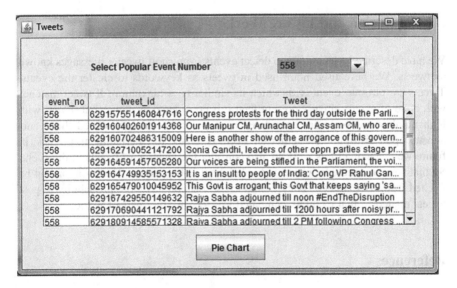

Fig. 3 Tweets corresponding to selected event

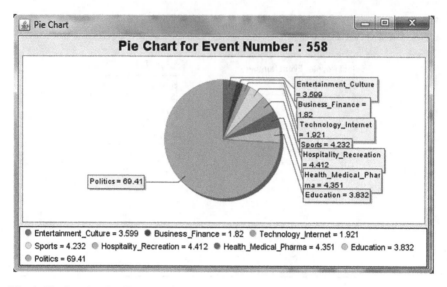

Fig. 4 Pie chart for classification of tweets corresponding to a particular event

5 Conclusion and Future Work

We have described a technique to detect events from short tweeter messages known as tweets. We have used noun used in tweets as keywords to cluster the events. Tweets are put into event with which its similarity is maximum. If tweet matches with none of events already present then it is considered as new event. Event with high number of tweets are considered as popular. We also classified the events depend upon distribution whether the event is related to entertainment, politics. In future we can extend this work to also consider the location of tweets in the event. So that we can conclude where particular event is happening. This location can be fetched from user tweets. As user from event location has more probability to post tweets related to that event as compared to users in other part of world.

References

1. H Sayyadi, M Hurst, and A Maykov. Event Detection and Tracking in Social Streams. In Proceedings of the Third International ICWSM Conference, pp. 311–314, 2009.
2. X. Zhou and L. Chen. Event detection over twitter social media streams. The VLDB Journal— The International Journal on Very Large Data Bases, 23(3), pp. 381–400.
3. W. Dou, X. Wang, W. Ribarsky and M. Zhou. Event Detection in Social Media Data. In IEEE VisWeek Workshop on Interactive Visual Text Analytics-Task Driven Analytics of Social Media Content, pp. 971–980, 2012.

4. H. Becker, M. Naaman and L. Gravano. Learning Similarity Metrics for Event Identification in Social Media. In Proceedings of the third ACM international conference on Web search and data mining, pp. 291–300, 2010.
5. A. Marcus, M. S. Bernstein, O. Badar, D. R. Karger, S. Madden and R. C. Miller. TwitInfo: Aggregating and visualizing microblogs for event exploration. In Proceedings of the SIGCHI conference on Human factors in computing systems, pp. 227–236, 2011.
6. H. Becker, M. Naaman and L. Gravano. Learning Similarity Metrics for Event Identification in Social Media. In Proceedings of the third ACM international conference on Web search and data mining, pp. 291–300, 2010.
7. M. Mathioudakis and N. Loudas. TwitterMonitor: Trend detection over the twitter stream. In Proceedings of the 2010 ACM SIGMOD International Conference on Management of data, pp. 1155–1158, 2010.
8. J. Weng, Y. Yao, E. Leonardi and F. Lee. Event Detection in Twitter. In Proceedings of the 5th International AAAI Conference on Weblogs and Social Media, pp. 401–408, 2011.
9. S. Unankard, X. Li and Mohamed A. Sharaf. Emerging event detection in social networks with location sensitivity. World Wide Web, pp. 1–25, 2014.
10. J. Sankaranarayanan, H. Samet, Benjamin E. Teitler, Michael D. Lieberman and J. Sperling. TwitterStand: News in Tweets. In Proceedings of the 17th ACM sigspatial international conference on advances in geographic information systems, pp. 42–51, 2009.
11. A. Guille and C. Favre. Event detection, tracking, and visualization in twitter: a mention anomaly based-approach. Social Network Analysis and Mining 5, no. 1, pp. 1–18, 2015.

DMDAM: Data Mining Based Detection of Android Malware

Abhishek Bhattacharya and Radha Tamal Goswami

Abstract Mobile malwares have been rising in scale as Android operating system enabled smart phones are getting popularity around the world. To fight against this outburst of Android malwares, different static and dynamic malware detection methods have been proposed. One of the popular methods of static detection technique is permission based detection of malwares through AndroidManifest.xml file using machine learning classifiers. However, the comparison of different machine learning classifiers on different data sets has not been fully cultivated by existing literatures. In this work we propose a framework which extracts the permission features of manifest files, generates feature vectors and uses different machine learning classifiers of a Data Mining Tool, Weka to classify android applications. We evaluate our method on a set of total 170 applications (100 benign, 70 malwares) and results show that highest TPR rate is 96.70 % while accuracy is up to 77.13 % and highest F1 score is 0.8583.

Keywords Android · Malwares · Static detection · Classification · Feature reduction

1 Introduction

The Android operating system has become soft target for attackers as the market share of Android has increased. Moreover, Android applications are easy targets for reverse engineering, which is a specific characteristic of Java applications in

Abhishek Bhattacharya (✉)
Department of Computer Science & Engineering, Institute of Engineering & Management,
Kolkata, India
e-mail: abhishek.bhattacharya@iemcal.com

R.T. Goswami
Department of Computer Science & Engineering, Birla Institute of Technology, Mesra, India
e-mail: rtgoswami@bitmesra.ac.in

© Springer Science+Business Media Singapore 2017
J.K. Mandal et al. (eds.), *Proceedings of the First International Conference
on Intelligent Computing and Communication*, Advances in Intelligent Systems
and Computing 458, DOI 10.1007/978-981-10-2035-3_20

general, and which is often abused by malicious attackers, who attempt to embed malicious program into benign applications, hence creating subspecies of existing malware. Android also permits the users to download apps from untrusted third party markets. To protect mobile users from rigorous threats of malwares, various approaches have been proposed. Static analysis is based on source code assessment looking at apprehensive patterns. Though some static analysis approaches have been successful, different obfuscation techniques have evolved. Dynamic analysis which is also known as behavior-based detection is a method which involves running the apk in a separate environment to analyze its run time logs. In this paper, we have proposed a data mining based malware detection framework for Android devices. Summering, our main contributions in this paper are:

- We described the process of extracting permissions from .apk files.
- We applied featured reduction concepts and Information Gain (IG) was used to rank extracted features and created data sets using those selected features.
- We performed empirical validation using machine learning classifiers and made the comparison of performances of different Weka based machine learning classifiers on different data sets.

We collected 170 samples of diverse categories of apks from different Android market. The collection consists of 100 benign and 70 malware apps. The main categories of benign apps are system tools, entertainment, news, music and audio and games etc. Malware apks are downloaded from Contagio malware dump. Our benign apk files were collected from Google Play Store. The reminder of this paper is organized as follows: Sect. 2 details the related works. Section 3 describes feature engineering and detection methods. Section 4 shows the empirical validations. Finally, Sect. 5 shows the ways to future works.

2 Related Work

One of the most significant tools that can be used for malware detection is Machine Learning techniques on static features that are extracted from apk files. Burguera et al. [2] proposed Crowdroid-an approach that analyzed the behavior of the applications through clustering techniques in Malware detection and they considered two categories of Android applications: tools and business. The results indicated a positive indication of using machine learning techniques in Android Malware. Permlyzer [4] discussed the usage of permission through both static and dynamic analysis. In [1], apks' permissions were extracted from manifest file and especially <uses-permission> tag was used for this purpose. Here Euclidian, Cosine and Manhattan distances were considered. Average accuracy obtained was 85 % through Manhattan distance and using Euclidian and Cosine distances they obtained 87.57 % and 90 % accuracy respectively. Here also comparison of performances of other built in classifiers of Weka was not included. In [6], a proactive machine learning approach, based on Bayesian classifier was proposed. This model was built by extracting features from 1000 samples and overall 90 % accuracy was achieved. A Java implemented custom package manager was used for extracting the features from Androidmanifest.XML files. Ryo et al. [5] introduced a model in which they used only the manifest files to detect malwares. During the experiment they had taken 365 samples and achieved overall correct detection ratio as 90.0 %. Apart from considering only permissions, they also considered Intent filter (action), Intent filter (category) and process name in every manifest file. But the comparison of performances of classifiers was not included in that work. Hyunjae et al. [10] analyzed malicious behaviors and permissions to increase detection accuracy. Finally they showed detection and classification performance as 98 % and 90 %, respectively. Aswani et al. [11] used BNS, KL, RS, KO and MI as feature selection techniques and BNS achieved highest classification accuracy (93.02 %).

3 Feature Engineering

In this work, we proposed a method for detecting Android malwares by comparing the performances of different machine learning algorithms.

3.1 Feature Extraction

In this section, we reviewed different feature sets that we have considered for this malware detection model. We collected those features from AndroidManifest.xml files which are packed with .apk files. To achieve that objective we first extracted the permissions using Androguard tool [9]. The structure for declaring permission in AndroidManifest.xml is shown in below.

<uses-permission android: name = "String"/>

These permissions are generally requested by any application during installation in mobile devices. We processed the manifest files searching for that <uses-permission> tag and extracted the permission strings. After processing all 100 benign and 70 malwares we created feature vector of permissions. The values of each selected features are stored as binary number where 0 and 1 represents presence and absence of that permission in the feature vector for that application.

3.2 Preprocessing

In our work, we carried out feature elimination to remove irrelevant features. It is actually carried out to select a subset of significant installation time features to use in model construction. Leaving out important attribute or keeping irrelevant attributes may cause confusion for mining algorithms. After elimination of unimportant features, only 83 common features from both malware and benign samples are considered. The main goal of this phase is to find a minimum set of features such that the resulting probability distribution of data class is as close as possible to the original distribution created with all features. Information Gain (IG) was applied to our model as a feature ranking tool to our combined feature set and at first we selected top 10 features from total 83 combined and common features (both malware and benign) according to their Information Gain scores. Similarly we applied Information Gain to select top 5 and top 2 features from the original dataset. Finally we prepared 4 datasets having different number of features (83, 10, 5 and 2 respectively). We discarded any feature with IG score 0.

4 Empirical Validation

To evaluate our method, we used the datasets described in Sect. 4 which are composed of permissions of 170 sample Android applications. Top ranked features were selected according to their IG score to create ensemble features and Weka tool was used to analyses the evaluation of proposed model.

4.1 Evaluation Metrics

We evaluated the model by measuring the following parameters-

- **True Positive Ratio (TPR)**: TPR = TP/(TP + FN) where TP represents the quantity of benign applications properly classified as benign and FN represents the quantity of benign applications incorrectly classified as malware.
- **False Positive Ratio (FPR)**: FPR = FP/(FP + TN) where FP represents the quantity of malware which are inaccurately classified as benign and TN represents the quantity of malware which are truly classified as malware.
- **Accuracy**: It determines the percentage of the total quantity of predictions is correct.
- **F1-Measure**: It computes the harmonic mean of Precision (P) and True Positive Ratio (TPR). F1= (2 * P * TPR)/(P + TPR). F1 score 1 indicates good performance in classification of minority data [8].
- **Area under ROC Curve**: It sets up the relation between TPR against FPR. It shows the predictive power of classifier.

4.2 Classifiers

One of the most engaging features of machine learning algorithms is that they improve their ability to discriminate normal behavior from anomalous behavior with experience. K-fold cross-validation was used for evaluating the results of a numerical analysis which generates an independent dataset. Using K = 10 folds in cross validation means 90 % of data is utilized for training purpose and 10 % for testing in each fold test. To visualize the classification performance of the models, we constructed the Receiver Operating Characteristic (ROC) curve. From the result it is depicted that RandomForest classifier provides the best value in AUC. AUC values closer to 1 denotes better classifier. Result shows that for dataset 1 highest AUC of 0.817 in RandomForest classifier is seemed to be the most predictive of all. For dataset 2,NaiveBayesMultinomial classifier provides the best value in AUC (0.722). The best AUC for dataset 3 is obtained with PART classifier (0.723). On dataset 4, best AUC of 0.667 is obtained with SMO classifier. Figure 1 shows the TPR results for four different datasets with different feature settings. ZeroR classifier provides the best TPR for all datasets. Figure 2 shows that average True positive rate (TPR) score is highest (96.6952) in dataset 4 with top 2 selected features followed by dataset 3 with top 5 selected features. The obtained result is undoubtedly better than that of [6]. It also clearly depicts the benefit of intelligently selecting a reduced feature set. The average TPR of dataset 4 and dataset 3 of our model are better than that of [7]. More over, Fig. 3 shows that dataset 2, dataset 3 and dataset 4 which are formed based on Information Gain score of attributes, produces better average TPR than that of datasets which are formed using Mutual Information, Pearson correlation coefficient and T test of Wang [8]. Average TPR of both dataset 3 (96.1904 %) and dataset 4 (96.6952 %) of our study produce better performances than average TPR (93.43 %) with dataset 1 of [13] and average TPR (88.23 %) with dataset 2 of [13]. The average TPR percentage (96.2 %) in J48 classifier of our framework produces better result than average TPR (93 %) of Alazab's framework [10]. The average TPR percentage (92.4 %) in NaiveBayes classifier of our framework produces better result than average TPR percentage (91 %) of Alazab's framework [10] (Fig. 4).

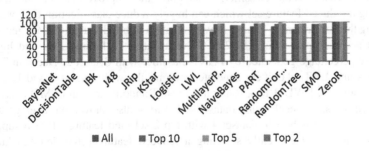

Fig. 1 Comparison of TPR results of four different datasets

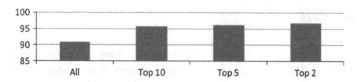

Fig. 2 Average true positive rate

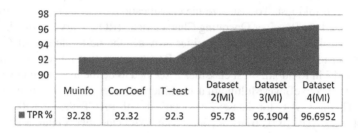

Fig. 3 Comparison of our TPR with MI, T, CorrCoef of Wang's TPR

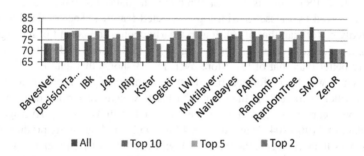

Fig. 4 Comparison of different accuracy with four different datasets

Figure 5 demonstrates the comparison of different F1 score with all four datasets using different Weka classifiers. It shows that dataset 3 obtains significantly higher F1 score (0.952) using Kstar classifier, followed by F1 score (0.86869) with dataset 4 using Decision Table classifier. It supports the improvement of intelligently selecting a reduced feature set which can hugely reduce classification time without affecting the performance. Figure 4 reveals that though individually SMO generates highest accuracy percentage (81.295 %) with dataset 1 followed by 2nd highest accuracy percentage generated by PART classifier with dataset 2. Decision Table, Logistics and LWL classifiers generate same results with dataset 3 and Decision Table, Logistics, LWL, J Rip, Random Forest, Random Tree and SMO classifiers generated the same result using dataset 4. Figure 4 also shows that average accuracy performance is best in dataset 4 with top 2 selected features. It also supports the advantage of intelligently selecting a reduced feature set which can hugely reduce classification time. PUMA [12] achieved 86.41 % accuracy, and

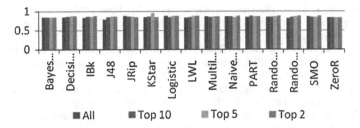

Fig. 5 Comparison of different F1 scores with four datasets

DroidpermissionMiner achieved 81.56 % accuracy with Mutual information (MI) as feature selection tool. In our framework SMO with full set of attributes generates 86.09 % accuracy which is quite comparable with [12].

5 Conclusion and Future Works

In this paper, we presented a framework using machine learning algorithms for classifying android applications whether they are malware or benign. Our model was evaluated using ROC curve (AUC), TPR and F1 scores. Results showed that highest TPR rate was 96.70 % while accuracy was up to 77.13 % and highest F1 score achieved was 0.8583. Unlike prior research in this field, in this paper we presented the comparison of performances of different classifiers of Weka in terms of TPR, AUC and F1 scores. Moreover, unlike any other prior studies, in this paper we made the comparative studies of different feature reducts, created by Information gain as feature ranking technique. Machine learning-based detection approaches are having two limitations: they have high false alarm rates and determining what features should be learned in the training phase is an intricate task [3]. Through proper investigation it can be shown that because of sparseness of feature vector, detection ratio of may be inaccurate [8]. Moreover, a number of malwares request same permissions that are also requested by benign applications. However, the comparisons that we made may not be always meaningful because different studies use different data sets. In future, we would like train classifiers with larger datasets to get more accurate classification.

References

1. Borja Sanz, Igor Santos, Xabier Ugarte-Pedrero, Carlos Laorden, Javier Nieves, and Pablo Garcia Bringas, "Instance-based Anomaly Method for Android Malware Detection", SECRYPT, SciTePress, pp. 387–394, 2013.
2. I. Burquera, U. Zurutuza, and S. Nadjm-Tehrani, "Crowdroid: behavior-based malware detection system for Android," in Proc. the 1st ACM workshop on Security and privacy in smartphones and mobile devices, 2011, pp. 15–26, 2011.

3. K. Allix, T. F. D. A. Bissyande, J. Klein, and Y. Le Traon,, "Machine Learning-Based Malware Detection for Android Applications: History Matters!," Technical Report, University of Luxembourg, 2014.

4. Wei Xu, Fangfang Zhang, Sencun Zhu, "Permlyzer: Analyzing permission usage in Android applications", In Proc. of IEEE International Symposium on Software reliability Engineering (ISSRE), pp. 400–410, 2013.

5. Ryo Sato, Daiki Chiba and Shigeki Goto, "Detecting Android Malware by Analyzing Manifest Files", Proceedings of the APAN – Network Research Workshop, pp. 1–9, 2013.

6. S. Y. Yerima, S Sezer, G. McWilliams," Analysis of Bayesian classification-based approaches for Android malware detection", IET, vol. 8, issue 1, pp. 25–36, 2014.

7. Paolo Rovelli, Ýmir Vigfússon, "PMDS: Permission-Based Malware Detection System", Information Systems Security Lecture Notes in Computer Science vol. 8880, pp. 338–357, 2014.

8. Wei Wang, Xing Wang, Dawei Feng, Jiqiang Liu, Zhen Han, Xiangliang Zhang, "Exploring Permission-Induced Risk in Android Applications for Malicious Application Detection", IEEE Transactions on Information Forensics and Security, vol. 9, issue 11, pp. 1869–1882, 2014.

9. https://code.google.com/p/androguard.

10. Hyunjae, Kang, Jae-wook Jang, Aziz Mohaisen, Huy Kang Kim, "Detecting and Classifying Android Malware Using Static Analysis along with Creator Information", International Journal of Distributed Sensor Networks vol. 2015, pp. 1–9, 2015.

11. A. M. Aswini, P. Vinod, "Android Malware Analysis Using Ensemble Features", Security, Privacy, and Applied Cryptography Engineering Lecture Notes in Computer Science vol. 8804, pp. 303–318, 2014.

12. Borja Sanz, Igor Santos, Carlos Laorden, Xabier Ugarte-Pedrero, Pablo Garcia Bringas, Gonzalo Álvarez, "PUMA: Permission Usage to Detect Malware in Android", International Joint Conference CISIS'12-ICEUTE'12-SOCO'12 Special Sessions, Advances in Intelligent Systems and Computing, vol. 189, pp. 289–298, 2013.

13. Zarni Aung, Win Zaw, "Permission-Based Android Malware Detection", International Journal Of Scientific & Technology Research, vol. 2, issue 3, pp. 228–234, 2013.

Comparative Study of Parallelism on Data Mining

Kartick Chandra Mondal, Sayan Bhattacharya and Anindita Sarkar

Abstract Today's world has seen a massive explosion in various kinds of data having some unique characteristics such as high-dimensionality and heterogeneity. The need of automated data driven techniques has become a necessity to extract useful information from this huge and diverse data sets. Data mining is an important step in the process of knowledge discovery in databases (KDD) and focuses on discovering hidden information in data that go beyond simple analysis. Traditional data mining methods are often found inefficient and unsuitable in analyzing today's data sets due to their heterogeneity, massive size and high-dimensionality. So, the need of parallelization of traditional data mining algorithms has almost become inevitable but challenging considering available hardware and software solutions. The main objective of this paper is to look at the need and limitations of parallelization of data mining algorithms and finding ways to achieve the best. In this comparative study, we took a look at different parallel computer architectures, well proven parallelization methods, and programming language of choice.

Keywords Data mining · Parallelism · Parallel computing · Java

1 Introduction

Data mining is the process of analyzing raw data from very large databases to turn them into useful and previously unknown information which helps in finding out interesting patterns, trends and relationships within data. It is an important step in the process of knowledge discovery in databases (KDD) [1]. Thus, it enables businesses

K.C. Mondal (✉) · Sayan Bhattacharya
Department of Information Technology, Jadavpur University, Kolkata, India
e-mail: kartickjgec@gmail.com

Sayan Bhattacharya
e-mail: s4sayan.bhattacharya@gmail.com

Anindita Sarkar
School of Mobile Computing, Jadavpur University, Kolkata, India
e-mail: sarkar.anindita5@gmail.com

© Springer Science+Business Media Singapore 2017
J.K. Mandal et al. (eds.), *Proceedings of the First International Conference
on Intelligent Computing and Communication*, Advances in Intelligent Systems
and Computing 458, DOI 10.1007/978-981-10-2035-3_21

to improve quality of service, increase revenue and cut down the operational cost. The application of data mining is widespread ranging from scientific applications, analysis of financial data, retail and telecommunication industries to biological data analysis, commercial computing and many more.

Today's world has seen a massive explosion in various kinds of data such as high-dimensionality and heterogeneity [2]. So, data mining applications have to deal with huge data sets from different sources [3]. This can be relational databases, flat files or other sources and can be spread over multiple systems or platforms. Analyzing such huge amount of data involves some serious amount of computation. Even with the most sophisticate data mining algorithms and advanced hardware, finding patterns and relationships in such data becomes pretty time consuming at times. The need of parallelization of existing data mining algorithms and high performance parallel computing is therefore becoming an essential part of the solution [4, 5].

Consider the scenario of an important data mining task called association rules mining [3]. Here, it is essential to identify certain hidden patterns, trends and relationships within the data which can be used to predict likely outcomes. The input data sets for association rule discovery are often found to be enormous, diverse and having high dimensionality which affects the performance and efficiency of the mining process. Sometimes, it can take hours or even days to produce a complete set of result. However, often such processes are executed under specific performance requirements where it is rather expected to be interactive and should produce results within a fixed time span. To meet the performance requirements and speed up the execution time of such algorithms, some kind of parallelism becomes an almost inevitable solution [4]. In most of the cases, a data mining algorithm has to work upon existing transactional databases which are often found to be in the form of parallel database and are spread over multiple sites [6]. To perform the association rule discovery in a serial fashion, all of these have to be brought to a single site or computer which becomes practically impossible due to the huge expense involved in doing so. Thus, parallelism becomes the obvious choice in such scenarios.

Parallelism mostly helps in improving performance and enables to achieve scalability amongst other numerous benefits offered. Firstly, it allows to do same work in less time [4]. The trick is, a task always takes lesser time to execute if it is executed concurrently rather than in a serial fashion. With the advancement of technology, computer hardware has become more powerful and less expensive. Old single processor systems have been quickly replaced by their multi-processor counterparts which no longer suffers from memory and CPU speed limitations. Parallelism mainly offers the following advantages in a multi processor environment:

- The capabilities of a multi-processor system can be fully utilized [5].
- It offers the option to break a huge and complex problem into relatively simpler and smaller sub-problems, assigning each task to a separate processor.
- Separate results produced by concurrent execution of sub-problem which can be combined to get the final result which reduces the processing time and improves performance.

Besides these advantages, parallelism also comes with a set of limitations, specially in terms of its implementation challenges and complexity. Without proper implementation, parallelism may suffer from problems like improper communication, synchronization issues and performance overhead. The creation of new threads and processes involves memory allocations, initializations and management activities which are quite expensive. Without proper designing, these costs may be even higher. Also, communication between these task, threads and processes and maintaining synchronization also becomes a serious issue. If these are not handled properly the algorithm will fail to produce correct results [6]. It may sound very simple to break a task into smaller sub-problems, run each one concurrently and combine the result but in reality everything comes with a performance overhead. Breaking down tasks and combining results should be lossless considering the same input dataset to make sure it does not affect the overall functioning of the algorithm. Also, a poorly designed parallel algorithm might fail to utilize a multi processor system properly having issues like improper data distribution and task duplication [7].

However, parallelism is certainly not the only approach to achieve better performance. There are also a number of other approaches which can be used as alternate solution or in combination with parallelism. Some of these methods are code optimization, restricting the search space, sampling etc. In [4], presents a detailed analysis of these techniques and their potential benefit/limitations over parallelism.

This paper is organized as follows: Sect. 2 gives an overview of different parallel computer architectures available and their potential benefits and shortcomings along with different parallelism techniques and quick comparison between them. We are considering Java based parallelism techniques in this paper. In Sect. 3, we have discussed about the potential advantages of Java and compare it with other traditional programming languages like Fortran, C etc. with some existing state-of-the-art parallelization architectures and libraries. Finally, a conclusion has been drawn of the topic presented here.

2 Parallel Computer Architectures

The first and foremost requirement of parallelism is the availability of more than one processor in the system. A multiprocessor architecture is absolutely necessary for running parallel algorithms where the very fundamental is to divide the workload of processing among multiple processors thereby speeding up the processing and reducing execution time [5, 7]. Multiprocessor systems are the backbone of parallelism and are also known as parallel computer architectures. Parallelism is the technique to distribute the workload of a data mining algorithm over multiple processors. There exists different techniques to achieve that, which can be broadly classified in two major categories: Data Parallelism, Task or Control Parallelism. In brief, data parallelism refers to execution of the same set of operations/instructions on small subsets of a large data set at the same time [8, 9]. It focuses on distributing the data set across different nodes in a parallel computing environment. On the other hand, control parallelism refers to the concurrent execution of different operations/instructions on the same data set. It focuses on distributing the execution processes/tasks across different nodes in a parallel computing environment.

In this section we have discussed different parallel computer architectures and the basic properties of each along with their advantages and disadvantages. The parallelism technique to be implemented for a particular system greatly depends on the underlying hardware and system architecture. Broadly, there exists two major variant of multiprocessor systems known as tightly coupled and loosely coupled. A combination of both or hybrid system is also available as shown in Fig. 1. So, there are primarily three types of parallel computer systems available: Shared Memory Multiprocessor (SMP) System, Distributed System, Hybrid SMP Cluster System. In Table 1, a comparative analysis of advantages and disadvantages of these system have been shown.

A tightly coupled system consists of more than one processor sharing the common memory using a common bus which is also used to access the I/O system. The processors are capable of executing independently and may have their private cache memory. The shared memory can be used for message passing and communication between the processors [6, 10]. Such systems are particularly suitable for parallel execution where different tasks/processes can execute a different set of code called

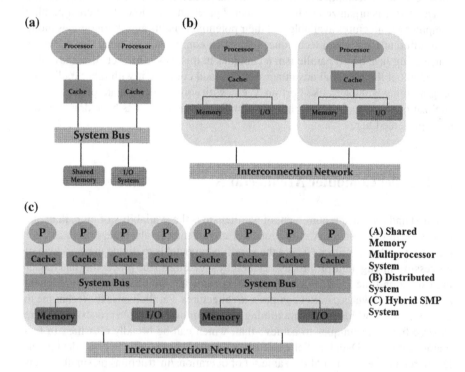

Fig. 1 Parallel computer architectures

Table 1 Advantages and disadvantages of parallel computer architectures

Architecture	Advantages	Disadvantages
Shared memory multiprocessor system	Easy to implement parallel programs, has direct and equal access to the system memory, does not need any extra mechanism to transfer data or information between processing units	High cost, less bandwidth available per processor that can limit scalability and can hamper the system performance, failure of one can lead to the failure of the overall system
Distributed system	Provides better scalability, much more robust and fault tolerant than their shared memory counterparts, system continues to work even if some computational nodes fail, relatively low cost than the tightly coupled ones [5]	Needs special care and attention in the application level to maintain communication between processing nodes [6], transferring information between different processors is difficult as no common shared memory is used here
Hybrid SMP cluster systems	Combines the best of both systems and is very much suitable for parallel program execution	Costly and difficult to setup

task or control parallelism. Figure 1a is an example architectural overview of shared memory multiprocessor system.

On the other hand, loosely coupled system consists of more than one autonomous computer. Each system in the network will have its own local memory and I/O system and they appear to the external user as a single system. The computers share system resources through the interconnected network. Such systems are also known as distributed systems [6]. Figure 1b is an example of distributed systems [10].

A combination of these above two is known as a hybrid system. Recently, it became very popular and is known as hybrid SMP cluster systems [6]. It is a more advanced variant and can be looked as a loosely coupled system made out of multiple shared memory multiprocessor systems coupled together tightly. The hybrid SMP systems are connected through a high speed network [10]. Figure 1c shows an architectural overview of hybrid SMP cluster systems.

3 Parallelism in Java

As discussed earlier, in todays world parallel computing has become an absolute necessity in domains which involve complex, extensive and time-consuming computations and data mining is no exception. Historically, the primary choice for implementation of parallel programming has been traditional programming languages like

Table 2 Features based comparison of well-known programming languages

Language	Object oriented	Functional	Procedural	Generic	Reflective	Portable	Security
Fortran	Yes		Yes	Yes		Limited	Limited
C			Yes				Limited
C++	Yes	Yes	Yes	Yes			Limited
JAVA	Yes	Yes	Yes	Yes	Yes	Yes	Secure

C and Fortran. Java was rather considered inefficient for computation-extensive tasks as Java programs run much slower than C or Fortran [11]. But, despite its performance, Java has emerged as the preferred language for parallelism, especially for its "Write Once Run Anywhere" facility and well rich feature set which are lacking in traditional programming languages like C, C++ and Fortran. A feature-wise comparison of these well known and widely used programming languages is shown in Table 2.

The main objective of this section is to present advantages of Java and some existing parallelization techniques which can give directions for parallel data mining algorithm development. We limit our discussions to only those libraries which are quite generic in nature and not tied to any specific algorithm. These approaches can be applied to parallelize any data mining algorithm to improve its performance. In cases, where these methods are not suitable, they can well be used as a baseline and can be extended to develop new parallelization engines of choice as per the demand of the target algorithm.

3.1 Advantages of Java

Complex data mining applications are often found to be deployed across different organizations. Thus, data mining applications involve a number of different platforms and traditional programming languages are often found in-sufficient to handle such heterogeneity. Java's portability feature provides a comprehensive solution to this problem [11]. The well-defined thread model of Java allows portable parallel programming across different platforms with better performance. Being an object-oriented language Java programs are modular and easily maintainable. Besides, Java provides a number of other additional advantages such as its ease of use, network-centric features, security, memory management etc.

Network programming in Java is a lot easier than other languages due to its vast network-centric features [12]. It is an ideal network programming language providing facilities such as TCP/IP based socket programming, object serialization and Remote Method Invocation (RMI) which easily streamed across a communication link and invoke methods on a remote object located in a different address space on the network as if the object exists locally.

Security considerations are very important from a parallel programming viewpoint. Different parties/organizations participating in a parallel computing environment requires their systems to be safe from theft of data and malicious attacks. Java has been designed with special care to its safety and security features. Several features have been included at the language level to implement security. As discussed in [11], some of these features include exclusion of pointers, ensuring type safety checking, providing access control for variables and methods, automatic garbage collection, inclusion of final classes and methods etc. Java uses a security model known as the sandbox model [11] which allows user to run even untrusted code safely.

3.2 Parallel Java Architectures and Libraries

In this section, we are going to present a comparative analysis of some of the recent parallel architectures and libraries available. These available architectures can be broadly classified in two major categories: Automatic Parallel Architecture and Semi-Automatic Parallel Architecture. An automatic parallelization framework [13] is generally capable of parallelize the input program without any need of modification to the implementation logic of the framework itself. Such implementations are very intensive in nature as they have to incorporate huge analyzing and processing logic to be able to understand and interpret the input program and decide on the approach to parallelize it accordingly. Development of such framework is quite complex as there is no further scope of run time alteration. However, it does not require an expert to run the framework as there is no need of any framework modification depending on input program.

A semi-automatic parallelization framework requires some kind of manual intervention or modification of code/logic of the implementation based on the type of input program to be parallelized. One advantage of such framework is that they are relatively easy to implement as the designer has to focus on the core logic of the framework as further alterations can be made during run time depending upon nature of input program. But the major disadvantage of this, it may suffer from problems like incorrect output and/or high running time if the input program is not properly analyzed and framework modifications are not properly made. A detail study of these architectures have been presented in the Table 3.

Table 3 Comparison of different parallel architecture and libraries

Cite	Type	Approach	Targeted for	Implementation
[14]	Automatic	Trace-based parallelization. Traces (hot paths of execution in a program) are used as units of parallel work	Data-parallel applications	Java based implementation using Jikes RVM
[15]	Automatic	Thread spawning. The system achieves parallelism by creating new asynchronous threads for each new method invocation	Java programs which uses pointer-based dynamic data structures and recursion	Java based
[16]	Automatic	Extract instruction signatures (read from/write to memory etc.) to infer data dependencies between instructions. Create a set of tasks based on the data dependencies and execute them in parallel on multiple cores of the processor using a work-stealing algorithm [17]	Sequential Java programs	Java based
[18]	Automatic	Collect trace information on-the-fly during program execution to recompile methods dynamically for parallel execution. Also uses important features of JVM like multi-thread execution, run-time sampling, on-demand recompilation etc.	Java programs	Java based implementation using Jikes RVM
[19]	Semi-Automatic	Operates through two major components—a wrapper generator and a parallelization engine using techniques based on distributed parallel method execution, code migration and bytecode transformation	Computation extensive Java programs with loosely-synchronous tasks and to an application without source code	Java based
[20]	Semi-Automatic	Using an API consisting of two Python decorators. One identifies functions that should be parallelized and the other marks functions which are free of side-effects. The whole system consists of 3 components translation, scheduling and distribution	Sequential Python programs on multi core, cloud or cluster systems	Python based

4 Conclusion

In this paper, we have discussed about the need of parallelism of data mining algorithms from both operational and performance point of view. However, despite a lot of advantages offered by parallelization, there are also a number of limitations and implementation implications it has which we have tried to point out in course of our analysis. A parallel computer architecture is the primary need for implementing parallelization in any system. Therefore, we tried to give an account of different parallel computer architectures available in todays world. It is very important to analyze a data mining application thoroughly in terms of multiple factors and parameters before determining the choice of parallelism for the application. This not only requires good understanding of the application itself but also require in depth knowledge of benefits and shortcomings of the parallelization methods. From the implementation point of view, a programming language is needed to develop parallel applications for the system. We have suggested Java as the language of choice and discussed about some exclusive features it offers in support for parallel programming. Finally, we have covered some state-of-the-art parallelization architectures and libraries available today which can be taken as guidelines for implementing Java based parallelization for data mining algorithms.

References

1. M. Stonebraker, R. Agrawal, U. Dayal, E.J. Neuhold and A. Reuterr *DBMS Research at a Crossroads: The Vienna Update.* 1993: In Proc. of the 19th VLDB Conference, Dublin, Ireland
2. M.S. Chen, J. Han and P.S. Yu *Data mining: An overview from database perspective.* December 1996: IEEE Transactions on Knowledge and Data Eng.
3. R. Agrawal, T. Imielinski and A. Swami *Mining association rules between sets of items in large databases.* 1993: In Proc. of 1993 ACM-SIGMOD Int. Conf. on Management of Data, Washington, D.C.
4. A.A. Freitas and S.H. Lavington *Mining Very Large Databases with Parallel Processing.* 1998: Kluwer Academic Publishers
5. M.J. Zaki *Parallel and Distributed Association Mining: A Survey* Rensselaer Polytechnic Institute
6. F. Stahl, M.M. Gaber and M. Bramer *Scaling up Data Mining Techniques to Large Datasets Using Parallel and Distributed Processing*
7. S. Paul *Parallel and Distributed Data Mining* Karunya University, Coimbatore, India
8. T.G. Lewis *Data parallel computing: an alternative for the 1990s.* September 1991: IEEE Computer, 24(9)
9. W.D. Hillis and L. Steele Jr. *Data parallel algorithms.* December 1986: Comm. ACM, 29(12)
10. A. Kaminsky *Parallel Programing in Java.* Presented at the CCSCNE 2007 Conference April 20, 2007
11. L. F. Lau, A. L. Ananda, G. Tan, W. F. Wong *JAVM: Internet-based Parallel Computing Using Java.* School of Computing, National University of Singapore
12. E.R. Harold *Java Network Programming.* 1997: O'Reilly and Associates
13. U. Banerjee, R. Eigenmann, A. Nicolau, D. Padua *Automatic program parallelization.* Proceedings of the IEEE 81(2), 211–243 (1993)

14. B.J. Bradel, T.S. Abdelrahman *Automatic Trace-Based Parallelization of Java Programs.* Edward S. Rogers Sr. Department of Electrical and Computer Engineering, University of Toronto, Toronto, Ontario, Canada M5S 3G4

15. B. Chan and T.S. Abdelrahman *Run-Time Support for the Automatic Parallelization of Java Programs.* Department of Electrical and Computer Engineering, University of Toronto, Toronto, Ontario, Canada M5S 3G4

16. J. Rafael, I. Correia, A. Fonseca, B. Cabral *Dependency-Based Automatic Parallelization of Java Applications.* University of Coimbra, Portugal

17. R.D. Blumofe, C.E. Leiserson *Scheduling Multithreaded Computations by Work Stealing.* J. ACM 46(5), 720–748 (1999)

18. Y. Sun, W. Zhang *On-line Trace Based Automatic Parallelization of Java Programs on Multi-core Platforms.* Department of ECE, Virginia Commonwealth University

19. P.A. Felber *Semi-Automatic Parallelization of Java Applications.* Institut EURECOM 06904 Sophia Antipolis, France

20. S.C. Mller, G. Alonso, A. Amara, A. Csillaghy *Pydron: Semi-Automatic Parallelization for Multi-Core and the Cloud.* Proceedings of the 11th USENIX Symposium on Operating Systems Design and Implementation, October 68, 2014, Broomfield, CO

Brief Review on Optimal Suffix Data Structures

Kartick Chandra Mondal, Ankur Paul and Anindita Sarkar

Abstract Suffix tree is a fundamental data structure in the area of combinatorial pattern matching. It has many elegant applications in almost all areas of data mining. This is an efficient data structure for finding solutions in these areas but occupying good amount of space is the major disadvantage of it. Optimizing this data structure has been an active area of research ever since this data structure has been introduced. Presenting major works on optimization of suffix tree is the matter of this article. Optimization in terms of space required to store the suffix tree or time complexity associated with the construction of the tree or performing operation like searching on the tree are major attraction for researcher over the years. In this article, we have presented different forms of this data structure and comparison between them have been studied. A comparative study on different algorithms of these data structures which turns out to be optimized versions of suffix tree in terms of space and time or both required to construct the tree or the time required to perform a search operation on the tree have been presented.

Keywords Suffix array · Suffix cactus · Suffix tree · Pat tree

1 Introduction

Many potential applications in language processing uses suffix tree as an extremely powerful data structure. This data structure is constructed for a given text in such a way, that every node in the tree correlates to a suffix of the text. It helps in pat-

K.C. Mondal (✉) · Ankur Paul
Department of Information Technology, Jadavpur University, Kolkata, India
e-mail: kartickjgec@gmail.com

Ankur Paul
e-mail: paulankur.143@gmail.com

Anindita Sarkar
School of Mobile Computing, Jadavpur University, Kolkata, India
e-mail: sarkar.anindita5@gmail.com

tern matching in a long string. By virtue of this capability, it plays a crucial role in some text compression techniques as it allows repeated sequence of characters to be represented explicitly. Suffix tree have been used in softwares to detect plagiarism [1]. This is done by concatenating the strings involved and looking for sufficiently long strings that occur in more than one of them. Suffix trees were first introduced by P. Wiener in 1973 in his paper [2]. Donald Knuth characterized this as "Algorithm of the year 1973". McCreight (1976) and Ukkonen (1995) [3] have simplified the construction of suffix tree. Thereafter, this has been an active area of research and there have been a number of different algorithms proposed to construct and maintain a suffix tree [4].

Suffix tree has profound application in the field of bioinformatics where there is a need for large strings of DNA or proteins analysis with no defined boundary. It is also used in the electronic storage and retrieval of information in large documents such as encyclopedia and other reference works which requires the use of searching systems that are efficient both in time and storage requirements [5]. Memory requirement and response time largely depends on the choice of data structure and algorithm used for processing the large documents. For a small document where processing time is not a matter to concern, by scanning the text input a query can be answered. However, for a very long text when multiple searches are expected, performing a pre-processing step to build a data structure for speeding up the search is worthy. Suffix tree is a solution to this.

A major bottleneck in the application of suffix tree is the large amount of memory needed. Optimizing the various aspects of suffix tree remains an open area of research. The optimization may be done to make the suffix tree more time optimal or space optimal or both. In this article, various optimization techniques have been described that may be applied on a suffix tree to make it more efficient. A comparative analysis of them have been represented here. This article would be a nice starting point for beginners wishing to explore more on suffix tree.

In the next section, some important terminology have been given to understand the optimal structure of suffix tree. After that, a comparative study on these suffix tree have been presented. At the end, we have concluded our study on optimization of suffix tree.

2 Terminology

2.1 Suffix Tree

Suffix Tree for a given collection of strings is defined to be the tree containing nodes, which represent the strings S arranged in such a manner that all the nodes from root to an intermediate node have same prefix for the string associated with that intermediate node [6–8]. Here, the root is associated with an empty string as shown in Fig. 1. Following are the characteristics of a suffix tree:

Fig. 1 Suffix tree for
"ABCBCB"

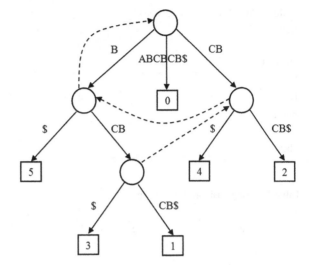

1. Except root, at least 2 children are associated with every internal node.
2. Any edge of the tree is a part of the non-empty substring of S.
3. If two edges are divided from an internal node, then the two string labels won't have the same starting character.
4. All the string labels can be found on different paths from root to leaf.

2.2 Suffix Link

In a complete suffix tree, all internal non-root nodes have a suffix link to another internal node. If the path from the root to a node spells the string "Sa", where "S" is a single character and "a" is a string (possibly empty), it has a suffix link to the internal node representing "a" [7]. The dotted lines in Fig. 1 represent the suffix links in the suffix tree.

2.3 Suffix Array

For a given string, sorted array of all suffixes are called a suffix array. This is a space efficient data structure as compared to that of suffix tree [4, 9]. Suffix array may also be represented by the elements which are starting positions of the suffixes sorted in lexicographical order (Table 1). Considering the string "abcbcb", a sentinel $ is added to this to represent the end of string. It may be represented in the form as shown in Table 2. All possible suffixes of the string "abcbcb" sorted lexicographically is

Table 1 Sorted suffixes for "abcbcb"

Suffix	i
$	7
abcbcb$	1
b$	6
bcb$	4
bcbcb$	2
cb$	5
cbcb$	3

Table 2 String positions for text "abcbcb"

i	1	2	3	4	5	6	7
S[i]	a	b	c	b	c	b	$

Table 3 Suffix array for "abcbcb"

i	1	2	3	4	5	6	7
A[i]	7	1	6	4	2	5	3

shown in Table 1. Each element of the suffix array may be represented as the starting position of each suffix for a given text. Table 3 shows the suffix array obtained for the text "abcbcb" where each element of the array being the starting position of the corresponding suffix.

2.4   LCP (Longest Common Prefix) Array

LCP array is a supplementary data structure to the suffix array. Size of the longest common prefixes of the sorted suffix array [5, 10] are stored in it. The complete suffix array with the suffixes is shown in Table 4. Successive suffixes are compared by lexicographically to construct the LCP array (L) for determining the longest common prefix. The array is shown in Table 5.

3   Optimized Suffix Structure

Several researches have been done to optimize suffix tree. These optimization techniques can be categorized as Space Optimization, Time Optimization & Hybrid or Mixed Optimization. Under this section, our aim is to give a brief overview of the

Table 4 Complete suffix array for "abcbcb"

i	1	2	3	4	5	6	7
A[i]	7	1	6	4	2	5	3
1	$	a	b	b	b	c	c
2		b	$	c	c	b	b
3		c		b	b	$	c
4		b		$	c		b
5		c			b		$
6		b			$		
7		$					

Table 5 LCP array for "abcbcb"

i	1	2	3	4	5	6	7
L[i]	0	0	1	1	3	2	2

optimized structures of each category and compare them in a common platform. In Table 6, we have summarized different optimal data structures, algorithms and their special features and applications that have been discussed so far.

3.1 Space Efficient Structure

One of the crucial factors that is needed to be taken care in application of suffix tree is the large amount of space required by suffix tree. Alternate data structures which are more economical in space have been discussed here. However, the operations and performance factors remain close or same as suffix tree.

Suffix Cactus

Suffix Cactus is an alternative data structure to suffix tree and suffix array. Size and search performance of suffix cactus lies between suffix tree and suffix array. It is a compact version of suffix tree or suffix array augmented with extra information. A suffix cactus is formed from a suffix tree by concatenating the internal node with any of its children. The concatenations are called the branches of suffix cactus [11]. The name 'cactus' comes from the way the branches start in the middle of other branches. Figure 2 shows the suffix cactus for the text, cabacca$.

In order to implement this structure, we require 2 arrays: SUFFIX and DEPTH. SUFFIX is the basic suffix array that we have discussed earlier. The DEPTH array

Table 6 Comparison on optimized suffix data structures

Algorithm or data structure	Special Features	Advantages	Application
Space efficient			
Suffix cactus [11]	Internal nodes of suffix tree are concatenated with one of its children	Space required is much less when compared to that of Suffix Tree but more than Suffix Array	Regular expression matching, string matching
LC Trie [12, 16]	Two levels of compression: Path Compression and Level Compression are applied on Suffix Tree to obtain LC Trie	Compact data structure which requires much less space when compared to that of Suffix Tree	Pattern matching from large documents such as books, encyclopedias and DNA sequences
Time efficient			
Stellar [13]	Association between nodes bind through suffix-link and tree-edges under them are used to traverse recursively	Significantly improved search performance on a suffix tree when the suffix tree is stored in disk	Used for performing search on real DNA sequences which is not possible to be stored in main memory
Balanced Indexing Structure [14]	Balanced Search Tree on the suffixes. Supports indexing when text is online	Constructs suffix tree in linear time	Real Time Indexing for applications where text changes dynamically
Space and Time efficient			
Las Vegas Type Suffix Tree [15]	Odd Tree of suffixes at odd positions is constructed first. Even Tree is constructed similarly. The trees may be merged using two ways: Sequential Approach and Parallel Approach	Work optimal parallel algorithm to construct suffix tree using linear space	Dictionary matching
Augmented Suffix Array [5]	Suffix Tree is pruned to contain only first suffix of each segment. LCP of each segment is stored in another array. Combined data structure forms Augmented Suffix Array	For very large texts, most part of the data structure can be stored in secondary memory without compromising search operation efficiency	Operation on large texts

Fig. 2 Suffix cactus for the text, cabacca

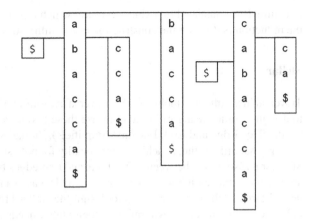

stores information about the branching depth of each branch. The same holds for the performance in many applications, such as regular expression matching.

LC Trie (Path and Level Compressed Suffix Tree)

LC Trie (Level Compressed Trie) [12] is a compressed version of a suffix tree. The LC-trie is very easy to implement. The compression works well to reduce the space requirement to such an extent that for n leaves tree, $\lceil loglog(n) \rceil$ bits are sufficient. Construction time for such level compressed tree is O(n log k) and space complexity is O(n), where text and alphabet sizes are n and k, respectively.

This trie has two variant: Path Compressed and Level Compressed.

1. **Path Compression**: It reduces the search cost of the binary tries. For branching from an internal node, indexing is used for indicating the character. This kind of tree is also known as **Patricia tree**.
2. **Level Compression**: It is used to reduce the size of the Patricia tree. If the ith highest level of the tree is complete against the (i + 1)th level, then the ith level is replaced by a single node with degree 2^i. This process iterates in top-down way to get a structure for adopting the distribution of the input. The data structure obtained is referred to as LC-trie or Level Compressed trie.

3.2 Time Efficient Structure

Optimization with respect to time complexity is an important factor to analyze as far as suffix trees are concerned. Making time economical suffix trees may be done by reducing the tree construction time or by reducing operation time over the tree. Stellar is a time economical algorithm where searching cost is reduced. Another data structure, BIS (Balanced Indexing Structure) which is also time economical but here the reduction is done on the construction of the suffix structure.

Stellar

Traversal in a suffix tree includes edge traversal and suffix link traversal [13]. The aim to design a complete algorithm to optimize these two traversals gives rise to Stellar (Suffix-Tree Edge and Link Locality AmplifieR). Stellar is a linear-time, top-down strategy that utilizes the, to achieve high locality for both suffix-links and tree-edges. Structural association between suffix-links and tree-edges below associated subtrees are used to achieve high locality among them. The association between nodes connected through these two are used to design the stellar [13]. If a node is inspected only once using the top-down traversal, then Stellar needs linear time complexity in the size of the suffix tree.

Balanced Indexing Structure

It is always been difficult to develop a real-time construction algorithm for suffix tree. Presented in this section is a different data structure, Balanced Indexing Structure (BIS), which needs O(logn) time to construct a suffix tree for every single input symbol [14]. Let, $S_1, \ldots, S_n$ are the suffixes of S length n. Basic elements for each suffix S_i, BIS has n nodes. Constructing the BIS is done through a recursive algorithm. The cost of addition of text in worst case is O(log n), where size of the text to be inserted is n. Three data structures are incorporated in the BIS nodes are mentioned as follows:

1. A lexicographically ordered list of suffixes.
2. A lexicographic node ordering in a balanced search tree on suffixes.
3. Suffixes using a textual ordering.

3.3   Space and Time Efficient Structures

This section discusses different structures which make suffix trees both space and time economical.

Las Vegas Type Suffix Tree

A simplified suffix tree of Las Vegas Type construction algorithm which yields an O(log n) time, O(n) space and O(n) work complexity described in [15]. There are two approaches to this method: sequential approach and parallel approach. The depth of all refinement nodes in the trees is determined by DFS to deduce their lengths. Complete tree construction procedure include coupled–DFS merging along with final DFS refinement procedure which needs O(n) time. Whereas, in the parallel approach, the linear ordering of the suffixes makes its complexity to O(log n) time.

Augmented Suffix Array

A very efficient structure for string processing with respect to execution time is suffix tree whereas suffix array is memory efficient but requires more execution time. It is observed that suffix tree data structure spent more time on visiting nodes near to root but memory is required more for the nodes near to leaf. When suffix array is augmented by pruned suffix tree, it is known as augmented suffix array [5]. Augmented suffix array requires $O(|w| + \log(m) + k)$ time for pattern search of length w in text t with the array divided in segments of length m. It requires about $n\log(n) + 4n\log(n)/m + O(n)$ memory bits for storing the tree.

4 Conclusion

Various optimization techniques of suffix tree, with respect to space, time as well as both space and time have been discussed. The key observation is that optimizing one of the factors, say space efficiency may lead to giving away the efficiency of the time in exchange. Moreover, suffix trees need to be present in main memory in order to be highly efficient. However, for large texts like dictionary or DNA strings, it is not possible to accommodate such suffix structures in main memory. As a solution, alternate data structures and algorithms belonging to suffix family have been introduced which make it possible to perform the necessary operations like search and insert in linear time when the data is disk resident.

Fields where search time is not that important as space required to store the suffix tree, the time efficiency may be given away to some extent and the structure made more space efficient. This is specially used in applications developed for hand held devices where memory is limited. Thus it is the need of the application based on which the optimization factor is applied. However, it is still remains an open area of research to get a suffix tree which takes linear space and operable in linear time and is resident of main memory. Applying the optimization techniques described in this article to any suffix tree based algorithm to make it further efficient is an important future scope in this field.

References

1. M. Kay, *Suffix Trees*, Stanford University, 2004.
2. P. Weiner, *Linear Pattern Matching Algorithms*, SWAT 73 Proceedings of the 14th Annual Symposium on Switching and Automata Theory, Pages: 1–11, 1973.
3. R. Giegerich and S. Kurtz, *From Ukkonen to McCreight and Weiner: A Unifying View of Linear-Time Suffix Tree Construction*, Algorithmica 19, Pages: 331–353, 1997.
4. L. M. S. Russo, G. Navarro and A. L. Oliveira, *Fully-Compressed Suffix Trees*, LATIN, Vol. 4957, LNCS, Pages: 362–373, 2008.
5. L. Colussi and A. De Col, *A time and space efficient data structure for string searching on large texts*, Inf. Process. Lett. 58(5), Pages: 217–222, 1996.
6. M. Farach, *Optimal Suffix Tree Construction with Large Alphabets*, FOCS, IEEE Computer Society, Pages: 137–143, 1997.
7. S. Kurtz, *Reducing the Space Requirement of Suffix Trees*, Softw. Pract. Exper. 29(13), Pages: 1149–1171, 1999.
8. R. Kolpakov, G. Kucherov, T. Starikovskaya, *Pattern Matching on Sparse Suffix Trees*, in Data Compression, Communications and Processing (CCP), 2011 First International Conference on, pp. 92–97, 2011.
9. L. Wang, K. Huang, J. Zhang, J. Yao, *A Complete Suffix Array-Based String Match Search Algorithm of Sliding Windows*, in 2012 Fifth International Symposium on Computational Intelligence and Design (ISCID), vol. 2, pp. 210–213, 2012.
10. J. Shun, *Fast Parallel Computation of Longest Common Prefixes*, in SC14: International Conference for High Performance Computing, Networking, Storage and Analysis, pp. 387–398, 2014.
11. J. Karkkainen, *Suffix Cactus: A cross between Suffix Tree and Suffix Array*, CPM, Pages: 191–204, 1995.
12. A. Andersson and S. Nilsson, *Efficient Implementation of Suffix Trees*, Softw. Pract. Exper. 25(2), Pages: 129–141, 1995.
13. S. J. Bedathur and J. R. Haritsa, *Search-Optimized Suffix-Tree Storage for Biological Applications*, HiPC, Vol. 3769, LNCS, Springer, Pages: 29–39, 2005.
14. A. Amir, T. Kopelowitz, M. Lewenstein, N. Lewenstein, *Towards Real-Time Suffix Tree Construction*, SPIRE, Vol. 3772, LNCS, Pages: 67–78, 2005.
15. M. Farach and S. Muthukrishnan, *Optimal Logarithmic Time Randomized Suffix Tree Construction*, ICALP, Vol. 1099, LNCS, Pages: 1–18, 1995.
16. T. Yang, Z. Mi, R. Duan, X. Guo, J. Lu, S. Zhang, X. Sun, B. Liu, *An ultra-fast universal incremental update algorithm for trie-based routing lookup*, in 2012 20th IEEE International Conference on Network Protocols (ICNP), pp. 1–10, 2012.

Demalvertising: A Kernel Approach for Detecting Malwares in Advertising Networks

Prabaharan Poornachandran, N. Balagopal, Soumajit Pal,
Aravind Ashok, Prem Sankar and Manu R. Krishnan

Abstract From search engines to e-commerce websites and online video channels to smartphone applications, most of the internet applications use advertising as one of their primary source of revenue generation. Malvertising is the act of distributing malicious software to users via advertisements on websites. The major causes of malvertisement are the presence of hundreds of third party advertising solutions and the improper verification of ads at the publisher's site. Moreover, smartly tailored advertisements are placed which exploit a browser's bugs and vulnerabilities to infect user with malicious software. In this paper, we highlight loopholes in the currently applied advertising policies and the vulnerabilities that are exploited to attack customers by serving malicious ads on user applications. The major contribution of the authors is a framework developed to identify malicious advertisements at the publishers' end. It is based on two types of analyses. The first type of analysis involves static analysis of the advertisement's source code. The other type is the behavioral analysis of the advertisements done in a secure sandboxed environment to detect any malicious activity. We extracted a total of 9 features from

Prabaharan Poornachandran (✉) · N. Balagopal · Soumajit Pal · Aravind Ashok · Prem
Sankar · M.R. Krishnan
Amrita Center for Cyber Security Systems and Networks, Amrita Vishwa
Vidyapeetham, Amritapuri Campus, Kollam, India
e-mail: praba@amrita.edu

N. Balagopal
e-mail: balagopaln89@gmail.com

Soumajit Pal
e-mail: soumajit@am.amrita.edu

Aravind Ashok
e-mail: aravindashok@am.amrita.edu

Prem Sankar
e-mail: premsankar@am.amrita.edu

M.R. Krishnan
e-mail: manurk@am.amrita.edu

© Springer Science+Business Media Singapore 2017
J.K. Mandal et al. (eds.), *Proceedings of the First International Conference
on Intelligent Computing and Communication*, Advances in Intelligent Systems
and Computing 458, DOI 10.1007/978-981-10-2035-3_23

15,000 advertisements and classified it using a trained one class SVM classifier. Our result shows that 53 % of the suspicious ads contain dubious iFrames while 69 % of them perform redirections followed by drive by download 18 % with very low false positive and false negative rates.

Keywords Malvertising · Malware analysis · Web advertising · SVM classifier · Static analysis · Dynamic analysis

1 Introduction

Web advertising has become a billion dollar business, supported by large multilayer ad network infrastructure. These web advertisements have become a major source of revenue for a site which publishes advertisements. Even though web advertising has multiple advantages, it has its own disadvantages too. Majority of the ads are distributed as third party content to different publisher websites. So, for the publisher to control over the website content has become limited. Many of the advertising networks cache the browsing behavior of users to serve targeted ads, which is a threat to one's privacy. As far as cyber security is concerned, the negative side of web advertising is that malvertising can serve as a very effective tool in hands of cybercriminals to steal information or cause potential damage to a user. Since advertising on mobile applications has become a widespread way of getting customers, the new target of adversaries have been through this medium. Social networking sites such as Facebook and Twitter have been misused as a platform for spreading malware [1]. Compared to other malware delivery mechanisms, a small fraction of malicious ads can cause major damage [2]. Once the damage is done, the advertiser can easily remove the malicious ad from the compromised network without any trace.

Dasient research states that 97 % of the Fortune 500 websites are at a high risk of getting infected with malware due to external partners [3]. On 3rd January 2014, malvertising attacks were carried out by Yahoo ads. Malicious ads were served by ads.yahoo.com [4]. The malicious iFrames in the ads redirected the users to some infected files that were hosted on third party servers. As a result even without clicking on an infected ad the visitors to the malicious ad were redirected to an exploit kit. Another well-known incident is the malvertising incidence that happened to the New York Times. A fake anti-virus scanner was found in their homepage that attempted the website users to install the rogue anti-virus scanner [5]. In 2010, the research released by Dasient indicated that 1.3 million malicious ads are viewed per day that accounts to 59 % for Drive-by download and fake security software contributed to 41 % [3]. According to Cisco, advertisements are 182 times more likely to serve malware than a smut website [6]. Symantec reports that some massive malvertisements are leading to ransomwares attacks [7].

Detection of these malicious ads and identification of its corresponding attack types are critical for a secure browsing experience.

2 Online Advertisement: Threats

With the help of large multi-layer ad network infrastructure, display ads are delivered to the user through advertisers, publishers and advertising network. In this online advertising scenario, advertisers are the creators of advertisement. Publishers on behalf of advertisers publish the ads on their web pages. The role of advertising network is to bring advertisers and publishers together. Audiences are the users who visit publisher pages and receive ad contents.

Ad-syndication Model

This business model increases the chance of posting malicious content on a big publisher's web site. Well-known trusted ad network domains may outsource to smaller, newer and perhaps not-so-trusted ad domains. Overall 75 % of the landing sides that serves malicious ads use multiple level of Ad Syndication [2]. Instead of compromising a popular website by exploiting the vulnerabilities in their underlying software, attackers find it easier to attack their ad network whose security practices may not be on the same level [8] (Fig. 1).

Dynamic Delivery of Ads

Dynamic delivery of ads is one of the features that are misused by cyber criminals. HTML code snippets are provided that are used in conjunction with normal websites so that it can be used to embed advertisements. Based on user or content characteristics, content of advertisements can change dynamically. The issue with this is the integrity of content is not easy to determine that is shared among different domains.

Flash Ads

Malvertisements can take the form of flash programs. Attackers take advantage of flash program's ability to embed the business logic directly to the ad that can be

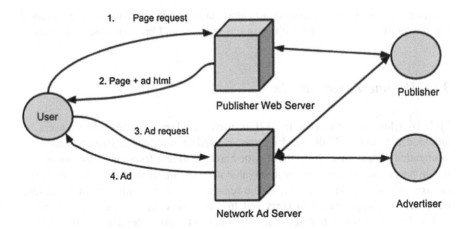

Fig. 1 Online advertising network

even used to evade detection [9]. The malware sometimes looks like ordinary flash file, but it may contain encrypted redirect function which circumvent the publisher while uploading the ad. Once the ad is published in a website, the flash file can launch the malicious redirect based on the business logic embedded in it.

QR Codes
QR codes have become very popular in consumer advertising. The QR code and its content cannot be deciphered by the human eye [10]. An infected QR code may contain a link to a malicious website that can infect the device with a Trojan.

Loopholes in Ad-policies
There are many reasons for malvertising to take place at the publishers' site. One major reason is the loopholes in the ad policies. Initially an attacker (advertiser) posts benign ads and once the trust is established between the advertiser and the publisher, the advertiser, in this case the attackers, start posting malicious ads right under the nose of the publisher. Another challenge for large publishers is that, at any given time, they run large number of ad tags. Monitoring all the ads being served is a great challenge.

Pop-up Ads Malvertisements can also take the form of random pop ups while browsing a page saying that you are a lucky visitor and you won an expensive gift.

Reputation Based Systems
Publishers usually rely on reputation based systems. Criminal organizations convince the publishers by providing fake documentation and the advertisers can create fake ad agencies to make the publishers believe them as legitimate ones. So reputation based systems is not an ideal solution for preventing malvertisement [11].

3   Attack Methodologies

Attackers use a variety of methods to carry out malvertising. Most of the methods use the loopholes and vulnerabilities in the browser and browser based languages.

3.1   IFrame Based Attacks

Link Hijacking as discussed in [12] is used in advertising to redirect users to unintended websites. Malformed iFrames is used to automatically redirect users to unintended pages. This makes the iFrame barely visible. Moreover, if the width and height properties are set to a negative value it makes the iFrame invisible. Attacker carryout iFrame based attack due to the reason that the web advertisement servers need not be compromised to carry out this type of attack. Attackers can easily embed hidden iFrames that serve malvertisements while interacting with a legitimate user.

The recent incidents of malvertising attacks via Yahoo ads [8] and the video sharing websites YouTube and DailyMotion shows the use of highly effective iFrame based web attacks on larger online communities. In the DailyMotion malvertising incident, an invisible iFrame has redirected the websites' users to the exploit serving a fake antivirus malware [13].

3.2   Malicious JavaScript Codes

eval(): The JavaScript *eval* function language construct is very dangerous and it's evil is discussed in these works [14–16]. In some scenarios, ad scripts use a series of string concatenations to construct URLs or JavaScript functions and it is passed to *eval*. The purpose of doing this is to make the origin of ad hard to trace and the attackers can dynamically choose ads for display.

Whitespace randomization: Obfuscation and code packing techniques are used by attackers to evade detection of the malicious JavaScript ads [17]. White space randomization is used as a JavaScript obfuscation technique. Attacker can scatter whitespace characters throughout their code, taking advantage of the fact that JavaScript ignores whitespaces. Even though it changes the static patterns, it will not change the semantics of the JavaScript and is helpful for the attacker to carry out attack [8]. Most of the security technologies use content matching to detect obfuscation not white space randomization. Hence, signature based detection techniques for obfuscation doesn't fare well.

Presence of Shell code: Shell codes obfuscation techniques using *unescape* has been used to distribute malware. In the work done by [18] describes the use of *unescape* function to allow shell code in JavaScript based malwares. Shell codes embedded inside JavaScript ads which are used for the manipulation of URL encoded strings. Presence of escaped characters and *unescape* functions are also discussed obfuscation mechanisms [19]. Hence presence of such functions and shell code in an advertisement is classified as suspicious in our system. Attackers also use string manipulation techniques, like manipulating comments in JavaScript codes [8]. This technique can make the analysis hard for a researcher and the effectives of the malicious ad can be maintained for a longer period of time.

The later sections present the methodology, design and implementation work of our system to detect malvertisements.

4   Our Methodology

In our proposed malvertisement detection framework for the publishers, we perform two types of analysis as shown in Fig. 2.

Static Analysis: The system analyses the advertisement code and looks for the attack methodologies mentioned in the above section. The presence of hidden

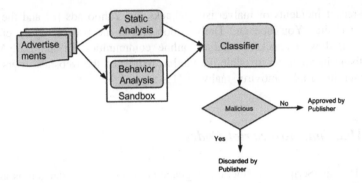

Fig. 2 High level system flow diagram

iFrames, *eval* functions etc. are taken as features and these feature sets is sent it to a supervised classifier.

Behavior analysis: In the behavior analysis section the dynamic properties of the advertisements are checked in a sandboxed environment. The advertisement is opened in browser environment and the dynamic properties of the OS like registry changes, processes spawned etc. and the network is monitored. These properties are considered as the second set of features and sent to the supervised classifier.

5 Design and Implementation

In the static analysis module, the system will analyze the source code of the advertisement. Within the advertisement it performs the following analysis:

a. **Bad iFrame Detection**: In our detection model, once an iFrame is found, it is classified based on its size. If the size of iFrame is small, null or negative then the system will classify that iFrame as suspicious. For example as shown below, if the iFrame has width = 1 and height = 1, definitely it cannot serve the purpose of an advertisement, Instead it can be used for carrying out attacks. Small iFrame can make the exploit invisible to naked eye.

```
<iframe width="1" height="1" frameborder="0" scrolling="no" marginheight="0"
marginwidth="0" src="http://bad-network.com/getbadfile.php">
</iframe>
```

For large iFrames, the presence of object tag is checked. Since attackers use Object tags to embed malicious scripts, the presence of object tag is classified as suspicious.

b. **Unescape Function Detection**: The static analysis also involves checking for the presence of *unescape* function. The presence of JavaScript *unescape* function with a large amount of escaped data could potentially indicate the attempt to inject a large amount of shell code or malicious JavaScript. This could help the attacker to take control of the system through browser vulnerabilities.

c. ***Shell Code Detection***: The system also checks for the presence of any VB script or shell code that can be used as an anti-detection mechanism used by attackers. The other analysis in the static module includes ***whitespace randomization*** checking and presence of ***eval*** function.

d. ***Dynamic analysis***: The dynamic analysis checks the system behaviour by running the advertisements i.e. by visiting the page with the ads using multiple web browsers in a sandboxed environment. Cuckoo is an open source malware analysis framework. We use the cuckoo sandbox to visit the pages using Internet Explorer, Google Chrome and Mozilla Firefox web browsers. We use these as they are some of the popular web browsers. Also, choosing different web browsers and OS for inspecting one page increases the chances of detecting the malicious advertisements. The system records network traffic trace, registry changes, memory dumps of malware process, files being created, deleted and downloaded by the advertisement during its execution in the sandboxed OS snapshot running the advertisement. These records are used as features and passed to the one class SVM classifier. The idea of the behaviour analysis is that an advertisement is not supposed to make file changes or spawn processes or even make suspicious redirects. Hence any such events are taken as a suspicious feature.

Classification Algorithm

A feature vector F is made based on these static and dynamic analyses which are sent to a supervised classifier S. Each element f_i of the feature set represents a Boolean value, which indicates the presence or absence of an attack method found out in the static analysis module. All the f_i are fed as inputs to a trained One Class SVM classifier S [20] where it is categorizes the advertisements to a class C which is either malicious or normal.

One class classifiers have been extensively used for outlier detection [21], novelty detection [22] and concept learning [23, 24]. In one class SVM classifier, we train a model based on the data from a particular known class. The objective is to define a boundary around the known class using support vectors such that any data outside the boundary will be considered as outliers or the not-known class.

6   Evaluation of the System

We crawled more than 15,000 ad impressions from Alexa's top websites for analysis. Out of that we manually verified 14,627 to be normal legitimate ads and the remaining 373 as malicious. We trained our one class classifier with the 9 features extracted from these 14,627 legitimate ads.

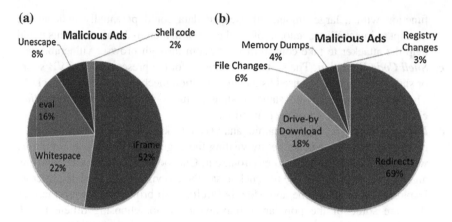

Fig. 3 Pie chart of malicious ads based on **a** Static analysis features and, **b** dynamic analysis features

On cross-validation for testing the accuracy of the system, our model detected 362 of the 373 ads as malicious with 3 false positives and 1 false negative. Out of the 362 ads, 193 ads contained suspicious iFrames; 78 ads had the whitespace randomization in the code greater than the threshold; in 57 advertisement's code, the system was able to detect the presence of potentially dangerous *eval* function; 28 ads had *unescape* function with a large amount of escaped data and the remaining 6 ads had shell codes embedded inside their codes. The statistical results is shown in the Fig. 3 given below.

7 Conclusion

Legitimate use of online advertising is necessary for the web economy as it enables each party to profit from the system. Since when the cyber criminals started using online advertisement for carrying out malicious activities, malvertisement attacks are unfortunately turning into a more serious problem for online advertisement business model. As malvertising targets the known vulnerabilities of operating system and browser, the softwares in the system should be up to date. Compared to other malware delivery mechanisms, a small malicious ad campaign can cause major damage. We have come up with a system that is a combination of static and dynamic analysis that could effectively detect malvertisements with very low false positives and false negatives. We hope our system helps in mitigating the malwares spread across the internet via advertisements and provide safe and secure browsing environment for the users.

References

1. Yan, Guanhua, et al. "Malware propagation in online social networks: nature, dynamics, and defense implications." Proceedings of the 6th ACM Symposium on Information, Computer and Communications Security. ACM, 2011.
2. Mavrommatis, Niels Provos Panayiotis, and Moheeb Abu Rajab Fabian Monrose. "All your iframes point to us." (2008).
3. Ariana (2010). Q1'10 web-based malware data and trends. [ONLINE] Available at: http://blog.dasient.com/2010/05/q110-web-based-malware-data-and-trends.html. [Last Accessed 12 February 2014].
4. Larry Seltzer (2014). Yahoo serves malicious ads. [Online] Available at: http://www.zdnet.com/yahoo-serves-malicious-ads-7000024775/ [Last Accessed 15 May 2014].
5. Aimee Picchi (2009). Malvertising hits The New York Times. [Online] Available at: http://www.dailyfinance.com/2009/09/14/malvertising-hits-the-new-york-times/ [Last Accessed 15 May 2014].
6. Cisco Annual Security Report: Threats Step Out of the Shadows.
7. Symantec Corporation Internet Security Threat Report 2013: Volume 18.
8. Feinstein, Ben, Daniel Peck, and I. SecureWorks. "Caffeine monkey: Automated collection, detection and analysis of malicious javascript." Black Hat USA 2007 (2007).
9. Ford, Sean, et al. "Analyzing and detecting malicious flash advertisements." Computer Security Applications Conference, 2009. ACSAC'09. Annual. IEEE, 2009.
10. Maria Tucker, Cracking Mobile: A Guide to Mobile Malvertising.
11. Seifert, Christian, Ian Welch, and Peter Komisarczuk. "Identification of malicious web pages with static heuristics." Telecommunication Networks and Applications Conference, 2008. ATNAC 2008. Australasian. IEEE, 2008.
12. Zarras, Apostolis, et al. "The dark alleys of madison avenue: Understanding malicious advertisements." Proceedings of the 2014 Conference on Internet Measurement Conference. ACM, 2014.
13. Michael Mimoso (2014). Malicious ad on dailymotion redirect to fake AV attack. [ONLINE] Available at: http://threatpost.com/malicious-ads-on-dailymotion-redirect-to-fake-av-attack/103494. [Last Accessed 1 February 2014].
14. Richards, Gregor, et al. "The eval that men do." ECOOP 2011–Object-Oriented Programming. Springer Berlin Heidelberg, 2011. 52–78.
15. Cova, Marco, Christopher Kruegel, and Giovanni Vigna. "Detection and analysis of drive-by-download attacks and malicious JavaScript code." Proceedings of the 19th international conference on World wide web. ACM, 2010.
16. Chellapilla, Kumar, and Alexey Maykov. "A taxonomy of JavaScript redirection spam." Proceedings of the 3rd international workshop on Adversarial information retrieval on the web. ACM, 2007.
17. Li, Zhou, et al. "Knowing your enemy: understanding and detecting malicious web advertising." Proceedings of the 2012 ACM conference on Computer and Communications Security. ACM, 2012.
18. Curtsinger, Charlie, et al. "ZOZZLE: Fast and Precise In-Browser JavaScript Malware Detection." USENIX Security Symposium. 2011.
19. Seifert, Christian, Ian Welch, and Peter Komisarczuk. "Identification of malicious web pages with static heuristics." Telecommunication Networks and Applications Conference, 2008. ATNAC 2008. Australasian. IEEE, 2008.
20. Heller, Katherine, et al. "One class support vector machines for detecting anomalous windows registry accesses." Workshop on Data Mining for Computer Security (DMSEC), Melbourne, FL, November 19, 2003. 2003.
21. Gunter Ritter, María Teresa Gallegos: Outliers in statistical pattern recognition and an application to automatic chromosome classification, Pattern Recognition Letters 18(6): 525–539 (1997).

22. C. Bishop, "Novelty detection and neural network validation", Proc. IEE Conference on Vision and Image Signal Processing, pp. 217–222, 1994.
23. N. Japkowicz, C. Myers and M. Gluck, "A novelty detection approach to classification", Proc. of 14th IJCAI Conference, Montreal, pp. 518–523, 1995.
24. Soman, K. P., R. Loganathan, and V. Ajay. machine learning with SVM and other kernel methods. PHI Learning Pvt. Ltd., 2009.

Closure Based Integrated Approach for Associative Classifier

Soumyadeep Basu Chowdhury, Debasmita Pal, Anindita Sarkar and Kartick Chandra Mondal

Abstract Building a classifier using association rules for classification task is a supervised data mining technique called Associative Classification (AC). Experiments show that AC has higher degree of classification accuracy than traditional approaches. The learning methodology used in most of the AC algorithms is apriori based. Thus, these algorithms inherit some of the Apriori's deficiencies like multiple scans of dataset and accumulative increase of number of rules. Closed itemset based approach is a solution to the above mentioned drawbacks. Here, we proposed a closed itemset based associative classifier (ACFIST) to generate the class association rules (CARs) along with biclusters. In this paper, we have also focused on generating lossless and condensed set of rules as it is based on closed concept. Experiments done on benchmark datasets to show the amount of result it is generating.

Keywords Associative classifier · Class association rules · Closed itemset · Data mining

S.B. Chowdhury · Debasmita Pal · K.C. Mondal (✉)
Department of Information Technology, Jadavpur University, Kolkata, India
e-mail: kartickjgec@gmail.com

S.B. Chowdhury
e-mail: SoumyadeepBasuChowdhury@gmail.com

Debasmita Pal
e-mail: debasmita.p12@gmail.com

Anindita Sarkar
School of Mobile Computing, Jadavpur University, Kolkata, India
e-mail: sarkar.anindita5@gmail.com

© Springer Science+Business Media Singapore 2017
J.K. Mandal et al. (eds.), *Proceedings of the First International Conference on Intelligent Computing and Communication*, Advances in Intelligent Systems and Computing 458, DOI 10.1007/978-981-10-2035-3_24

225

1 Introduction

Knowledge extraction from databases has been gaining popularity in many of the major applications like marketing, manufacturing, fraud detections and telecommunications. This process is aimed to discover useful, hidden information from the huge amount of transactional raw data which includes data preparation and selection, data cleansing, data transformation, data mining and interpretation. The most important step in the process of KDD is data mining which extract meaningful patterns from cleaned and transformed data. Association Rule Mining (ARM) and Classification are the major data mining tasks which gain popularity in recent times. Recently, researchers have become interested in performing *Associative Classification (AC)* by merging the process of ARM and Classification together. The result of which is Class Association Rules (CARs) set.

Several research studies [1–5] show that AC is superior than other traditional classification approaches such as [6, 7]. The primary advantage of AC is that it produces very simple knowledges with "If-Then" rule that can easily be understood and interpreted. Moreover, using this approach additional useful hidden information are able to find which traditional methods are unable to detect. This new data insights help to upgrade the predictive accuracy of the classifier according to various experimental studies as shown in [5, 8–11]. The learning methodology used by most CAR algorithms test the correlations between attribute and class value of the training data set. It may result in redundant rules which may cause an exponential growth of rules when no suitable pruning method is invoked as shown in [12, 13]. This problem usually happens for very small minimum support threshold value or highly correlated input data.

Here, we propose our ACFIST (Associative Classifier using Frequent generalized Itemset Suffix Tree) algorithm that utilizes a close lattice based approach used in ARM. The concept of closed itemset in ARM minimizes the generation of candidate itemsets and consequently the overhead caused by the exhaustive search techniques like Apriori. Majority of the current CAR algorithms extract the highest frequency class connected with the itemset of the training data and ignore all other class labels. Nevertheless, there are many applications like online shopping cart, which may require generation of rules with multiple labels, giving the decision makers more alternatives to select from. Here, we have focused to extract the complete set of possible classes for each rule so that the end-user can use them in their business activities. We have also focused on generating lossless and condensed set of rules as it is based on closed concept. Experiments were conducted on six benchmark datasets to show the amount of result it is generating and time it needs to complete the process.

In the next section, we have explained our proposed approach in great detail. During the explanation, we have used an example to evaluate the algorithm simultaneously. In Sect. 3, we have presented the experimental results generated from our experiments. Conclusion and future scope of the work have been presented in the last section of this paper.

2 Methodology

2.1 Problem Statement

A special case of association rule mining known as associative classifier involves only class attribute as consequent of the rules. Such as, a rule X→Y, where X can be any set of disjoint attribute value pairs and Y is only a class attribute. The formal definition of the problem statement of associative classification is as: The training data set D has n distinct attributes $X_1, X_2, ..., X_n$ and Y is a list of classes where an attribute can be categorical or continuous and |D| is the number of rows. Any discretization method can be applied for continuous attributes. Our goal is to construct a classifier from set of rules generated D, as R: $X \to Y_1$, where X is a set of disjoint attribute value pairs and Y_1 is a class which belongs to Y. Here, the main task is to generate a set of rules for predicting the classes of previously unknown data as accurately as possible.

2.2 Proposed Approach

Our proposed algorithm has been termed as Associative Classifier using Frequent generalized Itemset Suffix Tree (ACFIST). Detail analysis of our proposed approach has been discussed in this section. ACFIST is divided into three different phases and several sub-phases under it. These phases of ACFIST algorithm is depicted in Fig. 1. Phase 1 looks for associations between attributes and class in the input data set. Phase 2 is used to generate the frequent closed itemsets (FCIs) from the association between attribute-value generated in the previous phase. These we can generate biclusters from these FCIs using the object id list present in the compact data structure. These phase equivalent to another algorithm FIST proposed in [14]. This is the reason we call our approach an integrated approach. After finding the complete set FCIs, third phase is used for rule sorting based on thresholds such as support, confidence. Output from phase 3 is the Class Association Rules for representing the classifier which is the ultimately goal of proposed approach. The general algorithmic flow of ACFIST is shown in Algorithm 1. The inputs to this algorithm are Source datasets named inputDB.csv, Class Attribute name, minimum support value and minimum confidence value.

In the following subsections, we have explained the three phases of this algorithm along with result execution on the example database shown in Fig. 2. The minimum support and minimum confidence taken for this example execution are 80 % and 80 %, respectively (Fig. 3).

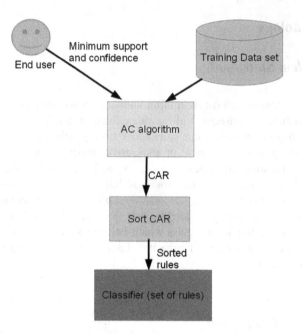

Fig. 1 Phases of associative classifier

TID	A	C	D	T	W	Class Label
1	1	1	?	1	1	Y
2	?	1	1	?	1	Y
3	1	1	?	1	1	Y
4	1	1	1	?	1	N
5	1	1	1	1	1	N
6	?	1	1	1	?	N

inputDB.csv

Fig. 2 Transaction database D1

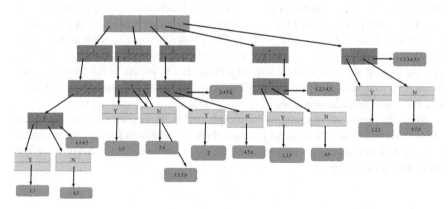

Fig. 3 FGIST after updating with all CARs

Algorithm 1 Associative Classifier Version Of FIST: General algorithm

---

Input: inputDB.csv, Class Attribute, *minsupport* Value, *minconfidence* Value
Output: Class Association Rules
  Phase 1: Parsing the input data set
  Phase 2: Mining frequent closed itemsets
    Phase 2.1: Create frequent Generalized Itemset Suffix-Tree (FGIST)
    Phase 2.2: Find frequent closed itemsets
  Phase 3: Generating Class Association Rules
    Phase 3.1: Update FGIST such that each path from root to leaf represents a CAR
    Phase 3.2: Generate CAR

Phase 1: Parsing the Input Dataset

In this phase, the source dataset e.g., *inputDB.csv* is taken as input and generates *inputDBFIST.csv* and *inputDBClass.csv* as output. Where inputDBClass.csv contains all class attributes along with their object lists and inputDBFIST.csv contains the entire source dataset except the class attribute. Basically, it divides the input dataset into these two parts. inputDBFIST.csv will later go as input to the phase 2 to generate the FCIs and inputDBClass.csv will be used in phase 3 to generate CARs. Algorithm 3 shows the function, CSVParse(), for parsing input dataset. The output of this phase to the example database is given in Fig. 4.

Phase 2: Creating Frequent Generalized Itemset Suffix-Tree

After the first phase, inputDBFIST.csv goes as input to the second phase. Output of this phase is the data structure called Frequent Generalized Itemset Suffix-Tree (FGIST) which stores all frequent closed itemsets along with their row-ids. Detail explanation of this phase is explained in [14] as we have extended their approach for generating CARs. Output of the example dataset is shown in Fig. 5.

Phase 3: Creating Class Association Rules

After creating final FGIST by updating and pruning each branch of the FGIST in Phase 2, the tree contains only frequent closed itemsets. The FGIST along with input-DBClass.csv containing class labels and their object list can be used to create *Class Association Rules*. The algorithm for creating CARs is given in Algorithm 2. The block diagram of generating CARs for the example dataset is shown in Fig. 6.

GenerateRules()

The pseudo-code of generateRule() function is given in Algorithm 4. This function is called from root of the tree to create *Class Association Rules* from FGIST. For each children under the root, it recursively calls another function *createRule()* as shown in Algorithm 5.

Algorithm 2 Phase 3: Creating CARs from FGIST.

Input: FGIST and inputDBClass.csv
Output: FGIST containing CAR
  begin
  for all Frequent Closed Itemset I with object list OID $\neq$ null **do**
   for all Class labels cl with object list OID $\neq$ null **do**
   if (I.OID $\cap$ clj.OID) $\neq$ null **then**
    Create class node named classNode, whose value is cl
    classNode.OID=I.OID $\cap$ cl.OID
    I.child=classNode
   end if
   end for
  end for
  end

CreateRule(Rule Prefix, HNode I)

This function is called for the first time by *geneateRules()* function with two arguments: An empty rule named *prefix* and reference towards the first node of the branch. When createRule() function is called, it receive two values corresponding to the following arguments: first, a rule set *prefix* corresponding to the frequent closed item(s) and class labels gathered by recursive calls performed till this node. This itemset is initialized to empty set at first call by the *generateRules()* function. Second, a ref-

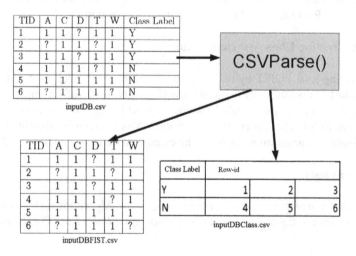

TID	A	C	D	T	W	Class Label
1	1	1	?	1	1	Y
2	?	1	1	?	1	Y
3	1	1	?	1	1	Y
4	1	1	1	?	1	N
5	1	1	1	1	1	N
6	?	1	1	1	?	N

inputDB.csv

CSVParse()

TID	A	C	D	T	W
1	1	1	?	1	1
2	?	1	1	?	1
3	1	1	?	1	1
4	1	1	1	?	1
5	1	1	1	1	1
6	?	1	1	1	?

inputDBFIST.csv

Class Label	Row-id		
Y	1	2	3
N	4	5	6

inputDBClass.csv

Fig. 4 Parsing input dataset using CSVParse()

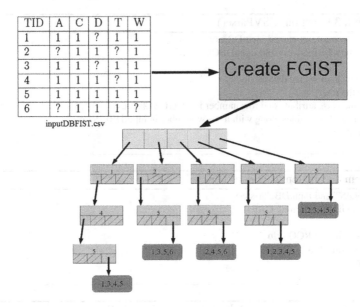

TID	A	C	D	T	W
1	1	1	?	1	1
2	?	1	1	?	1
3	1	1	?	1	1
4	1	1	1	?	1
5	1	1	1	1	1
6	?	1	1	1	?

inputDBFIST.csv

Fig. 5 FGIST of the example dataset

Fig. 6 Phase 3 execution for example dataset

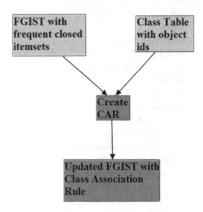

erence to a node named *I* corresponding to the first node of the branch currently processed. *createRule()* function creates rule whose antecedent starts with the node name *I* corresponding to the frequent closed itemsets. After completing all these steps shown in Algorithm 5, the updated suffix data structure containing only CARs for the given example is shown in Fig. 3.

Algorithm 3 Function: CSVParse()

Input: Source dataset named inputDB.csv
Output: inputDBFIST.csv and inputDBClass.csv
  begin
  for all row R **in** source dataset **do**
    Parse the row
    Store the class attribute and row number in inputDBClass.csv
    Store rest of the values along with it's row number in inputDBFIST.csv
  end for
  end

Algorithm 4 Function: generateRules()

Input: FGIST and inputDBClass.csv
Output: FGIST containing CAR
  begin
  for all Child C in ROOT **do**
    C.createRule({ },C)
  end for
  end

Algorithm 5 Function: createRule(Rule prefix, HNode I)

Input: Rule prefix and HNode I
Output: FGIST containing CAR
  begin
  Add I.item to prefix
  for all Child C1 of I **do**
    C1.createRule(prefix,C1)
  end for
  if I.OID $\neq$ null **then**
    for all Class label cl **do**
      T = { I.OID $\cap$ cl.OID }
      if T = null **then**
        Continue with next class label
      else
        classNode.item=cl
        classNode.OID=T
        I.child=classNode
      end if
    end for
  end if
  Remove I.item from prefix
  end

---

3   Experimental Results

We have conducted experiments to evaluate the number of rules and execution time of our proposed approach. Experiments were conducted using six data sets mentioned in Table 1 which are taken from UCI Machine Learning Repository. All the

Table 1 Data set used for experiments

Data set name	# Attributes	# Instances	# Class	# rules	Execution time (ms)
Breast cancer	32	569	2	608	172
Tic-tac-toe	9	958	2	308	344
Car	6	1728	4	124	125
Glass	10	214	7	41	32
Vehicle	18	946	4	40	31
Iris	4	150	3	4	15

Fig. 7 Graphical plot of execution time against size of dataset

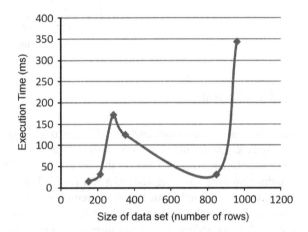

experiments were performed on a 2.93 GHz, Core 2 duo PC under Windows XP Operating System. In our experiments, minimum confidence is set to 80 % but choosing a value minimum support is more complex. This is because minimum support has a strong effect on the quality of the resulting classifier produced. If minimum support is set too high, those rules that cannot satisfy minimum support but have high confidences will not be included and also the CARs may fail to cover all the test cases. After few experiments, we set minimum support to 80 % in all the experiments reported below.

Figure 7 shows the graph which plots execution time against size of data set. Figure 8 shows the graph which plots total number of rules generated against size of data set.

Fig. 8 Graphical plot of
total no. of rules against size
of dataset

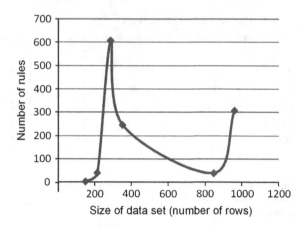

4 Conclusion

Integrating association rule and classification is gaining popularity as a major data mining task. Traditional classification techniques like decision tree or rule induction has less accuracy than ACFIST. In this paper, some problems related to associative classifier have been investigated and a new ACFIST algorithm is proposed as it's outcome. This algorithm has the following main features: it uses FIST [14] algorithm to generate frequent closed itemsets which then used to create class association rules. It uses confidence and support as measures to compare two rules. This algorithm gives list of all rules, which can be formed from one itemset and also support, confidence of those rules. This helps to prevent information loss and also helps end users to have an idea about the precedence of those rules.

In some applications, condensed representation of CAR is needed which can be generated by using rule pruning. *Database Coverage* is one such method for rule pruning. As a future development, we will incorporate this to our method. Also, we have not compare the performance of our algorithm with other state-of-art algorithm in this paper due to space limitation and since our approach is an updated version of FIST which has good time and memory complexity than other algorithm as shown in their paper [14]. But, as future scope, we will do the performance comparison after incorporating rule pruning to our approach.

References

1. B. Liu, W. Hsu, and Y. Ma. Integrating Classification and association rule mining. In KDD'98, New York, NY, Aug. 1998.
2. F. Thabtah, P.Cowling, Y. Peng. MCAR: multi-class classification based on association rule. Proceeding of the 3rd IEEE International Conference on Computer Systems and Applications (pp. 1–7). Cairo, Egypt.

3. X. Li, D. Qin and C. Yu. ACCF: Associative Classification Based on Closed Frequent Itemsets. Proceedings of the Fifth International Conference on Fuzzy Systems and Knowledge Discovery -. FSKD. pp. 380–384.
4. L. Wenmin, J.Han, J.Pei. CMAR: accurate and efficient classification based on multiple class-association rules. In ICDM' 01, pp. 369–376, San Jose, CA, Nov. 2001.
5. X.Yin and J.Han. CPAR: Classification based on predictive association rule. In SIAM International Conference on Data Mining SDM, pp. 369–376, 2003.
6. J. Quinlan. C4.5: Programs for machine learning. San Mateo, CA: Morgan Kaufmann, (1993)
7. D. Jensen and P. Cohen. Multiple comparisons in induction algorithms. Machine Learning 38(3), (pp. 309–338), 2000.
8. Y. Chien, Y. Chen. Mining associative classification rules with stock trading data A GA-based method. Knowledge-Based Systems23, 605–614, (2010).
9. X. Wang, K. Yue, W. Niu. An approach for adaptive associative classification. Expert Systems with Applications: An International Journal, Volume 38 Issue 9, 11873–11883, (2011).
10. C.H. Wu, J.Y. Wang, C.J. Chen. Mining condensed rules for associative classification. ICMLC 2012, 1565–1570, (2012).
11. C. Chen, D. Chiang. Improving the Performance of Association Classifiers by Class Prioritization. Journal of Computational Information Systems 8: 4, 1697–1712. (2012)
12. F. Thabtah. Review on Associative Classification Mining. Journal of Knowledge Engineering Review, Vol. 22:1, 37–65. Cambridge Press, (2007).
13. A. Veloso, W. Meira, M. Zaki, M. Goncalves, H. Mossri. Calibrated Lazy Associative Classification. Information Sciences: an International Journal, Volume 13 (181), 2656–2670, (2011).
14. K.C. Mondal, N. Pasquier, A. Mukhopadhyay, U. Maulik,S. Bandhopadyay. A New Approach for Association Rule Mining and Bi-clustering Using Formal Concept Analysis. In 8th International Conference, MLDM 2012, Berlin, Germany, July 13–20, 2012. pp. 86–101.

Approaches and Challenges of Big Data Analytics—Study of a Beginner

Ankita Roy, Soumya Ray and Radha Tamal Goswami

Abstract Big data analytics is a process of examining large set of data and extracting only the useful information out of it for further research. Data that is generated these days doesn't follow any particular structure. Data can be structured, un-structured or semi-structured. Data can be in form of text, image, video, live streams etc. It is a challenging job to handle such data for data mining, web mining and text mining or environmental research works. In this paper we have discussed about the step by step process of big analytics and the relevant challenges while applying on real life events. Also this paper provides a comparative study on the popular data mining algorithms which are generally used for big analytics.

Keywords Data mining · Big data · Big analytics · Algorithms · Applications of big data

1 Introduction

Data that is too large to store in traditional database is called big data. Big data can be of various size and dimensions. It has mainly three characteristics: volume, variety and velocity. Terms like petabyte, exabyte explain the volume of big data. It can be of various types, like: Structured, Semi-structured and Unstructured. Previously, structured data was used for data warehousing. And for unstructured data, first the data is extracted, transformed to structured data and then loaded in the warehouse. It was called ETL process. Some cases loading is done before trans-

Ankita Roy · Soumya Ray (✉) · R.T. Goswami
Department of Computer Science & Engineering, Birla Institute of Technology,
Mesra, Kolkata Campus, Kolkata 700107, India
e-mail: soumyaray@bitmesra.ac.in

Ankita Roy
e-mail: ankitasarkerroy@gmail.com

R.T. Goswami
e-mail: rtgoswami@bitmesra.ac.in

© Springer Science+Business Media Singapore 2017   237
J.K. Mandal et al. (eds.), *Proceedings of the First International Conference on Intelligent Computing and Communication*, Advances in Intelligent Systems and Computing 458, DOI 10.1007/978-981-10-2035-3_25

formation, which is referred as ELT process. In case of big data following these processes are impossible. The velocity of big data is unpredictable, so it is too difficult to store and analyze big data to extract the useful information.

In this paper we have done a thorough study of big data analytics and described the steps of the workflow. The implementation challenges of mining algorithms for the purpose of big data analytics are also explained briefly.

2   Related Work

Many algorithms are already defined for the analysis of large data sets. Many researchers also proposed advanced algorithms. Thabet Slimani proposed current trend in association rule mining and also compared the performance of different algorithms. He and his friends are working on Parallel Implementation of Classification Algorithms Based on Map Reduce. They have worked on unstructured data and tested their algorithm. A rapid distributed algorithm for data mining association rules is proposed by Cheun. D.W., by reducing the number of message passed. Revolution analytics offers a framework where parallel algorithms can be optimized within HADOOP. Karthik Katamba et al. working on asynchronous algorithms on map reduce [1]. These all researchers are testing different algorithms on rough data to find out only the useful information, pattern from huge data set, for future analytics.

3   Data Analytics

Every day huge data is generated by geological sensors, social media and satellites. If we want to predict a social or natural scenario from that huge data, first we have to fetch only the useful information out of it and then analyze that part to come to a conclusion. This process is called Data Analytics. High-performance analytics can be used for simpler and faster processing of only relevant data. We can apply high-performance data mining, predictive analytics, text mining, forecasting and optimization on big data.

3.1   Scale of Big Data and Scope of Big Data Analytics

In the year 2008, enterprise server systems in all over world have processed 9.57×1021 bytes of data [2] and in every 2 years from then, this number is expected to be at least doubled. Facebook operates on around five hundred TB user log and hundreds of TB image data every day. YouTube statistics says that every minute 100 h of video are uploaded and more than 135 thousand hours are watched [3]. In 2012, smartphones sold like never before and as a result almost 46 million mobile apps were downloaded, each app collecting more data. These statistics only gives an idea about the current existence of large data sets. Data can be structured (financial data, government statistics), semi-structured (sms, emails) or completely unstructured (audio, video files). Healthcare applications, social networking recommendation systems, business analysis operations are examples of those applications which require effective analysis large data sets.

3.2 Nature of Alalytics Work

The term Big data analytics broadly covers any decision making task, which is data driven. Analytics is used by both the corporate world and academic scientists and researchers. In business world, analytics team uses business information data like: customer, sales, profit etc. In academia, researchers analyze data to test hypothesis and then form their own theory. They have full control on the source data. But in both the cases, the overall analytical workflow is almost similar. Data analytics can be done through acquiring data, choosing proper architecture, shaping Data into that architecture, Coding and/or Debugging and finally reflecting and Iterating the result.

4 The Steps of Big Data Analytics and Related Issues

Analysis of big data is primarily divided into five interrelated steps. But every step has to face some major challenge handling the large data. Here we discuss the steps and relevant challenges (Fig. 1).

4.1 Acquire Data

The first challenge of Big data analytics is to find out the source. Many private companies and governments sell data into market for public interest. But the online available data does not follow any particular format. Some are not in even machine readable format. So it is really hard to use those data in practical work.

Fig. 1 Steps for big data analytics

4.2 Choose Architecture

Choosing the architecture is a very important step of this analysis work. The architecture has to be efficient by means of performance and cost. Terabytes of data we cannot store in a single machine's memory. We have to rely on distributed computing approach. We can use HADOOP to store our data, where commodity hardware will be involved, so it can be cost effective. Then we need to make a cluster of nodes, where our data will be stored divided into different blocks. We can very well use that data whenever required. Taking proper decision is highly important for the next steps.

4.3 Shape Data into Architecture

After we have got the proper dataset and platform, we need to upload the data into that platform. But we have to make sure that the data is uploaded in proper order, so that it will be compatible with the style of computation. So data has to be structured, partitioned and distributed before uploading.

4.4 Coding

For big data analytics, languages like R, Python, Pig can be used over HADOOP. For large data sets, instead of simple excel tools, cloud analysis is used, which is far more complicated than traditional desktop tools.

4.5 Reflecting and Iterating the Result

After successful completion of coding, debugging needs to be done, which follows reflecting and ultimately through iteration result is achieved. Tools like R, Python, Matlab provide desirable environment.

5 Desirable Algorithms

Many algorithms are designed to perform analysis on large data sets, but choosing the appropriate algorithm for analytic work depends on some factors, like size and structure of the data set. Platform is also an important factor for analytics as computation depends on parallel, distributed or real time approach. Following sub sections describe describe some of the algorithms, which can be useful for data mining jobs to be performed in big analytics.

5.1 Naïve Bayes Approach

Naïve Bayes is the simplest approach to be implemented. Here we use some prior probabilities, likelihood and posterior probabilities, so that we can classify the incoming data items. This is suitable for simple filtering and straightforward classification problems. Software packages like Apache Mahout, Weka support naïve bayes.

5.2 K-Means Clustering Algorithm

This is a method for estimating the mean (vectors) of a set of k groups. If the number of variables is large, then computation in k-means is faster than hierarchical clustering, provided k is small.

5.3 Apriori Algorithm

Apriori algorithm is mainly used over transactional database. It provides a solution for mining on frequent item set and learning of association rules. Using Apriori algorithm we can highlight the general trends in a database. This algorithm has powerful application in market research and recommendation systems.

5.4 Frequent Pattern Growth Algorithm

Frequent pattern growth algorithm is used to find frequent patterns and used in market analysis, retail etc. This algorithm never breaks long patterns of any transaction; rather it preserves whole information for FP mining. It reduces infrequent items and stores the remaining in descending order of frequency.

5.5 A Comparative Study of These Four Algorithms

See Table 1.

Table 1 Comparative study of data mining algorithms

	Naïve bayes approach	K-means clustering	Apriori algorithm	FP growth algorithm
Suitability	• Simple Filtering • Straightforward classification	• Small number of groups with large number of variables in each group	• Very large data • Breadth oriented search • Where time is not a constraint	• Finding frequent patterns • Vertical data format
Type of user	• Beginners • Expert Statisticians	• Technology savvy people with clear idea about their purpose	• Big data analysts • Data miners	• Technology savvy people with proper resources
Applications	• Text Classification • Spam filtering	• HADOOP • Sensus • Social Networks	• Market research • Recommendation system	• Retail supermarket like Amazon
Software support	• Apache Mahout • WEKA • ORANGE: for novice and experts	• Apache Mahout k-means currently supports: Collaborative filtering, Mean shift clustering, Fuzzy k-means • ELKI • MATLAB • R	• Java library SPMF	• Apache Mahout • MAFIA, it is an algorithm for mining maximal frequent item set for transactional database

6   Applications of Big Data Analytics

In Sect. 3.1 we have mentioned the scale and scope of big data analytics. Some industries use only text data in form of mails, chat conversations, tweets, social media postings etc. Some of the industries like mobile industry; they work on location and time data as well, to know the current position of the customer. These are most privacy sensitive types of big data. Retail and manufacturing companies use radio frequency identification data. Social networking data is very important for telecommunications and some other industries. Hence the application scope of big analytics is also increasing proportionately.

6.1   Social Media and Internet

In 2011, the Virginia earth quake or japan tsunami was experienced by many people seconds after they got tweets of the disasters. So, we can understand the speed and also the impact of the flow of information in social networks. Using big analytics in social networks we can easily understand the pattern of behavior, shape the flow of information, manage resources efficiently and predict situations beforehand. Internet itself is a source of web data. By proper application of big analytics, we can make efficient indexing of image and videos, so that searching will be easy and quick. This will provide new scope in search engines.

6.2　Government Organizations and Public Sector

All sectors are going digital now. The USA President issued a directive in 2012, named "Building a 21st Century Digital Government", demonstrating the flow of decision making to address critical data handling and the need of data analytics [4]. By 2015 they have closed 1000 data centers and moved almost 79 services to cloud. Some US Federal agencies also moved their public websites to Amazon's EC2. AWS GovCloud has been created for these commercial purposes [4]. Analytics have significant application in fraud detection also, for the purpose of giving social security of common people.

6.3　Health and Human Welfare

Medical data corresponds to EMRs and images like x-rays, USGs etc., which is growing rapidly in both record size and whole population coverage. Pharmaceutical data like drug molecules, their structures, bio molecular information; personal practice reports like dietary habits, patterns of exercise, activity records etc are also included in healthcare data. A survey by the McKinsey Global Institute [5] calculated that more than three hundred billion Dollars in value can be made every year by healthcare analytics. That means cost-benefits of big analytics here are highly appreciated. If genome structure, heredity, lifestyle, clinical and healthcare data can be analyzed properly, probability of critical illness can be predicted beforehand. That means, data quality and efficiency of analysis are critical in health informatics.

6.4　Business and Economy

Everyday a business enterprise collect large amount of data of various kinds. Customer relation data, inventory data, supplier data, store related data, market data, video feeds, customer profiles, customer preferences, financial data are the examples. They might be log data or images or video feeds or text data and also it may exceed exabytes. Since business analytics on such huge datasets are done in reputed companies, the infrastructure provided is well structured and integrated.

6.5　Experimental Processes and Research

Big data analysis has good applications in computing and experimental processes. Simulations, which were previously done with scaled data, now can be done using exascales. Also, exabytes of data is generated from high speed collision of sub-atomic particles. These data must be analyzed to prove scientific theorems.

7 Conclusion

This paper is a study of different aspects of Big Data analytics. Starting with the basic steps, we have covered scale and scope of big analytics, common applications and useful mining algorithms here. Since big data is an emerging trend in business and academia, they have to work together to invent new technologies, optimize old algorithms and advancement of analytical tools to support large data sets with different structures.

References

1. Karthik Kambatla, Naresh Rapolu, Suresh Jagannathan, Ananth Grama, "Asynchronous Algorithms in MapReduce" in: IEEE International Conference on Cluster Computing, CLUSTER, 2010.
2. P. Middleton, P. Kjeldsen and J. Tully, "Forecast: The Internet of Things, Worldwide," Gartner, 2013.
3. Hye-Chung Kum, Ashok Krishnamurthy, Ashwin Machanavajjhala, and Stanley C. Ahalt. "Social genome: putting big data to work for population informatics" In IEEE computer. January 2014. ICCC 2006. pp-1566–1569.
4. Mohammad Alizadeh, Albert G. Greenberg, David A. Maltz, Jitendra Padhye, Parveen Patel, Balaji Prabhakar, Sudipta Sengupta, Murari Sridharan, "Data center TCP (DCTCP)", in: Shivkumar Kalyanaraman, Venkata N. Padmanabhan, K. K. Ramakrishnan, Rajeev Shorey, Geoffrey M. Voelker (Eds.), SIGCOMM, ACM, 2010, pp. 63–74.
5. Kim H. Pries, Robert Dunnigan, "BIG DATA ANALYTICS: A practical guide for managers", 2015, ISBN: 13: 978-1-4822-3452-7 (eBook - PDF).
6. L. Atzori, A. Iera and G. Morabito, "The Internet of Things: A survey," Computer Networks, vol. 54, no. 15, pp. 2787–2805, 2010.
7. D. Zage, K. Glass, and R. Colbaugh." Improving supply chain security using big data. In International Conference on Intelligence and Security Informatics". IEEE, 2013. Marisa Paryasto is a resea.
8. Agrawal, R. and R. Srikant. "Fast algorithms for mining association rules", In Proceedings of the 20th VLDB Conference. Santiago, Chile. 1994. http://cs.stanford.edu/people/chrismre/cs345/rl/ar-mining.pdf. Accessed March 27, 2014.
9. T. Rathika, J. Senthil Murugan, "FP Tree Algorithm and Approaches in Big Data, in", Proceeding of International Journal of Innovative Research in Computer and Communication Engineering, September 2014, Vol. 2, Issue 9, pp. 5716–5721.
10. M. Vijayalakshmi, M. Renuka Devi, "A Survey of Different Issues of Different Clustering Algorithms used in Large Data Sets", International Journal of Advanced Research in Computer Science and Software Engineering, March 2012.
11. Michael Minelli, Michele Chambers, Ambiga Dhiraj, "Big Data, Big Analytics: Emerging Business Intelligence and Analytic Trends for Today's Businesses", published 2013 by John Wiley & Sons, Inc.

12. Joseph McKendrick, "Big Data, Big Challenges, Big Opportunities, IOUG Big Data Strategies Survey", IOUG, Sept 2012.
13. Q. He, F. Zhuang, J. Li and Z. Shi, "Parallel Implementation of Classification Algorithms Based on MapReduce", International Conference on Rough Set and Knowledge Technology, Proceeding, pp. 655–662.
14. L. Atzori, A. Iera and G. Morabito, "The Internet of Things: A survey," Computer Networks, vol. 54, no. 15, pp. 2787–2805, 2010.

A Novel MapReduce Based k-Means Clustering

Ankita Sinha and Prasanta K. Jana

Abstract Data clustering is inevitable in today's era of data deluge. k-Means is a popular partition based clustering technique. However, with the increase in size and complexity of data, it is no longer suitable. There is an urgent need to shift towards parallel algorithms. We present a MapReduce based k-Means clustering, which is scalable and fault tolerant. The major advantage of our proposed work is that it dynamically determines the number of clusters, unlike k-Means where the final number of clusters has to be specified. MapReduce jobs are iteration sensitive as multiple read and write to the file system increase the cost as well as computation time. The algorithm proposed is not iterative one, it reads the data from and writes the output back to the file system once. We show that the proposed algorithm performs better than an Improved MapReduce based k-Means clustering algorithm.

Keywords Davies-Bouldin index · MapReduce · Clustering · k-Means

1 Introduction

There is a remarkable growth in data generation from multiple sources such as social media, business enterprises, sensors and so on [1, 2]. The IDC Digital Universe study estimates that the amount of digital data will grow to 40 zettabytes by 2020 and this will be doubled each year [3]. Acquiring knowledge hidden in this huge amount of data is a big challenge. Clustering is a powerful unsupervised learning data mining technique which is very useful for this purpose [4]. However, as the size of data is very large, single machine is no longer suitable for clustering. The need is to make a gradual shift towards the use of multiple machines with

Ankita Sinha (✉) · P.K. Jana
Department of Computer Science and Engineering, Indian School of Mines,
Dhanbad, India
e-mail: ankitasinha051@gmail.com

P.K. Jana
e-mail: prasantajana@yahoo.com

© Springer Science+Business Media Singapore 2017
J.K. Mandal et al. (eds.), *Proceedings of the First International Conference on Intelligent Computing and Communication*, Advances in Intelligent Systems and Computing 458, DOI 10.1007/978-981-10-2035-3_26

247

distributed storage and distributed processing [5]. Clustering divides the data into groups called clusters, where objects in one cluster are more similar than the objects belonging to other clusters. One of the most popular and commonly used clustering algorithm is k-Means. It is very simple to implement and robust. However, there are some inherent weaknesses in k-Means clustering. For examples, the user has to specify the number of clusters before the start of the algorithm, which may lead to the resolution problem. The generated clusters are also very much sensitive to the selection of initial seeds [6]. Therefore, many algorithms [7, 8, 9], have been reported to improve the performance of the k-Means. However, all such algorithms have been developed for standalone systems and hence they are very slow to process large size data. Therefore many efforts has been made to implement k-Means on multiple machines [6, 10–13], some of which are also based on MapReduce.

In this paper, we propose a new MapReduce based k-Means clustering algorithm, which runs on multiple machine. However, our algorithm has the following advantages over the existing algorithms. It dynamically determines the number of clusters and hence it does not require the user to specify the number of clusters to be generated. Most importantly, it does not iterate over the map() and reduce() phases unlike other MapReduce based k-Means [6, 11, 12] and hence it is faster. The proposed algorithm is tested extensively through simulation and the results are compared with improved k-Means [12] to show the efficacy of the proposed algorithm.

In the recent years, various clustering algorithms have been developed which are based on MapReduce. Cui et al. [10] have reported a MapReduce based k-Means algorithm in which the iteration dependency of MapReduce jobs is taken into consideration to obtain a scalable algorithm. However, they use three MapReduce jobs for this purpose, which increase communication and I/O cost. In [11] another parallel k-Means algorithm has been presented, but no care has been taken for initial seed selection. Bahmani et al. [6] have also presented MapReduce based k-Means clustering algorithm where the cluster number has to be specified. The algorithm in [12] does not take the iteration dependence of MapReduce into account and calls both map() and reduce() functions multiple times before convergence. In [4] Weizhong Yan et al. have presented an algorithm which uses MPI to implement the power iteration clustering in a parallel environment. However it does not address the node failure which is inherently handled by Hadoop. Work has also been done for GPU based clustering. For example, G-DBSCAN [14] is the parallel implementation of density based clustering algorithm DBSCAN which is about 100 times faster than the CPU implemented sequential DBSCAN algorithms. The algorithm in [15] presents a GPU based parallel computing for large scale data clustering. The performance gain was 30–60 times in GPU than on a 3 GHz CPU implementation. GPUs' disadvantage lies in its limited non-orthogonal instruction set and programming model. On the other hand, our proposed algorithm is iteration independent and thus faster. It solves the over-resolution problem due to automatic selection of cluster number in the Reduce phase.

The organization of this paper is as follows. An overview of k-Means and the proposed algorithm are presented in Sect. 2. The experimental results are described in Sect. 3 followed by the conclusion and future work in Sect. 4.

2 Proposed Work

2.1 An Overview of k-Means

k-Means is a partition based clustering algorithm. It chooses the initial cluster centers, i.e., seeds randomly, and then iteratively assigns the data points to them until convergence. Each point is added to the nearest center, and in each iteration a more optimal center is selected [6, 10–12]. The pseudo code in Algorithm 1 explains the working of the basic k-Means.

Algorithm 1: k-Means

Step 1: Randomly select k centre from the data set.
Step 2: Calculate distance between each data point to the centres selected.
Step 3: Assign each point to its nearest centre.
Step 4: Calculate new centre

$$v_i = \left(\frac{1}{c_i} \sum_{i=1}^{c_i} x_i \right)$$

Step 5: Repeat Step 3 & Step 4 until no change.
Step 6: Stop

2.2 Proposed Algorithm

The basic idea of our proposed algorithm is as follows. The initial data is divided into small chunks or blocks and distributed over the individual machines (nodes). The distribution of the data is performed by internal mechanism of Hadoop and no prior knowledge was available regarding the data sets [16, 17]. Each node works in parallel to process the data points assigned to it. MapReduce is a parallel programming paradigm with two phases map() and reduce(). The output of one map() task is independent of the output of other map() tasks [17]. Each map() generates k clusters thus k centroids. The centroids are chosen as the representative which are merged in the reduce() phase on the basis of dynamically calculated threshold value. Finally, the data points in a particular cluster are mapped to the merged clusters generated to get the final output.

In map() phase, k-Means algorithm is run on each node in parallel. The output of the map() is set of clusters. The clusters generated by one node may be similar to the cluster generated by other nodes. Therefore, one of the biggest challenge is to merge the output of map() phase in the reduce() phase. In conventional k-means the number of clusters has to be specified prior to the start of the algorithm. This may lead to the problem of over-resolution i.e. the obtained number of clusters is more than the actual number of clusters. In our proposed work we handle this problem by intelligently selecting the number of clusters. The algorithm itself decide the actual optimal number of clusters. Moreover, the proposed work is non iterative as the initial read and final write back to the HDFS (Hadoop Distributed File System) is done only once.

The outputs of the map() phases are merged in reduce() based on a threshold value. Threshold value is calculated based on the following equation.

$$\tau = \frac{1}{n^2} \sum_{i=1}^{n} \sum_{j=1}^{n} d_{ij} \quad \text{for} \quad i \neq j \tag{1}$$

Here n is the total number cluster centers generated by each map() method. We consider the average distance between two cluster centers as the threshold value. If the distance between two or more cluster centers is less than the threshold, they are merged to form a single cluster. Algorithms 2 and 3 show the working of map() and reduce() phases respectively.

Algorithm 2: map (key, value)

```
Input : Data set and initial k
Output: Set of cluster
Step 1: Select k random centres centre[k] from D_n
Step 2: Centroid = Centre
Step 3:for i =1to n do
3.1:for j=1 to k
3.2:   Distance[j]=calculateDistance(Data[i],Centroid[j])
endfor
3.3: minDistance =min{Distance[j]
3.4: Add D[i] to Centroid[j] with minDistance
endfor
```

$$\text{Step 4: NewCentroid}[i] = \frac{\sum (data \in cluster_i)}{\text{number of elements in } cluster_i}$$

```
Step 5:if (Centroid != NewCentroid)
             Centroid = NewCentroid goto Step 2
endif
Step 6:stop
```

```
                      Algorithm 3 : reduce (key, value)

Input : set of centroids generated by map()
Output: Final set of cluster
Step 1:for each i in Dₙ'
for each j in Dₙ'
if i is equal to j then
                    add i to merge[i]
continue
endif
if distance(i,j)<= τ then
                    add j to merge[i]
                    remove j from Dₙ'
endif
endfor
```

$$Centroid = \frac{\sum merge_i}{|merge_i|}$$

```
                remove i from Dₙ'
endfor
Step 2: stop
```

3 Experimental Results

The proposed algorithm was simulated on Cloudera Hadoop (CDH4) [18] following MapReduce programming paradigm. We also executed Improved k-Means clustering algorithm [12] for the sake of comparison.

3.1 Datasets

We used two data sets to evaluate the performance of our proposed MapReduce based k-Means algorithm. We generated a synthetic well separated data set in Matlab R2013a. The raw and clustered data generated by running our algorithm are shown in Fig. 1. The algorithm was simulated by varying the initial k from $k = 5$ to $k = 20$. In all the cases the proposed algorithm generates 5 clusters. This shows that the algorithm runs perfectly for different values of k and solves the over-resolution problem.

We also tested our proposed method on the real world data set (Iris) which is available at [19]. The Iris flower data set or Fisher's Iris data set is a multivariate data set introduced by Ronald Fisher in his 1936 paper. There are 150 instances of data in four dimensions. As it is multidimensional data, we have not shown the clustering results visually. However, we judge its performance using Davies-Bouldin validity index described later in Sect. 3.3 (Fig. 2).

(a)   **(b)**

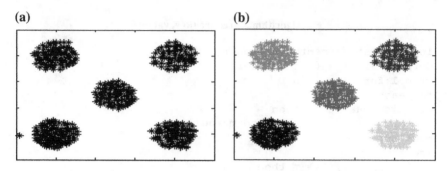

Fig. 1 Clustering results of the proposed algorithm. **a.** Before clustering. **b.** After clustering

Fig. 2 Final number of
clusters generated

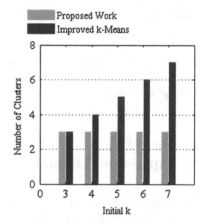

3.2 Performance Evaluation

In this section, we describe the experimental results run on the setup described in the
beginning of this section. We have extensively compared our proposed work with
another MapReduce based k-Means clustering algorithm presented in [12]. We show
in Fig. 3 the comparison of the two on the basis of the rounds each of the algorithm
takes to converge. We have taken the average over 10 instances. As can be seen from
the results the number of rounds are approximately same for both the algorithms.
Our proposed work is independent of the number of rounds the algorithm takes to
converge and reads and writes back to HDFS only once. Whereas in the algorithm
presented in [12] the number of reads and write back to HDFS is equal to the number
of rounds, hence IO cost in [12] is much higher than proposed work.

Secondly, the number of clusters generated by [12] is dependent on the k value
provided at the start of the algorithm. In contrast to that our proposed work detects
the number of clusters automatically. The value of k was varied from $k = 3$ to 7. In
all cases the final number of clusters generated was found to be 3 for iris data for

Fig. 3 Number of rounds
taken to converge

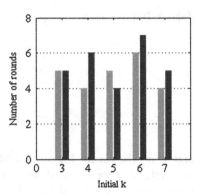

our proposed work whereas for Improved k-Means it is always equal to k as depicted in Fig. 2. From the experimental results it is deducible that our claim that the proposed work free from the problem of over-resolution is true.

3.3 Cluster Validation

Performance of clustering algorithm can be measured in terms of validation indices. Davies-Bouldin index [20] is a widely used validation metric to measure the quality of the cluster. The lower the value of the DBI the better is the cluster quality. Lower DBI indicates the cluster has low intra-cluster distances and high inter-cluster distances. DBI is defined as follows:

$$DBI = \frac{1}{k} \sum_{i=1}^{k} \max_{i \neq j} \left(\frac{\sigma_i + \sigma_j}{d(c_i, c_j)} \right) \qquad (2)$$

Here, k is the number of clusters, c_m is the centroid of cluster m, σ_m is the average distance of all points in the cluster m to centroid c_m and $d(c_i, c_j)$ is the distance between centroids c_i and c_j.

Table 1 indicates the comparison of DBI value obtained by varying the number of k provided in the map phase run for our proposed algorithm and Improved k-Means respectively. As we can see from the results the DBI values obtained for different k values provided initially is approximately the same for our proposed

Table 1 DBI for small k

k (map)	DBI for proposed work	DBI for improved k-Means
3	0.2963318	0.30497122
4	0.2725026	0.32628764
5	0.2599847	0.35453211
6	0.2895517	0.37244492
7	0.2760805	0.40839967

Table 2 DBI for large k

k (map)	DBI for proposed work
10	0.2982702
15	0.2438739
20	0.2357644
25	0.2900837
50	0.2530135
75	0.2725032

algorithm whereas for Improved k-Means the DBI value is increasing with k. For each value of k we executed our proposed algorithm for 10 times and mean of the result obtained is quoted here. However, with some error the number of final clusters generated varied in few cases. Moreover, to validate our claim that there will not be any problem of over-resolution for our proposed algorithm, the initial value of k was taken to be very large (up to $n/2$). In such cases as well the final number of clusters generated was found to be 3 only, which is the optimal number of clusters in iris data [21]. Also the DBI value was found to be in the same range for very high values of k, as shown in Table 2.

4 Conclusion and Future Work

In this paper, we have presented a MapReduce based k-Means clustering algorithm, which dynamically determine the number of clusters unlike the basic k-Means and also the other MapReduce based k-Means clustering algorithms present in literature. Iteration dependence of MapReduce programs is also taken into consideration. The experimental results show that our clustering algorithm works efficiently on real-world as well as synthetic data sets. The proposed solution provides a novel method for parallel data clustering.

MapReduce is not capable to handle the applications which need their processing online. In future we would like to implement such features in our algorithm such that it can process large scale data in real time without degrading the quality of clusters. We would further like to implement our work on GPU and CPU clusters.

Acknowledgments This research work is supported by Council of Scientific and Industrial Research (CSIR), New Delhi, India. The authors are grateful to CSIR for the financial assistance provided to carry out the research work.

References

1. Chen, CL Philip, and Chun-Yang Zhang: Data-intensive applications, challenges, techniques and technologies: A survey on Big Data. *Information Sciences* 275: 314–347 (2014)
2. DT Editorial Services, Big Data Black Book, Dreamtech Press (2015)

3. Gantz, John, and David Reinsel.: The digital universe in 2020: Big data, bigger digital shadows, and biggest growth in the far east: IDC iView: IDC Analyze the Future 2007, 1–16 (2012)
4. Fahad, Adil, et al.: A survey of clustering algorithms for big data: Taxonomy and empirical analysis., *IEEE Transactions on Emerging Topics in Computing* 2.3, 267–279 (2014)
5. Shirkhorshidi, Ali Seyed, et al.: Big data clustering: A review. *Computational Science and Its Applications–ICCSA*, Springer International Publishing, 707–720 (2014)
6. Bahmani, Bahman, et al.: Scalable k-Means++. *Proceedings of the VLDB Endowment* 5.7, 622–633 (2012)
7. Huang, Xiaohui, Yunming Ye, and Haijun Zhang.: Extensions of kmeans-type algorithms: A new clustering framework by integrating intra cluster compactness and intercluster separation. *Neural Networks and Learning Systems, IEEE Transactions on* 25.8, 1433–1446 (2014)
8. Maldonado, Sebastián, Emilio Carrizosa, and Richard Weber: Kernel Penalized K-means: A feature selection method based on Kernel K-means. *Information Sciences* 322, 150–160 (2015)
9. Zhong, Caiming, et al.: A fast minimum spanning tree algorithm based on K-means. *Information Sciences* 295, 1–17 (2015)
10. Cui, Xiaoli, et al.: Optimized big data K-Means clustering using MapReduce. *The Journal of Supercomputing* 70.3, 1249–1259 (2014)
11. Zhao, Weizhong, Huifang Ma, and Qing He.: Parallel k-Means clustering based on MapReduce. *Cloud Computing*, Springer Berlin Heidelberg, 674–679 (2009)
12. Anchalia, Prajesh P.: Improved MapReduce k-Means Clustering Algorithm with Combiner. *Computer Modelling and Simulation (UKSim), 2014 UKSim-AMSS 16th International Conference on Computer Modelling and Simulation*, IEEE (2014)
13. Kumar, Jitendra, et al.: Parallel k-means clustering for quantitative ecoregion delineation using large data sets. *Procedia Computer Science* 4, 1602–1611 (2011)
14. Andrade, Guilherme, et al.: G-DBSCAN: A GPU accelerated algorithm for density-based clustering. *Procedia Computer Science* 18, 369–378 (2013)
15. Cui, Xiaohui, Jesse St Charles, and Thomas Potok: GPU enhanced parallel computing for large scale data clustering. *Future Generation Computer Systems* 29.7, 1736–1741 (2013)
16. Dean, Jeffrey, and Sanjay Ghemawat.: MapReduce: simplified data processing on large clusters. *Communications of the ACM* 51.1, 107–113 (2008)
17. Apache. Apache hadoop. http://hadoop.apache.org
18. Cloudera: http://www.cloudera.com
19. UCI Machine Learning Repository: https://archive.ics.uci.edu
20. Davies, David L., and Donald W. Bouldin.: A cluster separation measure. *IEEE Transactions on Pattern Analysis and Machine Intelligence* 2, 224–227(1979)
21. Traganitis, Panagiotis A., Konstanti14nos Slavakis, and Georgios B. Giannakis: Sketch and Validate for Big Data Clustering. arXiv preprint arXiv:1501.05590 (2015)

Supplier Selection in Uncertain Environment: A Fuzzy MCDM Approach

Sobhan Sarkar, Vishal Lakha, Irshad Ansari and Jhareswar Maiti

Abstract This paper addresses a critical issue of selection of supplier occurred in supply chain of a manufacturing company. As there are lot more criteria present for decision making of suitable supplier selection among many, it becomes more challenging task for any company to make as this decision is entangled with company's profit and time. So, to address this problem, this paper proposes a multi-criteria decision making (MCDM) method using Decision Making Trial and Evaluation Laboratory (DEMATEL) based on Analytic Network Process (ANP), i.e., DANP, with fuzzy Vise Kriterijumska Optimizacija I Kompromisno Resenje (FVIKOR) to judiciously select suppliers based on important criteria and to point out interrelationships among dimensions and criteria in SCM by Network Relationship Map (NRM) for this company. Furthermore, the ranking is supported by sensitivity analysis.

Keywords DEMATEL-ANP · Fuzzy VIKOR · Network Relationship Map (NRM) · Supplier selection · Sensitivity analysis

Sobhan Sarkar (✉) · Vishal Lakha · Irshad Ansari · Jhareswar Maiti
Department of Industrial & Systems Engineering, Indian Institute of Technology, Kharagpur 721302, India
e-mail: sobhan.sarkar@gmail.com

Vishal Lakha
e-mail: vishallakha@gmail.com

Irshad Ansari
e-mail: irshad.ans.786@gmail.com

Jhareswar Maiti
e-mail: jhareswar.maiti@gmail.com

© Springer Science+Business Media Singapore 2017
J.K. Mandal et al. (eds.), *Proceedings of the First International Conference on Intelligent Computing and Communication*, Advances in Intelligent Systems and Computing 458, DOI 10.1007/978-981-10-2035-3_27

1 Introduction

Suppliers are integral part of an organization. Appropriate suppliers with high capability and competitiveness can ensure the considerable amount of reduction in operational costs and improve the level of quality of the end products of any company. Thus, supplier selection is a vital managerial issue for any company. In this present study, a problem associated with supplier selection of a welding company has been addressed with proposed solution of implementing hybrid multi-criteria decision making (MCDM) methodology i.e., Decision Making Trial and Evaluation Laboratory (DEMATEL) based on Analytic Network Process (ANP), i.e., DANP, with fuzzy Vise Kriterijumska Optimizacija I Kompromisno Resenje (FVIKOR) (see the flowchart in Fig. 1) with a view to not only help the company to decide the best supplier but also to assess properly the interrelationships among dimensions and criteria in order to increase supply chain (SC) efficiency overall. To address the problem initially, an extensive literature review was done that helped to figure out twelve important criteria and six dimensions regarding supplier evaluation in SC which, in turn, were verified by industry experts.

In past, many researchers have used various techniques for supplier selection. Of them, very few but important studies are discussed here as detailed illustration of related literature review is beyond the scope of this paper. Chiou et al. used the issue of green competencies to solve the green supplier selection problem using Analytic Hierarchy Process (AHP) [1]. A fuzzy analytic network process (ANP) with multi people decision making environment has been used by Buyukozkan and Cifci for the selection of sustainable suppliers [2]. Dekker et al. explored that the important green supply chain management (GSCM) practices include the environmental factors in supplier selection, maintenance and development [3]. Govindan et al. described the initiatives of sustainable SC and proposed a model of fuzzy multi-criteria approach for the supplier selection based on triple bottom line (TBL) approach (i.e., economic, environmental and social considerations) [4]. With the help of fuzzy MCDM and grey theory applied to a case study, Tseng and Chiu had shown in their model that how a case company targeted to select the green suppliers in order to meet their GSCM practices [5]. Sarkis and Dhavale proposed a hybrid model combining Bayesian network (BN) and Monte Carlo Markov Chain (MCMC) simulation to get the ranking of sustainable suppliers [6]. Mazdeh et al. presented, in their paper, a solution for supplier selection under single item dynamic lot sizing problem, which is based on Fordyce-Webster Algorithm [7]. Green supplier selection model

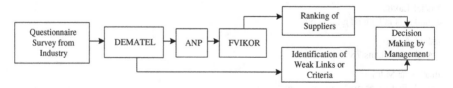

Fig. 1 Flowchart of proposed methodology

encompassing environmental and economic criteria had been proposed by Hashemi et al. [8]. Another study, in line, that dealt with prioritizing green supplier also proposed a method that deploys fuzzy ANP technique (Galankashi et al. [9]). Rajesh and Ravi proposed Grey Relational Analysis (GRA) model to find out resilient suppliers and also demonstrated the comparisons of GRA with AHP and ANP that could further validate this model [10].

The remainder of this paper is organized as follows: In Sect. 2, brief description of methodology used is given. The problem of the company is stated in Sect. 3. Section 4 illustrates results and discussion. Finally, some conclusions are drawn and the scopes for future study are discussed in Sect. 5.

2 Methodology

This paper only outlines the three methodologies used namely, DEMATEL, ANP, and FVIKOR. For elaborate discussion, which is beyond the scope of this paper, authors are requested to go for further studies [9, 11–13].

2.1 DEMATEL

The DEMATEL process can be briefly summarized as follows (for detailed study, refer to Hsu et al. paper [11]):

Step 1: *Calculate the average matrix or Initial Influence Matrix A.*

$$A = [a_{ij}]; \quad a_{ij} = \frac{1}{H} \sum_{k=1}^{H} x_{ij}^k \tag{1}$$

where i and j = 1, 2, …, n, k = 1, 2, …, H; H = Total number of experts.

Step 2: *Calculate the normalized direct influence matrix D*

$$S_1 = \max \left(\max_{1 \le i \le n} \sum_{j=1}^{n} a_{ij}, \max_{1 \le j \le n} \sum_{i=1}^{n} a_{ij} \right); \quad D = \frac{A}{S_1} \tag{2,3}$$

Step 3: *Compute the total relation matrix (TC)*

$$TC = [t_{ij}] = \sum_{i=1}^{\alpha} D^i = D(I - D)^{-1}, \text{ as } \lim_{k \to \alpha} D^k = [0]_{n \times n} \tag{4}$$

Here, $D = [d_{ij}]_{nxn}$, $0 \leq d_{ij} < 1$, $0 \leq (\sum_i d_{ij}, \sum_j d_{ij}) < 1$, and at least one column sum $\sum_j d_{ij}$ or one row sum $\sum_i d_{ij}$ equals to 1.

Here, 'r' and 's' vectors are the (n × 1) vectors representing the sum of the rows and that of the columns of the matrix TC, respectively. They are as follows:

$$r = [r_i]_{nx1} = \left(\sum_{j=1}^n t_{ij} \right)_{nx1} \qquad s = [s_j]'_{1xn} = \left(\sum_{i=1}^n t_{ij} \right)'_{1xn} \qquad (5,6)$$

[Here, superscript implies it to be the Transposition]

Here, the sum $(r_i + s_j)$ gives an index representing the total effects both given and received by the i-th factor, and the difference $(r_i - s_j)$ represents the net effect, the i-th factor contributes to the system. The i-th factor is net causer and net receiver when $(r_i - s_j)$ is positive and negative, respectively.

Step 4: *Fix the cut-off value and obtain the Network Relationship Map (NRM):*
Cut off/Threshold value (α) is generally selected by experts in respective domain. Here, α is taken 0.20. Based upon this, NRM can be drawn to show the interrelationships within dimensions and criteria (Figs. 3 and 4).

2.2 Analytic Network Process (ANP)

In this paper, ANP is only outlined briefly below in steps [11]:

Step 1: *Create an unweighted supermatrix*
From DEMATEL, the TC matrix is obtained. Then, normalization is performed. Here, $TC = [t_{ij}]_{nxn}$ an $TD = [t_{ij}]_{mxm}$ d are calculated by criteria and dimensions (here, clusters), respectively. Then, normalization of TC is performed to get the ANP weights for dimensions or clusters by utilizing TD. Then, each column will sum in order to get the normalized form. A new matrix TC^α is derived by normalizing TC by dimensions (clusters). Then, unweighted supermatrix is calculated by Eq. (7), which is based on transposing the normalized influence matrix TC^α by clusters. Here,

$$W = (TC^\alpha)' \qquad (7)$$

Step 2: *Calculation of the weighted supermatrix* (W^α)
To derive W^α matrix, each column will sum for normalization. Now, the total influence matrix TD is normalized and a new matrix TD^α is obtained.

Step 3: *Limit the weighted supermatrix*

To obtain the weights of all criteria, the following calculation is performed.

$$\underset{g \to \infty}{Lim} (W^\alpha)^g; \text{ where g is any number of power.} \tag{8}$$

2.3 Fuzzy VIKOR

This method is explained briefly as follows [12, 13]:

Step 1: Calculate the aggregated fuzzy values of alternatives rates and then defuzzify it by best non-fuzzy performance (BNP) calculation through centre of area (COA) method [12].

Step 2: *Determination of the worst f_j^- and best f_j^* values for each criterion.*

$$f_i^* = \max_i (f_{ij}) \text{ and } f_i^- = \min_i (f_{ij}) \tag{9}$$

Step 3: *Calculate S_i and R_i.*

$$S_i = \sum_{j=1}^n \frac{w_j(f_i^* - f_{ij})}{(f_i^* - f_i^-)}; \ R_i = \max_i \left(\frac{w_j(f_i^* - f_{ij})}{(f_i^* - f_i^-)} \right) \tag{10}$$

Step 4: *Compute Q_i.*

$$S = \max_i(S_i); \ S^* = \min_i(S_i); \ R^- = \max_i(R_i);$$

$$R^* = \min_i(R_i); \ Q_i = \{ \frac{v(S_i - S^*)}{(S^- - S^*)} \} + \{ \frac{(1-v)(R_i - R^*)}{(R^- - R^*)} \} \tag{11}$$

where S_i, R_i are considered as distance rate of i-th alternative to positive and negative ideal solution, respectively and w_j, v, and $(1 - v)$ are the weights for j-th criterion, maximum group utility, and individual regret, respectively.

Step 5: Create the ranking of alternative suppliers by sorting S, R, and Q values in an ascending order.

Step 6: A compromise solution is proposed that implies that alternative $(A^{(1)})$ is ranked first based on minimum Q value if the two conditions (C1 and C2) are met i.e., (i) **C1:** It is acceptable advantage showing that $Q(A^{(2)}) - Q(A^{(1)}) \geq (\frac{1}{m-1})$, where m is number of alternatives, and $A^{(2)}$ is an alternative having second position in ranking list by Q; and (ii) **C2:** It is acceptable stability. From S or/and R values, $A^{(1)}$ must be the best ranked.

This solution is stable within a range of decision making process i.e., 'with veto (v less than 0.5)', or 'with consensus, v is nearly equal to 0.5', or 'by majority rule, v is greater than 0.5'. If anyone of the abovementioned conditions is not met, then compromise solutions are proposed. It consists of: (1) $A^{(1)}$ and $A^{(2)}$ if condition C2 is not met, or (2) $A^{(1)}$, $A^{(2)}$, ..., $A^{(M)}$, if condition C1 is not met; then with the help of relation $Q(A^{(M)}) - Q(A^{(1)}) < \left(\frac{1}{m-1}\right)$ for highest value of M, $A^{(M)}$ is computed.

3   Problem Statement of the Company

This study addressed a supplier selection problem of a welding industry. The company, under study, has been experiencing a serious threat related to suppliers in terms of their various factors such as process capability, delivery issue, geographical location, quality of product, costing, service level, lead time issue etc. Out of them mentioned, lead time is a serious issue. Most of the time, suppliers undervalue the scheduled time to deliver their product, that in turn, leads delay in manufacturing cycle of products. Under such circumstances, this company has been trying to figure out initial basic feasible solution of ranking their enlisted suppliers to exploit them properly to increase the overall productivity of the supply chain.

Data Collection: By questionnaire survey, the data were collected (in linguistic form; very low, low, medium, high, very high, and excellent) from five experts from the company under study. In this paper, six dimensions i.e., level of quality (D_1), delivery (D_2), risk (D_3), cost (D_4), service level (D_5), environmental collaboration (D_6), and twelve criteria i.e., ingredient consistency (C_1), process capability

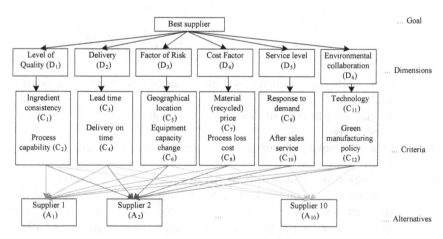

Fig. 2 The hierarchy of supplier selection

(C_2), lead time (C_3), delivery on time (C_4), geographical location (C_5), equipment capacity change (C_6), material (recycled) price (C_7), process loss cost (C_8), response to demand (C_9), service level after sales (C_{10}), technological issue (C_{11}) used for recycling processes, and green manufacturing policy (C_{12}) are considered. There are ten suppliers denoted by A_1, A_2, ..., A_{10} required to be evaluated based on criteria as decided by the experts. The total hierarchical structure with ten suppliers (i.e. alternatives), twelve criteria, and six dimensions are shown in Fig. 2 in order to select the best supplier (goal).

4 Results and Discussion

(i) From the DEMATEL method, delivery schedule (D_2) is the most important dimension with the value of 3.951, while environmental collaboration (D_6) is the least important dimension with the value of 2.369 (Fig. 3). So, there is an urgent need for the company to get involved with the environment-friendly activities. In contrast to the importance, risk (D_3), cost (D_4), service level (D_5) and environmental factors (D_6) dimensions are the net causes and are classified in the cause group. On the other hand, quality (D_1) and delivery (D_2) are the net receivers and are classified in effect group based on $(r - s)$ values. Service level is found to have the greatest direct impact on the other dimensions. On the other hand, risk has very less impact on others. Delivery is found out to be a factor, which is the most affected by the other factors. For dimensions and criteria, NRMs are displayed in Figs. 3 and 4. From DANP method, the weights obtained of all criteria $(C_1, C_2, ..., C_{12})$ are 0.0835, 0.0820, 0.0836, 0.0827, 0.0799, 0.0857, 0.0607, 0.1073, 0.0890, 0.0773, 0.0835, and 0.0851, respectively (by Eq. 8 performed by MATLAB coding).

Fig. 3 Network relationship map of the dimensions based on the threshold value, $\alpha = 0.20$

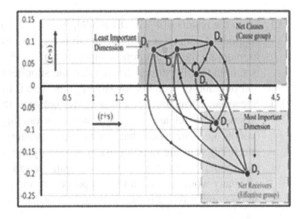

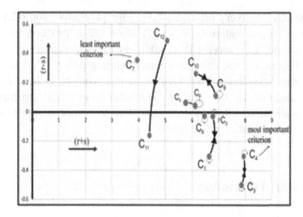

Fig. 4 Network relationship map of the criteria within dimensions, $\alpha = 0.20$

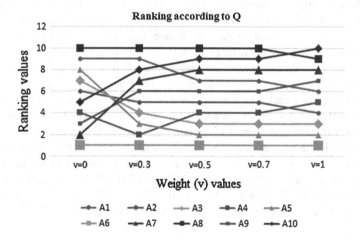

Fig. 5 Results of sensitivity analysis

(ii) From the results of FVIKOR model, A_6 (−0.7712) is found out to be the best supplier. A_8 supplier with rank tenth is required to work hard on its principle to maintain the production in supply chain. In Fig. 5, the sensitivity analysis is shown that illustrates the fact that the supplier (A_6) is the best through all combination of weights for strategy i.e., $v = 0$ to 1.0.

(iii) Spearman correlation co-efficient (R) between the result of ranking of our proposed model and that of company's ranking is found to be equal to 0.6606, i.e., our model has a strong relationship (as R lies in between 0.60 and 0.79) with the ranking list produced by the expert in this company and this model is found to be 66.06 % valid based on this expert judgment. If company would stress on this proposed ranking of their suppliers, they might attain competitive advantage over the previous selection they maintained.

Like other studies, there are some limitations in our present study. One of the limitations is that number of experts is only five whereas it could have been more for better judgment. Furthermore, sensitivity analysis could be done with more number of weight values to properly investigate the change in ranking of suppliers.

5 Conclusion

In this study, our proposed fuzzy hybrid method i.e., DANP and FVIKOR, which has much more potential towards the initial solution for the problem, has been implemented for the initial solution of the problem. In future study, any MCDM technique such as Technique for Order of Preference by Similarity to Ideal Solution (TOPSIS), multi-objective optimization on the basis of ratio analysis (MOORA) or multi-objective optimization on the basis of simple ratio analysis (MOOSRA) or ELimination Et Choix Traduisant la REalité (ELECTRE) or any other new MCDM approach could be deployed to investigate for better ranking in the supply selection framework as the future studies.

Appendix

See Figures 3, 4 and 5.

References

1. Chiou, C. Y., Hsu, C. W., & Hwang, W. Y., "Comparative investigation on green supplier selection of the American, Japanese and Taiwanese Electronics Industry in China," *Proceeding of 2008 IEEE International Conference on Industrial Engineering and Engineering Management, IEEM 2008*, pp. 1909–1914 (2008).
2. Buyukozkan, G., & Cifci, G., "A novel fuzzy multi-criteria decision framework for sustainable supplier selection with incomplete information," *Computers in Industry*, vol. *62*, pp. 164–174 (2011).
3. Dekker, R., Bloemhof, J., & Mallidis, I., "Operations Research for green logistics - An overview of aspects, issues, contributions and challenges," *European Journal of Operational Research*, vol. *219*, no. 3, pp. 671–679 (2012).
4. Govindan, K., Khodaverdi, R., & Jafarian, A., "A fuzzy multi criteria approach for measuring sustainability performance of a supplier based on triple bottom line approach," *Journal of Cleaner Production*, vol. *47*, pp. 345–354 (2013).
5. Tseng, M., & Chiu, A. S. F., "Evaluating fi rm' s green supply chain management in linguistic preferences," *Journal of Cleaner Production*, vol. *40*, pp. 22–31 (2013).
6. Sarkis, J., & Dhavale, D. G., "Supplier selection for sustainable operations: A triple-bottom-line approach using a Bayesian framework," *International Journal of Production Economics* (2014).

7. Mazdeh, M. M., Emadikhiav, M., & Parsa, I., "A heuristic to solve the dynamic lot sizing problem with supplier selection and quantity discounts," *Computers & Industrial Engineering*, vol. *85*, pp. 33–43 (2015).
8. Hashemi, S. H., Karimi, A., & Tavana, M., "An integrated green supplier selection approach with analytic network process and improved Grey relational analysis," *International Journal of Production Economics*, vol. *159*, pp. 178–191 (2015).
9. Galankashi, M. R., Chegeni, A., Soleimanynanadegany, A., Memari, A., Anjomshoae, A., Helmi, S. A., & Dargi, A., "Prioritizing Green Supplier Selection Criteria Using Fuzzy Analytical NetworkProcess," *Procedia CIRP*, vol. *26*, pp. 689–694 (2015).
10. Rajesh, R., & Ravi, V., "Supplier selection in resilient supply chains: a grey relational - analysis approach," *Journal of Cleaner Production*, vol. *86*, pp. 343–359 (2015).
11. Hsu, C.-H., Wang, F.-K., and Tzeng, G.-H., "The best vendor selection for conducting the recycled material based on a hybrid MCDM model combining DANP with VIKOR," *Resources, Conservation & Recycling,* vol. *66*, pp. 95–111 (2015).
12. Rostamzadeh, R., Govindan, K., Esmaeili, A., & Sabaghi, M., "Application of fuzzy VIKOR for evaluation of green supply chain management practices," *Ecological Indicators*, vol. *49*, pp. 188–203 (2015).
13. Sarkar, S., & Sarkar, B., "A New Way to Performance Evaluation of Technical Institutions: VIKOR Approach", *Proceeding of 2014 Global Sustainability Transitions: Impacts and Innovations*, pp. 209–216 (2014).

Intelligent Computing for Skill-Set Analytics in a Big Data Framework—A Practical Approach

Sirisha Velampalli and V.R. Murthy Jonnalagedda

Abstract Over the last few decades there is considerable increase in number of students both in Traditional as well as Online education. Especially students are showing utmost interest to learn from advanced online systems (Moretti in EDM, 2014, [1]) such as Intelligent Tutoring systems, Massive Open Online Courses (MOOC) and Virtual learning environments. These all systems are generating huge amount of Data. Now it is crucial to handle Big Data in Education. If Big Data in Education is handled properly by applying Big Data Analytics Techniques and Tools then some Intelligent Patterns can be retrieved which helps to improve Education process. In this paper Hadoop Framework (White in The definitive guide, O'Reily Media, 2009, [2]) is used to handle and process data. Analytics is applied by taking Resumes Data which are the most useful and commonly available Educational Data for Analysis and various valid Skill Inferences are drawn. Further Performance Analysis for Experiments is done by comparing Nondistributed Environment and Distributed Hadoop cluster by increasing the number of nodes. Stepwise experimentation is provided in Appendix with screenshots.

Keywords Analytics · Big data · Education · Hadoop · MapReduce · Skill

1 Introduction and Motivation

Voluminous and variety of data is generated rapidly from all sectors including Banking sector, Telecom sector, Mobiles, Social Networking, Online shopping, Education and Health Care. Discovering valid and hidden patterns from these Big data sources is quite useful task. Hadoop Distributed File System (HDFS) can efficiently store huge data sets where as Mapreduce can process and used for analysis of such huge

S. Velampalli (✉) · V.R.M. Jonnalagedda
Department of CSE, University College of Engineering, JNTUK 533003, Kakinada, India
e-mail: sirisha.velampalli@gmail.com

V.R.M. Jonnalagedda
e-mail: mjonnalagedda@gmail.com

© Springer Science+Business Media Singapore 2017   267
J.K. Mandal et al. (eds.), *Proceedings of the First International Conference on Intelligent Computing and Communication*, Advances in Intelligent Systems and Computing 458, DOI 10.1007/978-981-10-2035-3_28

data sets. User Knowledge Modeling is one of the key Application area in Educational Data Mining [3, 4] and Learning Analytics. Identifying skill set is quite a worthy task because Industries hire the students according to their requirement. In prior to Recruitment process if College Management or concerned Authorities are able to give valid report to Placement authorities about Skill set of their students then a prior estimate can be made by them, whether existing skills are sufficient or any training on novel courses is needed. Through this research Placement process becomes very easy and students as well as College Authorities can estimate which skills are required further and know about existing Skill set.

2 Methodology

The Methodology followed for the entire experimentation can be explained in a step-wise manner as follows:

Step 1:   Required Resumes to be processed say $R1, R2, \ldots Rn$ of all formats are taken.

$$[allformats.pdf, .doc, .jpeg, .png...] \tag{1}$$

Step 2:   Resumes of all formats are converted to Textual format (.txt) for easy processing.

$$[.pdf, .doc, .jpeg, .png- > Textfiles] \tag{2}$$

Step 3:   Textual format files are stored in Hadoop Distributed File System (HDFS).

$$[R1, R2, \ldots Rn- > HDFS] \tag{3}$$

Step 4:   Now content of all Text files are merged in HDFS.

$$[MergeR1, R2 \ldots Rn] \tag{4}$$

Step 5:   Preprocessing is done on merged content to process required data and remove unwanted data.

Step 6:   Finally Mapper and Reducer functions are applied on data and valid patterns are retrieved.

$$[ProcessingR1, R2 \ldots Rn- > Mapreduce] \tag{5}$$

3 Big Data Framework

In Big Data Framework the data will be divided and mapped into Distributed systems. Apache Spark, GraphLab, HPCC Systems (High Performance Computing Cluster), Dyrad, Hadoop [5] can be used for storing and processing Distributed Data. In this paper Hadoop is used. The programming model [6] MapReduce popularized by Google, which is used for parallel processing of large data sets. Processing flow for MapReduce is shown in Fig. 1. Hadoop [7] is a MapReduce based framework for processing huge data sets. Hadoop has two subsystems namely a Distributed file system called HDFS and MapReduce paradigm. Architecture of Hadoop is shown in Fig. 2. In HDFS architecture, the master node runs Namenode and worker nodes run

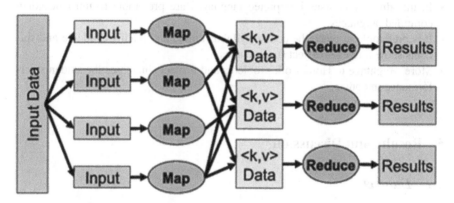

Fig. 1 MapReduce computation flow

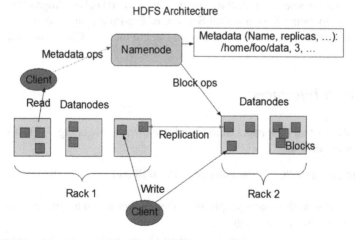

Fig. 2 HDFS architecture

Data Nodes, which manage local data storage and other information. For controlling and scheduling tasks, Hadoop has JobTracker, which assigns roles to the worker nodes as Mapper or Reducer task by initializing TaskTrackers.

4 Related Work

Capability Matrix [8] and sum-scores can be used to estimate stduent skill knowledge. Performance of student skill knowledge can be compared using [9] clustering methods. Papers [10, 11] applied Learning Analytics for improving Computer Science course quality and drawn the following inferences:

- In introductory courses interpreted languages are preferable to introduce than compiled languages.
- It is preferable to place the course syllabus online and readily available and students show interest to select such courses.
- More weightage to Homework problems and Projects increases the students performance in course.

5 Results and Discussion

5.1 Dataset

Resumes of various formats are taken for Experimentation. We restricted our Dataset to only Computer Science student Resumes as we want to analyse Programming Languages and Operating Systems Skill Set. Lot of Pre-Processing is done on Data to extract valid inferences and remove unused Data to increase Computational Performance.

5.2 Result Inferences:

Final Analysis of all Languages Skill Count is shown clearly in Fig. 3. The following inferences are drawn:

- It is inferred that "C" is the most common skill and "perl" is the least common skill.
- It is also inferred that many people are skilled in Procedural languages than Object Oriented and Scripting Languages.

Experimentation is done by extracting various Operating Systems skills and analysis is shown in Fig. 4.

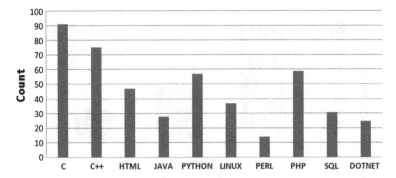

Fig. 3 Language skill set statistics

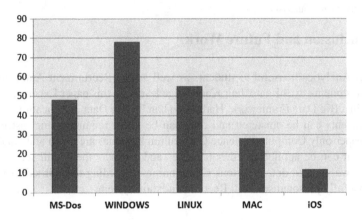

Fig. 4 Operating systems statistics

- It is inferred that Windows Operating Systems is the most common skill and iOS is the least common skill.

5.3 Performance Analysis

Experiments are done in a Distributed Hadoop Framework by increasing the number of nodes. There is drastic increase in performance from Non-distributed to Distributed Framework. The performance analysis is shown in Fig. 5.

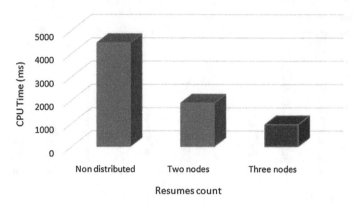

Fig. 5 Performance analysis

6 Conclusion and Future Work

Skill Set Analytics is useful to students as well as Job Recruitment Authorities. It helps the students to estimate their Knowledge levels. In this paper Experimentation is done in a Big Data Framework, Hadoop to handle Big Data in Education. Similar skills students can be clustered and they can be given required same instruction. In this paper only Computer Science Curriculum Resumes are taken and Analysed, similarly it can be applied to other branches and estimate their Skill sets as well. In future we want to detect various Communities according to their skill sets and develop a fully personalised skill Recommendation system.

Fig. 6 Screen shot for converting all files into text format

7 Appendix

Methodology followed in entire experimentation is shown stepwise clearly with the help of screenshots (Figs. 6, 7, 8, 9, and 10).

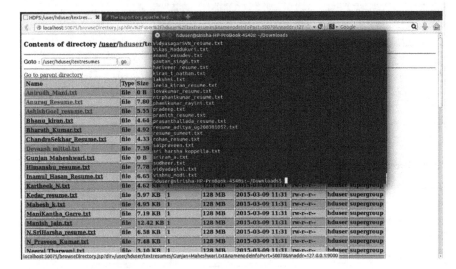

Fig. 7 Screen shot for placing text files in HDFS

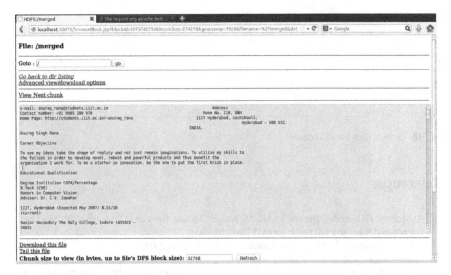

Fig. 8 Screen shot for merged text content

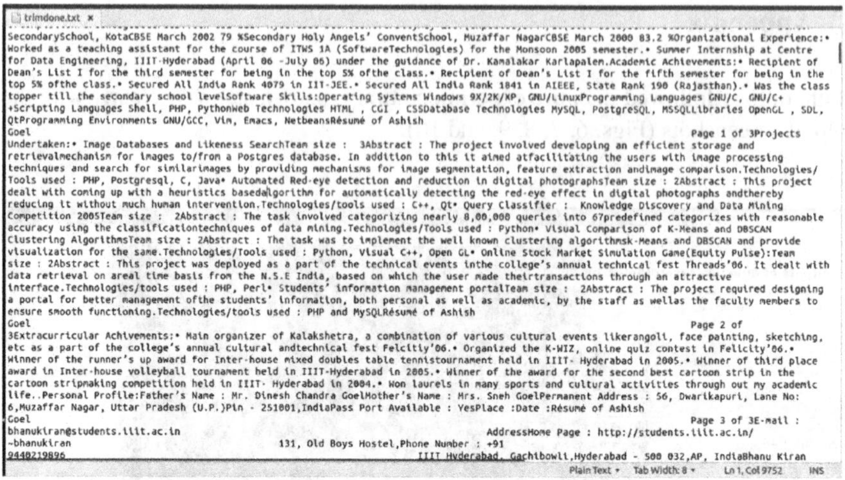

Fig. 9 Screen shot for trimmed content

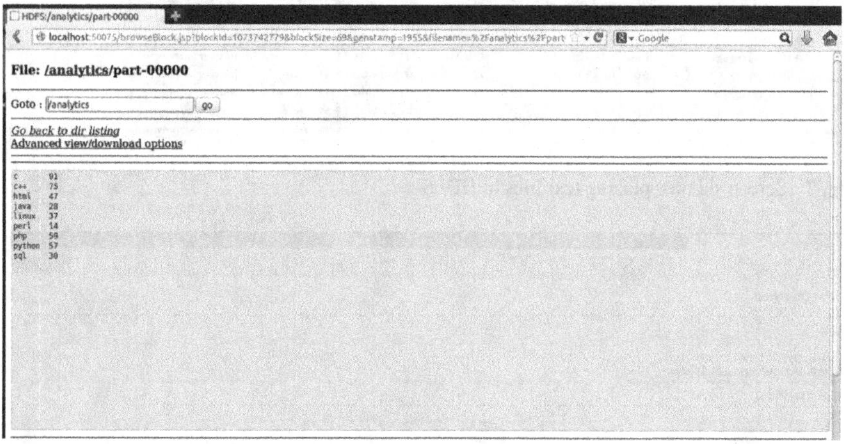

Fig. 10 Screen shot for skill count

References

1. A. Moretti, J. Gonzalez-Brenes, and K. McKnight. Mining the web to leverage collective intelligence and learn student preferences. EDM (2014)
2. White, T. Hadoop: The Definitive Guide. O'Reilly Media (2009)
3. http://www.educationaldatamining.org
4. Baker, R.S.J.d.,Yacef, K. The State of Educational Data Mining in 2009: A Review and Future Visions. J. Educational Data Mining, I (1), 3–17 (2009)
5. Kai Hwang, Geoffrey C. Fox, Jack J. Dongarra: Distributed and Cloud Computing Book: From Parallel Processing to the Internet of Things. Morgann Kaufmann (2012)

6. J. Dean and S. Ghemawat: MapReduce: Simplified data processing on large clusters, Proc. 6th Symposium on Operating Systems Design and Implementation (OSDI 04), pp. 137–150 (2004)
7. Apache Hadoop, http://hadoop.apache.org
8. Ayers, E, Nugent, R, Dean N: Skill Set Profile Clustering Based on Student Capability Vectors Computed from Online Tutoring Data. Educational Data Mining 2008: 1st International Conference on Educational Data Mining, Proceedings (refereed). R.S.J.d. Baker, T. Barnes, and J.E. Beck (Eds), p. 210–217, Montreal, Quebec, Canada (2008)
9. Elizabeth Ayers, Rebecca Nugent and Nema Dean: A Comparison of Student Skill Knowledge Estimates, EDM (2009)
10. A. Moretti, J. Gonzalez-Brenes, and K. McKnight. Towards data-driven curriculum design: Mining the web to make better teaching decisions. EDM (2014)
11. A. Moretti, J. Gonzalez-Brenes, and K. McKnight. Mining student ratings and course contents for computer science curriculum decisions. EDM (2014)

Implication of Performance Appraisal Process on Employee Engagement Mediated Through the Development and Innovation Oriented Culture: A Study on the Software Professionals of IT Companies in West Bengal

Manas Kumar Sanyal, Soma Bose Biswas and Rana Ghosh

Abstract In the face of stiff global competition and fast changing technology, the sustainability of the software companies undoubtedly depends on their contented and committed intellectual capital. To attract and retain best talents from the market, the industry emphasizes on strategic and innovative HR practices. As the driving force of the industry is its' workforce, maintenance and management of skills and potentialities of the individual employees' and working teams are considered as the instrument to retain talents in the organizations and also to continue to provide quality service to the customers. Performance management hence plays a central role in these organizations. Continuous monitoring and management of individual competences therefore always be in focus for the HR department. The paper investigates the performance assessment process and its' implication on employee engagement mediated though an HRD culture conducive of growth and innovation in some IT/ITes/BPO companies of West Bengal.

Keywords Performance–process factor · Innovation oriented culture · Employee development · Goal setting · Monitoring

M.K. Sanyal
Engineering, Technology and Management, Kalyani University, Kalyani Nadia 712345, West Bengal, India
e-mail: manas_sanyal@rediffmail.com

M.K. Sanyal
BA Department, Kalyani University, Kalyani Nadia 712345, West Bengal, India

S.B. Biswas (✉)
Department of Business Admnistration, Narula Institute of Technology, Agarpara, Kolkata 109, India
e-mail: soma_mba@rediffmail.com; somabbiswas@gmail.com

Rana Ghosh
Guru Nanak Institute of Dental Science and Research, Kolkata, India
e-mail: ranaghosh@rediffmail.com

© Springer Science+Business Media Singapore 2017
J.K. Mandal et al. (eds.), *Proceedings of the First International Conference on Intelligent Computing and Communication*, Advances in Intelligent Systems and Computing 458, DOI 10.1007/978-981-10-2035-3_29

1 Introduction

The explosion of knowledge-based industry is a reality even in the developing nation like India. The Indian knowledge resources are now one of the largest contributors towards providing IT solutions to their global customers. The young knowledge capitals of the organizations are aspiring for continuous growth and autonomy at their work. The ease of access to the information continuously generates quest and that leads to innovation. Innovation at different technical levels, changes the demands of society very fast. As a result to combat the external threats like competitive market and fast changing technology, the organizations also need people who are accommodative to change. The organizations are fostering and also capitalizing creative and inimitable idea of their human capital. Continued and technological growth has also shortened the life span of any organizations. So, only developing a talented workforce through training and development of individual employees' only is not the solution but to continue with sustainability in the market for long term, the organization should be flexible to change through restructuring its processes, business models, technology used compatible to the changes in market preferences [3, 18, 21, 36]. The country is set to position herself as one of the large supplier of young engineers and skilled workforce to the World by 2020 [6]. The age group of workforce population less than 35 has augmented from 353 million in 2001 to 430 million in 2011 [50]. The phenomenal escalation of software industry has accelerated the growth pace of Indian economy. Large number of proficiently trained young job seekers with proficiency in English language and mathematical aptitude makes India a hot destination to invest for the foreign multinationals [33]. Information Technology is defined as the management of raw data into meaningful, workable instructions though a protocol of transmuting the machine language, storing, transmitting and securely retrieving the information through a specific software and hardware applications [43, 47, 48]. Indian professionals are creating real heights of their proficiency in IT, in India as well as abroad.

The software industry has created a diverse image than the other traditional industries of India, in different aspects like procurement of highly talented and educated workforce, structure of employment, work culture and organizational functionalities, transnational identity, cultural and social identity of the industry, implementation of strategies in their work [54].

Cross cultural clients and the use of sophisticated techniques has differentiated software industry from the others. The demand of this profession is high adaptability and readiness to change. Organizations require reducing the operating time and costing for project development as well as project delivery time maintaining the quality of service. This increases the importance of knowledge workers and proper cultural capital in the software industry. Consequently, identification, maintenance and management of technology-fit—skills, socially and culture-fit human capital is a vital requisites for any organization irrespective of its size for proper functioning and sustainability in long run [14].

To improve individual performances and thereby attain the organizational goal, IT industry spends a huge attention over their HRM practice in the organization. Big as well as small companies' gives thrust to the strategic man management process and integrate the imperatives to the all human resource functions. At one side, the software companies are dealing with transnational clients and hence facing challenges by fast technological changes causing swift skill obsolesce. On another side being a project based industry the movements of the employees are fast across the teams dealing with diverse skill sets required for the projects even across the globe, different project leaders, entirely different challenges posed by the technology and client's demand. Adaption of the fast change hence is the most important for the efficient organizational performance. A strategic performance management system integrated to the broad organizational objective is always considered as a feasible solution to problems arising out of change.

The people dimension of this talent based industry has always been a major concern of the management professionals for effective and efficient organizational performance. IT is a skill intensive industry. Tschang [53], examined the skill set required for the IT organization, and reported two main skill sets: (1) Product Development and (2) Business Development skill. Basic and advanced technical skills, system skills like project management skills and innovative skills are within the first skill set. The IT companies are facing challenges to maintain a reasonable talent pool. According to [49] Attrition has increased from 13 % in 2010 to 19 % in 2011 in the Indian IT sector. It was reported by data quest survey, 2011 that during the year, almost 70 companies had reported significant changes in their senior (C-level) management level, including CEOs/MDs [49]. Employees often changes jobs and join to the company's competitors for better salary, prospect and career growth. The reason is that in spite of a large number of manpower supply at the junior level, there is a dearth of talented and skilled professional at the middle level or at the top level, which increase the scope of poaching and case of attrition [10]. According to Dataquest survey to identify the best employers in IT industry, India, it was found that long term development scope for employees and creative HR practices are vital tool to retain their employees.

To build up a competitive talent management strategy, individual performance management strategy must be developed and it must be integrated with the development policy, promotion policy, reward policy and career planning. Various components of talent Management which should be integrated together to create a high performing team, are: recruitment, selection, performance management, mentoring, individual development, leadership development, career planning, and recognition and reward [25, 46]. But most unfortunately, performance results are treated as assessment of one's ability [42] and concentration of HR department lies more on assessment of performance rather on management of performance.

The paper tries to study the performance appraisal practices in various software companies and the employees' attitude towards the transparent appraisal practices related to appraising performances of the individual. The paper tries to study the effect of goal setting, evaluation process of performance of individual; and also the feedback system on learning and development and hence establishing innovation promoting culture.

2 Review of Existing Literatures

2.1 Assessment of Employee Performance and Its' Implication

The holistic approach to achieve organizational excellence, identifying the rational factors contributing to peak performance through measurement of individual and team performances and integrating it to organizational strategic requirement through developing the capabilities of the workforce and thus providing a sustained competitive advantage in any equation delivering value to the customers and society is generally defined as Performance management systems [4]. The structured and bias free method of quantification and administration of individual performances towards managing the proficiencies at individual level, cohesion between team members and organizational performance in the market has drawn attention [24] of the researchers. Performance appraisal is being used as powerful tool of performance management to improve skill, knowledge, ability and personality of the individual [22], though there are conflicts relating its' process, and mechanism to encourage inter-individual comparison (for promotions and salary decision), intra-individual comparison (training need identification, feedback mechanism) simultaneously [39]. Even due to several researches, it is still not be possible to make the appraisal system unbiased. Hitherto the process in many cases failed to achieve its' objective and became mere annual ritual by the supervisors measure performance of employee against the roles, objectives and expectations typically fixed by the top level management [12]. The quest to develop a bias free integrated and comprehensive employee assessment process is always in focus and that resulted in multidimensional research studies in this field: some researchers investigated to develop different exclusive techniques to assess employee performance [45] like: people capability maturity model [30], EVA model, 360° appraisal method [11] some researchers suggested to train the rater on 'how to evaluate' to improve the quality of assessment [13] and to reduce bias while assessing employees' performances [38, 52] etc. The concept of performance appraisal have now become strategic [40], emphasising on comprehensive management of performances at individual levels shifting from traditional approach of only "measurement the performances" [15]. Though the implications of appraisal in HR practices are also manifold [29] including feedback management and mentoring [19, 27, 31] training need identification or decisions related to promotional, career planning, salary and increments, reward administration etc., [4, 16] but the process is highly criticized about its efficient administration and utilization. The sophistication in using the appraisal techniques to appraise the employee performance in IT companies is diverse from organization to organization like depending on the size of the companies.

2.2 Goal Setting and Innovation

Performance appraisal being a part of management works in a systematic form that follows a cycle like performance planning, performance expectation and acceptance of performance goal communication, performance monitoring and feedback, performance evaluation and implementation in HR decision making and appraisal feedback [14]. Software companies seek to achieve the desired result through setting the goal mutually with the subordinates; regular monitoring and mentoring of employees' daily activities towards achieving the result [9] and integrating organizational—team—individual performances. To achieve excellence, these companies are found to nurture an HRD climate conducive to innovation. Centrally managed appraisal programs categorize the organizational objectives after identifying the maturity level of the organization. The organizational objective is set according to the strategies made at the top management level and then en route to the individual echelon through the goals set jointly. Goal setting process promotes participation in decision making [35] and hence also facilitates social learning [8]. The result based approach [4] promotes attainment of the business result yielded by employees through, concrete, measurable, work achievements judged against fine-tuned targets or goals set mutually by the subject and the assessor. Bandura [37], proposed that the management should work as a facilitator to facilitate employees to attain their individual goals best by directing their own endeavors towards achieving the organizational objective. The result/outcome is expected to generate through a climate of mutuality based on common understanding and a process of achievement of goals/targets set and evaluated periodically jointly by management and subordinate.

The self goal setting process endorses self monitoring of the goal approaches to achieve result, develops self efficacy [7] among the employees and generates the innovation stimulating behavior [23]. The knowledge of results [34] generates confidence among the employees and develops the attitude among them for future plan of performances.

Hypothesis 1 Goal Setting positively influence employee development and innovation culture.

2.3 Performance Monitoring and Management Support to Facilitate Employee Development and Innovation

Extensive researches have proven the fact that one of the effective ways to develop the skill of the employees is undoubtedly continuous monitoring of peoples' performance. Traditionally, appraisal system is used as a coercive tool to examine the employees' performances [2] but modern theories propounds the it should encourage a continuous and open discussion with the job incumbent about the hindrances and obstacles faced while performing and probable scope to improve [51]. The researchers have provided enough authentications about the fact that the perceived organizational support enhances the affective attachment with the organization and hence increase their commitment to the organizations [26]. The modern organizations are more relying on the mentorship scheme to expedite organizational learning and improve the performances of workers at operative level. Kram and Isabella [32], advocates for peer relationship as a source of development with mentoring practices. Prompt Guidance, management support to complex and critical works can enhance the morale of the employees, and thus contribute to organizational growth.

Hypothesis 2 Monitoring employee performances positively influence employee development and innovation culture.

2.4 Effect of Performance Evaluation on Learning and Development and Innovation

A proper and thorough evaluation of employee performances can help the job incumbent to get accurate feedback about the areas to develop, which surely help the employees in his further career. MindTree consulting integrated performance management system to organizational values and objectives through CLASS (caring, learning, achieving, sharing and social conscience) model [1]. The success lies in measuring the performance accurately and optimized use of vital resources and thus help the business organizations to produce result. Evaluation of performances were integrated to defining goal, finally planning and management of performances, performance counselling [41]. A fair, transparent and equitable evaluation procedure increase faith of the employees in the process and hence reduces bias. Folger et al. [20], addressed the political approach of assessing employee performance and its conjectures, and suggested a due process to overcome the short coming emphasizing on practice of fair and distributive justice.

Hypothesis 3 Evaluation of employee performances positively influence employee development and innovation culture.

2.5 Effect of Feedback on Learning and Development and Innovation

One of the major importance of the appraisal system manifests in identifying the area of development of the appraisee. A tailor made performance management strategically anchored to manage the performances at individual level, team level and organizational level, integrated with prompt feedback, continuous guiding, responsibility based performance appraisal, development feedback, connected with manpower planning, nurturing team work, creating operating line excellence [1]. A proper feedback can help the employees to provide their best at work place. It is found from recent researches [17], that positive feedback on employee performance are more accepted than the negative feedbacks during the appraisal interview. Meriac et al. [28], have proposed that a distinct judgement and feedback on employee performance which can help to develop the employee performances rather concentrating on only regular, prompt and futuristic feedback mechanism to improve employee performance [44].

Hypothesis 4 Feedback about the employee performances positively influence employee development and innovation culture.

A seven factor model depicting the relation between performance appraisal process factors and employee engagement is proposed as in Fig. 1.

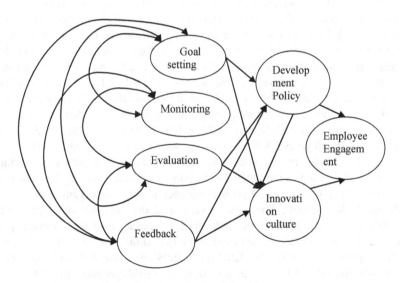

Fig. 1 Seven factor model describing inter factor relationship

3 Objectives

1. To study the linkage between goal setting, employee development and innovation encouraging culture
2. To study the association between performance monitoring, employee development and innovation encouraging culture
3. To study the association between performance evaluation, employee development and innovation encouraging culture
4. To study the association between feedback, employee development and innovation encouraging culture
5. To study the influence of employee performance assessment system on employee engagement mediated through the organizational employee development and innovation oriented culture

4 Methodology Used

To study the research objectives the data are collected from 19 Kolkata based software companies those are members of NASSCOM Kolkata (NASSCOM 2014). The data are stratified into different level as: small/medium companies, major Indian software companies and transnational firms having foreign partnerships in the business. To obtain the result survey method was used and around 900 feedback forms were circulated through mail or personally out of which 769 filled up questionnaires obtained. The data were selected after rejection of 78 questionnaire as they were partly filled up. The data were further filtered on the basis at least one of appraisal sessions attended by the respondents and 58 questionnaires were discarded. The data set finally consisted of total 633 sample with 256 employees in the junior grade (40.4 %), 183 employees at middle level (28.9 %), 130 at managers rank (20.5 %) along with 60 (10.1 %) non IT executives. The data is distributed in terms of 65.9 % male and 34.1 % female participant. The age group is distributed as: age less than 26 years is 31.8 %, More than 26–below 31 years—42, 16.9 %: 31–below 36 years, and 36–40 years 5.8 % and the respondents in age group above 40 years is 2.5 %. 35.3 % of the respondents were found between 1–<2 years of working experience, 43.2 % are in the range 2–<5 years, respondents having 5 years–<10-years were 18.1, 3.3 % over 10 years experience in the same organization.

Research Instrument: The paper investigated the influence of several independent variables along with other independent variables as gender, work experience, age of the respondents obtained from the administration of questionnaire which are gathered from the validated set of questions from the works on performance appraisal [5] and Organizational commitment (Meyer and Allen 1991). The paper studies the effect of these independent variables on the dependent variable i.e., employee engagement/loyalty mediated through development policy and innovation culture. The items were scaled in a five point Likert type scale with agree–disagree poles (5 = strongly agree, 1 = strongly disagree) to conclude the study.

Table 1 Chi-square tests showing an association between performance appraisal system and development and innovation oriented culture

		Employee strength	Employee development	Innovation
Goal setting	For all organizations	99.606**		123.472**
	Above 15,000	50.194**		75.539**
	Below 15,000	52.657**		76.023**
Monitoring	For all organizations	74.775**		112.704**
	Above 15,000	41.988**		68.707**
	Below 15,000	37.327**		72.466**
Evaluation	For all organizations	73.223**		110.539**
	Above 15,000	39.806**		11.854**
	Below 15,000	50.890**		66.865**
Feedback	For all organizations	97.425**		159.306**
	Above 15,000	62.624**		83.936**
	Below 15,000	47.452**		106.290**

Data Source Primary **significance level 0.01 (2-tailed)

Analysis and Finding: It was found that the selected 19 IT companies vary in size, business focus, customer base, etc., the selected IT companies in West Bengal, the data are divided on the basis of their workforce size and clustered in two groups: employee strength above 15000 and below 15000 (NASSCOM, "Members", 2012). To study the opinions of the respondents, separate questions were administered describing the association between the performance appraisal process variables and their influence on employee development as well as on innovation oriented culture. A chi-square testing is done in these two different groups. The hypothesis 1–4 is tested, which establishes the association of performance appraisal process (goal setting, performance monitoring, evaluation system, feedback process) with learning and development and innovative climate in the system irrespective of the size of the companies. The Pearson chi-square was calculated with p value 0.000 which is less than 0.05 (Table 1), resulting in rejection of the null hypothesis and accepting the alternative hypothesis (H1 to H4) that the employees of the selected IT companies accepted that the performance appraisal process factor is significantly associated with employee development and innovation orientation in organizations of all size.

The correlations are confirmed between the variables (Table 2):

To understand the effect of appraisal imperatives on employee engagement, an exploratory and confirmatory factor analysis is conducted. To comprehend the reliability of the data, the Cronbach's Alpha is calculated as 0.937 for 41 independent variables obtained from the administration of questionnaire, which is adequate to progress further. Seven factors extracted show a positive and significant association with their observed variables with very high factor loadings. The varimax rotated component for seven factors is found as 66.703 % of the total variance. The factors are: Goal setting, performance monitoring, evaluation, and

Table 2 Correlation of performance appraisal process variables with employee development and innovation oriented culture

	Mean	SD	Goal setting	Employee development	Innovation orientation
Goal setting	3.20	0.839	1		
Employee development	3.58	0.847	0.337**	1	
Innovation orientation	3.48	0.828	0.285**	0.253**	1
Monitoring	3.58	0.847	1		
Employee development	3.48	0.824	0.237**	1	
Innovation orientation	3.45	0.860	0.248**	0.253**	1
Evaluation	3.58	0.847	1		
Employee development	3.48	0.824	0.256**	1	
Innovation orientation	3.36	0.804	0.243**	0.253**	1
Feedback	3.58	0.847	1		
Employee development	3.48	0.824	0.316**	1	
Innovation orientation	3.36	0.804	0.294**	0.253**	1

**Correlation shows significance at 0.01 level (2-tail). *Data Source* Primary

feedback, Innovation reinforcing culture, employee development, and employee engagement. To establish the objective of the study, a confirmatory factor analysis is done between the seven factors (Goal setting, performance monitoring, evaluation, and feedback, Innovation reinforcing culture, employee development, and employee engagement). The results are as shown:

Number of observed variable	41
Sample moment values	861
Parameters estimated	97
DF	764
χ^2	1370.076
Probability	0.000

Model fitting information	
RMSEA	0.035
NFI	0.841
CFI	0.889

The RMSEA shows that the model is moderately fit and NFI, CFI values confirms that the data is sufficient to describe the model.

The correlations between the factors extracted are also showing a significant and positive correlation (Table 3). The correlation matrix establishes a strong relationship between the factors as obtained from factor analysis. Empowerment agreement establishes a strong relations innovation oriented culture and

Table 3 Standardized regression estimates

	Goal setting	Monitoring	Evaluation	Feedback	Employee development	Innovation culture
Clear communication of corporate plan and business goals	0.754**					
Self acceptance of the goal	0.793**					
Mutual goals setting	0.736**					
Acceptance of individual *KRA criteria*	0.710**					
Discussions over *constraints to achieve KRA's*	0.737**					
Scope to assert for support to understand the KRA's	0.795**					
Promoting performance through *Goal setting process*	0.740**					
Communication of changing intermediary goals and expectations by top management	0.682**					
Regular review of goals	0.695**					
Continuous mentoring plan		0.726**				
Prompt guidance		0.762**				
Immediate resource support		0.693**				
Scope of *self evaluation*			0.659**			
Fairness in evaluation			0.763**			
Efficiency in evaluation			0.620**			
Evaluation of *behavior beyond job description*			0.714**			
Equity in evaluating performance			0.738**			
Review of *development plan*			0.701**			
Reliable feedback system				0.608**		
Proper attention and time spent in review discussions				0.764**		
Careful performance review discussions				0.650**		
Open communication in review discussions				0.737**		
Identification of individual *developmental need*				0.681**		

(continued)

Table 3 (continued)

	Goal setting	Monitoring	Evaluation	Feedback	Employee development	Innovation culture
Counseling on mistakes				0.692**		
Exercise SWOT analysis for future project				0.758**		
Identification *of competency area*				0.729**		
Promotion of continuous learning					0.780**	
Encouraging self learning					0.584**	
Focus on skill based mentoring					0.709**	
Regularity in *mentoring and guidance*					0.748**	
Scope of long term *development*					0.750**	
Positive rating on *innovation*						0.786**
Practice of encouraging the Innovative Ideas						0.774**
Management's belief in innovation at lower level						0.713**

Data Source Primary **significance level 0.01

Table 4 Standardized Regression estimates among the constructs

	Development	Innovation
Goal setting	0.173**	0.138**
Evaluation method	0.181**	0.109**
Feedback	0.299**	199**
Employee development	0.0	164**
Innovation		

Data Source Primary **Regression weights: significance level 0.01 (2-tailed)

empowerment agreement (0.427). Reward scheme and empowerment agreement shows influence on employee engagement.

The following table shows the standardized regression estimations (Table 4).

The covariances within the factors are as follows (Table 5).

The inter construct correlations are (Table 6).

Regression weights of observed variables are showing a significant and high association with their constructs. The regression estimates of the observed variable for the latent "Employee engagement" are also very high: *Membership of the organization* (0.744**), *Generating a feeling of motivation* to see the success of the company (0.875**), *Employees involvement* in the decisions that is related to them (0.832**), *Reference to friends* to join this company (0.798**), *Desiring to serve*

Table 5 Covariance among the constructs

	Goal setting	Monitoring	Evaluation	Feedback
Goal setting				
Monitoring	0.191**			
Evaluation	0.266**	0.206**		
Feedback	0.217**	0.179**	0.021**	

Data Source Primary **Regression weights: significance level 0.01 (2-tailed)

Table 6 Inter construct correlations

	Goal setting	Monitoring	Evaluation	Feedback
Goal setting				
Monitoring	0.368**			
Evaluation	0.541**	0.502**		
Feedback	0.517**	0.511**	0.514**	

Data Source Primary **Regression weights: significance level 0.01 (2-tailed)

this company till retirement (0.799**), *Comfortable to share feelings, opinion within team* (0.823**). The regression weights shows the relationship with employee development policy and innovation oriented culture as 0.466** and 0.281** (Fig. 2).

It is evident from the above discussions that the specifications are confirmatory. The relationship between the observed variables and the latents are defined and the observed variables are sufficient to describe the latent variables. It can be concluded that the model fairly and accounts for the variables observed in the data. The

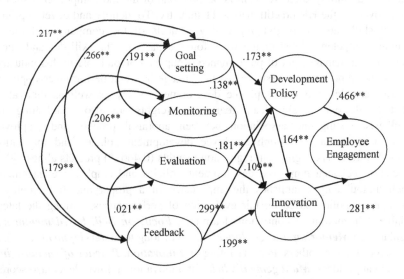

Fig. 2 Seven factor model with regression weight and covariance values

regression weights are significant at p value less than 0.05 among all factors with employee engagement which proves the significant relationship of all the factors with employee engagement. Calculating the total effects of the six factors on employee engagement, it is understood that employee loyalty is highly influenced by the innovation oriented culture which is mostly influenced by the employee development practices in the organizations.

5 Findings

After referring a comprehensive number of the literatures, it is evident that proper and fair implementation of performance appraisal system undoubtedly functional in an organization to create a work force with inimitable competence set. The IT industry has differentiated itself from the traditional companies in many ways such as its' project based nature, multinational clients, multilevel projects including offshore assignments, fast changing technology, result orientation, cross functional teams and swift changes through projects and changes in work teams. The industry is also suffering from external threats of large number of competitors, quality seeking customers, and stringent Governmental laws. The industry is also unique with its large numbers of young Executives at the top level also. The young dynamic, talented and growth seeking knowledge worker group is delivering value to the world with an unmatched skill set. The industry is facing two fold problems with their workforce: 1. A dearth of right skill at senior and middle level, 2. Retention of rightly skilled employees at middle level and top level. Smaller organizations are facing the problem in more severe notes. Hence retention of right skill set and employee engagement is become one of the most important issues to ponder over for the HR practitioners at IT industry. To identify and develop proper job—fit skill sets for ongoing/upcoming projects of the company and the to maintain competence level of the work force at par with the skill demand, measurement and management of performance occupies a pivotal role in this industry. Testing the hypothesis has established the linkage between performance appraisal process variables and the employee development policy as well as innovation oriented culture. The significant regression weights are also establishing the hypothesis, testing the association between appraisal process and employee engagement mediated through employee development policies and innovation oriented culture. The significant regression weights (0.466**) predicts almost 50 % association between employee development policies and employee engagement, which undoubtedly establishes the importance of a proper and fair employee development policy based on fair evaluation of performances. Among the latent variable, "Promotion of continuous learning", *Focus on Skill based mentoring*, Regularity in *Mentoring and guidance*, Scope of long term *development* have got high scores than the others. Positive rating on *Innovation, Practice of encouraging* the Innovative Ideas, *Management's belief* in innovation at lower level also scored a high estimate. Selection of the employees for increments, rewards, or identification of skill sets for next projects are decided based on previous performance data. These issues are generally settled in appraisal feedback sessions. During the field study, it is apparent from the interviews that while deciding personnel for the next projects or rewards and incentives, fair and equitable decisions are hardly made. In many cases the choice of people are biased by the line managers. The scope of innovation and empowerment is obviously low in the small companies as impressions gathered during in depth interviews of the people. The feedback system is also criticized by many of the respondents from big MNCs as well as SMEs. The normalization process in the appraisal system is found to be one major source of distrust among the executives of the big MNCs. Highly complex activities and organizational practices has created many lacunas of distrust and profligate power play in the companies. This environment creates distrust and they are real cause for most deserving and potential employees.

A causal linkage between the appraisal process variables, employee development policies, innovation oriented culture and employee engagement is found to be significant. It is also clear from the regression estimates that innovation oriented culture and employee development policies are exhibiting a consequential as well as positive effect on faith among the employees, The respondents in all companies irrespective of their size have clearly indicated a positive impact of fair and equitable appraisal process facilitating employees' long term development and encourages innovation among the work force can enhance the organizational performance and can help to create a committed set of work force.

6   Implications and Conclusion

It is evident from the above discussion that though the experiential variables are showing a high association with the latent variables and significant relationship together but the regression weights of appraisal process factor is showing somewhat less values on employee development and innovation culture. This implies a lack of coordination between appraisal process and the development and innovation encouraging HRD policy. As understood from the in depth study, that the problem is inherent in its structure and mode of operation. The increased power politics some time worsen the hard competition even more for the employees. Appraisal is looked as a coercive tool and approach of reprimand. As gathered from the direct interview of the employees and the HR professionals from the software companies' understudy, it is implicit that, policy wise, the monitoring of employees' performances in these companies are continuous in nature, involving the line managers/team leads and most of the HR payoffs are related to appraisal outcomes. Due to the high mobility of the employees through different kinds of projects and teams experiencing different skill requirement and also different types of leadership practice high adaptability and high flexibility is a requisite to sustain on the job for the employees. On the part of the employer it is also a challenge to assess the employee performance and manage administrative formalities on the basis of the results. The organizations reported a quarterly system of appraisal to assess the employee performance. Hence the appraisal process should be administered more critically and with prudence. The team leader should be trained to handle the appraisal data and in man management more rigorously. Continuous monitoring and prompt guidance can develop a sustained group of work force. The organizations can follow a cascading model of mentoring at all levels of work force. The process should start from top level management. Step I: Identification of requisite set of competences according to the organizational requirements. Step II: Clustering of similar competency sets based on the importance to the organization, complexity in nature, similarity and/or uniqueness Identification of competency sets at all of work force. Step III: Mapping the Competency sets of the employees at all level. Step IV: Identification of the best performers for each of the cluster. Step IV: Assigning the role of mentor for the group whose competency set is aimed to develop. The process should be continued in a continuous manner. Best mentor award can create interest among the line managers and can motivate them to spend time and effort creating a best pool of subordinates. Subordinate development can be a criterion of supervisory appraisal system which will create a mutual dependency and can reduce the bias.

A congenial environment of mutuality, faith, role dependency, management of esteem and positive competition can surely provide a positive and synergy effect to create a high performing team. Practice of clear communication from the immediate supervisory levels, leaders' position power, and a positive attitude of line managers are always favorable to provide productive results. Proper attention and accounting of the organizational citizenship behavior can augment the employee engagement intention with the organizations.

References

1. Agarwala, T. (2007). *"Strategic Human Resource Management"*. New Delhi: Oxford University Press.
2. Anthony, R.N. (1965). *"Planning and control systems: A framework for analysis"*. Graduate School of Business Administration, Harvard University: Boston.
3. Armstrong, M. (2010). *""Armstrong's Handbook of Reward Management Practice improving performance through reward"*. Kogan Page: New Delhi.
4. Armstrong, M., & Baron, A. (2002). *"Performance Management: The new realities"*. New Delhi: Jaico Publishing House.
5. Armstrong, M., & Brown, D. (2006). *"Strategic Reward: Making it Happen"*. London, UK: Kogan Page.
6. Arora, A., & Gambardella, A. (2005). *"From Underdogs to Tigers: The rise and growth of the software industry in Brazil, China, India, Ireland and Israel"*. New York: Oxford University Press.
7. Bandura, A. (1988). "Self-regulation of motivation and action through goal systems". In V. Hamilton, G.H. Bower, & N.H. Frijda (Eds.), *Cognitive perspectives on emotion and motivation* (pp. 37–61). Dordrecht, Netherlands: Kluwer Academic Publishers.
8. Bandura, A. (1977). *"Social learning theory"*. Prentice-Hall: Englewood Cliffs. N.J.

9. Bandura, A. (1978). "The self system in reciprocal determinism". *American Psychologist. Vol 33(4)*, 344–358.
10. Bhatnagar, J. (2007). "Talent Management strategy of employee engagement in Indian ITES employees: Key to retention". *Employee Relations. Vol. 29. No. 6.*, 640–663.
11. Bhattacharyya, D.K. (2007). *"Human Resource Research Methods"*. New Delhi: Oxford University Press.
12. Bohlander, G., & Snell, S. (2007). *"Human Resource Management"*. New Delhi: Change learning India.
13. Borman, W.C. (1979). "Format and Training effects on Rating accuracy and Rater Errors". *Journal of Applied Psychology. Vol. 64. No. 4*, pp. 410–421.
14. BoseBiswas, S. (2014). *"Performance Appraisal System for Executives: A case study in some selected companies in West Bengal"*. Kalyani. India: The University of Kalyani.
15. Bourne, M., Mills, J., Wilcox, M.N., & Platts, K. (2000). "Designing, Implementing and Updating Performance Measurement Systems". *International Journal of Operations and Production Management. Vol. 20. No. 7*, 754–771.
16. Cleveland, J.N., Murphy, K.R., & William, R.E. (1989). "Multiple Uses of Performance Appraisal: Prevalence and Correlates. *Journal of Applied Psychology. Vol. 74*, 130–135.
17. Culbertson, S.S., Henning, J.B., & Payne, S.C. (2013). "Performance appraisal satisfaction: The role of feedback and goal orientation". *Journal of Personnel Psychology, 12(4)*, 189–195.
18. Dover, P. A., Lawler, W., & Hilse, H. (2008). "Creating an entrepreneurial mindset at Infineon technologies: The Infineon-Babson global manager development programme". *Journal of Change Management 8 (3–4)*, 265–77.
19. Fleenor, J.W., & Prince, J.M. (1997). *"Using 360-degree Feedback in the Organizations: An Annoted Bibliography"*. North Carolina: Centre for Creative Leadership.
20. Folger, R., Konovsky, M.A., & Cropanzano, R. (1992). "A Due Process Metaphor for Performance Appraisal". *Research in organizational Behavior. Vol. 14*, 129–177.
21. Frahm, J., & Brown, K. (2007). "First steps: Linking change communication to change receptivity". *Journal of Organizational Change Management. 20(3)*, 370–87.
22. Ghosh, A.K. (2013). "Employee Empowerment: A Strategic Tool to obtain Sustainable Competitive Advantage". *International Journal of Management. Vol. 30. No. 3. Part 1*, 95–107.
23. Gliem, J.A., & Gliem, R.R. (2003). *"Calculating, Interpreting, and Reporting: Cronbach's Alpha Reliability Coefficient for Likert-Type Scales"*. Midwest Research to Practice Conference in Adult, Continuing, and Community Education.
24. Hartog, N.D., Paul, B., & Jaap, P. (2004). "'Performance Management: A model and Research Agenda". *Applied Psychology: An international review. 53(4)*, 556–569.
25. Heinen, S.J., & O'Neill, C. (2004). "Managing talent to maximize performance". *Employment Relations Today. Vol. 31., p. 2*.
26. Hutchison, S., & Sowa, D. (1981). "Perceived Organizational Support". *Journal of Applied psychology. Vol. 71. No. 3*, 500–507.
27. Ilgen, D.R., & Feldman, J.M. (1983). "Performance Appraisal: A Process Focus". In *Research in Organizational Behavior* (pp. 5, 141–197). Greenwich: CT: JAI Press.
28. John, P., Meriac, C., Allen, G., & Therese, M. (2015). "Seeing the Forest but Missing the Trees: The Role of Judgments in Performance Management". *Industrial and Organizational Psychology, 8*, 102–108.
29. Judge, T.A., & Ferris, G.R. (1993). "Social context of performance evaluation decisions". *Academy of Management Journal. Vol 36*, 80–105.
30. Khatri, K.G. (2006). *"Performance Management—People CMM level 2 Process Area—CSC in India's experience"*. New Delhi: Computer Sciences Corporation India.
31. Kim, J.S., & Hamner, W.C. (1976). "Effect of Performance Feedback and Goal Setting on Productivity and Satisfaction in an Organizational Setting". *Journal of Applied Psychology. Vol. 61, No. 1*, 48–57.
32. Kram, K.E., & Isabella, L.A. (1985). "Mentoring Alternatives: The role of peer relationship in career development". *Academy of Management Journal. Vol. 28. No. 1*, 110–132.

33. Lane, S. (2001). *"Offshore software development: Localisation, globalisation and best practices in the evolving industry"*. Boston, USA: Aberdeen IT Services Practice: Aberdeen Group.
34. Locke, E.A., Cartledge, N., & Koeppel, J. (1968). "Motivational effects of knowledge of results: A goal-setting phenomenon". *Psychological Bulletin, Vol 70(6)*, 474–485.
35. Locke, E.A., Shaw, K.N., Saari, L.M., & Latham, G.P. (1980). "Goal setting and task performance: 1969–1980". *Psychological bulletin*, 1–90.
36. Malik, K.A., & Bakhtawar, B. (June 2014). "Impact of appraisal system on employee performance: A comparison of permanent and contractual employees of Pakistan Telecommunications Company Limited (PTCL)". *European Scientific Journal*, 98–109.
37. McGregor, D. (1957). "An uneasy look at performance appraisal". *Harvard Business Review, Vol 35 No. 3*, 89–94.
38. Murphy, K.R., & Balzer, W.K. (1989). "Rater Errors and Rating Accuracy". *Journal of Applied Psychology, 74, 619–624*, 619–624.
39. Murphy, K.R., & Cleveland, J.N. (1995). *"Understanding Performance Appraisal: Social, Organizational, and Goal based Perspectives"*. Thousand Oaks: CA: Sage.
40. Murphy, K.R., Balzer, W., Kellem, K., & Armstrong, J. (1982). "Effects of Purpose of Rating on Accuracy in Observing Teacher Behavior and Evaluating Teacher Performance". *Journal of Educational Psychology. Vol. 76*, 45–54.
41. Pareek, U., & Rao, T.V. (2006). *"Designing and Managing Human Resource System"*. New Delhi: Oxford and IBH Publishing Co. Pvt. Ltd.
42. Pfeffer, J. (1994). *"Competitive advantage through people: unleashing the power of the workforce"*. Boston, M.A: HBS Press.
43. Proctor, K.S. (2011). *"Optimizing and Assessing Information Technology: Improving Business Project Execution"*. John Wiley & Sons.
44. Pulakos, E.D., Mueller, H.R., Arad, S., & Moye, N. (2015). "Performance management can be fixed: An on-the-job experiential learning approach for complex behavior change". *Industrial and Organizational Psychology: Perspectives on Science and Practice. Vol 8*, 51–76.
45. Ramya, T.V. (2005). "Normalization of Performance Ratings and its Practice in Indian Software Organizations". *IIMB management review, December*.
46. Scheweyer, A. (2004). *"Talent Management Systems: Best Practices in Technology Solutions for Recruitment, Retention and Workforce Planning*. New York: Wiley.
47. Schuler, R. (1989). "Strategic Human Resource Management and Industrial Relations". *Human Relations. Vol. 42*, 157–84.
48. Schunk, D.H. (1989). " Social cognitive theory and self-regulated learning". In B.H. Zimmerman, & D.H. Schunk, *Self-regulated learning and academic achievement: Theory, research, and practice* (pp. 83–110). New York: Springer-Verlag.
49. Sharma, O. (2011). *"DQ CMR Best Employers 2011: Reaching out to retain"*. Retrieved from Data quest: http://dqindia.ciol.com/content/top_stories/2011/111092201.asp.
50. Shivakumar, G. (2013). *"India is set to become the youngest country by 2020"*. New Delhi: THE HINDU.
51. Simons, R. (1990). "The role of management control systems in creating competitive advantage: New perspectives". *Accounting, Organizations and Society. Vol. 15 (1/2)*, 127–143.
52. Smither, J.W., & Reilly, R. (1987). "True Inter correlation among Job Components, Time Delay in Rating and Rater Intelligence as Determinants of Accuracy in Performance Ratings". *Organizational Behavior and Human Decision Processes. 40*, pp. 369–391.
53. Tschang, T. (2001). *"The Basic Characteriostic of skills and Organizational Capabilities in the Indian Software Industry"*. Tokyo: ADB Institute.
54. Upadhya, C., & Vasavi, A.R. (2006). *"Work, Culture and Sociality in the Indian IT Industry: A Sociological Study"*. Bangalore: School of Social Sciences, National Institute of Advanced Studies.

Part V
Device System and Modeling

Estimation of MOS Capacitance Across Different Technology Nodes

Sarita Kumari, Rishab Mehra, Amit Krishna Dwivedi
and Aminul Islam

Abstract This paper presents an in-depth analysis of NMOS capacitances across various technology nodes and device parameters which are extracted for different operating regions namely accumulation, cutoff, saturation and triode, while keeping the aspect ratio same for each transistor. Since MOS capacitances are the key parameters for estimating process development, material selection and device modeling, this paper enlists their variation with gate-to-source voltage (V_{GS}) while keeping drain-to-source voltage (V_{DS}) constant. This paper also aims to present the impact of capacitance variation on device performance that includes operating speed, power consumption, delay product and so on. The simulations results have been extensively verified using HSPICE simulator @ various technology nodes.

Keywords MOS capacitance · Triode · Accumulation · Saturation · Gate · Switching

1 Introduction

The gate capacitance is an important parameter for CMOS technologies in terms of device characteristics and modeling. Various performance attributes are directly related with the MOS capacitance. Hence, an extensive study of MOS capacitance and its reliance on various device parameters and technology nodes is essential [1].

Sarita Kumari · Rishab Mehra (✉) · A.K. Dwivedi · Aminul Islam
Department of Electronics and Communication Engineering, Birla Institute of Technology,
Mesra, Ranchi 835215, Jharkhand, India
e-mail: rishabmehra03@gmail.com

Sarita Kumari
e-mail: ksarita92@gmail.com

A.K. Dwivedi
e-mail: amit10011.13@bitmesra.ac.in

Aminul Islam
e-mail: aminulislam@bitmesra.ac.in

© Springer Science+Business Media Singapore 2017
J.K. Mandal et al. (eds.), *Proceedings of the First International Conference
on Intelligent Computing and Communication*, Advances in Intelligent Systems
and Computing 458, DOI 10.1007/978-981-10-2035-3_30

With the increasing demand of compact devices, circuit miniaturization and reduction in operating bias, precise and accurate analysis of the circuit parameters is required. For understanding the basic characteristics of MOS capacitances, the intrinsic capacitances are separated from the parasitic capacitances, overlap capacitances and junction capacitances [2]. The overlap capacitances are determined geometrically by the gate and source/drain overlap areas. Large parasitic overlap and intrinsic capacitances are one of the main limitations for device application in the high frequency domain [3]. Apart from this, when the MOS is operated in the accumulation region, the capacitances between the substrate and source/drain are termed as junction capacitances [4, 5]. Under normal operating conditions, these capacitances are nonreciprocal in nature since the behavior of the MOSFET charging does not solely depend upon the two terminals between which the capacitance is formed [6].

Several models have been already reported in the literature for extraction of gate and overlap capacitances of the MOS. A two terminal p-substrate MOS structure has been utilized in [7] in which the poly-silicon depletion and quantum effects are neglected for simplification purposes. In [8], capacitance measurements have been performed on strained Si on relaxed $Si_{0.8}Ge_{0.2}$ buffer and then modelled using solutions of Poisson-Schrodinger equations. Statistical variations in gate capacitance of subthreshold MOSFET have been analyzed using 65-nm benchmarking BSIM4 model. Similarly, in [9] a novel model for gate capacitance has been proposed which includes temperature effects and takes into account both charge quantization and poly-Si depletion. However, in this paper the MEYER's model of gate capacitance has been utilized for accurate and precise determination of capacitance values and their variation with gate-to-source bias (V_{GS}) is studied keeping the drain-to-source voltage (V_{DS}) fixed at a particular value.

MOS capacitance model along with quantitative descriptions of gate capacitances is stated in Sect. 2. Simulation results and discussions along with the variation of capacitances with different technology nodes are mentioned in Sect. 3. Effects of capacitance variation on device performance have been also reported in the same section. Finally, concluding remarks are made in Sect. 4.

2 MOSFET Capacitance Model

The various capacitances formed in an NMOS transistor are shown in Fig. 1 and the MEYER's model used for their extraction is shown in Fig. 2.

Fig. 1 Various capacitances formed in an NMOS transistor

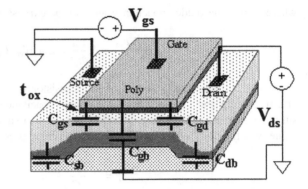

Fig. 2 Capacitance model of a MOSFET

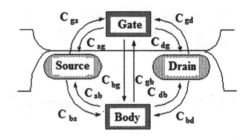

2.1 Gate-to-Bulk Capacitance (C_{GB})

In the accumulation region,

$$C_{GB} = cap \tag{1}$$

In the cutoff region,

$$C_{GB} = \frac{cap}{\sqrt{\left[1 + 4.\frac{V_{GS} + V_{SB} - V_{FB}}{\gamma^2}\right]}} \tag{2}$$

In the saturation region,

$$C_{GB} = \frac{G^+ .cap}{\sqrt{\left[1 + 4.\frac{\gamma(V_{SB} + PHI)^2 + V_{SB} + PHI}{\gamma^2}\right]}} \tag{3}$$

In the triode region,

$$C_{gb} \approx 0 \tag{4}$$

where, cap is the oxide capacitance and can be expressed as

$$cap = \frac{A.\varepsilon_{ox}}{t_{ox}} = A.\frac{\varepsilon_{ox}}{t_{ox}} = L_{eff} \, W_{eff} \, C_{ox} \tag{5}$$

Here, effective channel length of the transistor is denoted by L_{eff}, W_{eff} is the effective width and C_{OX} expresses the oxide capacitance per unit area. PHI is the surface potential under strong inversion and G^+ is a smooth factor having a fixed value. γ is the body effect coefficient and can be expressed as

$$\gamma = \frac{\sqrt{2q\varepsilon_{si}n_{sub}}}{C_{ox}} \tag{6}$$

where, q is the oxide charge, n_{sub} is the substrate doping concentration and permittivity of silicon is ε_{si}. Also, V_{SB} and V_{FB} are source-to-bulk and flat band voltages respectively.

2.2 Gate-to-Source Capacitance (C_{GS})

In the accumulation region,

$$C_{GS} \approx 0 \tag{7}$$

In the cutoff region,

$$C_{GS} = CF5.cap + \frac{cap.(V_{GS} - V_{TH})}{0.75PHI} \tag{8}$$

In the Saturation region,

$$C_{GS} = CF5.cap \tag{9}$$

In the triode region,

$$C_{GS} = CF5.cap.\left\{ 1 - \left[\frac{V_{Dsat} - V_{DS}}{2(V_{Dsat} + V_{SB}) - V_{DS} - V_{SB}} \right]^2 \right\} \tag{10}$$

where, V_{Dsat} is the saturation drain voltage and $CF5$ is the modified MEYER control for the capacitance multiplier for C_{GS} in the saturation region.

2.3 Gate-to-Drain Capacitance (C_{GD})

In the accumulation and cutoff region,

$$C_{GD} \approx 0 \tag{11}$$

In the saturation region,

$$C_{GD} = CF5.cap.D^+ \tag{12}$$

In the triode region,

$$C_{GD} = CF5.cap.\max\left\{D^+, 1 - \left[\frac{V_{Dsat} - V_{DS}}{2(V_{Dsat} + V_{SB}) - V_{DS} - V_{SB}}\right]^2\right\} \tag{13}$$

Here, D^+ is another smooth factor with fixed value.

3 Simulation Results and Discussions

The MOS capacitances are extracted @ various technology nodes (250, 180, 130, 90, 65 and 45 nm) using HSPICE simulator. Since the MOS capacitances are a function of the transistor aspect ratio (W/L ratio) [3], the aspect ratios at all technology nodes are kept constant during analysis.

The parametric details regarding the capacitance estimation are tabulated in Table 1 and the capacitance versus V_{GS} plot for different technology nodes are shown in Figs. 3, 4, 5, 6, 7, and 8.

Following observations can be made from these figures:

- In the accumulation region, the gate-to-body capacitance (C_{GB}) is essentially constant up to twice of flat band voltage ($2\varphi_f$) and then decreases exponentially. The gate-to-source capacitance (C_{GS}) and gate-to-drain capacitance (C_{GD}) are almost equal and negligible in magnitude.

Table 1 Parameters for capacitance estimation @ T = 25 °C

Technology nodes (nm)	Channel length (L) (μm)	Channel width (W) (μm)	V_{DS} (V)	V_{GS} (V)
250	30	30	0.65	−3.3 to 3.3
180	30	30	0.64	−3.3 to 3.3
130	30	30	0.60	−2.5 to 2.5
90	30	30	0.65	−2.5 to 2.5
65	30	30	0.43	−2.5 to 2.5
45	30	30	0.44	−2 to 2

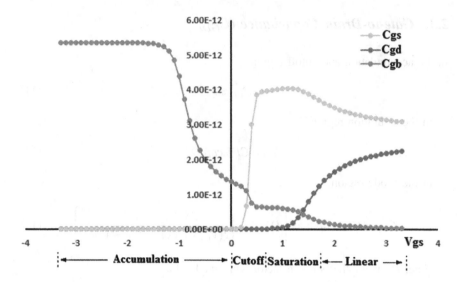

Fig. 3 Capacitance versus V_{GS} plot @ 250 nm technology node

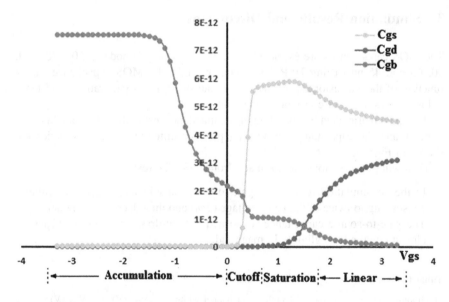

Fig. 4 Capacitance versus V_{GS} plot @ 180 nm technology node

- In the cutoff region, C_{GB} continues to decrease while C_{GS} begins to rise steeply and reaches a maximum value. However, C_{GD} continues to remain negligible.
- In the saturation region, C_{GB} as well as C_{GS} become effectively constant and C_{GD} begins to increase towards C_{GS}.

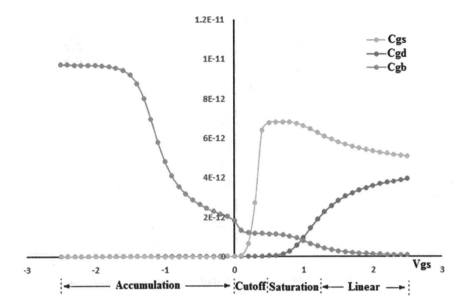

Fig. 5 Capacitance versus V_{GS} plot @ 130 nm technology node

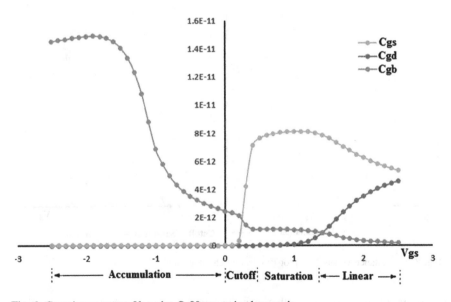

Fig. 6 Capacitance versus V_{GS} plot @ 90 nm technology node

- In the triode region, C_{GB} decreases further and becomes almost negligible. Apart from this, C_{GS} decreases while C_{GD} continues to increase. These two capacitances approach each other and become essentially equal if the gate-to-source bias is increased further.

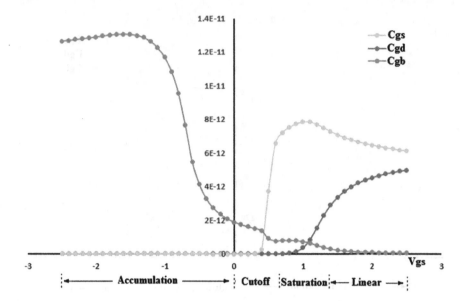

Fig. 7 Capacitance versus V_{GS} plot @ 65 nm technology node

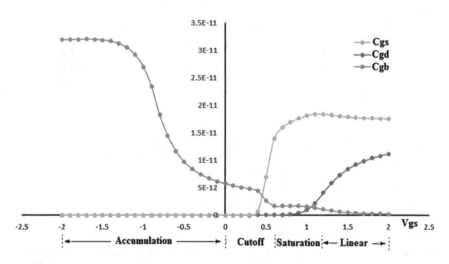

Fig. 8 Capacitance versus V_{GS} plot @ 45 nm technology node

Some of the inferences that can be drawn as the technology generation decreases:

- The gate-to-source capacitance (C_{GS}) becomes essentially linear in the saturation region for 65 and 45 nm technology nodes whereas it is constant in case of other technology nodes.

- The capacitance values increase with decrease in the technology node since a very thin gate oxide separates the gate and source/drain electrode which increases the effects of intrinsic and parasitic capacitances tremendously [4].
- The difference in C_{GS} and C_{GD} in the triode region increases with decrease in the technology node. This is due to the decrease in the difference between V_{Dsat} and V_{DS} since the operating bias decreases with decrease in the technology generation.
- The gate-to-bulk capacitance (C_{GB}) increases considerably in comparison to C_{GS} and C_{GD} due to decrease in the oxide thickness t_{OX} with different technology nodes.

The dynamic response of a MOS transistor is dependent on the time taken by the intrinsic and parasitic capacitances of the device and the extra capacitances introduced by the interconnecting lines to charge (discharge). Hence, with increase in device capacitance the average propagation delay (t_p) increases and hence the circuit switching and speed decreases.

For an inverter with transistors operating in the saturation region, t_p can be expressed as [10]

$$t_p = \frac{1}{2}\left(t_{PHL} + t_{PLH}\right) = \frac{C_L}{2V_{DD}}\left[\frac{1}{K_P} + \frac{1}{K_N}\right] \tag{14}$$

where, t_{PHL} and t_{PLH} represents the t_P during high to low and low to high transitions respectively. K_N and K_P are the device process parameters for the NMOS and PMOS transistors and C_L is the load capacitance.

This load capacitance primarily consists of the gate capacitance of fan-out gate, drain diffusion capacitance of the output transistors and the interconnect capacitances. Therefore, the load capacitance C_L increases with decreasing technology node due to the increase in gate capacitance and drain diffusion capacitances.

4 Conclusion

A precise and accurate extraction of MOS capacitance is presented across various operating regions @ different technology nodes, keeping the aspect ratios constant. Effect of capacitance variation on device switching characteristics has been also studied. All simulation results have been done in the HSPICE simulator using MEYER's model gate capacitances and demonstrate that the MOS capacitance mainly depends upon the bias voltage and region of operation.

References

1. R. Chau, Datta S., M. Doczy, Doyle B., Kavalieros J., M. Metz, "High-κ/metal-gate stack and its MOSFET characteristics," *IEEE Electron Dev. Lett.*, vol. 25, no. 6, pp. 408–410, 2004.
2. Y. Taur, Ning T. H., "Funds. of modern VLSI devices," *Cambridge university press*, 2009.
3. I. Saad, Riyadi M. A., F. M. N. Zul A., R. Ismail, "Reduced parasitic capacitances analysis of nanoscale vertical MOSFET," *Semiconductor Electronics (ICSE) IEEE Int. Conference on*, pp. 25–29, Jun. 2010.
4. Y. Ohkura, Toyabe T., H. I. R. O. O. Masuda, "Analysis of MOS capacitances and their behavior at short-channel lengths using ac device simulator," *IEEE Trans. Comput.-Aided Design Integr. Ckts Syst.*, vol. 6, no. 3, pp. 423–430, 1987.
5. Rabaey J. M., Chandrakasan A. P., Borivoje N., "Digital integr. ckts," vol. 2, *Englewood Cliffs: Prentice hall*, 2002.
6. D. E. Ward, "Charge-based mod. of cap. in MOS T," *Integr. ckts. labs*, June 1981.
7. J. He, X. Xi, M. Chan, K. Cao, Niknejad A., Hu C., "A physics based analytical surface pot. and cap. model of MOS operation from acc. to dep. region," *Proc. Nanotech NSTI*, vol. 2, pp. 302–305, Feb. 2003.
8. F. Gilibert, D. Rideau, F. Payet, F. Boeuf, E. Batail, M. Minondo, Jaouen H., "Strained Si/SiGe MOS cap. modeling based on band structure analysis," *Solid-State Dev. European Research 35th Conf.* (ESSDERC 2005) *Proc.*, pp. 281–284, Sep. 2005.
9. L. Larcher, P. Pavan, Pellizzer F., Ghidini G., "A new mod. of gate cap. as a simple tool to extract MOS param.," *IEEE Trans. Electr. Dev.*, vol. 48, no. 5, pp. 935–945, 2001.
10. A. K. Dwivedi, P. Abhijeet, R. Mehra, A. Islam, "Versatile Noise Suppressed Variable Pulse Voltage Controlled Oscillator," *Int. Journal of App. Engg. Research (IJAER)*, vol. 10, no. 20, pp. 18633–18638, May. 2015.

Cross-Coupled Dynamic CMOS Latches: Scalability Analysis

Rishab Mehra, Sarita Kumari and Aminul Islam

Abstract This paper primarily focuses on the power dissipation of cross coupled CMOS dynamic latches and also takes the technology scalability of the design into account. Mainly 3 topologies namely the Cascade Voltage Switch Logic (CVSL), Dynamic Single Transistor Clocked (DSTC) and Dynamic Ratio Insensitive (DRIS) have been investigated. A comparative study is provided which validates the suitability of the above latches for high-speed low power applications. Further, a brief account regarding the use of these latches for the design of high speed edge triggered flip-flops is also provided. The simulations results have been extensively verified on SPICE simulator using TSMC's industry standard 180 nm technology model parameters and the technology scalability is tested with 22 nm predictive technology model developed by Nanoscale Integration and Modeling (NIMO) Group of Arizona State University (ASU).

Keywords Power dissipation · Switching · Latch · Scalability · Delay

1 Introduction

Latches and flip-flops are the most important components of sequential as well as microprocessor based systems. They primarily decide the clock frequency of most digital circuits since their delay needs to be taken into consideration while selecting a suitable clock cycle [1, 2]. Hence, a major portion of the overall circuit speed and power is affect by the type of topology of flip-flops and latches used in the design.

Rishab Mehra (✉) · Sarita Kumari · Aminul Islam
Department of Electronics and Communication Engineering, Birla Institute of Technology,
Mesra, Ranchi 835215, Jharkhand, India
e-mail: rishabmehra03@gmail.com

Sarita Kumari
e-mail: ksarita92@gmail.com

Aminul Islam
e-mail: aminulislam@bitmesra.ac.in

© Springer Science+Business Media Singapore 2017   307
J.K. Mandal et al. (eds.), *Proceedings of the First International Conference
on Intelligent Computing and Communication*, Advances in Intelligent Systems
and Computing 458, DOI 10.1007/978-981-10-2035-3_31

Several analysis involving the delay, power and leakage characteristics of flip-flops are available in literature. A novel analysis of leakage and delay characteristics of 19 (4 categories) nanometer flip-flop topologies is presented in [1, 2]. Apart from this, the delay-area and delay-leakage tradeoffs have also been reported. In addition, the work also involves the impact of layout parasitics and estimates the energy dissipation which is contributed by the static, dynamic and short circuit parameters. A similar analysis involving double edge triggered flip-flops is seen in [3]. Double edge triggered flip-flops (DETs) reduce the clocking dissipation without any compromise in speed and can be operated with a clock frequency half of that of the conventional single edge triggered flip-flops. Variability analysis of 14 single edge triggered flip-flops is reported in [4, 5]. Several studies involving the effect of process, temperature and supply variations on setup and hold times of flip-flops, propagation delay, delay sensitivity etc. have also been carried out.

This paper provides an accurate and precise analysis of the performance of dynamic CMOS latches in terms of average power dissipation, EDP, PDP, propagation delay and delay variability. Further, the technology scalability of the design is also validated. Apart from this, an overview regarding the design of single edge triggered flip-flops using the master-slave configurations of the above latches is also reported. The rest of the paper is organized as follows. A brief literature review of the Cascade Voltage Switch Logic (CVSL), Dynamic Single Transistor Clocked (DSTC) and Dynamic Ratio Insensitive (DRIS) latches is provided in Sects. 2, 3 and 4 respectively. Their performance in terms of the above design metrics is looked upon in Sect. 5 and the effect of technology scaling is also discussed. Further, the design of master-slave single edge-triggered flip-flops is provided in Sect. 6. Finally, concluding remarks are given in Sect. 7.

2 Cascade Voltage Switch Logic (CVSL) Differential Latches

The cascade voltage switch logic (CVSL) computes both true and complementary outputs from the true and complementary input values using a pair of NMOS pull-down networks (n-type) and a pair of PMOS pull-up networks (p-type) [6]. The two types of CVSL latches are shown in Fig. 1. The CVSL latches have a speed advantage over the conventional True Single Phase Clocking (TSPC) latches but are sensitive to transistor aspect ratios (particularly the p-type latch) [7]. They are also sensitive to temperature and process variations (Tables 1 and 2).

The n-type latch has a potential speed advantage since all of the logic switching is performed by the NMOS transistors, leading to a reduction in the input capacitance. This can be intuitively seen from Table 3 depicting the propagation delay (t_p) for the p-latch to be nearly 2.5 times of that of the n-latch.

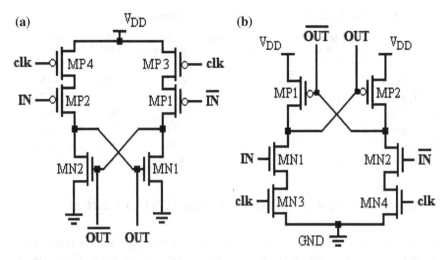

Fig. 1 a CVSL-p-type latch, b CVSL-n-type latch

Table 1 Specifications for CMOS n-type latches @ different technology nodes

Technology node (nm)	PMOS channel length (L_p) (nm)	PMOS channel width (W_p) (nm)	NMOS channel length (L_n) (nm)	NMOS channel width (W_n) (nm)	V_{DD} (V)	Temperature (°C)
22	2 × 22	4 × 22	2 × 22	4 × 22	0.8	25
180	2 × 180	4 × 180	2 × 180	4 × 180	1.8	25

Table 2 Specifications for CMOS p-type latches @ different technology nodes

Technology node (nm)	PMOS channel length (L_p) (nm)	PMOS channel width (W_p) (nm)	NMOS channel length (L_n) (nm)	NMOS channel width (W_n) (nm)	V_{DD} (V)	Temperature (°C)
22	2 × 22	8 × 22	2 × 22	1 × 22	0.8	25
180	2 × 180	8 × 180	2 × 180	1 × 180	1.8	25

Table 3 Design Metrics for CMOS latches @ 180 nm technology node

Latch	Propagation delay $(t_p) \times 10^{-9}$ (s)	Delay variability (a.u.) (σ/μ)	Power dissipation $\times 10^{-7}$	PDP $\times 10^{-15}$	EDP $\times 10^{-24}$
CVSL n-type	1.99	0.062	3.84	0.76	1.53
CVSL p-type	4.99	0.192	5.80	2.95	15.61
DSTC n-type	1.98	0.060	4.08	0.81	1.62
DSTC p-type	4.30	0.157	5.81	2.53	11.30
DRIS n-type	2.03	0.048	2.33	0.47	0.95
DRIS p-type	3.08	0.043	2.51	0.77	2.39

Table 4 Design Metrics for CMOS latches @ 22 nm technology node

Latch	Propagation delay (t_p) $\times 10^{-10}$ (s)	Delay variability (a.u.) (σ/μ)	Power dissipation $\times 10^{-9}$	PDP $\times 10^{-18}$	EDP $\times 10^{-27}$
CVSL n-type	6.33	0.77	6.21	5.61	13.5
CVSL p-type	8.17	0.41	4.73	4.28	5.31
DSTC n-type	6.39	1.06	6.19	6.21	26.2
DSTC p-type	11.90	0.15	4.67	5.67	7.22
DRIS n-type	3.89	0.09	2.96	1.15	0.44
DRIS p-type	9.25	0.05	2.30	2.13	1.98

3 Dynamic Single Transistor Clocked (DSTC) Latches

The two clocked transistors in the CVSL p-type and n-type latches (Fig. 1) are merged into a single transistor so as to improve the power efficiency. This design is called the Dynamic Single Transistor Clocked (DSTC) latch (Fig. 2). However, a charge sharing takes place in the latched state between the two input transistors which conduct simultaneously during input transition. This charge cannot be recovered in the dynamic latches leading to timing issues and large propagation delays, especially for the p-type latch [7]. This can be verified from Tables 3 and 4 in which the DSTC p-type latch has the second largest t_p at 180 nm technology node and the largest at 22 nm technology node.

Hence, the n-type latch due to better switching characteristics and immunity to transistor sizing can be used and cascaded in any form to design high-speed positive or negative edge-triggered flip-flops. However, the same is not true for the p-type latch due to its speed bottlenecks.

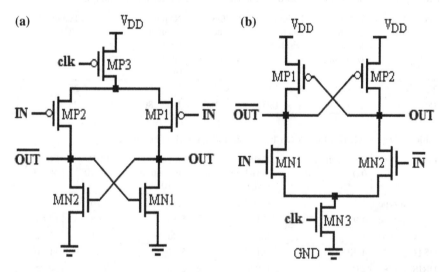

Fig. 2 a DSTC p-type latch. **b** DSTC n-type latch

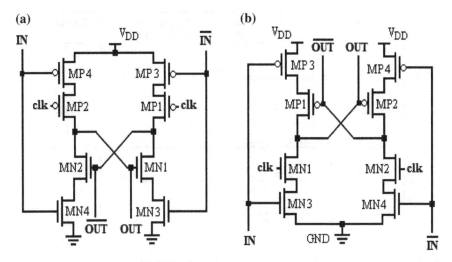

Fig. 3 **a** DRIS p-type latch. **b** DRIS n-type latch

4 Dynamic Ratio Insensitive (DRIS) Differential Latches

The Dynamic Ratio Insensitive (DRIS) latches (Fig. 3) are the most insensitive to transistor sizing and aspect ratios. Hence, they are the most robust out of all the dynamic latch topologies discussed so far. This can be seen from the Tables 3 and 4 in which the DRIS n-type latch has the least delay variability at both 22 nm and 180 nm technology node followed by the DRIS p-type latch.

Apart from this, the DRIS n-type and p-type latches have the least amount of power dissipation as compared to other latches (Tables 3 and 4) at both 22 nm and 180 nm technology nodes. Hence, verifying the technology scalability of the design. Such characteristics are useful for the design of high-speed low-power flip-flops which is discussed in the following sections.

5 Simulation Results and Discussions

The transistor sizing and other specifications used for simulation are tabulated in Table 1 (for n-type latches) and Table 2 (for p-type latches). The channel lengths and widths are selected so as to provide an optimum value of propagation delay t_p. Further, the n-type topologies were relatively insensitive to W_p/W_n ratios and work fine for equal widths. However, the p-type latches required the W_p/W_n ratio to be at least 6 for proper circuit functioning. This was not a necessity for the DRIS latches; however the aspect ratios were kept same for the sake of comparison.

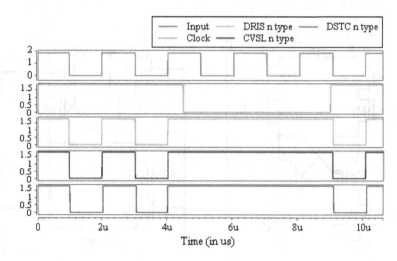

Fig. 4 Output characteristics of n-type latches @ 180 nm technology node

The output characteristics of the 6 dynamic CMOS latches at 180 nm technology node are shown in Fig. 4 (for n-type latches) and Fig. 5 (for p-type latches). The n-type latches have an active high clock input since the clock is applied to the gate of NMOS transistors. Similarly, the clock is active low for the p-type latches due to the clocking of PMOS transistors.

The above CMOS latch topologies were examined for various design metrics such as propagation delay t_p, delay variability, power dissipation, power delay product (PDP) and energy delay product (EDP). The results are tabulated in Table 3. Further, a similar analysis is carried at 22-nm technology so as to verify the scalability of the above designs, keeping the sizing ratios constant. The results for the same are tabulated in Table 4.

As seen, the CVSL p-type and DSTC p-type latches provide the maximum amount of delay as compared to the other latches, whereas, the CVSL n-type and DSTC n-type latches have minimum value of t_p. Also, the n-type latches along with DRIS p-type latch are far more robust than the remaining 2 topologies. Power dissipation is least for the DRIS latches followed by CVSL n-type latch, whereas, it is maximum for the DSTC p-type latch.

The circuit functionality remains fine even when the MOS is scaled down to 22 nm; however some variation in the performance of the latches is observed. The DRIS n-type latch provides the least delay, validating its sizing independent characteristics. However, the delay increases radically for the DRIS p-type latches as compared to the other p-type latches. Despite this, the DRIS n-type and p-type latches continue to remain the most robust and also have the least power dissipation. The overall circuit performance improves as compared to 180 nm which validates the scalability of the designs.

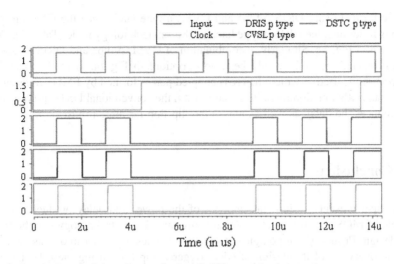

Fig. 5 Output characteristics of p-type latches @ 180 nm technology node

6 Design of High-Speed Edge Triggered Flip-Flops

Any of the discussed latches can be cascaded or combined suitably so as to form master-slave type edge triggered flip-flops (Fig. 6). A general design convention involves cascading of p-type and n-type latches. A high speed ratio insensitive latch

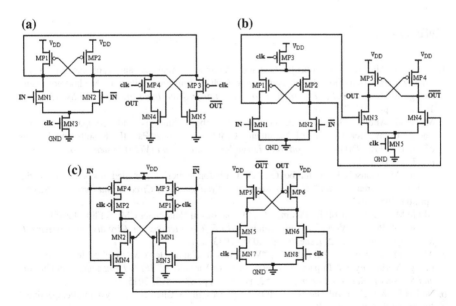

Fig. 6 Designs of edge triggered flip-flops using dynamic latches

can be formed using DRIS p-type and CVSL n-type latch since the CVSL n-type latch is faster than the DRIS n-type latch at 180 nm technology node (Table 3). This is shown in Fig. 6c. Similarly, a combination of CVSL p-type and DRIS n-type at 22 nm technology node would be a good option. In Fig. 6b, a modified DSTC p-type latch is used in which the cross coupled pair is formed by PMOS transistors. This makes it's functioning quite different from the conventional DSTC p-type latch and reduces the delay as well as makes the flip-flop far more robust [7, 8].

7 Conclusions

This paper presents an in-depth analysis of the power and delay product of cross coupled CMOS dynamic latches and also investigates the technology scalability of the design. Primarily 6 topologies of dynamic latches are taken into consideration and an overview of the design of edge triggered flip-flops using these latches are provided. Through rigorous examination using SPICE circuit simulator, it is observed that the DRIS n-type and p-type latches are the most power efficient at both 180 and 22-nm technology node. Also, these latches are more robust and insensitive to transistor sizing as compared to other latches. The DRIS n-type latch also has the least PDP and EDP despite technology scaling. However, the least delay at 180-nm technology node is provided by the DSTC n-type latch. The DSTC p-type latch provides maximum delay at 22-nm technology node and the CVSL p-type type is the speed bottleneck at 180-nm technology node.

References

1. Massimo Alioto, Elio Consoli, Gaetano Palumbo, "Analysis and Comparison in the Energy-Delay-Area Domain of Nanometer CMOS Flip-Flops: Part I—Methodology and Design Strategies," *IEEE Transactions on Very Large Scale Integration (VLSI) Systems*, vol. 19, no. 5, pp. 725–736, 2011.
2. Massimo Alioto, Elio Consoli, Gaetano Palumbo, "Analysis and Comparison in the Energy-Delay-Area Domain of Nanometer CMOS Flip-Flops: Part II—Results and Figures of Merit," *IEEE Transactions on Very Large Scale Integration (VLSI) Systems*, vol. 19, no. 5, pp. 737–750, 2011.
3. Alioto, M., Consoli E., Palumbo G., "DET FF topologies: A detailed investigation in the energy-delay-area domain," *IEEE International Symposium on Circuits and Systems (ISCAS)*, pp. 563–566, 2011.
4. Alioto Massimo, Consoli E., Palumbo G., "Variations in Nanometer CMOS Flip-Flops: Part I— Impact of Process Variations on Timing," *IEEE Transactions on Circuits and Systems I: Regular Papers*, vol. 62, issue: 8, pp. 2035–2043, 2015.
5. Alioto M., Consoli E., Palumbo G., "Variations in Nanometer CMOS Flip-Flops: Part II— Energy Variability and Impact of Other Sources of Variations," *IEEE Transactions on Circuits and Systems I: Regular Papers*, vol. 62, issue: 3, pp. 835–843, 2015.
6. Neil H.E. Weste, David Harris, "CMOS VLSI Design: A Circuits and Systems Perspective," *Pearson Education Inc.*, 2005.

7. Jiren Yuan, Christer Svensson, "New Single Clock CMOS Latches and Flip-flops with Improved Speed and Power Savings," *IEEE Journal of Solid State Circuit*, vol. 32, no. 1, pp. 62–69, January 1997.
8. IEEE Conference PublicationsJ. Yuan and C. Svensson. "High speed CMOS circuit technique," *IEEE Journal of Solid State Circuit*, vol. 24, pp. 62–70, Feb. 1989.

Cross-Coupled Dynamic CMOS Latches: Robustness Study of Timing

Rishab Mehra, Swapnil Sourav and Aminul Islam

Abstract This paper presents an in-depth analysis of the propagation delay of dynamic CMOS latches and its variability when subjected to process, voltage and temperature (PVT) variations. Three basic topologies namely the cascade voltage switch logic (CVSL), dynamic single transistor clocked (DSTC) and dynamic ratio insensitive (DRIS) have been investigated for robustness and switching characteristics. The extensive analysis provides well-defined guidelines for selection of variation-aware CMOS latches used in digital logic design. All simulations have been performed on 180 nm TSMC industry standard technology node using SPICE circuit simulator.

Keywords Robust · Propagation delay · Switching · Latches · Supply variations

1 Introduction

Latches are the basic building blocks of synchronous digital circuits and have a substantial impact on circuit speed and power dissipation [1]. Typically, 80 % of the total clock energy is dissipated by them [2] and they affect the performance energy trade-off at the chip level. A system's robustness is primarily dependent upon sequential circuits and hence most modern—designs which are subjected to large process/voltage/temperature (PVT) variations, require sufficient margins to prevent timing violations [3–5].

Rishab Mehra (✉) · Swapnil Sourav · Aminul Islam
Department of Electronics and Communication Engineering,
Birla Institute of Technology, Mesra, Ranchi 835215, Jharkhand, India
e-mail: rishabmehra03@gmail.com

Swapnil Sourav
e-mail: meec10032142@bitmesra.ac.in

Aminul Islam
e-mail: aminulislam@bitmesra.ac.in

© Springer Science+Business Media Singapore 2017
J.K. Mandal et al. (eds.), *Proceedings of the First International Conference on Intelligent Computing and Communication*, Advances in Intelligent Systems and Computing 458, DOI 10.1007/978-981-10-2035-3_32

Various delay and variability analysis of digital circuits have already been reported in literature. A thorough variation-aware comparison of 14 topologies of single-edge triggered flip-flops is reported in [6] and sensitivity of each flip flop (FF) topology to process variations is studied. Further, the impact of process variations on performance and hold time is also analyzed. The same FFs have been scrutinized for voltage, temperature and clock skew variations in [7]. Further, variability of energy and impact of technology scaling has also been reported. Both [6] and [7], use a simulation setup involving the FO4 configuration. A similar analysis for different full adder circuits is reported in [8] and delay sensitivity is analytically modeled with respect to supply variations. In [9], static timing analysis (STA) of digital circuits in the presence of supply voltage and ground variations is carried out. Apart from this, the propagation delay of static and dynamic CMOS logic is studied under process variations in [3]. Yet, to the best of our knowledge, no researcher has done the propagation delay and delay variability of dynamic CMOS latches (Fig. 1) with cross coupled configurations and differential outputs.

In view of the above, this paper provides an extensive analysis of the propagation delay and delay variability of dynamic CMOS latches (Fig. 1) with cross-coupled configurations and differential outputs.

The second part of this work is structured as follows. The impact of process, voltage and temperature (PVT) variations on the propagation delay (t_p) of the

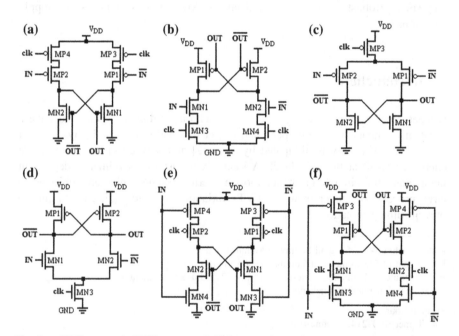

Fig. 1 a CVSL p-type. b CVSL n-type. c DSTC p-type. d DSTC n-type. e DRIS p-type. f DRIS n-type

dynamic latches is studied in Sects. 2, 3 and 4 respectively. Dependence of delay variability on the above parameters is analyzed in Sect. 5. Comparative results and the concluding remarks are made in Sect. 6.

2 Simulation Setup

International Technology Roadmap for Semiconductors (ITRS) identifies the key technical requirements and sets near and long term objectives for the semiconductor industry [2]. It projects variations in device parameters such as channel length (L), channel width (W), channel doping concentration (NDEP), oxide thickness (tOX), threshold voltage (Vt) and supply voltage (VDD), up to a range of ± 10 %. For greater accuracy of results, we run Monte Carlo simulations of 2000 iterations by varying the above device and environmental parameters and generating different SPICE model files for each set of parameters for considered circuits using at 180 nm TSMC industry standard model parameters [10]. The various process parameters including L, NDEP, tOX, and Vt are assumed to have independent Gaussian (or Normal) distributions with a 3σ variation of 10 %, as mentioned above.

3 Impact of Process Variations on Timing

A simple model can be used for defining the propagation delay t_p [3, 11]:

$$t_p = \frac{C_L}{I_{DS}} \cdot \frac{V_{DD}}{2} \tag{1}$$

$$I_{DS} = k \cdot W \cdot (V_{DD} - V_t)^\alpha \tag{2}$$

where, I_{DS} is the current provided by the output transistor of width W to the load capacitance C_L, k is a technology dependent transconductance coefficient, V_{TH} is the threshold voltage, α is the velocity saturation index ranging between one and two and V_{DD} is the supply voltage.

From (1), the delay is primarily dependent upon variations in the load capacitance C_L and the current I_{DS}. In particular, C_L accounts mainly for the gate capacitance of the fan out gate, interconnect capacitance and the output drain diffusion capacitance. Variations in C_L are primarily dependent on variations in the effective channel width, the channel length, and the oxide thickness. Among these, the most important parameter is the effective channel length [3]. Since the complementary states of the differential pair are highly dependent on the switching of the opposite MOS, t_p is studied with respect to the sizing of the MOS opposite to that whose drain gives the output state. The dependence of t_p on the length and width of the opposite transistor is shown in Figs. 2 and 3 respectively.

Fig. 2 Dependence of propagation delay on channel length of the opposite transistor

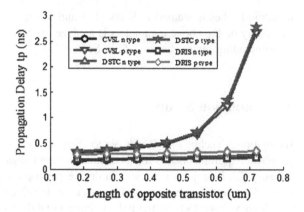

Fig. 3 Variation of propagation delay with the width of the opposite transistor

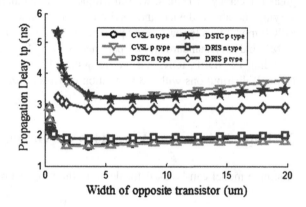

The CVSL p-type and the DSTC p-type latches are most sensitive to sizing variations and large changes in channel length and width may cause circuit malfunctioning or lead to timing violations. The DRIS n-type latch has the maximum immunity against channel length variations whereas the DSTC n-type latch provides the least delay for variations in width. The propagation delay for almost all latches decreases up to an optimized value of transistor aspect ratio and increases hence forth.

4 Impact of Supply Voltage Variations on Timing

Supply voltage (V_{DD}) variations are an important parameter for delay and its dependence is studied by varying the V_{DD} and V_{SS} values by $\pm 10\%$ around 1.8 V and 0 V respectively. The characteristic plots are shown in Figs. 4 and 5.

Fig. 4 Dependence of propagation delay on V_{DD}

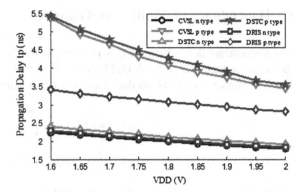

Fig. 5 Variation of propagation delay with V_{SS}

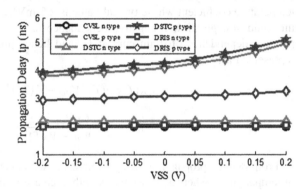

The impact of voltage variations on a generic parameter 'y' can be characterized using a sensitivity factor defined as [6]:

$$S_{V_{DD}}^{y} = \frac{\partial y}{\partial V_{DD}} \cdot \frac{V_{DD}}{y} \tag{3}$$

Hence, the sensitivity of delay t_p to V_{DD} can be found as [7, 8]:

$$S_{V_{DD}}^{t_p} = \frac{\partial t_p}{\partial V_{DD}} \cdot \frac{V_{DD}}{t_p} \approx 1 - \alpha \frac{V_{DD}}{V_{DD} - V_t} \tag{4}$$

A similar relationship can also be derived for VSS by making suitable adjustments in the I_{DS} equation. From (1) and (2), it is also clear that the delay is proportional to $V_{DD}/(V_{DD} - V_t)^{\alpha}$ and hence decreases with increase in the supply voltage. The opposite is true for VSS values which can be verified from the plots. The CVSL p-type and DSTC p-type latches are yet again most sensitive to VDD variation while other latches exhibit a fairly constant response.

5 Dependence of Timing on Temperature Variations

The temperature dependence of t_p is primarily due to variation in mobility (μ) and threshold voltage (Vt) of the MOSFETs which in turn affects the drain current I_{D-}. These dependences can be modelled using the following [10, 12]:

$$\mu(T) = \mu(T_o) \left(\frac{T_o}{T}\right)^m \tag{5}$$

$$V_t(T) = V_t(T_o) - K(T - T_o) \tag{6}$$

where, T_o is the reference temperature (300 K), m is the mobility temperature exponent whose typical value is 1.5 (ideally 1) and K is the threshold voltage temperature coefficient whose typical value is 2.5 mV/K. Hence, both mobility and threshold voltage show a negative dependence on temperature. Finally, the temperature dependence of drain current I_{D-} and hence propagation delay t_p can be represented as:

$$I_{DS}(T) \propto \mu(T) \left(V_{DD} - V_t(T)\right)^\alpha \tag{7}$$

$$t_p \propto \frac{C_L V_{DD}}{\mu(T) \left(V_{DD} - V_t(T)\right)^\alpha} \tag{8}$$

From (8), the propagation delay t_p increases with decrease in mobility (increase in temperature), whereas, it decreases with decrease in threshold voltage (increase in temperature). For supply voltages of around 2 V and threshold voltages of about 0.7 V, the impact of mobility degradation is more dominant and hence delay increases with increase in temperature. However, for V_{DD} values less than 1 V the impact of threshold voltage decrease surmounts the mobility degradation leading to overall positive temperature dependence [12]. This can be verified from Fig. 6 for

Fig. 6 Temperature dependence of propagation delay

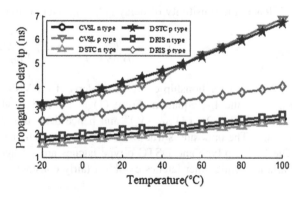

which the supply voltage is equal to 1.8 V. Again, the CVSL p-type and DSTC p-type latches are most sensitive to temperature variations and provide large amounts of delay at elevated temperatures. The DSTC n-type latch provides the minimum amount of delay when subjected to temperature variations followed by CVSL and DRIS n-type latches.

6 Variability Analysis of Dynamic Latches

A variability analysis of the dynamic CMOS latches is carried out at optimized aspect ratios (least delay) and its dependence on the above parameters (VDD and temperature variations) is studied. The relationship between delay variability and supply voltage variations (V_{DD} and V_{SS}) is shown in Fig. 7.

It can be seen that the CVSL p-type and DSTC p-type latches are the most sensitive to supply voltage variations and are the least robust as compared to the other latches. The delay variability decreases with increase in V_{DD} due to increase in the current I_D (from (2)) which is responsible for better switching characteristics. The opposite is true for V_{SS} as I_D increases with decrease in V_{SS} (more negative). A similar analysis is carried out with respect to temperature and the obtained results are shown in Fig. 8.

It can be seen that the delay variability increases with increase in operating temperature of the latch. This is justified from (7), since the drain current exhibits negative temperature dependence for circuits operating at around 2 V [12]. Hence, leading to degradation in device performance and switching. Yet again, the CVSL p-type and DSTC p-type latches are the most sensitive to temperature variations. Whereas the DRIS p-type and n-type latches are the most robust.

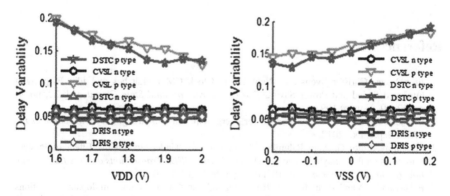

Fig. 7 Dependence of delay variability on VDD (*left*) and VSS (*right*)

Fig. 8 Temperature
dependence of delay
variability

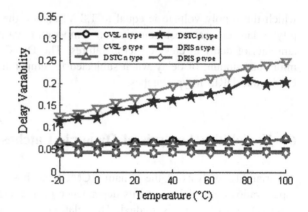

7 Conclusions

This second part of the paper presents a precise and accurate analysis of propagation delay (t_p) and delay variability which is carried out for 6 CMOS dynamic latches with cross coupled topologies and their responses when subject to process, supply and temperature variations are examined. All designs and their simulation results have been modeled using the industry standard 180 nm TSMC model parameters and the claimed theory is justified using SPICE simulator. The Cascade Voltage Switch Logic (CVSL) p-type latch is found to be the least robust and efficient. Through rigorous examination, it is found to be most sensitive to PVT variations and also provides maximum propagation delay. Hence, it is least suitable for high speed and variation-aware applications. The Dynamic Ratio Insensitive (DRIS) n-type latch provides minimum delay and is also least sensitive to aspect ratios. Moreover, the DRIS p-type latch is found to be the most robust. Also, the DSTC n-type latch provides the best tradeoff between circuit delay and variability.

References

1. Jiren Yuan, Christer Svensson, "New Single Clock CMOS Latches and Flip Flops with Improved Speed and Power Savings," *IEEE Journal of Solid State Circuit*, vol. 32, no. 1, pp. 62–69, January 1997.
2. B. Dally, "Architectures and circuits for energy-efficient computing," *keynote speech.at the CICC*, September 2012.
3. Massimo Alioto, Gaetano Palumbo, Melita Pennisi, "Understanding the Effect of Process Variations on the Delay of Static and Domino Logic," *IEEE Transactions on Very Large Scale Integration (VLSI) System,* vol. 18, no. 5, pp. 697–710, 2010.
4. F. Hassan,W. Vanderbauwhede, F. Rodríguez-Salazar, "Impact of random dopant fluctuations on the timing characteristics of flip-flops," *IEEE Transactions on Very Large Scale Integration (VLSI) System, vol. 20, no. 1, pp. 157–161, 2012.*

5. M. Alioto, G. Palumbo, M. Pennisi, "A simple circuit approach to reduce delay variations in domino logic gates," *IEEE Transactions on Circuits and Systems I: Regular Papers*, vol. 59, no. 10, pp. 2292–2300, 2012.
6. Alioto M., Consoli E., Palumbo G., "Variations in Nanometer CMOS Flip-Flops: Part I— Impact of Process Variations on Timing," *IEEE Transactions on Circuits and Systems I: Regular Papers*, vol. 62, issue: 8, pp. 2035–2043, 2015.
7. Alioto M., Consoli E., Palumbo G., "Variations in Nanometer CMOS Flip-Flops: Part II— Energy Variability and Impact of Other Sources of Variations," *IEEE Transactions on Circuits and Systems I: Regular Papers*, vol. 62, issue: 3, pp. 835–843, 2015.
8. Massimo Alioto, Gaetano Palumbo, "Impact of Supply Voltage Variations on Full Adder Delay: Analysis and Comparison," *IEEE Transactions on Very Large Scale Integration (VLSI) Systems*, vol. 14, issue: 12, pp. 1322–1335, 2006.
9. Ahmadi R., Najm F.N., "Timing analysis in presence of power supply and ground voltage variations," *IEEE International Conference on Computer Aided Design (ICCAD)*, pp. 176–183, 2003.
10. Neelam Arya, Shweta Singh, Manisha Pattanaik, "Temperature Insensitive Design For Power Gated Circuits," *9th International Conference on Industrial and Information Systems (ICIIS)*, pp. 1–6, 2014.
11. Sakurai T., Newton A.R., "Alpha-Power Law MOSFET Model and its Applications to CMOS Inverter Delay and Other Formulas," *IEEE Journal of Solid-State Circuits,* vol. 25, issue: 2, pp. 584–594, 1990.
12. Kanda K., Nose K., Kawaguchi H., Sakurai T., "Design Impact of Positive Temperature Dependence on Drain Current in Sub-1-V CMOS VLSIs," *IEEE Journal of Solid-State Circuits*, vol. 36, issue: 10, pp. 1559–1564, 2001.

A Design of a 4 Dot 2 Electron QCA Full Adder Using Two Reversible Half Adders

Sunanda Mondal, Debarka Mukhopadhyay and Paramartha Dutta

Abstract Quantum Cellular Automata (QCA) is one of the latest upcoming technology. In the present extension we have proposed a configuration of full adder utilizing two reversible half adders. However the proposed design contains only majority voter gate and inverter gate. Moreover the dissipation energy, incidence energy and effective area have been evaluated and reported also. To the best of our knowledge such a design in the present form has not been reported in literature as yet.

Keywords QCA · MV gate · Logically reversible half adder · Full adder · Coulomb's principle

1 Introduction

CMOS is the king of digital industry for last few decades as it has high operating speed, low power consumption, efficient use of energy and high degree of noise immunity [1, 2]. But day to day as our demands increases the transistor density on a chip also grows exponentially. And while to be integrated at nano scale level CMOS technology faces lots of problem. Mainly lithography cost and off-state leakage current grows exponentially when its dimension decreases. QCA is a developing innovation which has most of the ingredients to overcome the problem of CMOS technology.

Sunanda Mondal (✉) · Paramartha Dutta
Department of Computer & System Sciences, Visva Bharati University, Santiniketan, India
e-mail: sund.mondal@gmail.com

Paramartha Dutta
e-mail: paramartha.dutta@gmail.com

Debarka Mukhopadhyay
Department of Computer Science, Amity School of Engineering and Technology,
Amity University, Kolkata, India
e-mail: debarka.mukhopadhyay@gmail.com

© Springer Science+Business Media Singapore 2017
J.K. Mandal et al. (eds.), *Proceedings of the First International Conference on Intelligent Computing and Communication*, Advances in Intelligent Systems and Computing 458, DOI 10.1007/978-981-10-2035-3_33

Using only majority voter (MV) gate and inverter gate every QCA circuit can be designed. Also in QCA power consumption is negligible and it works at high speed [3–5].

In case of irreversible circuit, during its operation some heat is dissipated for each lost bit. But reversible circuit allow to recover the information. As a result its power consumption is very low. The proposed design is verified and simulated using QCADesigner [6, 7]. As present form of the design is not available in the existing literature, hence we are unable to undergo any comparative study in this article.

The organization of the proposed design: The Sects. 1 and 2 describe QCA and its utility in reversibility in computational paradigm. While Sect. 3 indicates designing of full adder using reversible half adder, Sect. 4 indicates simulation result and Sect. 5 indicates architecture of full adder. The article is concluded at the end.

2 Preliminary Aspects of QCA

2.1 QCA Components in Brief

QCA is a square nano structure. QCA technology is built up in cell [8–10]. Four quantum dots make a cell. The free electrons are trapped into the diagonal position due to coulomb repulsion [11–13]. One of these two configuration represent binary 0 and other represent binary 1 as shown in Fig. 1a. These free electrons are used to store and transmit data. Conceptually a binary wire is an array of QCA cell where all cell have the same polarization. Here electrons don't move between the cells. As a result no power dissipates.

In QCA two logic gates are used as the fundamental gate—Majority voter gate and Inverter gate.

Majority voter logic is reflected in Eq. 1.

$$M(A, B, C) = AB + BC + CA \tag{1}$$

where A, B and C represent the inputs of the majority voter gate.

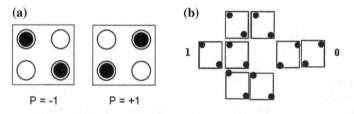

Fig. 1 **a** Polarity convention of QCA cell and **b** QCA inverter

Fig. 2 A QCA Majority voter gate

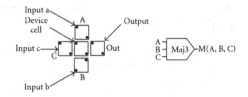

Cell polarization is changed by Inverter Gate as reflected in Fig. 1b. AND gate and OR gate can be formed using MV gate by feeding binary 0 and binary 1 through an input line respectively as illustrated in Fig. 2.

2.2 Advantages of QCA

This technology has many advantages.

1. No internal wiring is needed in QCA.
2. There is no current flowing between neighboring cell. Only when electron is lifted from their ground state enough energy is needed.
3. QCA cells are of very tiny size. It is of only 18 nm. So it requires comparatively less space to design a circuit.

2.3 QCA Clocking

QCA clocking is used mainly to supply power to the automaton and for controlling data flow direction. The clock of QCA consist of four phase—Switch, Hold, Release and Relax (Fig. 3). In Switch phase electron potential energy start to raise. In Hold phase electron potential energy is very high. During Release phase the electron potential energy begin to decrease and the electrons slowly start to be latched. During Relax phase electrons are completely localized that means it get its polarity. Also each clock cycle has four zones and each zone has four phases that is already described. Each clock zone has a 90° phase difference with its previous zone.

Fig. 3 QCA clocking

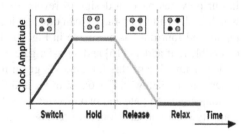

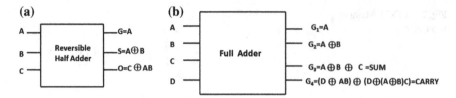

Fig. 4 **a** Block diagram of reversible half adder and **b** Full adder

2.4 Reversible Computing

In reversible computing information recovery is allowed that means we can recover input from available output [14, 15]. There should be a unique relationship between input vector and output vector. For example classical NOT operation is reversible that means from available output the input can be recovered and there is unique output vector for each specified input vector and vice-versa.

2.5 Advantages of Reversibility

There are many advantages of reversibility.

1. It consumes very low power to run a circuit that means it saves energy.
2. Its power dissipation is very low. In case of irreversible circuit, for a single bit of information the dissipated energy is $kT\ln 2$ where k represents the Boltzmann constant ($k = 1.3806505 \times 10^{-23}$ J/K) and T represents the operating temperature that is negligible but for the large number of bit this dissipated power is a headache that can be overcome by reversibility as reversible computation do not lose information.

3 Design and Implementation

In our previous paper a design of reversible half adder had been suggested by us which is shown in Fig. 4a. We can draw a block diagram of full adder that has four inputs and four outputs as shown in Fig. 4b. This full adder is designed using two reversible half adders [16] and a Ex-Or gate as shown in the Fig. 5.

The truth table of this full adder is given in Table 1. The circuit diagram of full adder is given below in Fig. 6a, b. In the addition section the circuit is composed of six majority voter gate and two inverter gate. The carry section is composed of nine majority voter gate and four inverter gate.

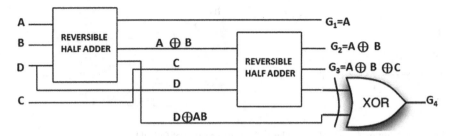

Fig. 5 Full adder using two reversible half adders

Table 1 Truth table of full adder

Input				Output			
D	A	B	C	G1	G2	G3	G4
0	0	0	0	0	0	0	0
0	0	0	1	0	0	1	0
0	0	1	0	0	1	1	0
0	0	1	1	0	1	0	1
0	1	0	0	1	1	1	0
0	1	0	1	1	1	0	1
0	1	1	0	1	0	0	1
0	1	1	1	1	0	1	1
1	0	0	0	0	0	0	0
1	0	0	1	0	0	1	0
1	0	1	0	0	1	1	0
1	0	1	1	0	1	0	1
1	1	0	0	1	1	1	0
1	1	0	1	1	1	0	1
1	1	1	0	1	0	0	1
1	1	1	1	1	0	1	1

4 Simulation Result

The proposed design and its outcome is verified by an open source simulator, QCA
Designer 2.0.3. In this layout G3 represent SUM and G4 represent CARRY. In the
bistable approximation we follow the parameters: QCA cell dimension $= 18$ nm,
clock high $= 9.800000e\text{-}022$J, clock low $= 3.800000e\text{-}023$ J, Dot diameter $= 5$ nm.
These are the default values of the mentioned parameters of the said simulator.
Figures 6b and 7 show the circuit design and the simulation diagram of proposed
Full Adder using Reversible Half Adder. The input A, B, C and D is given at 0th
phase of 0th clock. Output Sum is evaluated at phase '3' of 4th clock and Carry is
evaluated at phase '2' of 7th clock (Fig. 8).

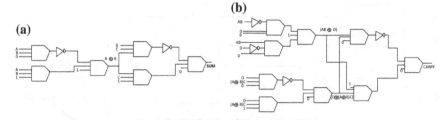

Fig. 6 **a** Addition section of full adder and **b** Carry section of full adder

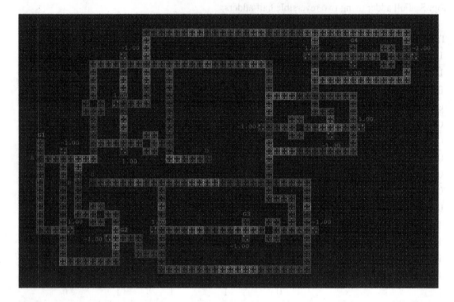

Fig. 7 Circuit layout of full adder

5 Architecture of Full Adder

In Table 2 E_m denotes the least amount of energy to drive the architecture which have N cells; E_{clock} denotes the clock energy supplied to the N cell architecture; E_{diss} is the dissipated energy from the N cell architecture; v_2 denotes the frequency of dissipation energy; τ_2 is the dissipation time to come to the relaxed state; v_1 is the frequency of incident energy; τ_1 denotes the required time to attain a higher quantum level n from a lower quantum level n_2; τ is the switching time between two different polarizations; t_p denotes the time required to propagate the signal through the entire architecture. In this table n represents the Quantum number, $\hbar$ symbolizes reduced Plank's constant, the mass of electron is denoted by m, a_2 denotes the area of a cell, N denotes the cells present in the architecture (in number), k is the total clock zones required for this architecture [6, 17]. Here, n = 10; $n_2 = 2$; m = 18 nm and N = 402.

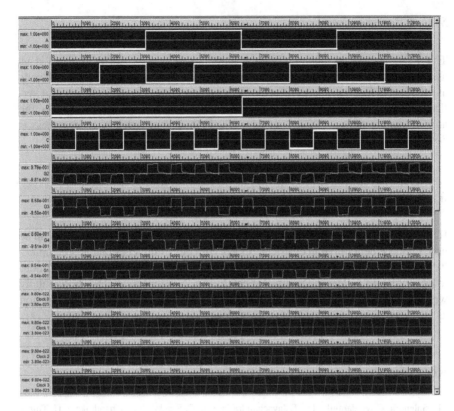

Fig. 8 Output of full adder

Table 2 Different parameters regarding the suggested full adder using two logically reversible half adders

Parameters	Values
$E_m = E_{clock} = \frac{n^2 \pi^2 \hbar^2 N}{ma^2}$	1.4962×10^{-17} J
$E_{diss} = \frac{\pi^2 \hbar^2}{ma^2}(n^2 - 1)N$	1.4813×10^{-17} J
$v_1 = \frac{\pi \hbar}{2ma^2}(n^2 - n_2^2)N$	2.1669×10^{16} Hz
$v_2 = \frac{\pi \hbar}{2ma^2}(n^2 - 1)N$	2.2346×10^{16} Hz
$\tau_1 = \frac{2ma^2}{\pi \hbar(n^2 - n_2^2)N}$	4.6147×10^{-17} s
$\tau_2 = \frac{2ma^2}{\pi \hbar(n^2 - 1)N}$	4.4748×10^{-17} s
$\tau = \tau_1 + \tau_2$	9.0896×10^{-17} s
$t_p = \tau + (k - 1)\tau_2 N$	5.4058×10^{-14} s

5.1 Effective Area

The 4 Dot 2 electron quantum cellular automata cells have square shape. Let us assume the size of a cell is a × a. The full adder is constructed using 402 such cells as reflected in Fig. 1a. So, the effective area regarding this design is 402 a^2 and vertically the coverage area is 31a and horizontally the coverage area is 36a. So, the area utilization ratio is 402:(31 × 36) = 402:1116. So the index of compactness is 36.021505.

6 Conclusion

Here we highlight a methodology of full adder using reversible half adder that can be used to implement different classical models with more efficiency and less complexity than its non reversible counterpart. As we use the logic of reversible half adder to design this full adder the number of cell increases automatically. So, horizontally we are working to decrease the number of cell and to minimize power consumption as far as possible.

References

1. Salendra. Govindarajulu, Dr. T. Jayachandra Prasad, C. Sreelakshmi, Chandrakala, U. Thirumalesh, "Energy-Efficient, Noise-Tolerant CMOS Domino VLSI Circuits in VDSM Technology", (IJACSA) International Journal of Advanced Computer Science and Applications, Vol. 2, No. 4, 2011.
2. Mingliang Zhang, Li Cai, Xiaokuo Yang, Huanqing Cui, and Chaowen Feng, "Design and Simulation of Turbo Encoder in Quantum-Dot Cellular Automata", IEEE Transactions On Nanotechnology, VOL. 14, NO. 5, September 2015.
3. G. C. S. Amarel and R. O. Winder, "Majority gate network", vol. 13. IEEE Transactions on Electronic Computers, 1964, pp. 4–13.
4. A. R. Meo, vol. 15. IEEE Transactions on Electronic Computers, 1966, pp. 606–618.
5. Kianpour, M., Sabbaghi-nadooshan, R., "A Novel Quantum-Dot Cellular Automata X-bit x 32-bit SRAM", Very Large Scale Integration (VLSI) Systems, IEEE Transactions on, PP. 1–1, Issue: 99, 14 April 2015.
6. Paramartha Dutta, Debarka Mukhopadhyay, "New architecture for flip flops using quantum-dot cellular automata", vol. 249. Springer, 2013, pp. 707–714.
7. "Qcadesigner home page http://www.atips.ca/projects/qcadesigner."
8. R. R. M. Nakahara and A. Saitoh, in Mathematical Aspects of Quantum Computing, vol. 1. World Scientific, Singapore, 2007.
9. G. H. B. C. S. L. A. O. Orlov, I. Amlani and G. L. Snider, "Science", vol. 277. American Association for the Advancement of Science, 1997, pp. 928–930.
10. P. M. S. C. H. J. W. H.S. P. Wong, D. J. Frank and J. J. Welser, vol. 87. IEEE, 1999, pp. 537–570.
11. R. L. P. Kaye and M. Mosca, "An Introduction to Quantum Computing", Oxford University Press, 2007.
12. K. N. R. Farazkish, M. R. Azghadi and M. Haghparastr, "World applied sciences journal", vol. 6, 2008, pp. 793–802.

13. T. F. T. Oya, T. Asai and Y. Amemiya, "A majority logic device using an irreversible single electron box", vol. 2. IEEE Trans. Nanotecnology, 2003, pp. 15–22.
14. Paramartha Dutta, Debarka Mukhopadhyay, Sourav Dinda, "Designing and implementation of quantum cellular automata 2:1 multiplexer circuit", vol. 25. International Journal of Computer Applications, 2011.
15. D. Tougaw and C. S. Lent, "Logical devices implemented using quantum cellular automata", vol. 75. AIP Publishing, 1994, pp. 1818–1825.
16. Sunanda Mondal, Debarka Mukhopadhyay, Paramartha Dutta, "A Novel Design Of a Logically Reversible Half Adder using 4 Dot 2 Electron QCA", NCCCIP-2015, PP. 123–130, May 2–3, 2015.
17. Debarka Mukhopadhyay, Paramartha Dutta, "A study on energy optimized 4 dot 2 electron two dimensional quantum dot cellular automata logical reversible flip-flops", Microelectronics Journal, Volume 46, Issue 6.

Secure Data Outsourcing in the Cloud Using Multi-secret Sharing Scheme (MSSS)

Amitava Nag, Soni Choudhary, Subham Dawn and Suryadip Basu

Abstract In the last few years, the demand of data outsourcing in the cloud has enormously increased. However, there are some significant barriers to cloud computing adoption. One of the most significant barriers of outsourcing data in the cloud computing is its security issues especially when sensitive data such as personal profile, financial records, medical reports, etc. are outsourced on cloud servers. Therefore, strong security measures are essential to protect data within the cloud. Recently, secret sharing schemes have been applied to protect sensitive data in cloud storage. In this paper, we propose a (t, n) multi secret sharing scheme (MSSS) for secure data outsourcing in the cloud. In the proposed scheme, t secrets are divided into n shares, which are distributed to n cloud servers (one share per server) and a trusted user can recover all secrets by combining shares from at least t servers. The analysis shows that the proposed scheme can provide secure and efficient data outsourcing system in the cloud.

Keywords Data outsourcing · Cloud storage · Secret sharing · MSSS · Trusted user

Amitava Nag (✉)
Department of IT, Academy of Technology, Hooghly 721212, India
e-mail: amitavanag.09@gmail.com

Soni Choudhary · Subham Dawn · Suryadip Basu
Department of CSE, Academy of Technology, Hooghly 721212, India
e-mail: choudhary.soni.0909@gmail.com

Subham Dawn
e-mail: knightdawnxp@gmail.com

Suryadip Basu
e-mail: suryadip.basu@yahoo.com

© Springer Science+Business Media Singapore 2017   337
J.K. Mandal et al. (eds.), *Proceedings of the First International Conference on Intelligent Computing and Communication*, Advances in Intelligent Systems and Computing 458, DOI 10.1007/978-981-10-2035-3_34

1 Introduction

Cloud computing is achieving increasing popularity in the last few decades as it provides a set of versatile IT services over the internet. It is a technique of usage of various servers over the internet, often known as cloud. In cloud computing, the client only needs internet access and any standard personal computer to access the privileges. In recent years, with evolution of easily available internet, cloud computing has become a very impressive tool in the technical world. The services such as data storage, applications, software are available to the local devices through internet. It can be available as online site of a company or offline storage and computing facilities [1] such as Google Drive, Apple's iCloud, Amazon's cloud Drive, Microsoft's SkyDrive, and Samsung's S-Cloud service. It can be used in different fields such as marketing, education, multimedia, e-commerce as shown in Fig. 1.

Traditionally, the cloud service includes several servers which are maintained and managed by a semi-trusted third party know as cloud provider [2]. However, outsourcing data to a third-party administrative control needs to be well protected especially when data owner shares their sensitive information on cloud servers [3]. Thus to preserve privacy of users' data in the cloud, the data must be encrypted before it is stored in the cloud server. Storing sensitive data in common encrypted forms is a conventional approach in which only the trusted users can download the encrypted files from the cloud and decrypt them with appropriate keys [4]. But the high complexity of encryption-based approach has severely limited their practical use [5]. Moreover, encrypted files are stored in a single storage. Therefore if the server which contains the encrypted files does not act properly due to an attack or some other reasons, then data user cannot recover the original secret. In this paper, we propose a threshold secret sharing-based approach for outsourcing sensitive data to cloud servers. The benefit of using threshold secret sharing for secure data outsourcing in the cloud is that, even if some cloud servers do not work properly,

Fig. 1 Application of cloud computing across many industry centres

the data user can still recover the secret if certain number of servers work appropriately.

2 Related Works

Secret Sharing Schemes were introduced independently by Shamir [6] and Blakely [7]. The basic idea in Secret Sharing [6] is to divide a secret data into several shares in such a way that certain subsets of the shares can recover the original secret. The advantage of secret sharing is its "fault-tolerance" and "availability". For instance, if one share is lost or damaged by an adversary, other shares can reconstruct the secret. Thus, nowadays secret sharing is being widely used in cloud computing [5, 8]. Several recent works [8] have proposed in the cloud based data outsourcing using Shamir's secret sharing scheme instead of traditional encryption [1, 8].

Cloud computing systems are organized as multiple servers and data storage devices which store distributed data of multiple users [9]. The shares of secret are stored at multiple servers. In order to reconstruct the secret, the shares are collected from distributed servers. Security of these data in cloud host is very much important and must be assured. The external attacks in a cloud can be resisted by service provider's authentication. The internal attacks which are imposed by anyone inside the system can also be resisted with the help of key. The inside intruder can access the data only if the key is available to him. In this way the data is safe from both internal as well as external attacks. But due to any physical damage to the server or due to any natural disaster, the cloud data may be lost. So cloud computing can be mixed with secret sharing to provide more security of data. If due to any natural disaster any single share is lost or damaged, then the rest of the shares would still be available. As long as threshold shares are available, the secret can be reconstructed. In [10], Nojoumian and Stinson used secret sharing scheme in cloud computing. Takahashi and Iwamura [11] have proposed a new mechanism of secret sharing for cloud computing which reduces the amount of share and thus suitable for cloud systems.

Shamir's (t, n) secret sharing scheme [6] is one of the most widely used techniques for secret sharing. In this scheme n shares of original secret are created. To reconstruct the original secret a minimum of t shares must be collected and any $t - 1$ or fewer shares are not sufficient to retrieve the secret. This is known as Threshold Scheme. Therefore, when secret sharing schemes are used in data outsourcing, fault tolerance property is achieved as even if $n - t$ shares are lost or damaged, the remaining t shares suffice to recover the information. Nevertheless, Shamir's (t, n) secret sharing scheme is not suitable to cloud systems because the size of the shares increases n times than the size of the secret. To solve this problem, we propose a (t, n) multi-secret sharing scheme in this paper which can be efficiently applied to cloud systems. In the proposed scheme, t secrets $\{S_1, S_2, ..., S_t\}$ are divided into n public shares $y_i (1 \leq i \leq n)$ and each share y_i is stored on the correspondent cloud server C_i among n cloud servers $\{C_1, C_2, ..., C_n\}$. Therefore, the proposed scheme reduces the size of the shares and thus it is suitable for cloud computing.

3 System Model

Our system model consists of three parties: (1) the data owner, (2) data user and (3) the cloud storage servers as shown in Fig. 2. The details of each party are described as follows:

1. Data Owner: The data owner (an enterprise or individual customer) stores a set of data files in the cloud. For security and confidentiality of the outsourced data in the cloud, data owners can encrypt their data before outsourcing.
2. Data Users: Only trusted data users are allowed to access outsourced data. Any user from cloud perimeter can download the data and any trusted data user has the privilege to decrypt the downloaded data with secret key.
3. Cloud Storage Servers: The cloud storage servers provide high-quality services to store and manage the outsourced data.

In this model, it is assumed that data owner can outsource data in a collection of cloud servers and only trusted data users can access it.

4 Proposed Scheme

This section deals with cloud security service using multi secret-sharing scheme. Assume that, there are t secret data files $S = \{S_1, S_2, ..., S_t\}$ to be shared by a data owner. The data owner D construct $n(t \leq n)$ encrypted shares $y_1, y_2, ..., y_n$ from t secret data files $S = \{S_1, S_2, ..., S_t\}$ using Shamir's (t, n) secret sharing schemes and stores $y_1, y_2, ..., y_n$ into n cloud servers $C_1, C_2, ..., C_n$ respectively. A trusted data user can reconstruct the original secrets by downloading t shares from any t servers. The complete procedure of the proposed scheme is illustrated in Fig. 3.

cloud server

files

Owner

data user

Fig. 2 System model

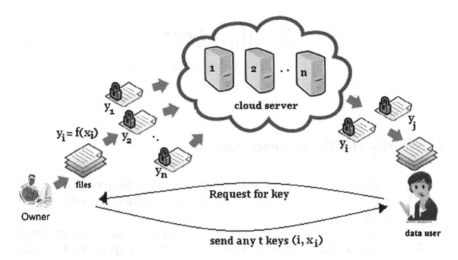

Fig. 3 The detail illustration of the proposed scheme

The processes of share generation and actual secret recovery are described below:

Algorithm 1: Share Construction

1. The data owner D first generates n random numbers x_1, x_2, ..., x_n.
2. Then D chooses a polynomial $f(x)$ of degree $t - 1$ randomly and a prime number p:

$$f(x) = a_0 + a_1 x + a_2 x^2 + \cdots + a_{t-1} x^{t-1} \bmod p \qquad (1)$$

in which $a_0 = S_1$, $a_1 = S_2$, ..., $a_{t-1} = S_t$
3. D computes the public shares $y_1 = f(x_1)$, $y_2 = f(x_2)$, ..., $y_n = f(x_n)$ and stores y_1, y_2, ..., y_n into cloud servers C_1, C_2, ..., C_n respectively.
4. D keeps (i, x_i) secret for $i = 1$ to n.

Algorithm 2: Secret reconstruction

A trusted user will recover the secrets $S = \{S_1, S_2, ..., S_t\}$ using the following procedure:

1. The user first requests the data owner for the private shares (i, x_i).
2. D first verifies whether the user is trusted or not. If the user is trusted, then D provides any t private share (i, x_i) secretly to that trusted user.
3. The trusted user downloads t public shares from t cloud servers. The basic rule for downloading these t shares is based on the public share y_i is downloaded from ith server.
4. The trusted user can uniquely determine the polynomial $f(x)$ from all t x_i as follows:

$$f(x) = \sum_{i=1}^{t} y_i \prod_{j=1, j \neq i}^{t} \frac{x - x_j}{x_i - x_j} \bmod p$$

$$= S_1 + S_2 x + S_3 x^2 + \cdots + S_t x^{t-1} \bmod p \tag{2}$$

5 Security and Performance Analysis

This section discusses the security and the performance of the proposed scheme. Table 1 compares the performances of the proposed scheme and other related schemes [9, 12].

Data confidentiality: The public shares stored in the cloud servers are generated by Shamir's secret sharing scheme. Thus each share is in encrypted form and cannot reveal any information.

Data correctness: The recovery phase of the proposed scheme is based on Lagrange interpolation polynomials. Thus if any user downloads $(t - 1)$ public shares and obtain $(t - 1)$ correspondent xi values from the data owner, he cannot determine the $(t - 1)$ degree polynomial $f(x)$ and so cannot obtain any information about the secrets.

Attack by intruder: A potential intruder can capture the public shares from the cloud. Since the data owner uses n random numbers $x_1, x_2, ..., x_n$ to generate n shares and all random numbers are kept secret, so it is not possible to determine the $(t - 1)$ degree polynomial $f(x)$ based on Lagrange interpolation polynomials without the values of $x_1, x_2, ..., x_n$. Therefore the proposed scheme is computationally secure from an attacker inside or outside the cloud perimeter.

Fault tolerant property: The main benefit of (t, n) secret sharing scheme over traditional encryption is that even if some shares are lost or damaged, the system can still recover the original secret if certain number of appropriate shares are available (known as fault tolerant property). The proposed scheme uses (t, n) secret sharing and thus has fault tolerant property. On the other hand Yoon et al. [12] used traditional encryption technique which is a single point of failure and hence does not have fault tolerant property. Although Liu et al. [9] proposed (n, n) secret sharing scheme, but it does not support fault tolerant property as it needs all shares to recover the original secret.

Table 1 Performance comparisons

	Data confidentiality	Data correctness	Secure from attacker	Fault tolerant property
Liu et al. [9]	Yes	Yes	Yes	No
Yoon et al. [12]	Yes	Yes	Yes	No
Proposed	Yes	Yes	Yes	Yes

6 Conclusion

In this paper, we propose multi-secret sharing scheme in order to protect the privacy and security of outsourced data in the cloud storage systems. Our multi-secret sharing approach can reduce the amount of shares with respect to the number of secrets. Also, it prevents untrusted users from obtaining the original secrets even though public shares are available in the cloud servers. Moreover, the proposed scheme preserves data confidentiality and correctness. The performance analysis indicates that the proposed scheme is suitable for cloud computing.

References

1. Ching-Nung Yang; Jia-Bin Lai, "Protecting Data Privacy and Security for Cloud Computing Based on Secret Sharing," in Biometrics and Security Technologies (ISBAST), 2013 International Symposium on, vol., no., pp. 259–266, 2–5 July 2013.
2. Kaiping Xue; Peilin Hong, "A Dynamic Secure Group Sharing Framework in Public Cloud Computing," in Cloud Computing, IEEE Transactions on, vol. 2, no. 4, pp. 459–470, Oct.-Dec. 1 2014.
3. Mazhar Ali, Kashif Bilal, Samee Khan, Bharadwaj Veeravalli, Keqin Li, Albert Zomaya, "DROPS: Division and Replication of Data in the Cloud for Optimal Performance and Security", IEEE Transactions on Cloud Computing, no. 1, pp. 1, PrePrints, doi:10.1109/TCC.2015.2400460.
4. Xiong, Jinbo, Ximeng Liu, Zhiqiang Yao, Jianfeng Ma, Qi Li, Kui Geng, and Patrick S. Chen. "A secure data self-destructing scheme in cloud computing." Cloud Computing, IEEE Transactions on 2, no. 4 (2014): 448–458.
5. Hadavi, Mohammad Ali, et al. "Security and searchability in secret sharing-based data outsourcing." International Journal of Information Security (2015): 1–17.
6. A. Shamir, "How to share a secret", Communications of the ACM. 22, (11), pp. 612–613 (1979).
7. G.R Blakely, "Safeguarding cryptography keys," in Proc. of AFIPS National Computer Conference, vol. 48, pp. 313–317, 1979.
8. Security Limitations of Using Secret Sharing for Data Outsourcing, Jonathan L. Dautrich, Chinya V. Ravishankar, Lecture Notes in Computer Science Volume 7371, 2012, pp 145–160.
9. Yanjun Liu, Hsiao-Ling Wu, Chin-Chen Chang, "A Fast and Secure Scheme for Data Outsourcing in the Cloud", TIIS 8(8): 2708–2722 (2014).
10. Nojoumian, M.; Stinson, D.R., "Social secret sharing in cloud computing using a new trust function," in Privacy, Security and Trust (PST), 2012 Tenth Annual International Conference on, vol., no., pp. 161–167, 16–18 July 2012.
11. Takahashi, S.; Iwamura, K., "Secret Sharing Scheme Suitable for Cloud Computing," in Advanced Information Networking and Applications (AINA), 2013 IEEE 27th International Conference on, vol., no., pp. 530–537, 25–28 March 2013.
12. Yoon, M., Jang, M., Shin, Y. S., & Chang, J. W. A Bitmap based Data encryption Scheme in Cloud Computing. International Journal of Security and Its Applications, vol. 9, no. 5, pp. 345–360, 2015.

Design of a BCD Adder Using 2-Dimensional Two-Dot One-Electron Quantum-Dot Cellular Automata

Kakali Datta, Debarka Mukhopadhyay and Paramartha Dutta

Abstract A full adder is designed which in turn is used as the building block to design a BCD adder. We have also compared the design of two-dot one-electron QCA BCD adder with the existing four-dot two-electron QCA BCD adder variant. The analysis of the proposed design justifies its effectiveness, in respect of energy utilization, compactness and stability.

Keywords Majority voter · Adder · Ripple carry adder · BCD adder

1 Introduction

QCA, proposed by Lent and Tougaw [6], promises an efficient technology, with high computation speed, small size and long lifetime. This does way with the CMOS technology drawbacks.

Two-dot one-electron QCA [3] as it is more advantageous over four-dot two-electron QCA. As in a cell there is only two dots and one electron, the total number of electrons participating in an architecture is almost halved, if not less. Therefore energy required is less. Also, there is no ambiguous configuration as in the case of four-dot two-electron [5]. Moreover, Coulomb's repulsion principle is obeyed while passing information from one cell to the next, thereby reducing the wiring complexity.

Kakali Datta (✉) · Paramartha Dutta
Department of Computer & System Sciences, Visva Bharati University,
Santiniketan 731235, West Bengal, India
e-mail: kakali.datta@visva-bharati.ac.in; kakalidatta@hotmail.com

Paramartha Dutta
e-mail: paramartha.dutta@gmail.com

Debarka Mukhopadhyay
Department of Computer Science, Amity School of Engineering & Technology,
Amity University, Kolkata 700156, West Bengal, India
e-mail: debarka.mukopadhyay@gmail.com

© Springer Science+Business Media Singapore 2017
J.K. Mandal et al. (eds.), *Proceedings of the First International Conference on Intelligent Computing and Communication*, Advances in Intelligent Systems and Computing 458, DOI 10.1007/978-981-10-2035-3_35

The following Sect. 2, we have discussed the cell structures of two-dot one-electron QCA, the QCA basic gates and clocking. In Sect. 3.1, we proposed the design of a full adder which acts as the basic block of the BCD adder. We propose a design of a BCD adder in Sect. 3.2. In Sect. 4 we have considered calculations based on potential energy for verifying the outputs. Design analysis has been done in Sect. 5. We have also compared the proposed two-dot one-electron architecture with an existing four-dot two-electron architecture in Sect. 6. Finally, in Sect. 7, the energy and power parameters are calculated, followed by conclusion in Sect. 8.

2   Overview of Two-Dot One-Electron QCA

Here, the QCA cells are demonstrated in Fig. 1a, the electrons when placed in the dot below and above, represent binary '0' and '1' respectively. Similarly, as shown in Fig. 1b, the electrons when placed in the right and left dot, represent binary '0' and '1' respectively. They changes position following Coulomb's law.

Here, the clocking is quite not like the clocking mechanism of CMOS as it shows the direction in which the signal flows and the input signals are supplied with energy so that the signal can reach the output end [1].

In [3] clocking (Fig. 2) and the basic building blocks (Fig. 3) are discussed.

The majority voter output is

$$M(A, B, C) = AB + \overline{C}(A + B) \tag{1}$$

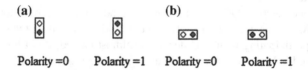

Fig. 1   The two-dot one-electron QCA cells

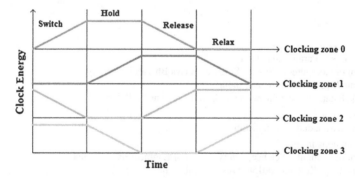

Fig. 2   The two-dot one-electron clocking

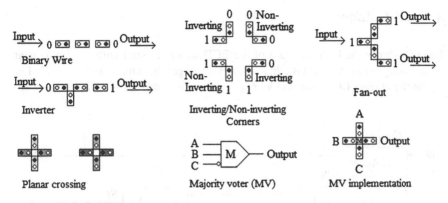

Fig. 3 Basic building blocks

(a) (b)

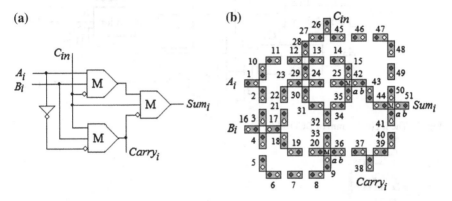

Fig. 4 The two-dot one-electron QCA implementation of a full adder

3 Designs Proposed

3.1 Full Adder

Bits A_i and B_i along with carry bit C_{in_i} is added by a full adder whose logic diagram along with its implementation are shown in Fig. 4. The sum and carry outputs are

$$Sum_i = A_i \oplus B_i \oplus C_{in} \tag{2}$$

$$Carry_i = A_i.B_i + A_i.C_{in} + B_i.C_{in} \tag{3}$$

3.2 BCD Adder

Figure 5 shows the logic diagram of the BCD adder. To start with, we implement 4-bit binary adder and add A_i, B_i and carry-in to get the sum S_i and carry C_i. Let S_{Bi} be the correct BCD sum and let C_{Bi} be the correct carry. If $A_i + B_i + C_{in_i} < 10$,

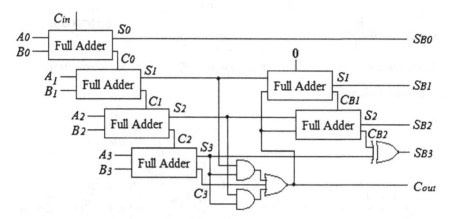

Fig. 5 The logic diagram of BCD adder

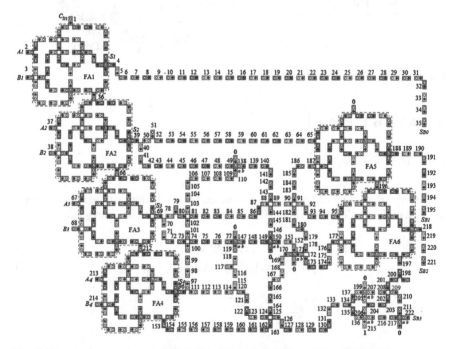

Fig. 6 The two-dot one-electron QCA BCD adder with full adders (FA) as building blocks

$S_{Bi} = S_i$ otherwise S_i must be corrected to change it to S_{Bi}. This correction can be made by adding 6 to S_i as the low-order 4-bits of $S_i + 6$ are equal to S_{Bi} whenever $A_i + B_i + C_{in_i} >= 10$. The output carry $C_{Bi}$is 1 whenever $C_i = 1$ or $S_i > 10$. In Fig. 5, we can see that a 4-bit ripple carry binary adder computes the sum S_i which passes through a second binary adder. It adds either 0 or 6 to it to form S_{Bi}. The two-dot one-electron QCA implementation of BCD adder is shown in Fig. 6.

Table 1 Output determination of full adder (FA)

Index of cell	Position of electron	Total potential energy	Remarks
1, 16, 26	–	–	Input cells A_i, B_i and C_{in} respectively
2–8	–	–	Same polarity as of 1 (Fig. 3d)
9	–	–	Opposite polarity of 8 (Fig. 3c)
10–15	–	–	Cell 15 has same polarity as of 1 (Fig. 3d)
17–20	–	–	Same polarity as of 16 (Fig. 3d)
21–25	–	–	Cell 25 has the polarity as of 17 (Fig. 3d)
27–33	–	–	Cell 33 has the polarity as of 26 (Fig. 3c)
34	–	–	Opposite polarity of 31 (Fig. 3b)
35	–	–	Opposite polarity of 34 (Fig. 3c)
36	a	-6.905×10^{-20} J	Electron latches at a as potential energy is least at a
	b	-0.294×10^{-20} J	
37–38	–	–	Same polarity as of 36 (Fig. 3a)
39	–	–	Opposite polarity of 15 (Fig. 3b)
40–41	–	–	Same polarity as of 43 (Fig. 3)
42	a	3.329×10^{-20} J	Electron latches at b as potential energy is least at b
	b	0.537×10^{-20} J	
43–44	–	–	Opposite polarity of 44 (Fig. 3b)
45–50	–	–	Opposite polarity of 10 (Fig. 3c)
51	a	3.329×10^{-20} J	Electron latches at b as potential energy is least at b
	b	0.537×10^{-20} J	

4   Output Energy State Determination

As simulator is not available till date, the outputs may be established by potential energy calculations and hence determined the outputs i.e. the positions of electrons where the potential energies are minimum [2].

In Table 1, we have obtained the potential energies of the QCA cells as numbered in Fig. 4b. The results obtained in Table 1 are used to deduce Table 2.

5   Analysis of the BCD Adder Design

An efficient QCA architecture must be stable and maximal area should be utilized. [4]. If all the input signals reach the majority voter gate with same clock and same strength and the output signal is with the same or next clock then the architecture is stable. These conditions are satisfied by our design.

Let us consider the dimension of a cell as p nm by q nm. Here, we require 505 which covers $505pq$ nm^2 of area and $(34p + 34q) \times (25p + 25q)$ nm^2 area is covered by the design. Hence, area utilization ratio is $101pq : 170(p + q)^2$. Number of crossovers here is 24.

6   Comparative Study

The comparative study of the BCD adder designs is done in the Table 3. In [5], the values of p and q are given to be 13 nm and 5 nm respectively.

7   Energy and Power Requirements

From [7], in an architecture with N cells, E_m is the energy to drive, E_{diss} is the dissipated energy, v_2 is the frequency of dissipated energy, τ_2 is the dissipation time, v_1 is the frequency of incident energy, τ_1 is the time to jump from quantum level n_2 to quantum level n, τ is time the cells require to jump between two consecutive polarizations, t_p is the propagation time, $v_2 - v_1$ is the differential frequency. These parameters are calculated in Table 4. Here we consider $n = 10$ and $n_2 = 2$, in course of the present deduction.

Table 2 Output state of two-dot one-electron QCA BCD adder

Cell index	Position of electron	Total potential energy	Remarks
1	–	–	Input C_{in}
2, 37, 67, 213	–	–	A inputs to FA$_i$, $i = 1, 2, ..., 6$ respectively
3, 38, 68, 214, 187, 177	–	–	B inputs to FA$_i$, $i = 1, 2, ..., 6$ respectively
4–35	–	–	Attains polarity of the sum of FA1 (Figs. 3c and 4)
36	–	–	Attains polarity of the carry of FA1 and C input to FA2
39–49	–	–	Attains polarity of the sum of FA2 (Figs. 3c and 4)
50	–	–	Attains Opposite polarity of 39 (Fig. 3b)
51–65	–	–	Attains Opposite polarity of 50 (Fig. 3b)
66	–	–	Attains polarity of the carry of FA2 and C input to FA3
69–77	–	–	Attains polarity of the sum of FA3 (Figs. 3c and 4)
78	–	–	Attains Opposite polarity of 69 (Fig. 3b)
79–95	–	–	Attains Opposite polarity of 78 (Fig. 3b)
96	–	–	Attains polarity of the sum of FA4 (Fig. 4)
97–109	–	–	Cells 109 Same polarity as of 96 (Fig. 3d)
110	–	–	Attains Opposite polarity of 109 (Fig. 3c)
111–114	–	–	Attains Opposite polarity of 96 (Fig. 3b)
115–117	–	–	Attains Opposite polarity of 114 (Fig. 3c)
118–119	–	–	Attains Opposite polarity of 116 (Fig. 3b)
120–135	–	–	Cells 120, 135 attain Opposite polarity of 114
136–137	–	–	Cell 137 attains polarity of 120 (Fig. 3d)
138	a	14.102×10^{-20} J	Electron latches at b as potential energy is least at b
	b	1.368×10^{-20} J	

(continued)

Table 2 (continued)

Cell index	Position of electron	Total potential energy	Remarks
139–146	–	–	Attains polarity of 138 (Fig. 3c)
147	a	14.102×10^{-20} J	Electron latches at b as potential energy is least at b
	b	1.368×10^{-20} J	
148–149	–	–	Attains polarity of 147 (Fig. 3a)
150	a	3.329×10^{-20} J	Electron latches at b as potential energy is least at b
	b	0.537×10^{-20} J	
151–152	–	–	Attains polarity of 150 (Fig. 3c)
153	–	–	Attains polarity of the carry of FA4 (Fig. 4)
154–170	–	–	Cells 154, 170 attain polarity of 153
171	a	-3.329×10^{-20} J	Electron latches at a as potential energy is less at a
	b	-0.537×10^{-20} J	
172–177	–	–	Attains polarity of 171 (Fig. 3d)
178–187	–	–	Cell 187 attains polarity of 171 (Fig. 3c, d)
188	–	–	Attains polarity of the sum of FA5 (Fig. 4)
189–195	–	–	Attains polarity of 188 (Fig. 3c)
196	–	–	Attains polarity of the carry of FA5 and C input to FA6
197–198	–	–	Attains polarity of the carry of FA6 (Figs. 3a and 4)
199–205	–	–	Cells 199, 204 attain Opposite polarity of 198
206	–	–	Attains Opposite polarity of 204 (Fig. 3b)
207	a	14.102×10^{-20} J	Electron latches at b as potential energy is least at b
	b	1.368×10^{-20} J	
208–211	–	–	Attains polarity of 215 (Fig. 3c)
212	–	–	Attains polarity of the carry of FA3 and C input to FA4
215	a	14.102×10^{-20} J	Electron latches at b as potential energy is least at b
	b	1.368×10^{-20} J	
216–217	–	–	Attains polarity of 215 (Fig. 3c)
218	–	–	Attains polarity of the sum of FA6 (Fig. 4)
219–221	–	–	Attains polarity of 218 (Fig. 3c)
222	a	3.329×10^{-20} J	Electron latches at b as potential energy is least at b
	b	0.537×10^{-20} J	

Table 3 Comparative study

	Proposed in [8]	Proposed in this paper
Cells required, N	1903 nos.	505 nos.
Area utilized	5494060 nm^2	275480 nm^2
Majority voter gate	25 nos.	25 nos.
Required energy	High (as 3806 electrons are concerned)	Less (as 505 electrons are concerned)

Table 4 Values of different parameters

Parameter	Value for design in [8]	Value for this design
$E_m = E_{clock} = \frac{n^2\pi^2\hbar^2 N}{2ma^2}$	7.084×10^{-17} J	1.802×10^{-17} J
$E_{diss} = \frac{\pi^2\hbar^2}{2ma^2}(n^2 - 1)N$	7.0139×10^{-17} J	1.784×10^{-17} J
$v_1 = \frac{\pi\hbar}{2ma^2}(n^2 - n_2^2)N$	2.0518×10^{17} Hz	5.219×10^{16} Hz
$v_2 = \frac{\pi\hbar}{2ma^2}(n^2 - 1)N$	6.411×10^{15} Hz	1.631×10^{15} Hz
$(v_2 - v_1) = \frac{\pi\hbar}{2ma^2}(n_2^2 - 1)N$	1.987×10^{17} Hz	5.056×10^{16} Hz
$\tau_1 = \frac{1}{v_1} = \frac{2ma^2}{\pi\hbar(n_1^2-1)}N$	4.873×10^{-18} s	1.915×10^{-16} s
$\tau_2 = \frac{1}{v_2} = \frac{2ma^2}{\pi\hbar(n_1^2-1)}N$	1.559×10^{-17} s	6.130×10^{-16} s
$\tau = \tau_1 + \tau_2$	1.608×10^{-16} s	6.322×10^{-16} s
$t_p\tau + (k - 1)\tau_2 N$	2.374×10^{-12} s	1.114×10^{-11} s

8 Conclusion

We have presented a full adder design, which is used to design a ripple carry adder. This ripple carry adder in turn is modified into a BCD adder. Analysis of the architectures has been made. Outputs of the proposed structure have been established by calculating potential energies. The energy and power requirements of the architecture have also been calculated. Last but not the least, a comparative study is also done.

References

1. Datta, K., Mukhopadhyay, D., Dutt, P.: Design of Logically Reversible Half Adder using 2D2-Dot 1-Electron QCA. Proceedings of the 4th International Conference on Frontiers in Intelligent Computing: Theory and Applications (FICTA-2015) pp. 379–389 (2015)
2. Datta, K., Mukhopadhyay, D., Dutta, P.: Design of a Binary to BCD Converter using 2-Dimensional 2-Dot 1-Electron Quantum Cellular Automata. National Conference on Computing, Communication and Information Processing 70, 153–159 (2015)
3. Datta, K., Mukhopadhyay, D., Dutta, P.: Design of n-to-2n Decoder using 2-Dimensional 2-Dot 1-Electron Quantum Cellular Automata. National Conference on Computing, Communication and Information Processing pp. 77–91 (2015)
4. Ghosh, M., Mukhopadhyay, D., Dutta, P.: A 2 dot 1 electron quantum cellular automata based parallel memory. In: Information Systems Design and Intelligent Applications, Advances in Intelligent Systems and Computing, vol. 339, pp. 627–636. Springer India (2015)
5. IV, L.R.H., Lee, S.C.: Design and Simulation of 2-D 2-Dot Quantum-dot Cellular Automata Logic. IEEE Transactions on Nanotechnology 10(5), 996–1003 (2011)
6. Lent, C., Tougaw, P.: A device architecture for computing with quantum dots. Proceedings of the IEEE 85, 541–557 (1997)
7. Mukhopadhyay, D., Dutta, P.: A Study on Energy Optimized 4 Dot 2 Electron two dimensional Quantum Dot Cellular Automata Logical Reversible Flipops. Microelectronics Journal, Elsevier 46, 519–530 (2015)
8. Shah, N., Khanday, F., Bangi, Z.: Quantum Cellular Automata Based Effcient BCD Adder Structure. Communications in Information Science and Management Engineering 2(2), 11–14 (2012)

Design Based New Coupling Metric for Component Based Development Environment

Jyoti Sharma, Arvind Kumar and M. Iyyappan

Abstract Software development has now days evolved into an extreme change that uses the best modules being run in various closed and open source software. The basic idea is to extract the best component so that it can be fitted into the ongoing software development process. For this reason reusable and best components are required. Merely selection of high cohesive components is not sufficient but we have to select the components which are having low coupling as well. In this proposal we have introduced the new coupling metric on a higher level that is at the package level.

Keywords Component · Complexity · Coupling · Cohesion · Coupling distance

1 Introduction

Component based software engineering is an advanced field of software engineering. Coupling and cohesion is having a direct impact on the complexity of the software. Different considerations has been proposed to reduce the complexity of the program by dividing them into functional modules [1]. During the development of any kind of software the design phase plays an important role and if the

Jyoti Sharma · Arvind Kumar (✉) · M. Iyyappan
Department of Computer Science and Engineering,
SRM University, Delhi-NCR, Sonepat 131029, Haryana, India
e-mail: k.arvind33@gmail.com

Jyoti Sharma
e-mail: kaushik.jyoti27@gmail.com

M. Iyyappan
e-mail: Iyyapan5mtech@gmail.com

© Springer Science+Business Media Singapore 2017   355
J.K. Mandal et al. (eds.), *Proceedings of the First International Conference on Intelligent Computing and Communication*, Advances in Intelligent Systems and Computing 458, DOI 10.1007/978-981-10-2035-3_36

designing phase of the particular software can be done easily then the coming corresponding phases in software development will get the good time as well.

Metrics are basically simply the measurement of a parameter. The design metrics introduced here is based on the concept that we are having a lot of number of packages in our code which may be an example or a commercial code. These packages are basically having the classes and classes are having the methods so the hierarchy which is obtained by using these three we will compute the design based metric in component based development environment.

CBSE is an approach which is used to enhance the reusability because reusability is a way to improve efficiency and productivity of software systems. Component based software systems are mainly constructed from the reusable components such as third party components and the commercial off the shelf components (COTS). As the rapid increase of software system size and complexity, it is very important to reduce the high software cost, time and complexity while increasing reliability, Performance and quality so on. Various graph based techniques has been proposed previously to indicate the complexity of the program [2].

In this paper we are using the concept of coupling distance to find the total coupling of components and our proposed approach will be used for calculating the total coupling of components and the average coupling of a project.

2   The Metric of Software

The software metrics are basically needed to measure several types of quality issues. Software Metrics provides a measurement for the software and the process of software production. Most of the existing metrics are applicable to small programs or components. Various number of measures has been designed previously and to measure such complexity measure some of the measures has been taken and it was observed that some of the complexity measures are not fulfilling the basic criteria so there is a need for designing of new metric [3]. There are some of the quality level attributes can be measured directly or indirectly like understand ability, modifiability, complexity and maintainability by using the software metrics and that is the reason that they are playing an important role in software engineering.

The coupling metric will have a direct impact on the complexity of software because if the coupling rate can be found between the components then selection of components will get easier for the development of software in component based development environment (CBSD). A model was designed previously in which design properties such as encapsulation and modularity are related to high level quality attribute such as flexibility and reusability [4]. We are considering the coupling metric by using the various source codes at the package level. In this research, we have presented a coupling metric for the various java codes by using the concept of coupling distance.

3   Coupling Computation at Package Level with Proposed Approach

For the computation of coupling we will consider four classes of a project named as online course portal system. For these different 4 classes we will compute the total coupling and at the end we will compute the average coupling for the project by using the proposed approach. A model of software quality has been proposed previously to indicate the quality in the software [5]. There are many complex systems in the world like the social system, physical system etc. Most of the systems are generally hierarchical structured like Fig. 1.

A large system is generally having the small submodules and so on.

3.1   Proposed Approach

For the new coupling metric using the concept of package in the component tree example we will follow the following number of steps.

Fig. 1 Ideal hierarchical system [6]

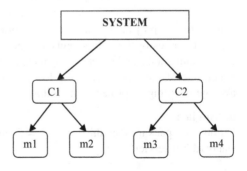

(1) Given an example or case study.

(2) Packages are considered as components, classes are considered as modules and methods are considered as submodules.

(3) Draw the graphical representation of components dependency with the component tree.

(4) Calculate the coupling distance and mostly the coupling distance is calculated by considering the height of lowest common ancestor in the graphical view of the component tree.

(5) Calculate the total coupling of components by using the below formula as:
Total coupling of component i = summation from j = 1 to M

$$\frac{\text{Coupling distance for i}}{N-1}$$

where i and j are the symbolic number for packages.

(6) In addition the Summation from j = 1 to M followed by

$$\frac{\text{Coupling of selected components}}{M}$$

where M is the total Number of components.

Now the average coupling of the project and by using those values we can deliver a quality based product.

3.2 Calculation of Total Coupling of Components with the Average Coupling of the Project

Considering a project of online course portal system in which there are various different classes or the components are there and these components are the application components and other components as well. Now considering the 4 classes in which one of them is acccou.java for the calculation of coupling metric with the average coupling of project is as follows:

acccou.java
Number of packages in this code is = 5 and the simple class is extending the standard httpservlet package.

```
{
Connection con=null;
{

    try
    {
                RequestDispatcher rd;
                        String st1="",st2="",st3="";
                int i,j=0,k=0;
                if((req.getParameterValues("sta"))!=null)
{
                        String id[]=req.getParameterValues("sta");
            int n=id.length;
            for(i=0;i<n;i++)
{
out.println("update upload set status='accepted' where upid="+id[i]+"");
s.executeUpdate("update upload set status='accepted' where upid="+id[i]+"");
con.commit();
                }
                    }
        }
        out.close();
        } }
```

The above code or example has been taken for the calculation by using the proposed metric given as [7].

Now calculation is as follows:

Here, total number of components in the given example = 5 as C1, C2, C3, C4, C5.
Total numbers of modules are 2 as M1 and M2 where M1 extends M2.
Total numbers of submodules are 1 named as SM1.

In the above code javax.servlet.http package is consisting of class http servlet and this class is consisting of method do get method.

Based on this internally defined hierarchy we will make the graphical representation of component tree with the component dependency [8].

accou is a subclass of httpservlet. The internal hierarchy can be shown as Fig. 2.

Fig. 2 Internal hierarchies
for package, classes and
methods [6]

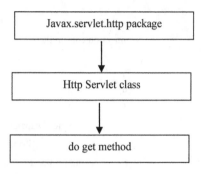

As defined in the previous work component dependency can be defined as:

Now the graphical representation with component dependency in the component tree will be as follows:

Component C1, C2, C3, C4, C5 all are directly related to the method M1 as shown below.

There was an idea of receiving the feedback or the input from the user side so that the software applications can be well understood [9]. Only component C5 is directly related to the module M2 because M2 is only the method or the module which is internally included in C5 component or package. A various number of models of software complexity has been proposed [10].

And finally this M2 is directly related to SM1 (Fig. 3).

Now the coupling distance for the various components has to be calculated as follows:

For C1 Coupling distance = 1
For C2 Coupling distance = 1

Fig. 3 Graphical
representation of component
tree with component
dependency [6]

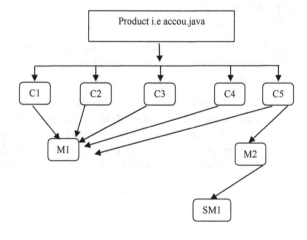

For C3 Coupling distance = 1
For C4 Coupling distance = 1
For C5 Coupling distance = 2 (according to previously defined hierarchy).

Therefore the calculation of total coupling of component I will be as follows:
Total coupling of component

$$i = \frac{1+1+1+1+2}{5-1}$$

$$= 6/4$$

$$= 1.5.$$

4 Results and Analysis

Similarly, we can calculate the average coupling of a project by calculating the coupling for the individual remaining classes (accreq.java, adlogin.java, aeexam.java) and it is shown below Table 1.

Average coupling of project can be calculated as:

$$\frac{1.5 + 1.5 + 1.4 + 1.6}{4} = 1.5.$$

Now from the above table we can analyze that no of packages in a class which are considered as components here are inversely proportional to value of total coupling of component I. Therefore by using the proposed coupling metric we are getting the different coupling value for the different component and the components having the low coupling value will be selected for the production of the software in component based development environment because always we required the low coupling and high cohesion. For example from these above class we will select the adlogin.java as a best class due to its low value and on an average by using these metrics the best project or the bunch of best components can be easily selected.

Table 1 Analysis for various classes

S. no	Class name	Number of components	Total coupling of component i
1	Accou.java	5	1.5
2	Accreq.java	5	1.5
3	Adlogin.java	6	1.4
4	Aeexam.java	4	1.6

4.1 Conclusions and Future Work

This paper proposes metrics for improving the quality of software components in CBSE environment based on a new strategy. During the optimal selection of components if the coupling is calculated by using the above metric then it will helps us to select the quality components for the development of software. Our results verified its integrity and operability. Some of the features of the metrics are:

1. These metrics can be used as a reusability factor for the various components.
2. These metrics can be used to evaluate the quality and efficiency of components. The coupling value calculated above helps to estimate the complexity of the software.

In this research we are calculating the average coupling of the project by just taking the four different classes with different number of components. The proposed approach can be applied to other projects also.

References

1. W. Stevens, G. Myers, and L. Constantine. Structured design, IBM Systems Journal, 13 (2):115–139, 1974.
2. T.J McCabe, A Complexity measure, IEEE Transactions on software Engineering, SE-2(4), December 1976, 308–320.
3. E. J. Weyuker, "Evaluating Software Complexity Measures." IEEE Transaction on software engineering, vol 14, pp 1357–1365, 1988.
4. Bansiya J. and Davis C. g.' "A Hierarchical model for object-oriented design quality assessment", IEEE Transactions on software Engineering, 28(1):4–17, 2002.
5. Dromey R.g., "A model for software product quality", IEEE Transactions on Software Engineering, 21(2):146–162, 1995.
6. L. Yu, Kai. C and Srini. R, "Multiple Parameter coupling metrics for layered component based software", University of Arkanas at little rock, USA, 2008.
7. Schildt Herbert (2008), 'The Complete Reference, Java', Seventh Edition, Tata McGraw Hill Publications.
8. A. Yadav and RA. Khan, "Class Cohesion Complexity Metric", Babasaheb Bhimrao Ambedkar University, ICCCT, 2011.
9. J.A Whittaker, "Software Invisible Users, IEEE Software, vol. 18, *pp. 84–88*, June 2001.
10. E Da-wei, The software complexity model and metrics for object oriented. School of computer Engineering, Jimei University, Xiamen, China, 16–18 April 2007, 464–469.

Scientific Workflow Management System for Community Model in Data Fusion

Boudhayan Bhattacharya and Banani Saha

Abstract The scientific experiments handle huge amount of data from various sources. The processing of data includes various computing stages along with their dependency pattern. The scientific workflow for data-intensiveness is used to model different processes. The scientific workflow paradigm integrates, structures and orchestrates services of heterogeneity and software design tools locally and globally to form scientific processes with complexity for enabling scientific discoveries. The Scientific Workflow Management System (SWfMS) deploys the scientific workflows for data-intensiveness by means of executing parallelism and the resources distributed in different infrastructures like grid and cloud. The community model is a data fusion methodology which is used to fuse data from various sources with multiplicity. The SWfMS for community model is used to describe the flow of data in various parts of the model and their corresponding working principle. This paper presents a data-intensive SWfMS for the community model.

Keywords Data fusion · Community model · SWfMS · Configuration generation workflow · Analysis campaign workflow

1 Introduction

The science with data-intensiveness is posing a huge challenge for the scientific workflows paradigm [1, 2]. The amount of data is expanding in leaps and bounds in each passing day. These data are generated by means of the modern experiments

Boudhayan Bhattacharya (✉)
Sabita Devi Education Trust – Brainware Group of Institutions,
398, Ramakrishnapur Road, Barasat, Kolkata 700124, India
e-mail: mailforboudhayan@gmail.com

Banani Saha
University of Calcutta, JD-2, JD Block,
Sector III, Salt Lake, Kolkata 700098, India
e-mail: bsaha_29@yahoo.com

© Springer Science+Business Media Singapore 2017
J.K. Mandal et al. (eds.), *Proceedings of the First International Conference on Intelligent Computing and Communication*, Advances in Intelligent Systems and Computing 458, DOI 10.1007/978-981-10-2035-3_37

and simulation results. The system also deals with the heterogeneousness and complex nature of data, their applications and the corresponding execution environments. A large-scale scientific experiment is usually based on scientific workflows to model data operations like inputting of data, processing of data, data analysis and aggregation of output data. Scientific workflow models are used to define the steps of data processing and their corresponding dependencies by means of Directed Graph or Directed Acyclic Graph [3]. The scientific workflow management works in distributed, homogeneous and non-homogeneous computing environments in organizational and geographical regions which process live data streams in gigabytes and archived and simulation data in petabytes available in different formats from various sources. To manage this kind of enactment, larger storage space and faster machines are required. Moreover, it is also scalable and diverse in terms of number of users, different applications, various data, resources for computation and technologies. There are several SWfMSs available in literature —Kepler [4, 5], VisTrails [6], Swift [7], Pegasus [8], Chiron [9], Taverna [10], Triana [11], Discovery Net [12], Yet Another Workflow Language (YAWL) [13], Business Process Execution Language (BPEL) [14], VIEW [15]. All of these models generate an architectural inference in which subsystems are developed through high-level organization. The organization of the paper is as follows— Sect. 2 outlines the Scientific Workflow Management Systems (SwfMS) along with requirements and life cycle of SWfMS, Sect. 3 explains the SWfMS architecture and Sect. 4 discusses the Community Model based on SWfMS.

2 Scientific Workflow Management System

The Scientific Workflow Management Systems (SWfMS) defines the infrastructure for setup, execution and monitor scientific workflow. The stages of SWfMS are as follows—Workflow script copying of previous stage, Changing file parameters, Running the modified script and Periodical checking of running jobs. If an error occurs, the background is searched in the different log files produced during the processing, the problem is solved and the steps are repeated until all the processing succeeds [16]. The SWfMS promotes production and standardization of different types of environment like Perl, Python or Shell scripts. The SWfMS was implemented successfully in different scientific applications [17, 18]. The experiments are managed based on the steps of the SWfMS for modeling of the processing steps as workflows with organized enactment [19].

2.1 Configuration Generation Workflow

The Configuration Generation Workflow is a collection of files which are used as inputs to the analysis campaigns. This collection of files is known as ensembles [16]. The ensemble is a collection of different configurations by a pre-defined order having different parameters. This workflow requires coordination of data fusion parameters, cluster parameters, grid parameters and cloud parameters. The Configuration forms a Markov chain sequence.

2.2 Analysis Campaign Workflow

The Analysis Campaign Workflow is a set of coordinated calculations which determines a set of specific data fusion quantities [16]. The analysis campaign workflow usually creates an intermediate data product. The intermediary calculations of each configuration are independent of each other.

The generic requirements of the SWfMS are Workflow Templates, Workflow Instances, Workflow Execution, Progress Monitoring, Workflow Execution History, Execution of Multiple Workflow Instances, Quality of Service Features, Stages in Input Data Files, Fault Tolerance, Data Provenance, Campaign Execution and Dispatch Campaigns [17, 18, 20] in sequence.

The scientific workflow is defined as a Directed Acyclic Graph (DAG) of the transitions of state from creating a process to completing a process [19, 21, 22]. The life cycle of scientific workflow phases involves Composition [21, 22] Deployment [19], Execution [21, 22] and Analysis [19, 22].

3 Architecture of Scientific Workflow Management System

The architecture of a SWfMS has five layers—Presentation layer, User Services layer, Workflow Execution Plan (WEP) Generation layer, WEP Execution layer and Infrastructure layer [4, 7–9]. The higher layers realize more concrete functionality by means of lower layers. The Presentation layer interacts with a user in SWfMS and the desired functions are realized at User Services layer. The Presentation layer is a tool for Interfacing with the user which interacts with users and SWfMSs at every stage of the life cycle of the scientific workflow. The User Services layer supports user functionalities like Monitoring and Steering of Workflow, Sharing of Workflow Information and Workflow Attribution Data Providing [23]. During workflow execution, workflow monitoring tracks the execution status and displays the required information [24]. The workflow monitoring verifies the result to prove a hypothesis [25]. The WEP generation layer produces a WEP in accordance with a design of scientific workflow based on three processes namely Refactoring of Workflow, Parallelization of Workflow and Optimization of Workflow. The WEP execution layer manages the execution of WEP through Scheduling of Workflow, Execution of Task and Robustness. The Scheduling Plan

(SP) is devised by SWfMS which use computing resources judiciously and prevents stalling of execution [26]. The infrastructure layer works as a platform for the SWfMS and the arrangement. This layer also maintains static or dynamic provisioning. The static provisioning provides resources for SWfMSs which are fixed, in workflow execution whereas dynamic provisioning enlarges or eradicates resources for SWfMSs vigorously [3].

4 Scientific Workflow Management System for Community Model

The community model [27–29] has been developed to apply the data fusion in different application areas like radar signal processing, image processing, data mining, decision support system etc. The Community Model paradigm (Fig. 1) is aimed to cater different application areas irrespective of environments. The different levels of community model ease out the burden from the master fusion filter by comparing the fused information using the reference sensor at each level. Each

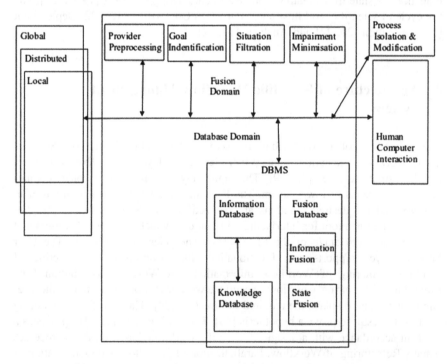

Community Model

Fig. 1 Community Model

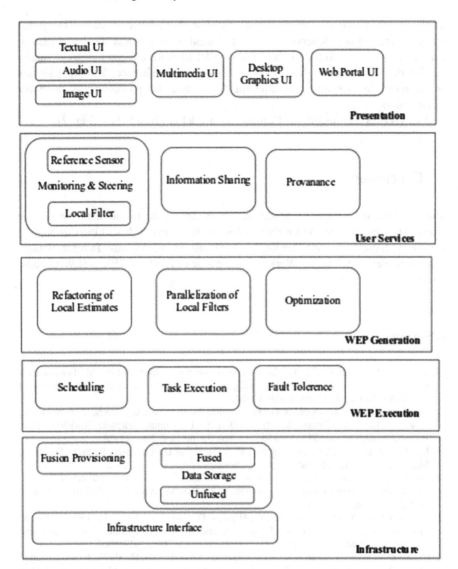

Fig. 2 SWfMS of Community Model

sensor has multiple levels in this model. These multiple levels are used for refinement of the sensors data collection. These levels use and execute different filtering algorithms for the fine-tuning of the data. The reference sensor feeds the data directly at each level to compare the fused data with its actual value. All filters with parallel organization, compare their respective sensor data with a common reference filter and calculates local states at each node. Thereafter, these locally estimated values are passed to the filters with parallel organization of the adjacent level and this process is repeated until the last (Nth) level of filter. The global estimates are defined depending on all the local estimates and the fused data in master fusion filter. All the filters in the parallel form have a common state due to the fact that all the filters share a system for common reference. The input signals are received by the filters at different levels and the processing is done for transmission.

The SWfMS architecture of Community Model is given below (Fig. 2).

5   Conclusion

The SWfMS of Community Model has been developed to facilitate the service oriented architecture (SOA) for data fusion implementation. It will then be applied to different SWfMS models like Kepler, VisTrails, Swift, Pegasus, Taverna, Triana, Discovery Net, Yet Another Workflow Language (YAWL), BPEL, VIEW to name a few.

References

1. Liew, C. S.: Optimization of the enactment of fine-grained distributed data-intensive workflows, Phd Thesis, Centre for Intelligent Systems and their Applications, School of Informatics University of Edinburgh, (2012).
2. Tsalgatidou, A., Athanasopoulos, G., Pantazoglou, M., Pautasso, C., Heinis, T., Grønmo, R., Hoff, H., Berre, A., Glittum, M., Topouzidou, S.: Developing Scientific Workflows from Heterogeneous Services, SIGMOD Record, vol. 35, no. 2, (2006), 22–28.
3. Liu, J., Pacitti, E., Valduriez, P., Mattoso, M.: A Survey of Data-Intensive Scientific Workflow Management, Journal of Grid Computing, Springer, (March 2015), 1–37.
4. Altintas, I., Berkley, C., Jaeger, E., Jones, M., Ludascher, B., Mock, S.: Kepler: Towards a Grid-Enabled system for scientific workflows, Grid Systems Workshop (GGF10) - The 10th Global Grid Forum, (2004).
5. Luda"scher, B., Altintas, I., Berkley, C., Higgins, D., Jaeger, E., Jones, M., Lee, E., Tao, J., Zhao, Y.: Scientific Workflow Management and the Kepler System, Concurrency and Computation: Practice and Experience, vol. 18, no. 10, (2006), 1039–1065.
6. Callahan, S., Freire, J., Santos, E., Scheidegger, C., Silva, C., Vo, H.: VisTrails: Visualization Meets Data Management, In Proc. of Special Interest Group on Management of Data Conference (SIGMOD'06), (2006), 745–747.
7. Zhao, Y., Hategan, M., Cliord, B., Foster, I., Laszewski, G. V., Nefedova, V., Raicu, I., Stef-Praun, T., Wilde, M.: Swift: Fast, reliable, loosely coupled parallel computation, In Proc. of IEEE Int. Conf. on Services Computing - Workshops (SCW), (2007), 199–206.
8. Deelman, E., Singh, G., Su, M.-H., Blythe, J., Gil, Y., Kesselman, C., Mehta, Vahi, G. K., Berriman, G. B., Good, J., Laity, A., Jacob, J. C., D. S. Katz, D. S.: Pegasus: A framework for mapping complex scientific workflows onto distributed systems. Scientific Programming, (2005), 13(3):219–237.

9. Ogasawara, E. S., Dias, J., Silva, V., Chirigati, F. S., de Oliveira, D., Porto, F., Valduriez, P., Mattoso, M.: Chiron: a parallel engine for algebraic scientific workflows, Concurrency and Computation: Practice and Experience, (2013), 25(16):2327–2341.

10. Oinn, T., Addis, M., Ferris, J., Marvin, D., Senger, M., Greenwood, R., Carver, T., Glover, K., Pocock, M., Wipat, A., Li, P.: Taverna: A Tool for the Composition and Enactment of Bioinformatics Workflows, Bioinformatics, vol. 20, no. 17, (2004), 3045–3054.

11. Majithia, S., Shields, M., Taylor, I., Wang, I.: Triana: A Graphical Web Service Composition and Execution Toolkit, In Proc. of IEEE International Conference on Web Services (ICWS'04), (2004), 514–524.

12. Rowe, A., Kalaitzopoulos, D., Osmond, M., Ghanem, M., Guo, Y.: The discovery net system for high throughput bioinformatics, Bioinformatics, vol. 19, no. 90001, (2003), 225–231.

13. van der Aalst, W. M. P., ter Hofstede, A. H. M., Kiepuszewski, B., Barros, A. P.: Workflow patterns, Distributed and Parallel Databases, vol. 14, (2003), 5–51.

14. Addis, M., Ferris, J., Greenwood, M., Li, P., Marvin, D., Oinn, T., Wipat, A.: Experiences with e-science workflow specification and enactment in bioinformatics, in e-Science All Hands Meeting 2003, S. Cox, Ed., (2003), 459–466.

15. Chebotko, A., Lin, C., Fei, X., Lai, Z., Lu, S., Hua, J., Fotouhi, F.: VIEW: a VIsual sciEntific Workflow management system, IEEE Congress on Services (July 9–13, 2007), 207–208.

16. Piccoli, L., Kowalkowski, J.B., Simone, J.N., Sun, X-H., Jin, H., Holmgren, D.J., Seenu, N., Singh, A.G.: Lattice QCD Workflows: A Case Study, IEEE Fourth International Conference on eScience, (2008), 620–625.

17. Podhorszki, N., Ludaescher, B., Klasky, S. A.: Workflow automation for processing plasma fusion simulation data, In Proc. of the 2nd workshop on Workflows in support of largescale science, Monterrey, CA. (2007), 35–44.

18. Deelman, E., et al.: Managing large-scale workflow execution from resource provisioning to provenance tracking: The CyberShake example, In Proc. of the Second IEEE International Conference on E-Science and Grid Computing (December 04–06, 2006), doi:10.1109/E-SCIENCE.2006.99.

19. GÖorlach, K., Sonntag, M., Karastoyanova, D., Leymann, F., Reiter, M. (2011) Conventional workflow technology for scientific simulation, in Guide to e-Science (Yang, X., Wang, L., Jie, W. eds.), Computer Communications and Networks, Springer London, 323–352.

20. Gil, Y. et al.: Examining the challenges of scientific workflows, Computer, Volume 40, no. 12, (Dec. 2007), 24–32.

21. Deelman, E., Gannon, D., Shields, M., Taylor, I.: Workflows and e-science: An overview of workflow system features and capabilities, Future Generation Computer Systems, (2009), 25 (5):528–540.

22. Mattoso, M., Werner, C., Travassos, G., Braganholo, V., Ogasawara, E., Oliveira, D., Cruz, S., Martinho, W., Murta, L. (2010) Towards supporting the life cycle of large scale Scientific experiments, In Pro of Business Process Integration and Management, 5, 79–82.

23. da Cruz, S. M. S., Luiza, M., Campos, M. Mattoso, M.: Towards a Taxonomy of Provenance in Scientific Workflow Management Systems, IEEE Computer Society, Congress on Services – I, (2009), 259–266.

24. Coalition, W. M.: Workflow management coalition terminology and glossary, (1999).

25. Costa, F., Silva, V., de Oliveira, D., Ocana, K. A. C. S., Ogasawara, E. S., Dias, J., Mattoso, M.: Capturing and querying workflow runtime provenance with prov: a practical approach. In Proc. of EDBT/ICDT Workshops, (2013), 282–289.

26. Bouganim, L., Fabret, F., Mohan, C., Valduriez, P.: Dynamic query scheduling in data integration systems, In Proc. of International Conference on Data Engineering (ICDE), (2000), 425–434.

27. Bhattacharya B., Saha, B.: Community Model - A New Data Fusion Filter Paradigm, American Journal of Advanced Computing (AJAC), Volume II Issue I, ISSN: 2368-1209, (January 2015), 25–31.

28. Bhattacharya B., Saha, B.: Community Model Architecture - A New Data Fusion Paradigm for Implementation, International Journal of Innovative Research in Computer and Communication Engineering (IJIRCCE), Volume 2 Issue 6, ISSN (Online): 2320-9801, ISSN (Print): 2320-9798, (June 2014), 4774–4783.
29. Bhattacharya B., Saha, B.: Signaling Cost Analysis of Community Model, Hotel Suryansh, Bhubaneswar, 3rd International Conference on Frontiers of Intelligent Computing: Theory and Applications (FICTA 2014), Advances in Intelligent Systems and Computing, Vol. 328, Springer, (2014), 49–56.

An Improved Stator Resistance Adaptation Mechanism in MRAS Estimator for Sensorless Induction Motor Drives

S. Mohan Krishna and J.L. Febin Daya

Abstract A comparative study of the conventional fixed gain PI and Fuzzy Logic based adaptation mechanisms for estimating the stator resistance in a Model Reference Adaptive System (MRAS) based sensorless induction motor drive is investigated here. The rotor speed is estimated parallely by means of a PI control based adaptive mechanism and the electromagnetic torque is also estimated to add more resilience. By considering the external Load torque perturbation as a model perturbation on the estimated stator resistance, the effects of the same on the estimated parameters are observed. The superiority of the Fuzzy based stator resistance adaptation mechanism is observed through detailed simulation performed offline using Matlab/Simulink blocksets. Furthermore, a sensitivity analysis of the stator resistance estimate with respect to load torque is also done to verify the effectiveness of the above concept.

Keywords Speed estimation · Adaptive control · Model reference · Machine model · Computational intelligence

Notation

i_{ds}^S, i_{qs}^S	d and q axis stator currents in stationary reference frame
$\hat{\psi}_{qrV}^S$, $\hat{\psi}_{drV}^S$	d and q axis Voltage model rotor flux linkages in stationary reference frame
$\hat{\psi}_{qrI}^S$, $\hat{\psi}_{drI}^S$	d and q axis Current model rotor flux linkages in stationary reference frame
L_r, L_m, L_s, σ	Rotor Magnetising and Stator inductance, Reactance
R_S, $\hat{R}_S$	Actual and Estimated Stator resistances
ω_r, $\hat{\omega}_r$, T_r	Actual and Estimated Rotor speeds, Rotor Time constant
K_p, K_I	Proportional and Integral gains

S. Mohan Krishna (✉) · J.L. Febin Daya
School of Electrical Engineering, VIT University, Chennai, India
e-mail: smk87.genx@gmail.com

© Springer Science+Business Media Singapore 2017
J.K. Mandal et al. (eds.), *Proceedings of the First International Conference
on Intelligent Computing and Communication*, Advances in Intelligent Systems
and Computing 458, DOI 10.1007/978-981-10-2035-3_38

1 Introduction

The indirect rotor flux oriented control implementation for a speed encoderless induction motor was significant in many ways. The presence of the speed encoder meant additional electronics, space constraints and cost. Besides, it would negate the inherent robustness of the induction motor. Consequently, many research efforts focused on the concept of sensorless speed estimation [1–4]. The speed estimation can be classified as one exploiting the concept of rotor spatial harmonics and the other depending on the machine model. Though the former is independent of machine parameters and considered as an accurate speed measurement, it introduces considerable measurement delays and cannot be used as a feedback signal for high performance drives. This led the researchers to focus more on machine model based speed estimation schemes, as they are relatively easy to implement, but the disadvantage was that, they were parameter dependent. Furthermore, the recent research focuses on machine model fed speed estimation mainly based on adaptive control. The classification of adaptive control based speed estimation schemes is shown in Fig. 1. Since the onset of sensorless control, parameter estimation has come to occupy considerable research space. Of all the parameters, identification of stator resistance is of significance in the condition and state monitoring of the drive. The speed estimation for a wide operating speed bandwidth is still a matter of concern and this led to the concept of online adaptation of certain critical motor parameters along with the speed. This was to avoid mismatching of the estimated and real values. Thermal parameters such as stator and rotor resistance were adapted online considering the important role they play during the operating range of the drive. As a result, many adaptation schemes were proposed for the same.

The different categories of stator resistance identification schemes were discussed by [5]. There are ones which relied purely on the measured or the reconstructed terminal quantities of the machine like stator voltages, currents etc. The stator resistance was obtained by making use of the machine model (Voltage and current model). The second category of identification schemes relied on an adaptive mechanism which would be based on a classical PI controller or Computational intelligence like Fuzzy Logic, Neural networks, ANFIS etc. A parallel stator resistance estimation scheme for a MRAS state estimator where the rotor fluxes are used to determine the error vector was implemented. Adaptation algorithms using

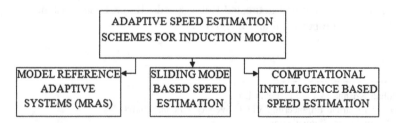

Fig. 1 Adaptive control based speed estimation schemes

full order Luenberger observer were also presented where the stator current error is input to the adaptation mechanism [6–8]. An instantaneous reactive power based algorithm was designed for joint state estimation of rotor resistance and speed with an extended Kalman filter and an Extended Luenberger Observer to suppress noise [9], but the above method, though more accurate and sensitive to model disturbances, occupied greater computational space and increased the complexity. Many algorithms have also been proposed for online tracking of rotor time constant, however, Stator resistance continues to be an influential parameter due to its dependence on the operating conditions, speed accuracy and stability of the drive system [10–14].

Existing research made use of mutual MRAS schemes involving parallel or sequential estimation of parameters or a comparison of different adaptation mechanisms. In this paper, a comparative analysis is performed on the conventional PI based and the Fuzzy Logic based stator resistance adaptation mechanism when subjected to load torque variations and change in the nominal value of stator resistance, with the rotor speed also being estimated parallely by means of a PI mechanism. In addition to it, the sensitivity of the stator resistance estimate for variations in the external load is also shown to indicate which one of the above gives an improved performance. The general configuration of the adaptation schemes is based on the Rotor Flux based MRAS and combines ideas of [5] and [14]. The work portrays the superiority and robustness of the Fuzzy logic based stator resistance identification mechanism over the classical PI scheme in the parallel state estimator. The electromagnetic torque is estimated using a mechanical model to verify the correctness and accuracy of the adaptation schemes. The model is validated by means of offline simulation using Simulink.

2 MRAS Based Simultaneous Parameter Estimation

MRAS form the backbone of adaptive control schemes. It has a high rate of adaptation and is easy to implement. The general configuration of MRAS is shown in Fig. 2. The reference and adjustable models are essentially voltage and current models fed from the machine terminals and the estimated quantity is obtained by an appropriate adaptation mechanism. The adjustable model depends on the unknown quantity. Depending on the output quantities of the models, the MRAS is characterized. Here, the configuration is based on rotor fluxes. The adaptation law makes use of a Lyapunov function to enable convergence.

The idea of Mutual Rotor flux based MRAS scheme is utilized here [15], where the roles of the two models are switched depending on the parameter to be estimated. The concept exploits the second degree of freedom, whereby the amplitude difference of the estimated rotor flux vectors is used to estimate the stator resistance. This identification scheme, in essence makes use of the speed estimate in obtaining the stator resistance and vice versa. The mechanism is shown in Fig. 3.

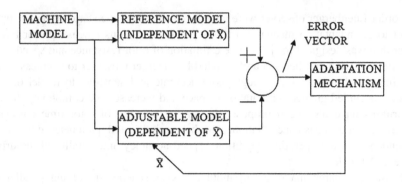

Fig. 2 General Configuration of MRAS

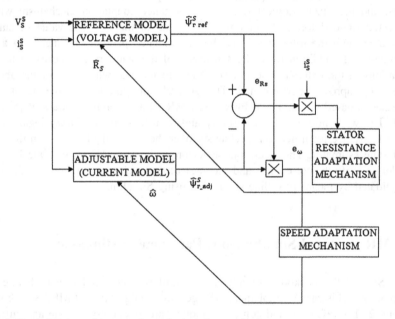

Fig. 3 Parallel stator resistance and speed adaptation mechanism

3   Structure of the Estimator

The equations and the respective mechanisms used for constructing the MRAS estimator depicted in Fig. 3 are as follows. The estimator is constructed in the stationary reference frame [16].

3.1 Reference Model/Adjustable Model for Speed Estimate/ Stator Resistance Estimate

$$\hat{\psi}_{qrV}^{s} = \frac{L_r}{L_m} \left[\int \left(V_{qs}^{s} - \hat{R}_s i_{qs}^{s} - \sigma L_s i_{qs}^{s} \right) dt \right] \tag{1}$$

$$\hat{\psi}_{drV}^{s} = \frac{L_r}{L_m} \left[\int \left(V_{ds}^{s} - \hat{R}_s i_{ds}^{s} - \sigma L_s i_{ds}^{s} \right) dt \right] \tag{2}$$

where, $\sigma = 1 - \frac{L_m^2}{L_s L_r}$.

3.2 Adjustable Model/Reference Model for Speed Estimate/ Stator Resistance Estimate

$$p\hat{\psi}_{qrI}^{s} = \left(\frac{-1}{T_r} \right) \hat{\psi}_{qrI}^{s} + \hat{\omega}_r \hat{\psi}_{drI}^{s} + \left(\frac{L_m}{T_r} \right) i_{qs}^{s} \tag{3}$$

$$p\hat{\psi}_{drI}^{s} = \left(\frac{-1}{T_r} \right) \hat{\psi}_{drI}^{s} + \hat{\omega}_r \hat{\psi}_{qrI}^{s} + \left(\frac{L_m}{T_r} \right) i_{ds}^{s} \tag{4}$$

3.3 Simultaneous Rotor Speed and Stator Resistance Adaptation Mechanism [9]

To ensure asymptotic stability of the system, the Popov's criterion for stability of Non Linear systems has to be satisfied:

$$S = \int_0^{t_0} \varepsilon^T W \, dt \geq -\gamma^2, \text{ for all } t_0 \tag{5}$$

where, $\varepsilon^T = [\varepsilon_{drI} \varepsilon_{qrI} \varepsilon_{drV} \varepsilon_{qrV}]$ and W is the non linear matrix given as:

$$W = \begin{bmatrix} -\Delta\omega_r \begin{bmatrix} 0 & -1 \\ 1 & 0 \end{bmatrix} & \begin{matrix} 0 & 0 \\ 0 & 0 \end{matrix} \\ \begin{matrix} 0 & 0 \\ 0 & 0 \end{matrix} & \frac{L_r}{L_m} \Delta R_S \begin{bmatrix} 0 & -1 \\ 1 & 0 \end{bmatrix} \end{bmatrix} \begin{bmatrix} \hat{\psi}_{drI} \\ \hat{\psi}_{qrI} \\ i_{ds} \\ i_{qs} \end{bmatrix} \tag{6}$$

where, $J = \begin{bmatrix} 0 & -1 \\ 1 & 0 \end{bmatrix}$, and $I = \begin{bmatrix} 1 & 0 \\ 0 & 1 \end{bmatrix}$ and $\Delta\omega_r = \omega_r - \hat{\omega}_r$ and $\Delta R_S = R_S - \hat{R}_S$;

The reduced form of the non linear matrix W is:

$$W = \begin{bmatrix} -\Delta\omega_r J & 0 \\ 0 & \frac{L_r}{L_m}\Delta R_S I \end{bmatrix} \begin{matrix} \hat{\psi}_{rl} \\ i_S \end{matrix} \qquad (7)$$

Equation (5) can be resolved into two inequalities given by:

$$S = X_1 + kX_2 \qquad (8)$$

where, $X_1 \geq -\gamma_1^2$ and $X_2 \geq -\gamma_2^2$ and $k = \frac{L_r}{L_m}$, γ_1 and γ_2 are positive real constants.

By exploiting the first inequality, the rotor speed is estimated. Here, the voltage and the current models are considered as the reference and adjustable models respectively. Therefore, we have:

$$X_1 = -\int_0^t \Delta\omega_r(\varepsilon_I^T J \hat{\psi}_{rl}) dt \qquad (9)$$

Here, $\varepsilon_I = \hat{\psi}_{rV}^S - \psi_{rl}^S$ and, on solving,

$$\varepsilon_I^T J \hat{\psi}_{rl}^S = \left[\hat{\psi}_{drV}\hat{\psi}_{qrl} + \hat{\psi}_{qrV}\hat{\psi}_{drl}\right] = e_\omega \qquad (10)$$

where e_ω is the rotor speed error. Using PI control based mechanism; the estimated rotor speed is obtained:

$$\hat{\omega}_r = \left\{K_P + \frac{K_I}{S}\right\}e_\omega \qquad (11)$$

$\hat{\omega}_r$ is the estimated speed. By utilizing the second inequality, the stator resistance is identified. Here, the roles of the reference and adjustable models are reversed. Therefore, we have:

$$X_2 = \int_0^{t_0} \Delta R_S \left(e_V^T i_S^S\right) dt \qquad (12)$$

where, $\varepsilon_V = \hat{\psi}_{rl}^S - \psi_{rV}^S$ and, on solving;

$$\varepsilon_V^T i_S^S = i_{ds}(\hat{\psi}_{drV} - \hat{\psi}_{drl}) + i_{qs}(\hat{\psi}_{qrV} - \hat{\psi}_{qrl}) = e_{Rs} \qquad (13)$$

where, e_{Rs} is the stator resistance error.

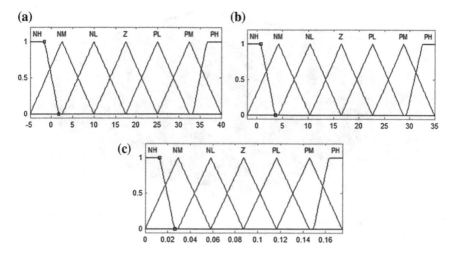

Fig. 4 Membership functions for the Fuzzy estimator, **a** Stator resistance error, **b** Change in Stator resistance error **c** Estimated Stator resistance

PI Based Stator Resistance Adaptation

$$\hat{R}_S = \left\{ K_P + \frac{K_I}{S} \right\} e_{Rs} \tag{14}$$

$\hat{R}_S$ is the estimated stator resistance.

Fuzzy Logic Based Stator Resistance Adaptation

The change in stator resistance error and the stator resistance error are taken as the inputs to the fuzzy estimator. The output is the estimated stator resistance and the fuzzification stage comprises of seven regions which are NH, NM, NL, Z, PL, PM, PH corresponding to negative high, negative medium, negative low, zero, positive low, positive medium and positive high. The triangular membership functions shown in Fig. 4 convert the inputs into corresponding fuzzy variables.

Table 1 Fuzzy rule matrix

		e_{Rs}						
		NH	NM	NL	Z	PL	PM	PH
Δe_{Rs}	NH	NH	NH	NH	NH	NM	NL	Z
	NM	NH	NH	NH	NM	Z	Z	PL
	NL	NH	NH	NM	Z	Z	Z	PM
	Z	NH	NM	Z	Z	Z	PM	PH
	PL	NM	Z	Z	Z	PM	PH	PH
	PM	Z	Z	Z	PM	PH	PH	PH
	PH	Z	PL	PM	PH	PH	PH	PH

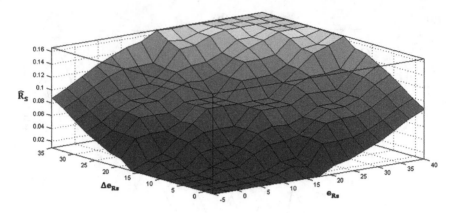

Fig. 5 Fuzzy control surface

The rule matrix is a 7 × 7 matrix having 49 rules shown in Table 1. The control range is between −5 and 40 for the stator resistance error, between −2 and 35 for the change in error and between 0 and 0.175 for the estimated stator resistance. The mamdani based interface is used and the fuzzy control surface is shown in Fig. 5.

3.4 Estimation of Electromagnetic Torque

$$T_e = \frac{3}{2}\frac{P}{2}\frac{L_m}{L_r}\left[\hat{\psi}_{drV}i_{qs}^s - \hat{\psi}_{qrV}i_{ds}^s\right]$$

(15)

4 Offline Simulation Results: Analysis and Discussion

The equivalent simulation model of the drive system along with the state estimator is constructed in MATLAB/SIMULINK and the tracking is observed and analyzed for both, the PI and the Fuzzy Logic based adaptation schemes. The model is run for the following cases:

(i)   Step Torque perturbation which is initially at no load and stepped up to rated load of 200 Nm after a fixed time interval
(ii) For half the rated load of 100 Nm.
(iii) At No Load, by doubling the nominal value of the actual Stator resistance of the motor.
(iv) Sensitivity analysis of estimated stator resistance with respect to Load.

4.1 MRAS Speed Estimator with PI Based Stator Resistance Adaptation Scheme

(i) See Fig. 6.
(ii) See Fig. 7.

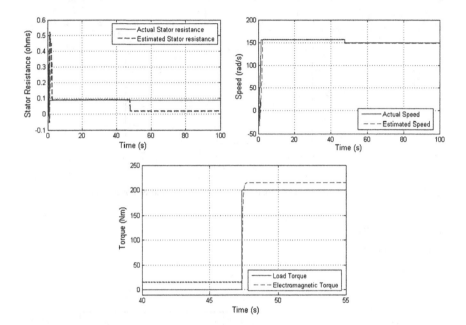

Fig. 6 Estimation of parameters at Step Torque perturbation

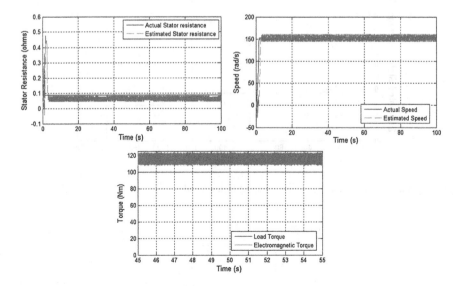

Fig. 7 Estimation of parameters at half the rated Load

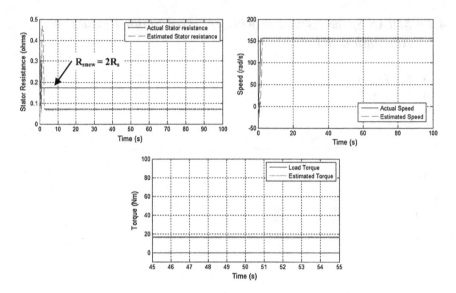

Fig. 8 Estimation of parameters at No load by doubling the actual Stator resistance

(iii) See Fig. 8.
(iv) See Fig. 9.

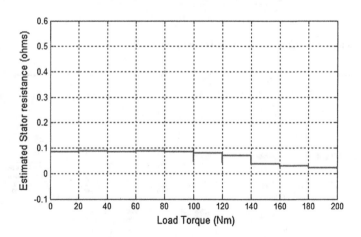

Fig. 9 Stator Resistance Estimate with respect to the Load

4.2 MRAS Speed Estimator with Fuzzy Logic Based Stator Resistance Adaptation Scheme

(i) See Fig. 10.
(ii) See Fig. 11.

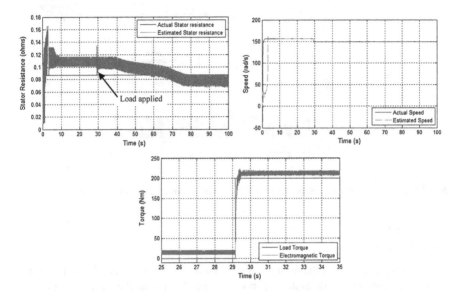

Fig. 10 Estimation of parameters at Step Torque perturbation

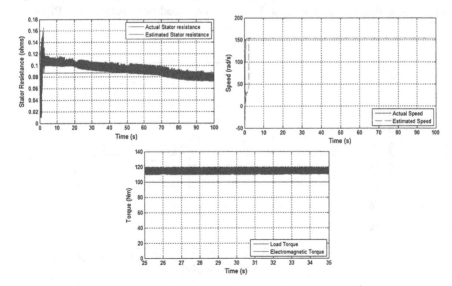

Fig. 11 Estimation of parameters at half the rated Load

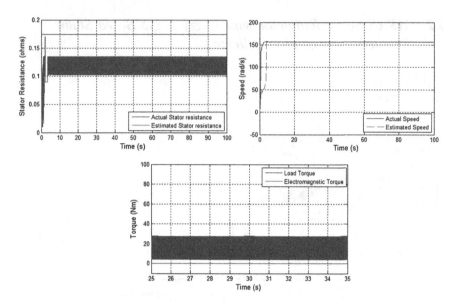

Fig. 12 Estimation of parameters at No load by doubling the actual Stator resistance

(iii)   See Fig. 12.
(iv)   See Fig. 13.

The specification of the motor considered is given in Appendix. The entire analysis is carried out in motoring mode at speeds around the base synchronous range. As it has been already explained, the stator resistance estimate depends on the speed estimate and vice versa. In all cases of Load perturbations, the PI based

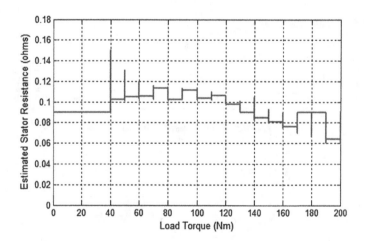

Fig. 13 Stator Resistance Estimate with respect to the Load

stator resistance estimate settles to a relatively lower value shown in Figs. 6 and 7 than the Fuzzy estimation scheme in Figs. 10 and 11. At the instance of transition from no load to full load, the mean of the instantaneous values of the flux levels come down, which can be attributed to lower settling of the estimated stator resistance in PI scheme and which, is not the case in Fuzzy Logic based scheme, where a constant flux level is maintained. Both the rotor speeds under all loading conditions converge to the actual value as can be seen in the figures. Since, for an ideal drive system, the load torque and the electromagnetic must be the same, the estimated electromagnetic torque in both the cases, more or less, follows the profile of the Load torque, but with a slightly higher magnitude to take into account the frictional and mechanical losses. The estimated torque is a function of the voltage model rotor fluxes and the d-axis and q-axis stator currents. However, q-axis component of the stator current (which is responsible for torque production) has inherent pulsations causing relatively more oscillations in the estimated parameters for fuzzy based adaptation as seen in Figs. 10, 11 and 12. When, the actual stator resistance is doubled ($R_{SNEW} = 2R_s$), though the Fuzzy estimated resistance in Fig. 12 settles to a higher value compared to the PI based scheme shown in Fig. 8. In both cases, it doesn't reach the reference value, due to decrease in the amplitudes of the flux levels as a result of increase in the voltage drop (due to increased value of the nominal resistance). For a ramp Load, the sensitivity of the estimated stator resistance is also studied and in Fig. 13, it can be observed that the Fuzzy based adaptation scheme is relatively more consistent than the PI based adaptation scheme shown in Fig. 9. This again, can be attributed to the amplitudes of the instantaneous rotor flux levels which are maintained almost constant in the former.

5 Conclusions

A comparative performance analysis between two Stator resistance adaptation schemes of Mutual MRAS based speed estimators have been presented for indirect vector controlled encoderless Induction motor drives. The concept utilizes the idea of simultaneous estimation of critical parameters such as the stator resistance and rotor speed, which play a vital role in the operating range of the drive system. The efficacy of Fuzzy Logic based adaptation mechanism is proved by the simulated results and by the susceptibility of the estimated stator resistance to changes in the external load. The above idea can be extended to closed loop tracking performance with variations in speed command.

Appendix

The motor model used in the simulation has the following ratings: A 50HP, three-phase, 415 V, 50 Hz, star connected, four-pole induction motor with equivalent parameters: $R_S = 0.087 \ \Omega$, $R_r = 0.228 \ \Omega$, $L_{ls} = L_{lr} = 0.8$ mH, $L_m = 34.7$ mH, Inertia, $J = 1.662$ kg m^2, friction factor $= 0.1$.

References

1. Marcetic, D.P., Krcmar, I.R., Gecic, M.A., Matic, P.R.: Discrete Rotor Flux and Speed Estimators for High-Speed Shaft-Sensorless IM Drives. IEEE T Ind Electron, 61, 3099–3108 (2014).
2. Smith, A.N., Gadoue, S.M., Finch, J.W.: Improved Rotor Flux Estimation at Low Speeds for Torque MRAS-Based Sensorless Induction Motor Drives. IEEE T Energy Conver, PP, 1–13 (2015).
3. Schauder, C.: Adaptive Speed Identification for Vector Control of Induction Motors without Rotational Transducers. IEEE T Ind App, 28, 1054–1061 (1992).
4. Zhao, L., Huang, J., Liu, H., Kong, W.: Second-Order Sliding-Mode Observer With Online Parameter Identification for Sensorless Induction Motor Drives. IEEE T Ind Electron, 61, 5280–5289 (2014).
5. Vasic, V., Vukosavic, S.N., Levi, E.: A Stator Resistance Estimation Scheme for Speed Sensorless Rotor Flux Oriented Induction Motor Drives, IEEE T Energy Conver, 18, 476–483 (2003).
6. Zaky, M.S.: A Stable Adaptive Flux Observer for a very low Speed-Sensorless Induction Motor Drive Insensitive to Stator Resistance Variations, Ain Shams Engg J, 2, 11–20 (2011).
7. Orlowska-Kowalska, T., Dybkowski, M.: Stator – Current-Based MRAS Estimator for a Wide Range Speed-Sensorless Induction-Motor Drive. IEEE T Ind Electron, 57(4), 1296–1308 (2010).
8. Kubota, H., Matsuse, K.: DSP-Based Speed Adaptive Flux Observer of Induction Motor, IEEE T Ind App, 29(2), 344–348 (1993).
9. Maiti, S., Chakraborty, C., Hori, Y., Minh C. Ta.: Model Reference Adaptive Controller-Based Rotor Resistance and Speed Estimation Techniques for Vector Controlled Induction Motor Drive Utilizing Reactive Power. IEEE T Ind Electron, 55(2), 594–601 (2008).
10. Umanand, L., Bhat, S.R.: Online Estimation of Stator Resistance of an Induction Motor for Speed Control Applications. In: Proc. Inst. Elect. Eng.-Elect. Power Applications, 142(2), pp. 97–103 (1995).
11. Tsuji, M., Chen, S., Izumi, K., Yamada, E.: A Sensorless Vector Control System for Induction Motors using Q-axis Flux with Stator Resistance Identification. IEEE T Ind Electron, 48, 185–194 (2001).
12. Guidi, G., Umida, H.: A Novel Stator Resistance Estimation Method for Speed-Sensorless Induction Motor Drives. IEEE T Ind App, 36, 1619–1627 (2000).
13. Raison, B., Arza, J., Rostaing, G., Rognon, J.P.: Comparison of Two Extended Observers for the Resistance Estimation of an Induction Machine. In: Proc. IEEE IAS Annual Meeting (2000).

14. Mir, S., Elbuluk, M.E., Zinger, D.S.: PI and Fuzzy Estimators for Tuning the Stator Resistance in Direct Torque Control of Induction Machines. IEEE T Power Electr, 13, 279–287 (1998).
15. Zhen, L., Xu, L.: Sensorless Field Orientation Control of Induction Machines based on a Mutual MRAS scheme, IEEE T Ind Electron, 45, 824–831 (1998).
16. Haron, A.R., Idris, N.R.N.: Simulation of MRAS-based Speed Sensorless Estimation of Induction Motor Drives using MATLAB/SIMULINK. In: Proc. First International Power and Energy Conference, pp. 411–415 (2006).

A New Implementation Scheme in Robotic Vehicle Propulsion Using Brushless DC Motor

Debjyoti Chowdhury, Arunabha Mitra, Santanu Mondal and Madhurima Chattopadhyay

Abstract This paper deals with the development of a propulsion system for a robotic vehicle using a permanent magnet Brushless DC (BLDC) motor with sensorless commutated drive. The proposed vehicle has four BLDC motor driven wheels, each having separate sensorless drive circuitry but all controlled by a single supervisory controller. The vehicle is capable of performing angular and linear displacements, ruled by a distantly located operator. A drive/break by-wire technology is utilized for operation of the wheels. In this work, a real time system with sensorless commutation is designed and implemented that utilizes a three phase inverter, a microcontroller and a motor speed feedback as drive circuitry. A suitable cost effective algorithm has also been developed to generate an appropriate six transistor switching sequence to commute the BLDC motor. The characteristics of the implemented drive give satisfactory outputs over a wide range of controlled speed variation from 330 to 2440 rpm. The effectiveness of the system so designed is demonstrated through the real time experimental data.

Keywords Brushless DC motor · Sensorless commutated drive · Back EMF · Robotic vehicle · Six switch inverter

Debjyoti Chowdhury (✉) · Arunabha Mitra (✉) · Madhurima Chattopadhyay
Heritage Institute of Technology, Kolkata, India
e-mail: djbabai.debjyoti@gmail.com

Arunabha Mitra
e-mail: arunabha.official@gmail.com

Madhurima Chattopadhyay
e-mail: madhurima.chattopadhyay@heritageit.edu

Santanu Mondal
Techno India, Salt Lake, India
e-mail: santanu_aec1984@yahoo.co.in

© Springer Science+Business Media Singapore 2017   387
J.K. Mandal et al. (eds.), *Proceedings of the First International Conference on Intelligent Computing and Communication*, Advances in Intelligent Systems and Computing 458, DOI 10.1007/978-981-10-2035-3_39

1 Introduction

The advantages of Brushless DC motors over brushed DC motors and induction motors due to permanent magnet rotor and electronic commutation technique makes it an obvious choice in electric vehicle applications [1]. The use of BLDC motors is expanding day by day in a wide range of applications especially in computerized equipments, space research programs, military support, transportation, production industries and household appliances. The main reason is that it holds high efficiency, torque, compactness over intact operating range and can be easily controlled due to proportional variation of torque with input voltage [2, 3]. Instead of using brushed DC motor with one central drive train propelling all four wheels in a Remote Operated Vehicle (ROV) [4], BLDC motors are mounted inside each wheel. Also, the electronic commutation by-wire technology is used instead of mechanical, hydraulic and pneumatic control systems. As a result, the weight of the system is reduced and the control over the vehicle becomes easier as well. The four wheel drive system of a commercial ROV is shown in Fig. 1. For robust traction control on various terrains, the ROV implements drive-by-wire analogy. The entire system comprises of a four wheel drive each incorporating a six-switch fed BLDC [5] motor and and a supervisory controller [6]. As found in literature, due to imperfections present in electronic commutation [7–9], each wheel drive tends to achieve different rotation speed [10]. The propulsion presented here in this paper tries to eliminate the foresaid problem by implementing an intelligent [11] supervisory controller to handle each individual wheel motor drive. This paper introduces the propulsion of a robotic [12, 13] vehicle using BLDC motor in Sect. 1. Then a description of the sensorless [14] drive system for BLDC motor is presented in Sect. 2. Next, Sect. 3 deals with real time implementation of the sensorless drive. The results obtained in real time simulation are nurtured in Sect. 4. Finally, Sect. 5 explores the scope for extension and modification of the present work.

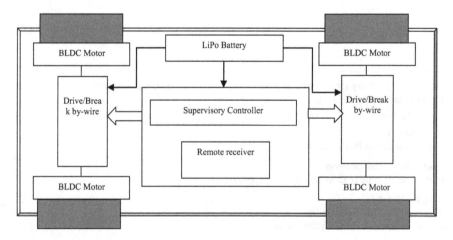

Fig. 1 Four wheel drive system for a Remote Operated Vehicle

Table 1 Implemented sequence for operation of switches

Rotor position (°)	Branch current			Switch	Closed	Digital sequence		
	I_a	I_b	I_c			D_1	D_2	D_3
0–60	+	−	Off	Q_1	Q_2'	1	0	Off
60–120	Off	−	+	Q_3	Q_2'	Off	0	1
120–180	−	Off	+	Q_3	Q_1'	0	Off	1
180–240	−	+	Off	Q_2	Q_1'	0	1	Off
240–300	Off	+	−	Q_2	Q_3'	Off	1	0
300–360	+	Off	−	Q_1	Q_3'	1	Off	0

2 Description of the Drive System

This section puts forward the idea of controlled rotation in a three phase BLDC motor using an external commutation circuit. The drive power is given to motor windings via a three phase inverter consisting of six Bipolar Junction Transistor (BJT) switches. The operation of the foresaid switches inside the driver follow a particular sequence which depends on the rotor position [15] as well as the rotor speed, shown by Table 1. The switching sequence is provided to the driver by a microcontroller, which takes a continuous feedback of the rotor position and the rotor speed. The rotor position is estimated by zero crossing the back EMFs [16] generated in the three windings of the BLDC motor. The rotor speed is measured and fed back to the microcontroller in order to verify the achievement of the desired rotational speed.

The block diagram of the real time model is described in Fig. 2. The reference speed is supplied with the help of a variable resistor. This analog value is converted into equivalent digital form by ADC in the microcontroller. Similarly, back EMFs [17] generated in the three windings are fed to the microcontroller, and the rotor

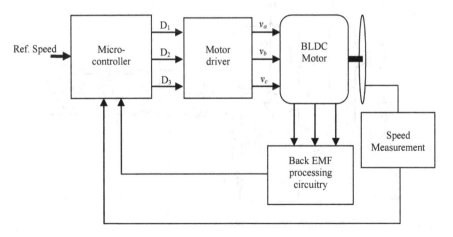

Fig. 2 Block diagram of the implemented hardware

speed is determined using a digital encoder. The microcontroller generates three
signals i.e., D_1, D_2 and D_3, which are supplied to the base terminals of the transistors used. The driver gives three phase supply to the motor i.e., v_a, v_b and v_c.

3 Real Time Implementation of BLDC Motor

The implemented hardware is built around a 2 W dual bridge motor driver
(L298 N) operated by an 8-bit Atmel microcontroller ATmega328p-pu. The three
phase six switch motor drive with reduced commutation lines by using digital NOT
gates connecting transistor bases is shown in Fig. 3.

A 2 W dual bridge motor driver (L298N) operated by an 8-bit Microcontroller
(i.e., Atmel ATMEGA328p-pu) has been used to drive the said motor as shown in
Fig. 4.

The configuration of the connections of the motor driver with the armature of the
motor is shown in Fig. 3. A diode is connected in parallel to each transistor to
provide snubbed protection to the designed system. Also, three similar resistors r_1,
r_2 and r_3 rated at 1 Ω (1 W) are connected for current protection. The designed real
time hardware with connections to the motor driver is shown in Fig. 4. An
appropriate algorithm considering a pre-defined transistor switching sequence
derived from Table 1 is written to the said microcontroller based commutation
system. The rotor speed is measured and fed back to the drive system by a photo
diode pair.

The ROV [18] drive system in order to maintain coordination among individual
motor speed controllers for four wheels, the proposed propulsion system makes use
of a supervisory controller. This supreme controller takes input from a user via a

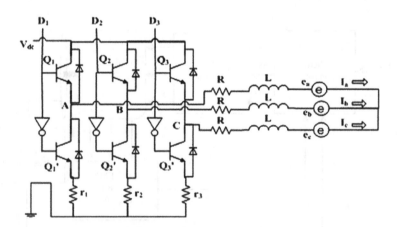

Fig. 3 Six transistor based three phase inverter

Fig. 4 The real time implemented hardware

wireless link to decide the desired rotational speed for every wheel's motor. The proposed algorithm of operation for the supervisory controller is described by a flow chart as shown in Fig. 5.

4 Hardware Simulation Results

This section puts forward the real time hardware simulation results of the designed Brushless DC motor drive.

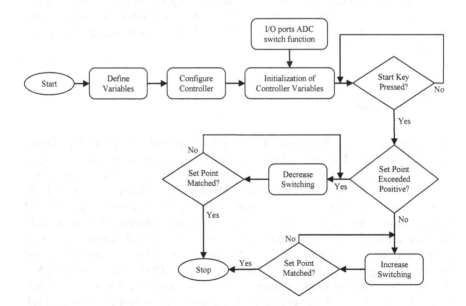

Fig. 5 Proposed algorithm of operation for the supervisory controller

(a)   (b)

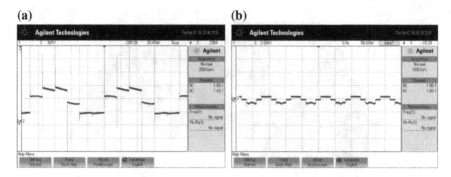

Fig. 6 a Back EMF generated in phase A. **b** Phase current supplied to phase A

Table 2 Parameters of the implemented BLDC Motor

Motor parameters	Values
Rated DC voltage	12 V
Rated current	110 mA
Stator resistance	1.7 Ω
Torque constant	14.5 mN m/A
Rated speed	5400 rpm
Life of working	50,000 h

The waveform traces for back EMF generated in phase A and current supplied are shown in Fig. 6. The technical specifications of the three ph. BLDC motor implemented in this work are displayed in Table 2. A rotor speed ranging from 330 to 2440 rpm has been achieved.

5 Conclusion

This paper proposes a Brushless DC motor based propulsion system for Remote Operated Vehicles. The intended scheme is innovative in terms of compactness, as the motor drive can be implemented right inside the wheel. In this work the rotor position is detected with the help of back EMF detection technique. The back EMF sensing technique is always advantageous as it requires minimum hardware and hence consumes less power thus making the system cost effective. This entire study has been carried out as a real time implemented hardware built around a low cost microcontroller. The output waveforms for back EMF and phase current obtained are given by Fig. 6 respectively in the previous section. The foresaid true back EMF detection is written in form of software algorithm inside each motor controller. Thus, allowing the flexibility to add another supervisory controller to change back

Fig. 7 Implementation of the
Remote Operated Vehicle

EMF detection according to the required load torque for varying terrain situations. The real time implementation of the proposed vehicle is shown in Fig. 7, on a custom designed robot chassis.

References

1. A. Tashakori, *Member IAENG*, M. Ektesabi, *Member IAENG* and N. Hosseinzadeh, "Modeling of BLDC Motor with Ideal Back-EMF for Automotive Applications", Proceedings of the World Congress on Engineering 2011 Vol II, WCE 2011, July 6–8, 2011, London, U.K.
2. Debjyoti Chowdhury, Madhurima Chattopadhyay and Priyanka Roy, "Modelling and Simulation of Cost Effective Sensorless Drive for Brushless DC Motor", International Conference on Computational Intelligence: Modelling, Techniques and Applications (CIMTA-2013), Procedia Technology 10 (2013), pp. 279–286.
3. Sharda Jaiswal, Debjyoti Chowdhury and Madhurima Chattopadhyay. "Performance Analysis of Sensored and Sensorless Drive of BLDC Motor Using Different Typs of DC/DC Converters In MATLAB/Simulink Platform", Proceedings of ITR International Conference, 06th April-2014, Bhubaneswar, ISBN: 978-93-84209-02-5.
4. M. Caccia, R. Bono, G. Bruzzone and G. Veruggio, "Bottom-following for remotely operated vehicles", Control Engineering Practice 11 (2003), pp. 461–470.
5. Behzad A and Rostami Alireza, "A novel starting method for BLDC motors without the position sensors", Energy Conversion Management 2009; 50(2):337–43.
6. Debjyoti Chowdhury and Madhurima Chattopadhyay, "Modeling and Real Time Implementation of Wireless Remote Controlled Vehicle", Second National Conference on Instrumentation and Control, 2013, pp 33.
7. S.A.K.H. Mozaffari Niapour, M. Tabarraie and M.R. Feyzi, "A new robust speed-sensorless control strategy for high-performance brushless DC motor drives with reduced torque ripple", Control Engineering Practice 24(2014)42–54.
8. Acarnley PP and Watson JF, "Review of position sensorless operation of brushless permanent-magnet machines", IEEE Trans. Ind.Electron 2006; 53(2):352–62.
9. Lai YS and Lin YK, "A new cost effective sensorless commutation method for brushless DC motors without using phase shift circuit and neutral voltage", IEEE Trans Power Electron 2007; 22(2):644–53.

10. Santanu Mondal, Arunabha Mitra, Debjyoti Chowdhury and Madhurima Chattopadhyay, "A New Approach of Sensorless Control Methodology for Achieving Ideal Characteristics of Brushless DC Motor Using MATLAB/Simulink", International Conference on Computer, Communication, Control and Information Technology (C3IT), February 2015, pp. 31.
11. R. A. Gupta, Rajesh Kumar and Ajay Kumar Bansal, "Artificial intelligence applications in Permanent Magnet Brushless DC motor drives", Published online: 25 December 2009 © Springer Science + Business Media B.V. 2009.
12. C. Canudas de Wit, "Trends in mobile robot and vehicle control", in Control Problems in Robotics, B. Siciliano and K. P. Valavanis, Eds. London, U.K.: Springer-Verlag, 1998, pp. 151–176. Lecture Notes in Control and Information Sciences 230.
13. M. Andersson, A. Orebäck, M. Lindström, and H. I. Christensen, "Intelligent Sensor Based Robotics", Heidelberg, Germany: Springer-Verlag, 1999. Ch. ISR: An Intelligent Service Robot, Lecture Notes in Artificial Intelligence.
14. Damodharan P, Sandeep R and Vasudevan K, "Simple position sensorless starting method for brushless DC motor", IEE Proc. Electro. Power Appl. 2008; 2(1):49–55.
15. Alireza R and Asaei Behzad, "A novel method for estimating the initial rotor position of PM motors without the position sensor", Energy Convers Manage 2009; 50(8):1879–83.
16. Shao JW, Nolan D and Hopkins T, "Improved direct back EMF detection sensorless brushless DC (BLDC) motor drives for automotive fuel pumps", IEEE Trans. Ind. Appl. 2003; 39 (6):1734–40.
17. Jiang Q and C. Hung "A new phase-delay-free method to detect back EMF zero crossing points for sensorless control of spindle motors", IEEE Trans. Magn. 2005; 41(7):2287–94.
18. Chokri Abdelmoula, Fakher Chaari and Mohamed Masmoudi, "A New Design of a Robot Prototype for Intelligent Navigation and Parallel Parking", Journal of Automation, Mobile Robotics & Intelligent Systems, 2009, Volume 3, No. 2.

Binary Fuzzy Goal Programming for Effective Utilization of IT Professionals

R.K. Jana, M.K. Sanyal and Saikat Chakrabarti

Abstract Utilization of IT professionals in software firms is an important problem in which vagueness is a very common factor. Most of these problems are multi-objective in nature and contain a large number of binary decision variables. In this work, a mathematical model is developed for allocating different categories of IT professionals, according to the requirement of various consultancy projects, as a binary fuzzy multiobjective programming problem. The considered objectives are effort maximization at all phases for executing the projects, and overall cost minimization of the firm. A binary fuzzy goal programming technique is applied to find the solution of the problem. A case example is presented based on the data collected from a software firm located at the Electronic Complex of Salt Lake, Sector-V, Kolkata, India and the effectiveness of the technique is demonstrated.

Keywords Multiobjective optimization · Fuzzy logic · Goal programming · Human resource planning · Information technology

1 Introduction

Human resources (HR) are the most valuable assets for any organization. This is true for software firms as well. There are different categories of employees in these firms. Information Technology (IT) professionals play an important role for the software firms. The existence and success of these firms depend largely upon the

R.K. Jana
Indian Institute of Social Welfare & Business Management, Kolkata, India
e-mail: rkjana1@gmail.com

M.K. Sanyal (✉)
University of Kalyani, Kalyani, India
e-mail: manas_sanyal@klyuniv.ac.in

Saikat Chakrabarti
Institute of Engineering & Management – Ashram Campus, Kolkata, India
e-mail: studymat.saikat@gmail.com

© Springer Science+Business Media Singapore 2017
J.K. Mandal et al. (eds.), *Proceedings of the First International Conference on Intelligent Computing and Communication*, Advances in Intelligent Systems and Computing 458, DOI 10.1007/978-981-10-2035-3_40

class of IT professionals employed and their proper utilization. They are highly paid as well. So, they must be effectively utilized for the benefit of the firms. It is very crucial to assign the most appropriate IT professionals according to their skills for successfully implementing the projects. At the same time, it must be taken into consideration that the cost involved in utilizing them is minimized. Using Taguchi's parameter design, a technique [11] is proposed to solve the IT professional selection problem for software projects. Zhou (2008) [14] proposed a technique for optimally allocating IT professionals for software projects. Also, a constraint-driven HR scheduling technique [12] is proposed by Xiao et al. (2008) for solving the HR scheduling problem for IT professionals.

The IT professional utilization problems are inherently multi-objective in nature. They involve complex and conflicting objective that are to be optimized simultaneously. Among several techniques, goal programming (GP) [3] has been used extensively [6] for solving multiobjective programming (MOP) problems. Unfortunately, the technique has not been used extensively for HR planning of software projects. It is used to allocate human resources for a health care organization [7]. A recent study [5] exploits GP for utilizing IT professionals in software firms. The problem is formulated as a GP problem with binary decision variables. Then the difference between the target value and the achievement function for all the goals is minimized.

The major shortcoming of the model is requirement of precise definition of target values of the goals. Narasimhan (1980) [10] proposed fuzzy goal programming (FGP) technique by combining fuzzy logic [13] and GP. It allows the goals to be defined imprecisely. The technique has been used to solve different types of imprecise MOP problems [9]. Till date, the literature reports the existence of only a few studies [4, 8] on application of FGP to allocate HR for business organizations. It is used to solve a multi-objective HR allocation problem [8]. In another study, FGP technique is employed for allocating tasks to employees in teamwork [4]. In particular, if the decision variables in a FGP problem are binary type then the problem is known as a binary FGP problem [2].

In this paper, a binary FGP model is formulated for effective utilization of IT professionals in software firms. Two goals are considered in this model. The first one is the cost goal and the second one is the effort goal. Due to the uncertainties of the decision-making environment, the goals are defined fuzzily. Triangular fuzzy numbers are used to define the target values of both cost and effort goals. Consideration of cost as a goal is always crucial for firms. The effort goal is equally important as it has relation to the completion of the projects in time. Also, the effort of individual IT professional is linked to rewards. Rewards are measured on a scale of 10. So, the goals are conflicting in nature. The Branch-and-Bound solver of LINGO 10 package is then used to obtain the most effective utilization of the IT professionals.

2 Fuzzy Subset, Membership Function, and Triangular Fuzzy Number

Fuzzy set theory was introduced by Zadeh (1965) [13] to extend the concept of ordinary set theory in which there is no partial belongingness of elements. On the other hand, fuzzy set theory allows both complete and partial belongingness of elements. Next, fuzzy subset, triangular fuzzy numbers, and triangular membership function are presented.

2.1 Fuzzy Subset

Suppose that Y is the collection of distinct objects and y is an element of Y. The fuzzy set $\tilde{P}$ in Y is then defined as follows:

$$\tilde{P} = \left\{ (y, \mu_{\tilde{P}}(y)) | y \in Y \right\} \tag{1}$$

where $\mu_{\tilde{P}}(y)$ is the membership function of y in $\tilde{P}$ that maps Y to the membership space $[0, 1]$.

2.2 Intersection of Two Fuzzy Sets

For any two fuzzy sets $\tilde{P}$ and $\tilde{Q}$, the intersection $\tilde{R} = \tilde{P} \cap \tilde{Q}$ is defined as:

$$\mu_{\tilde{R}}(y) = \min \left\{ \mu_{\tilde{P}}(y), \mu_{\tilde{Q}}(y) \right\}, y \in Y. \tag{2}$$

2.3 Triangular Fuzzy Number

A triplet $\tilde{T} = (t_1, t_2, t_3)$ is known as a triangular fuzzy number where t_1 denotes the least possible value, t_2 denotes the mean value, and t_3 denotes the highest possible value.

2.4 Triangular Membership Function

Let $\tilde{T} = (t_1, t_2, t_3)$ be a triangular fuzzy number. The triangular-shaped membership function $\mu_{\tilde{T}}(y)$ of $\tilde{T}$ is defined as (Fig. 1):

Fig. 1 Triangular
membership function

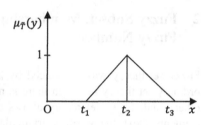

$$\mu_{\tilde{T}}(y) = \begin{cases} 0 & \text{if } y < t_1 \\ \frac{y - t_1}{t_2 - t_1} & \text{if } t_1 \le y \le t_2 \\ \frac{t_3 - y}{t_3 - t_2} & \text{if } t_2 \le y \le t_3 \\ 0 & \text{if } y > t_3 \end{cases} \tag{3}$$

3   Binary FGP Problem

In this section, a binary FGP problem [2] and its deterministic equivalent form are presented.

A classical MOP problem can be presented as:

$$\begin{aligned} \text{min: } &\{f_1(x), f_2(x), \ldots, f_M(x)\} \\ \text{subject to } &g_r(x) \le 0, \ r = 1, 2, \ldots, R \\ &x \ge 0, \end{aligned} \tag{4}$$

where x is vector of N decision variables; $f_m(x)(m = 1, 2, \ldots, M)$ are individual objectives, $g_r(x)(r = 1, 2, \ldots, R)$ are constraint functions.

In (4), if the decision variables are either 0 or 1, then it is classified as a binary MOP problem [9] and is defined as:

$$\begin{aligned} \text{min: } &\{f_1(x), f_2(x), \ldots, f_M(x)\} \\ \text{subject to } &g_r(x) \le 0, \ r = 1, 2, \ldots, R \\ &x = 0 \text{ or } 1. \end{aligned} \tag{5}$$

If it is possible to set target values for individual objectives in (5), then the corresponding binary GP problem [5, 6] takes the following form:

$$\begin{aligned} \text{min: } &\{h_1(n, p), h_2(n, p), \ldots, h_M(n, p)\} \\ \text{sucht hat } &f_m(x) + n_m - p_m = b_m, \quad m = 1, 2, \ldots, M \\ &g_r(x) \le 0, r = 1, 2, \ldots, R \\ &x = 0 \text{ or } 1 \\ &n, p \ge 0 \end{aligned} \tag{6}$$

where n is the vector of negative deviational variables $(n_1, n_2, \ldots, n_M)$, p is the vector of positive deviational variables $(p_1, p_2, \ldots, p_M)$, and $h_m(n, p)$ $(m = 1, 2, \ldots, M)$ are functional forms involving n and p.

The decision-making environment of IT professional utilization problem for software firms is highly uncertain. Consequently, defining precise target values for individual goals may not be an easy task. To capture the uncertainness of the decision-making environment and to provide flexibility to the decision-maker, the goals are expressed as fuzzy goals. The corresponding binary FGP problem can be written as:

Find x such that the following goals and constraints are satisfied:

$$
\begin{aligned}
& f_m(x) \cong b_m, \quad m = 1, 2, \ldots, M \\
& g_r(x) \le 0, r = 1, 2, \ldots, R \\
& x = 0 \text{ or } 1
\end{aligned}
\tag{7}
$$

where the symbol '$\cong$' denotes 'fuzzily equal to', and $f_m(x) \cong b_m$ indicates decision maker's partial satisfaction if $f_m(x)$ deviates slightly, up to some permissible tolerance values, from b_m.

Let b_m^L and b_m^U be the lower and upper limits of deviation of m-th goal in (7). Then following (3), the membership functions corresponding to the fuzzy goals can be constructed as follows:

$$
\mu_{f_m(x)}(x) = \begin{cases}
0 & \text{if } f_m(x) < b_m^L \\
\frac{f_m(x) - b_m^L}{b_m - b_m^L} & \text{if } b_m^L \le f_m(x) \le b_m \\
\frac{b_m^U - f_m(x)}{b_m^U - b_m} & \text{if } b_m \le f_m(x) \le b_m^U \\
0 & \text{if } f_m(x) > b_m^U
\end{cases}
\tag{8}
$$

Let $\mu_G(x)$ be the membership function corresponding to the fuzzy decision. Then using (8), it can be written as [1]:

$$
\mu_G(x) = \min\left\{ \mu_{f_1(x)}(x) \cap \mu_{f_2(x)}(x) \cap \ldots \cap \mu_{f_M(x)}(x) \right\}
\tag{9}
$$

Let, $\lambda_m = \max\left[\min\left\{ \mu_{f_1(x)}(x) \cap \mu_{f_2(x)}(x) \cap \ldots \cap \mu_{f_M(x)}(x) \right\} \right]$. Then the crisp equivalent of the binary FGP problem (7), providing the best compromised decision, is given by:

$$
\begin{aligned}
\max &: \lambda_m \\
\text{subject to} \quad & \mu_{f_m(x)}(x) \ge \lambda_m, m = 1, 2, \ldots, M \\
& g_r(x) \le 0, r = 1, 2, \ldots, R \\
& x = 0 \text{ or } 1.
\end{aligned}
\tag{10}
$$

4 IT Professional Utilization Model for Software Firms

In this section, we present a mathematical model for utilizing the IT professionals in software firms. The index, parameters, and decision variables are presented first.

4.1 Index, Parameters, and Decision Variable

The following notations are used to formulate the model:

i Index for IT professionals
j Index for IT professional category $(j = 1, 2, \ldots, J)$
I_j Number of IT professionals in category j $(I_1 + I_2 + \cdots + I_j = I)$
k Index for phase in a project $(k = 1, 2, \ldots, K)$
p Index for projects $(p = 1, 2, \ldots, P)$
c_{ijkp} Cost of employing the IT professional i of category j in phase k of project p
T_c Targeted cost of executing all the projects in time
e_{ijkp} Effort of the IT professional i of category j in phase k of project p
T_e Targeted effort for executing all the projects in time
M_{kp} Minimum number of IT professionals of specific category required to complete the task in time $(k = 1, 2, \ldots, K; p = 1, 2, \ldots, P)$
A_{kp} Available number of IT professionals of appropriate category for completing the task k of project p
x_{ijkp} IT professional i of category j involved in phase k of project p, a binary decision variable.

4.2 Goals of the Model

In our model, there are two goals that are presented below:

(i) Cost goal: The goal of the firm is to complete all the projects within the stipulated time by employing the IT professionals and by spending a targeted cost. The corresponding goal can be constructed as follows:

$$\sum_{p=1}^{P} \sum_{k=1}^{K} \sum_{j=1}^{J} \left\{ \sum_{i=1}^{I_j} c_{ijkp} x_{ijkp} \right\} \cong T_c. \tag{11}$$

(ii) Effort goal: To complete all the projects, the total effort of the IT professionals involved in different tasks of the projects should reach a targeted value. The corresponding goal can be constructed as follows:

$$\sum_{p=1}^{P} \sum_{k=1}^{K} \sum_{j=1}^{J} \left\{ \sum_{i=1}^{l_j} e_{ijkp} x_{ijkp} \right\} \cong T_e. \tag{12}$$

4.3 Constraints of the Model

The constraints of the model are presented as follows:

(i) Every task of the projects requires a minimum number of IT professionals to complete in time. So, the constraint can be written as:

$$\sum_{i=1}^{l_j} x_{ijkp} \geq M_{kp}, k = 1, 2, \ldots, K; p = 1, 2, \ldots, P. \tag{13}$$

(ii) The number of IT professionals assigned in phase t should be less than the available number of professionals of that category. So, the constraint can be written as:

$$\sum_{i=1}^{l_j} x_{ijkp} \leq A_{kp}, \ k = 1, 2, \ldots, K; p = 1, 2, \ldots, P. \tag{14}$$

(iii) If the IT professional i of category j is not involved in phase k of project p, then

$$x_{ijkp} = 0, \text{ for all } i, j, k, p. \tag{15}$$

If the IT professional i of category j is involved in phase k of project p, then

$$x_{ijkp} = 1, \text{ for all } i, j, k, p. \tag{16}$$

5 A Case Example

This case example has been taken from [5] in which the goals were precisely defined. Here, we have considered the imprecision of the decision-making environment and the goals are defined fuzzily. The required data for this example have been collected from a software firm. The firm is located at Salt Lake Electronic Complex, Sector-V, Kolkata, India. The firm does consultancy and also develops in-house products. Three broad categories of IT professionals – architect (A), developer (D) and quality engineer (Q) are employed for executing five different

Table 1 Involvement matrix

Category	Study	Design	Develop	Quality	Maintenance
A	×	×			
D			×		×
Q				×	×

Table 2 IT professionals requirement

Project	Study	Design	Develop	Quality	Maintenance
1	01	01	02	01	01
2	01	01	06	02	01
3	01	01	03	02	01
4	01	01	04	02	01

phases (i) study, (ii) design, (iii) development, (iv) quality assurance, and (v) maintenance for successfully completing the projects. If an IT professional is not involved in any project, then he will be involved in developing in-house products. The involvement of three categories of professionals in five phases is shown in Table 1 using the symbol '×'.

In the considered software firm, there are about 50 employees, among them 28 are IT professionals, and the rest are non-technical staffs. Out of the 28 professionals, 04 are architect, 16 are developer, and 08 are quality assurance engineer. The firm is, at present, planning to execute 04 upcoming projects. The IT professional requirement for the upcoming projects is shown in Table 2 [5].

The efficiency, in the scale of 10, of all the 28 IT professionals are as follows [5]: The efficiency of 04 architects are (8.4, 7.6, 8.5, 8.2). The efficiency of 16 developers are (8.1, 7.2, 7.8, 8.1, 8.8, 7.4, 7.7, 8.0, 7.4, 7.6, 8.2, 7.9, 7.5, 7.6, 7.8, 8.1). The efficiency of 08 quality assurance engineers are (7.3, 8.0, 7.6, 8.4, 7.9, 7.5, 7.7, 8.1). We have also collected data for cost to company for all the IT professionals, and time estimates on various tasks of the projects. It has been observed that the cost to company varies with the year of experience. Due to certain restrictions, these data are not shared here. The cost (in Rs. '00000) and the effort data, that are considered as triangular fuzzy numbers, are defined as (54, 60, 65) and (288, 300, 315), respectively.

6 Results and Discussion

We have solved the binary FGP problem using LINGO 10. Since there are binary variables, LINGO uses its Branch-and-Bound solver to find the solution. In the problem, there are 161 variables. Apart from the membership degree, all the 160 decision variables are binary in nature. The global optimal solution of the crisp

Table 3 Non-zero variables: FGP model

Non-zero variables							
x_{1112}	x_{2124}	x_{3233}	x_{4232}	x_{5351}	x_{7342}	x_{10231}	x_{15232}
x_{1122}	x_{2233}	x_{3344}	x_{4341}	x_{6231}	x_{7354}	x_{11234}	x_{16232}
x_{1233}	x_{2343}	x_{3352}	x_{4353}	x_{6341}	x_{8231}	x_{12232}	
x_{1343}	x_{3111}	x_{4113}	x_{5234}	x_{6351}	x_{8342}	x_{13232}	
x_{2114}	x_{3121}	x_{4123}	x_{5344}	x_{7234}	x_{9234}	x_{14232}	

Table 4 Assignment of IT professionals for FGP model

Project	Study	Design	Develop	Quality	Maintenance
1	A-3	A-3	D-6, 8, 10	Q-4, 6	Q-5, 6
2	A-1	A-1	D-4, 12, 13, 14, 15, 16	Q-7, 8	Q-3
3	A-4	A-4	D-1, 2, 3	Q-1, 2	Q-4
4	A-2	A-2	D-5, 7, 9, 11	Q-3, 5	Q-5

A: Architect, *D*: Developer, *Q*: Quality assurance engineer

model is obtained at 2591-th iteration. The solver takes just about 01 s to give the solution that contains 37 non-zero variables. The list of non-zero variables is presented in Table 3.

The corresponding assignment of IT professionals is shown is Table 4.

The maximum membership grade is obtained as 1.0. This indicates that both the goals have exactly reached to their target values. As a result, following the suggested assignment of IT professionals, the entire 04 project can be executed by spending a sum of Rs. 60 lac, and also the total effort required is 300.

6.1 Comparison with the GP Solution [5]

The GP model had 164 variables. Target values for the goal are same. LINGO solution was obtained at 3,031-th iteration with slightly longer time duration of 03 s. The assignment of IT professionals using GP is shown in Table 5.

The above table suggests that Architect 3 will be involved in Project 1 with the Study and Design phases. Similarly, Developer 7, 10 and 13 will be involved in the

Table 5 Assignment of IT professionals for the GP model

Project	Study	Design	Development	Quality	Maintenance
1	A-3	A-3	D-7, 10, 13	Q-2	Q-2
2	A-1	A-1	D-1, 4, 5, 8, 12, 16	Q-1, 8	Q-1, 8
3	A-4	A-4	D-3, 11, 14	Q-4, 6	Q-4
4	A-2	A-2	D-2, 6, 9, 15	Q-3, 7	Q-7

Development phase. Quality assurance engineer 2 will be involved in the Quality Assurance and Maintenance phases of Project 1. The utilization of the IT professionals for the remaining projects can also explained in the similar manner.

The results suggest that not a single goal was satisfied completely. The total cost for executing the entire 04 projects was calculated as Rs. 61.1238 lac, and the total effort required was calculated as of 293.7. The FGP approach provides better objective value for both the objectives. Therefore, it is clearly established that the binary FGP approach provides a better solution in comparison to the GP approach.

7 Conclusions

This paper addresses a very challenging problem for utilizing IT professionals in software firms. We have developed a mathematical model for such a problem. In the present literature, there are not an adequate number of mathematical models available. The problem has more than one objective and a large number of binary decision variables. We have used a binary FGP approach to find the best solution possible. The obtained solution recommends some change in terms of number of IT professionals to be utilized in different phases of execution of the projects in time. We have compared the solutions obtained with the GP solutions. It is observed that the binary FGP approach out performs the GP approach.

Acknowledgement The authors would like to thank Mr. R. N. Das, Project Leader, Dynamic Digital Technology Private Limited, Electronics Complex, Salt Lake, Sec-V, Kolkata-700091, India, for his support in this work.

References

1. Bellman, R. E., Zadeh, L. A.: Decision Making in a Fuzzy Environment, Management Sciences 17, B141–B164 (1970).
2. Chang, C.-T.: Binary Fuzzy Goal Programming: European J. of Operational Research 180(1), 29–37 (2007).
3. Charnes, A., Cooper, W.W.: Management Models and Industrial Applications of Linear Programming, Wiley, New York (1961).
4. Giannikos, I., Polychroniou, P.: A Fuzzy Goal Programming Model for Task Allocation in Teamwork, Int. J. of H.R. Development and Management 9(1), 97–115 (2009).
5. Jana, R. K., Chakrabarti, S.: A Goal Programming Approach to Human Resource Planning for Software Projects, Emerging Challenges for HR: VUCA Perspective, Eds.: U.K. Bamel, A. Sengupta and P. Singh, Emerald, 151–158 (2016).
6. Jones, D., Tamiz, M.: Practical Goal Programming, Springer, New York (2010).
7. Kwak, N.K., Lee, C.: A Linear Goal Programming Model for Human Resource Allocation in a Health Care Organization, J. of Medical Systems 21(3), 129–140 (1997).
8. Kwak, W., Shi, Y., Jung, K.: Human Resource Allocation in a CPA Firm: A Fuzzy Set Approach, Rev. of Quantitative Finance and Accounting 20(3), 277–290 (2003).

9. Lai, Y.-J., Hwang, C.-L.: Fuzzy Multiple Objective Decision Making: Methods and Applications, Springer, Berlin (1994).
10. Narasimhan, R.: Goal Programming in a Fuzzy Environment, Decision Sciences 11, 325–336 (1980).
11. Tsai, H.-T., Moskowitz, H., Lee, L.-H.: Human Resource Selection for Software Development Projects Using Taguchi's Parameter Design, European J. of Operational Research 151, 167–180 (2003).
12. Xiao, J., Wang, Q., Li, M., Yang, Y., Zhang, F., Xie, L.: Constraint-Driven Human Resource Scheduling Method in Software Development and Maintenance Process, Proc. of IEEE Int. Conf. on Software Maintenance, IEEE, pp. 17–26 (10.1109/ICSM.2008.4658050) (2008).
13. Zadeh, L. A.: Fuzzy sets. Information and Control, 8, 338–353 (1965).
14. Zhou, L.: A Project Human Resource Allocation Method Based on Software Architecture and Social Network, Proc. of 4th Int. Conf. on Wireless Communications, Networking and Mobile Computing, IEEE, 1–6 (10.1109/WiCom.2008.1749) (2008).

AVR Microcontroller Based Conference Presentation Timer

Sagar Bose, Soham Mukherjee, Sayak Kundu, Utpal Biswas and Mrinal Kanti Naskar

Abstract In this paper an implementation of an Atmel AVR based Conference Presentation Timer (CPT) is discussed which will serve as an automatic time keeper that keeps the track of time during presentation in a conference or seminar and alarms the speaker accordingly. Usually the duration of conferences vary therefore a provision of adjusting the total time is kept as well as a facility for changing the alarm time is also kept. However this paper elaborates a prototype which has further scopes for improvement.

Keywords Conference presentation timer · Atmel AVR · Automatic time keeper

1 Introduction

In most of the conferences a person is employed who alarms the author to complete his presentation within stipulated time. This process is monotonous, tedious and also accounts for distraction. The person engaged time-keeping many times miss the conference itself in spite of being interested in the conference matter. Our proposed

Sagar Bose (✉) · Soham Mukherjee · Sayak Kundu · M.K. Naskar
Jadavpur University, Jadavpur, India
e-mail: sgrbose@gmail.com

Soham Mukherjee
e-mail: sohamthedarkknight@gmail.com

Sayak Kundu
e-mail: sayakrnj@gmail.com

M.K. Naskar
e-mail: mrinaletce@gmail.com

Utpal Biswas
Kalyani University, Kalyani, India
e-mail: Utpal01in@yahoo.com

© Springer Science+Business Media Singapore 2017
J.K. Mandal et al. (eds.), *Proceedings of the First International Conference on Intelligent Computing and Communication*, Advances in Intelligent Systems and Computing 458, DOI 10.1007/978-981-10-2035-3_41

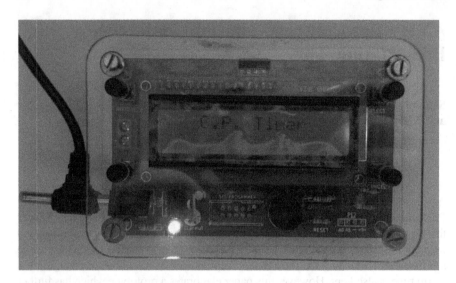

Fig. 1 Outlook of conference presentation timer

device i.e. Conference Presentation Timer (CPT) relieves the Time-Keeper of his duties and liabilities [3].

The CPT is pre-programmed to give a span of 15 min to present his/her lecture or presentation. This default settings of 15 min is usually set due to the fact that most of the conferences have a stipulated time period of 15 min. An alarm is sent off on the completion of 12 min to remind the presenter that his/her stipulated time is going to get over soon. However this total time of 15 min for presentation can be increased or decreased according to any particular conference. The moment of alarm can also be modified according to users' requirements.

As shown in the Fig. 1 of CPT the total time is displayed in the LCD screen. The buttons provided are for time adjustment as well as start and stop the timer. The buzzer is to signal the alarm and the end of a session. Its main control part is the Atmega8 Microcontroller which is programmed using MicroC.

The design of this proposed device is power efficient as a supply of +6 V DC or 4AA size batteries can power it up. Its portability is also notable as it can be easily carried in a pocket.

2   Device Hardware Description

The detailed schematic of the device is shown in Fig. 2. The schematic of the circuit is prepared using Cadsoft EAGLE PCB Design Software.

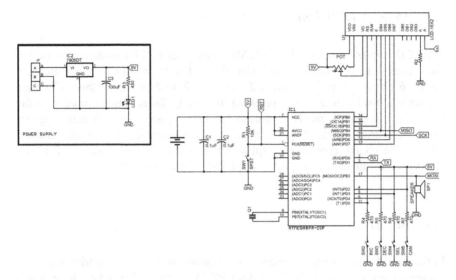

Fig. 2 Detailed schematic of CPT

2.1 Atmega8 Microcontroller Unit

Atmel ATmega8 used in this CPT is a 28-pin narrow DIP (Dual Input Package) IC. The AVR possesses a modified Harvard architecture with 8-bit RISC single-chip microcontroller. The low-power Atmel 8-bit AVR is chosen for this CPT because of this RISC-based microcontroller consists of 8 KB programmable flash memory, 512 K E^{2}PROM, 1 KB of SRAM and a 10-bit A/D converter. This available memory is used to store total time, alarm time after each time they are saved or changed. This MCU operates between 2.7–5.5 V which is supplied by 7805 Voltage Regulator and supports a throughput of 16 MPS with operating frequency at 16 MHz [1].

2.1.1 Advanced RISC Modified Harvard Architecture

The basic advantages of Atmega8 Microcontroller regarding its architecture are its vast 130 Instructions, 32 × 8 General Purpose Registers, Most instructions having One-clock Cycle Execution, and an On-chip Multiplier [1].

2.1.2 High Endurance Memory Segments

On the other hand regarding its high endurance nonvolatile memory it possesses 8 Kbytes of Self-programmable Flash memory, 512 Bytes of E^{2}PROM, 1 Kbyte Internal Static RAM, Data retention of 20 years at 85 °C and for 100 years at 25 °C [1].

2.2   16 × 2 LCD Unit

The LCD is interfaced with the AVR Microcontroller in 4 bit mode and is used to show Total time, Running time and Alarm time. The 16 × 2 LCD device used can show 16 characters per column and there are 2 such rows. In this LCD each character is displayed in 5 × 7 pixel matrix. The LCDs are programmed with the help of two registers namely command and data register. The command register stores the instructions that is given to the LCD as a command such as initialization, clearing the screen, controlling display, setting the cursor position, etc. The data register stores the data that is to be displayed on the LCD [2].

2.3   Buzzer

In the piezo buzzer sound is produced by principle of the piezoelectric effect. The buzzer is used as an alarm device in this CPT. When the CPT reaches alarm time it switches on the buzzer and a warning is issued. On reaching the final time the buzzer is switched on again and buzzes for two beeps to signal that the stipulated time is over.

The buzzer typically uses +5 V and has a very low power consumption.

2.4   Miscellaneous Items

The switches used in this device are Tact Switches that can be presses to start/stop or to set time. There are power LEDs to indicate that the device is in on state.

3   Working Algorithm

The microcontroller unit is programmed using Micro C PRO for AVR. The flowchart of the program is given in Fig. 3.

4   Future Scope for Development

Throughout this paper, we have presented the idea of a portable, dedicated device. The device may also be imagined as a handheld remote that can be used to control an already installed CPT in a seminar room. In that case the display unit will be a

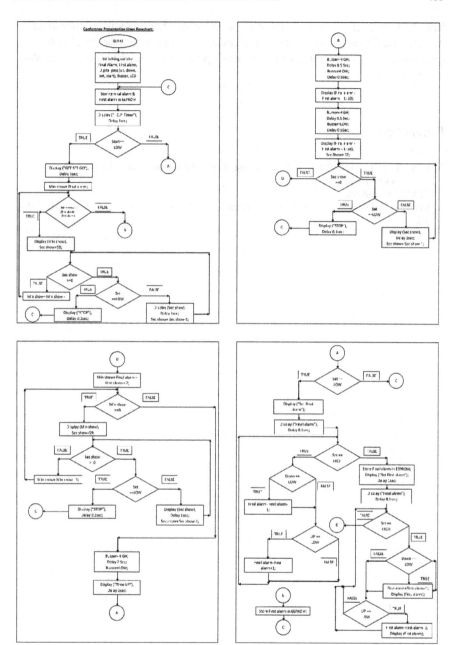

Fig. 3 Flowchart of working algorithm

large LCD or a LED matrix board along with a controller that will be controlled by the Remote. Extension of the dedicated device into a multipurpose one, i.e. to keep a provision for programming or to add a few peripherals may be an area to explore in near future.

References

1. Svendsli, O.J.: Atmel's Self-Programming Flash Microcontrollers. Atmel White Paper, 2464C–AVR–10/03.
2. X2 LCD Datasheet.
3. ATmega128 Datasheet, Web: http://www.atmel.com/dyn/resources/prod_documents/doc2467. pdf.

Part VI
Image Processing and
Pattern Recognition

A Person Identification System with Biometrics Using Modified RBFN Based Multiple Classifiers

Sumana Kundu and Goutam Sarker

Abstract In this present paper, we have designed and developed a person identification system with biometrics using modified Radial Basis Function Network based multiple classifiers. Three different classifiers using the same Modified RBFN with Optimal Clustering Algorithm, separately identify fingerprint, iris and facial images and the individual conclusions are fused together with programming based boosting. The conclusions from individual classifiers as well as the super-classifier performing fusion of conclusions are fuzzy in nature. Holdout method with Fuzzy Confusion Matrix is used to compute different performance metrics like accuracy, precision, recall and F-score. The different performance metrics are quite satisfactory. Also the learning and performance evaluation time with Fuzzy Confusion Matrix is low and affordable.

Keywords Fingerprint identification · Iris identification · Face identification · OCA · RBFN · Programming based boosting · Holdout method · Fuzzy confusion matrix · Precision · Recall · F-score

1 Introduction

Biometrics is the metrics related to human characteristics which can be used as a form of identification. Single biometric systems have limitations like high spoofing rate, high error rate, inter-class similarity and noise. Multimodal biometric systems prevail over some of these issues where biometric traits are acquired from multiple sources to recognize a person.

A PCA of symmetric sub-space model of neural network algorithm (SSA) for fingerprint recognition was presented in [1]. In an off-angle iris recognition system [2],

Sumana Kundu (✉) · Goutam Sarker
Computer Science and Engineering Department, NIT Durgapur, Durgapur, India
e-mail: kundu.sumana@gmail.com; sumana.kundu@yahoo.co.in

Goutam Sarker
e-mail: sarkergoutam@yahoo.co.in

© Springer Science+Business Media Singapore 2017
J.K. Mandal et al. (eds.), *Proceedings of the First International Conference on Intelligent Computing and Communication*, Advances in Intelligent Systems and Computing 458, DOI 10.1007/978-981-10-2035-3_42

the combination of LSEF and GC technique was used for iris segmentation and NeuWave Network was used for feature extraction. Another iris recognition method [3] was proposed based on the imaginary coefficients of Morlet wavelet transform. A face recognition application using Coiflet wavelets, PCA and neural network is described in [4]. In another face recognition application [5], the training patterns are learned incrementally. Here Gabor features and Zernike moment were used to extract features from face images. A fingerprint and iris feature-level fusion based recognition technique was proposed in [6] using conventional RBF neural network. Here iris features were extracted by block sum method and fingerprint features were by Haar wavelet method. Iris and face features were combined in [7] as new feature for representing persons which applied on modified PUM for recognition.

A multi-classification system based on simple bagging method was developed in [8]. In this system fingerprint, iris and face were used individually in OCA based modified RBFN classifier. Finally, an integrator concludes the decision of person identification based on simple voting logic or bagging method.

In this paper, we developed a multi-classification system using three different biometrics for person identification. There are three individual Optimal Clustering algorithm (OCA) based modified Radial Basis Function Network (RBFN) classifiers for Fingerprint, Iris and Face identification and also a Super-classifier which give the proper identification of person based on programming based boosting logic considering the result of three individual classifiers. Fuzzy confusion Matrix is used to evaluate the performances of all the classifiers of this person identification system.

2 Overview of the Multi-classification System

2.1 Preprocessing of Different Biometric Patterns

All the different biometric patterns, i.e. fingerprint, eye (iris) and face images of training and test databases have to be preprocessed [8].

2.1.1 Preprocessing of Fingerprint and Face Images

The preprocessing steps for fingerprint and face images are same. These steps are given below:

1. Conversion of RGB images into gray scale images.
2. Noise removal—To make noise free images 2D Median filter [8] have been used.
3. Image de-blurring—Blind deconvolution algorithm was used to de blur the images.
4. Background elimination.

5. Conversion of gray scale images into binary images.
6. Image Normalization—All images have been normalized into same and lower dimension.
7. Conversion of binary images into 1D matrix.

These 1D matrix file sets of fingerprint and face images were the inputs to the OCA of two individual classifiers.

2.1.2 Preprocessing to Extract Iris Images

In the preprocessing phase [8–10], iris images are extracted from eye images of training and test database. The necessary steps for this purpose are given below:

1. Conversion of RGB eye images into gray scale images.
2. Iris Boundary Localization—The radial-suppression edge detection algorithm [8–10] is used to detect the iris boundary. Then circular Hough Transformation is used to detect final iris boundaries and deduce their radius and center.
3. Extract the iris—The other parts of the eyes images such as eyelids, eyelashes and eyebrows have been removed to extract the iris [8].
4. Conversion into Binary images.
5. Image Normalization.
6. Conversion of binary images into 1D matrix.

This 1D matrix file set is the input to the OCA of respective classifier.

2.2 Theory of the Operation

The present system comprises of three individual classifiers for three different biometric features, i.e. fingerprint, iris and face identification. Each classifier consists of an OCA based modified RBFN [8, 11–14]. In three different classifiers, OCA formed clusters of various qualities of fingerprints of every person and angle (person-angle), various expressions of eyes (irises) of every person and two eyes (left and right) and different expressions of faces of every person and view (person-view) respectively. The mean (μ) and standard deviation (σ) of every cluster formed by OCA of three individual classifiers with approximated normal distribution output functions were used for every basis unit of RBFNs. Then Back propagation (BP) learning algorithm of RBFN, in different classifiers, classifies the "person fingerprint-angle" into "person fingerprint", the irises of two eyes (left and right) of each person into "person iris" and the "person-view" into "person" respectively. Finally a super-classifier concludes the final identification of the person combining the results of three individual classifiers based on programming based boosting method.

2.2.1 Identification Learning

There were three different training and test databases of fingerprint, iris and face for three individual classifiers. In the training fingerprint database, for each person's fingerprint, three different qualities of fingerprints and also 3 different angular (0°, 90° and 180°) fingerprints were there. In the training iris databases, for each person's iris (eye) patterns, three different expressions of individual left and right eye patterns were taken. Finally, in the training face database, for each person's face, three different expressions of face patterns and also three different angular views of face patterns i.e. frontal, 90° left and 90° right side view were taken. (Refer to Fig. 1).

After preprocessing, all different biometric patterns are fed individually as input to the different RBFNs of individual classifiers. When the networks learned all the different patterns (fingerprint, iris, face) of different qualities/expressions and of different angles or views for all different people, the classifiers are ready for identification of learned biometric patterns.

2.2.2 Identification Testing

The different test sets for testing of individual classifiers contains different person's (same person as training dataset) fingerprint, iris and face patterns of various qualities/expressions. These patterns are completely different from training sets. The test set to evaluate the performance of super-classifier contains pattern set (i.e. one fingerprint, iris and face image) for different persons of various qualities/expressions.

The different test patterns were fed to three different preprocessors of different classifiers. The preprocessed patterns were fed as inputs to the previously trained networks of three individual classifiers. After training, the networks of all the

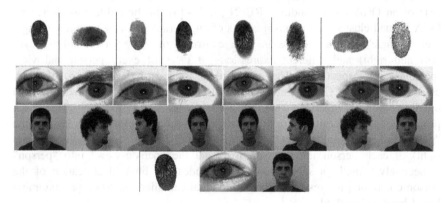

Fig. 1 Samples of few training and test patterns of different biometrics for single classifiers and a test set for super-classifier

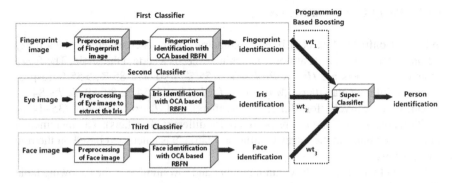

Fig. 2 Block diagram of multi-classification system for testing identification

classifiers gave high output values for known patterns and low output value for unknown patterns. A threshold value was required to distinguish between known and unknown patterns. The output value above threshold was considered as corresponding known pattern. In different overall output units, the BP networks produce different output activation. The normalized activation of each output unit represents the *probability* of belongingness of the input test pattern into the different classes. The test pattern is considered to belong to a class for which the normalized activation itself represents the *probability* of belongingness of that input test pattern into the particular classes. Finally super-classifier concludes the final identification of the person based on *programming based boosting method* considering the decisions of three individual classifiers. In this method, the weight assigned for each link is the *normalized accuracy* of that corresponding classifier. The *probability* of belongingness of the input test pattern for that corresponding class concluded by super-classifier by taking the *minimum* value of *probability* among three different classifiers. (Refer to Figs. 1 and 2)

3 Result and Performance Analysis

We have used the training and test databases for fingerprints samples from FVC 2004 database (http://www.advancedsourcecode.com/fingerprintdatabase.asp), Iris samples from MMU2 database (http://pesona.mmu.edu.my/~ccteo/) and Face samples from FEI database (http://fei.edu.br/~cet/facedatabase.html).

3.1 Performance Evaluation of the Classifiers

We use *Holdout method* with *Fuzzy Confusion Matrix* [15] to evaluate the performance of individual classifiers and super-classifier.

3.1.1 Fuzzy Confusion Matrix

In Fuzzy Confusion Matrix [15], each biometric pattern is considered to be a fuzzy member of any class, if its attribute matching exceeds a certain threshold. The fuzzy or graded members are *Highly Probable* (>55 %, >90 %, >85 %, >55 % respectively for different classifiers), *Moderately Probable* (45–54 %, 80–89 %, 75–84 %, 45–54 % respectively for different classifiers) and *Least Probable* (35–44 %, 55–79 %, 65–74 %, 35–44 % respectively for different classifiers). The Incremental value in each entry in the Fuzzy Confusion Matrix Σ is calculated based on the rule given below.

Let x and y represents the column number and row number in the following table (Refer to Table 1) respectively, where $x = 1$ and $1 \leq y \leq 3$.

For Actual class = Predicted class and for Actual class $\neq$ Predicted class,

$$\Sigma = [3 - (\text{Defuzzified Difference between Actual Class and Predicted Class})] \, \forall x, y.$$

where, the defuzzified values are: Highly Probable = 3, Moderately Probable = 2 and Least Probable = 1.

From the below mentioned Fuzzy Confusion Matrix (Refer Table 2), if we consider only two classes (say X and Y), then the accuracy, precision, recall and F-score are,

$$accuracy = \frac{A+D}{A+B+C+D} \qquad (1)$$

$$Precision = \frac{A}{A+B} \qquad (2)$$

Table 1 Incremental construction of fuzzy confusion matrix

		Actual class
Predicted class	Highly probable	+3
	Moderately probable	+2
	Least probable	+1

Table 2 Fuzzy confusion matrix (2 class)

			Actual class	
			X	Y
Predicted class	X	High	a_h	b_h
		Moderate	a_m	b_m
		Low	a_l	b_l
	Y	High	c_h	d_h
		Moderate	c_m	d_m
		Low	c_l	d_l

$$Recall = \frac{A}{A + C} \qquad (3)$$

$$F - score = \frac{2 * recall * precision}{recall + precision} \qquad (4)$$

where, $A = a_h + a_m + a_l$, $B = b_h + b_m + b_l$, $C = c_h + c_m + c_l$ and $D = d_h + d_m + d_l$.

The *accuracy* [8] of a classifier is the probability of its correctly classifying records in the test dataset. The *precision* of a classifier is the probability of records actually being in class X if they are classified as being in class X. The *recall* of a classifier is the probability that a record is classified as being in class X if it is actually belongs to class X and *F-score* is the harmonic mean of precision and recall.

3.2 Experimental Results

The proposed system was made to learn on a computer with Intel Core 2 Duo E8400, 3.00 GHz processor with 4 GB RAM and Windows 7 32-bit Operating System.

From Table 3, we can see the Fuzzy confusion matrix for super-classifier. The constructions of fuzzy confusion matrices are same for different classifiers. Each grid of fuzzy confusion matrix is further divided into three grids and each grid

Table 3 Fuzzy confusion matrix for super-classifier

			Actual class			
			Person 1	Person 2	Person 3	Person 4
Predicted class	Person 1	High	3	0	0	0
		Moderate	0	0	0	0
		Low	8	0	0	0
	Person 2	High	0	21	0	0
		Moderate	0	4	0	0
		Low	0	0	0	0
	Person 3	High	0	0	9	0
		Moderate	0	0	8	0
		Low	0	0	1	0
	Person 4	High	0	0	3	12
		Moderate	0	0	0	10
		Low	0	0	0	0
	Unknown	High	0	0	0	0
		Moderate	0	0	0	0
		Low	0	0	0	0

Table 4 Accuracy and learning time of the classifiers (in seconds)

Classifiers	Accuracy (%)	Training time	Testing time (single test sample)	Total time
First classifier (Fingerprint)	86.89	34.090	0.0526	34.1426
Second classifier (Iris)	85.86	15.796	0.0595	15.8555
Third classifier (Face)	95.00	160.007	0.0678	160.0748
Super-classifier	96.20	209.893	0.1803	210.0733

represents the fuzzy belongingness for the predicted class of any biometric pattern. From Table 4 we find that the accuracies of three different classifiers for three different biometric features (fingerprints, iris, and face) are 86.89, 85.86, 95.00 % and the accuracy of the super-classifier is 96.20 %. Thus, it is evident that the super classifier is efficient for person identification than considering single classifiers using single biometric features individually.

Also in Table 4, the multi classification system shows overall low testing time (<1 s) for different biometric patterns. In Table 5, precision, recall and F-score metrics clarify the performance of each class with holdout method. Hence, the proposed system shows improvement in terms of accuracy and testing time as compared to systems mentioned in the Sect. 1.

Table 5 Performance Measurement of the classifiers with holdout method

Performance evaluation metrics		1st classifier (Fingerprints)	2nd classifier (Iris)	3rd classifier (Face)	Super-classifier
Precision	Person 1	1.0000	0.7941	1.0000	1.0000
	Person 2	1.0000	1.0000	1.0000	1.0000
	Person 3	1.0000	1.0000	0.9310	1.0000
	Person 4	0.7241	0.6250	1.0000	0.8800
Recall	Person 1	0.8889	1.0000	0.9259	1.0000
	Person 2	0.7500	1.0000	0.8696	1.0000
	Person 3	0.8181	0.7778	1.0000	0.8571
	Person 4	1.0000	0.5556	1.0000	1.0000
F-score	Person 1	0.9412	0.8852	0.9615	1.0000
	Person 2	0.8571	1.0000	0.9303	1.0000
	Person 3	0.8999	0.8750	0.9643	0.9231
	Person 4	0.8399	0.5882	1.0000	0.9362

4 Conclusion

A biometric based person identification system using multiple classifiers with modified RBFN has been approached in the present paper. Person identification is more accurate and reliable because of the fact that the conclusions from different classifiers (using modified RBFN with OCA technique) acting on different biometrics of the different persons are fused or integrated together in the super-classification system with a methodology of programming based boosting. The fuzzy conclusions from the three individual classifiers as well as the super-classifier are more natural than hard or crisp conclusions. Performance evaluation in terms of accuracy, precision, recall and F-score of the present person identification system evaluated with Holdout method and Fuzzy Confusion Matrix is quite satisfactory. Also the learning and testing time is quite low and affordable.

References

1. Yu, C., Jian, Z., Bo, Y., Deyun, C.: A Novel Principal Component Analysis Neural Network Algorithm for Fingerprint Recognition in Online Examination System, Asia-Pacific Conference on Information Processing, pp. 182–186 (2009)
2. Moi, S., H., Asmuni, H., Hassan, R., Othman, R., M.: A Unified Approach for Unconstrained Off-Angle Iris Recognition, International Symposium on Biometrics and Security Technologies, pp. 39–44 (2014)
3. Lin, Z., Lu, B.: Iris Recognition Method Based on the Imaginary Coefficients of Morlet Wavelet Transform, Seventh International Conference on Fuzzy Systems and Knowledge Discovery, pp. 573–577 (2010)
4. Bhati, R., Jain, S., Maltare, N., Mishra, D., K.: A Comparative Analysis of Different Neural Networks for Face Recognition Using Principal Component Analysis, Wavelets and Efficient Variable Learning Rate, Int'l Conf. on Computer & Communication Technology, pp. 526–531 (2010)
5. Boughrara, H., Chtourou, M., Amar, C., B.: MLP Neural Network Based Face Recognition System Using Constructive Training algorithm, International Conference on Multimedia Computing and Systems, pp. 233–238 (2012)
6. Gawand, E., U., Zaveri, M., Kapur, A.: Fingerprint and Iris Fusion based Recognition using RBF Neural Network, Journal of Signal and Image Processing, 4(1), 142–148, (2013)
7. Lin, J., Li, J., Lin, H., Ming, J.: Robust Person Identification with Face and Iris by Modified PUM Method, International Conference on Apperceiving Computing and Intelligence Analysis, pp. 321–324 (2009)
8. Kundu, S., Sarker, G.: An Efficient Multiple Classifier Based on Fast RBFN for Biometric Identification, 2nd International Conference on Advanced Computing, Networking, and Informatics – Vol.1, Smart Innovation, Systems and Technologies 27, pp. 473–482 (2014)
9. Huang, J., You, X., Tang, Y., Y., Du, L., Yuan, Y.: A novel iris segmentation using radial-suppression edge detection, Signal Processing 89, 2630–2643 (2009)
10. Conti, V., Militello, C., Sorbello, F.: A Frequency-based Approach for Features Fusion in Fingerprint and Iris Multimodal Biometric Identification Systems, IEEE Transactions on Systems, Man, and Cybernetics—part c: Applications and reviews, 40(4), 384–395, (2010)

11. Sarker, G., Roy, K.: A Modified RBF Network With Optimal Clustering For Face Identification and Localization, International Journal of Advanced Computational Engineering and Networking, 1(3), 30 –35, (2013)
12. Sarker, G., Roy, K.: An RBF Network with Optimal Clustering for Face Identification, Engineering Science International Research Journal, 1(1), 70–74, (2013)
13. Sarker, G., Sharma, S.: A Heuristic Based RBFN for Location and Rotation Invariant Clear and Occluded Face Identification, International Conference on Advances in Computer Engineering and Applications, ICACEA – 2014, with IJCA, 30–36, (2014)
14. Kundu, S., Sarker, G.: A Modified Radial Basis Function Network for Occluded Fingerprint Identification and Localization, IJCITAE, Serial Publications, 7 (2), 103–109, (2013)
15. Sarker, G., Dhua, S., Besra, M.: An Optimal Clustering for Fuzzy Categorization of Cursive Handwritten Text with Weight Learning in Textual Attributes, International Conference ReTIS, pp. 6–11 (2015)

An Application of GIS Techniques Towards Pasture Land Use Studies

Urbasi Roy, Debasish Das and Mihir Bhatta

Abstract In fact the concept of land varies with the time and person concern. With the increasing pressure of population on the earth, the scientists are compelled to think about the land afresh. In this context, the present study involves an scientific investigation in the upper catchment area of Kangsabati watershed situated in the western part of Purulia district, bounded by latitude 23° 05′N to 23° 30′N and longitude 86° E to 86° 20′E. Tributaries of Kangsabati River are the main drainage in the area. In spite of moderate average annual rainfall (1446.4 mm) groundwater recharge is inadequate due to lack of permeability in the country rock. Present research involves planning pertaining to land use in a chronically drought-hit and degraded land. The main objective of this investigation is to identify suitable potential zones within the fallow or waste land for pasture development which will ultimately support a livestock rearing livelihood to the local rural and tribal poor. The different thematic layers of the area involving slope, drainage, lineament, surface water bodies, hydro-geomorphology and land use or land cover have been generated using SOI topographical sheets (73 I/3, 73 I/4 and 73 I/7), IRS-IB and IRS-P6 LISSIVMX satellite data aided by field verification for ground truth. The layers are analyzed in GIS environment and it reveals the potential pasture zones which are suitable for grazing by local livestock community. Generated digital maps in GIS environment have revealed many suitable areas for grazing as well as live stock watering.

Urbasi Roy · Debasish Das (✉) · Mihir Bhatta
Department of Environmental Science, University of Kalyani, Kalyani 741235, India
e-mail: ddas_kly@rediffmail.com

Urbasi Roy
e-mail: urbasi.roy81@rediffmail.com

Mihir Bhatta
e-mail: mihirbhatta@gmail.com

© Springer Science+Business Media Singapore 2017   425
J.K. Mandal et al. (eds.), *Proceedings of the First International Conference on Intelligent Computing and Communication*, Advances in Intelligent Systems and Computing 458, DOI 10.1007/978-981-10-2035-3_43

Keywords GIS techniques · Chronically drought-hit · Hard-rock · Pasture land-use

1 Introduction

Large number of people of Asia and African continents is involved in livestock rearing for their livelihood. India has 0.5 % of world's total grazing area but can carry 18 % of world's cattle population. As a result of poor availability of pasture and grazing land, animal either subsists on poor quality grasses available in the pasture and non-pasture lands or they are simply stall fed chiefly on crop residue.

In the present investigation, satellite based remote sensing and GIS techniques have been used for the planning of land-use in a chronically drought hit hard-rock terrain in the upper catchment area of Kangsabati [1], groundwater recharge is inadequate due to crystalline nature of the country rock and uneven relief of the terrain [2]. Hot summer and cold winter characterize the climate of the area under study. May is generally the hottest month with a mean daily maximum temperature of 40.3 °C and a mean daily minimum of 27.2 °C. January is the coldest month with the mean daily maximum temperature at 25.5 °C and the mean daily minimum at 12.8 °C. Agriculture is not an assured occupation in drought prone areas. Degraded pasture or grazing land occupies 12.07 sq km in the entire Purulia district of West Bengal [3]. Land use classification [4, 5], based on the present use of land is aimed to be recommended towards a better land classification (based on recommended use). Attributes for proposed land classification are related to pasture development mainly to provide an alternative livelihood, i.e. livestock rearing for the local rural and tribal people. Development of new surface water sources will efficiently manage the grazing distribution.

1.1 The Role of Remote Sensing and Geographical Information System (GIS)

With the recent availability of a wide range of sensors and there is considerable research in remote sensing that is being applied to rangeland or pastures. Over the past decades, remote sensing, geographical information system (GIS) and global positioning system (GPS) technologies have been integrated for detecting and mapping the distribution of noxious rangeland plants. Interpretation of Indian Remote Sensing Satellite (IRS) data can provide a dependable land use or land cover classification. Demarcation and identification of streams, paleo channels and other water bodies have been possible using satellite data in different scales [6, 7]. Use of GIS in land resource studies is at its early stage, but there have been many successful applications as well [8].

1.2 GIS and Its Application

Spatial information is always related to geographic space. Thus handling of spatial data involves the process of data acquisition, storage, analysis and output. Creation of spatial database in a GIS domain has become a very effective tool to aid and facilitate management and decision making. Data used in GIS often come in many types and are stored in different ways. Data sources are mainly obtained from digitization and scanning of paper maps and satellite images etc. Satellite data and SOI topo sheet data are very common input sources for GIS. In the present investigation GIS techniques involve the integrated and conjunctive analysis of multi-disciplinary spatial data within the same geo-referencing scheme. Aims of the present study have been identifying the suitable areas for livestock grazing including delineation of live-stock watering sites and health status of the animal as well as to create a digital data base revealing the above mentioned areas and sites.

2 Methodology

The flowchart of the methodology (Fig. 1a) involves data collection from survey of India toposheets (73 I/3, 73 I/4 and 73 I/7) where topo sheet no. 73I/3 in covers major part of the study area and geo-coded satellite imageries IRS1D of April 2001 and November 2000 through visual interpretation of standard FCC data aided by field verifications. Map has been derived from SOI toposheet using Wentworth method as described by Eadara and Karanam [9] also in the field slope angle and direction has been verified in some places using clinometer and GPS (model eTrex Garmin International, Inc). IRS P6 data were also consulted for better visual interpretability and then preparation of the thematic maps. Lineaments (fractures and joints) are placed in 1 m × 1 m grid and thus they offer lineament

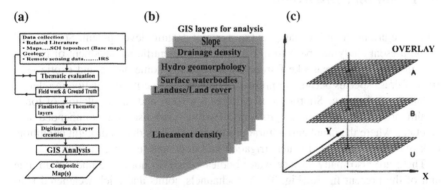

Fig. 1 **a** Flowchart of the methodology for GIS, **b** Thematic layers, **c** Thematic layers in overlay analysis (after [12])

number per grid is known as lineament density [10]. Layers (Fig. 1b) are finalized and subsequently digitization of different layers is done using TNT mips (Microimages) software (ver. 7.5) and through overlay analysis in GIS environment [11] composite maps are produced. In fact, thematic mapping is a task to produce a map for exploring, revealing, understanding and presenting the characteristics of a specialized phenomena and to communicate information (result) with a specific purpose to a certain audience by means of a numeric, graphical, visual, audio or by tactile representation [12].

FCC data (36.25 m spatial resolution) (Scale 1: 50000) dated 16th April 2001 (pre-monsoon) and 17th November 2000 (post-monsoon). IRS-P6 LISSIV MX data (5.8 m spatial resolution) (Scale 1: 25000) of December 2007 has also been studied for better visual interpretability towards post monsoon thematic information extraction. However, land use map has been generated using Survey of India (SOI) topographical sheet (no. 73 I/3, 73 I/4 and 73 I/7) and corresponding geo-coded satellite images. Preparation of final thematic maps, digitization of the thematic layers or maps and subsequent overlay analysis of the pertinent layers in GIS environment using TNT mips for delineating the suitable sites for stock watering in the produced composite map(s) (digital data base) has been accomplished [13]. The biochemical parameters of a grazing animal (e.g. goat) such as blood glucose, total RBC, total haemoglobin or Hb, packed cell volume or PCV, mean corpuscular volume, mean corpuscular haemoglobin or MCH and mean corpuscular haemoglobin concentration or MCHC and meteorological data such as ambient temperature (°C), relative humidity (%), average annual rain fall (mm) and elevation (mMSL) from previously published article [14] has been entered to the newly prepared digitized map as the non-spatial data or attributes. The data have been entered in such a manner so that, when the mouse (computing) has been ported and clicked on any selected point subsequently opening one or more windows containing the information about different non-spatial attributes [15].

3   Result and Discussion

Still there are no sound management systems of pasture development. So, the goal of the present study has been to move towards a scientific management of pasture land in drought hit areas like Purulia by introducing some native and exotic varieties of grass species because at present India is facing a critical situation in relation to land use planning. Shortages of pasture, firewood and fast depletion of forest wealth are assuming serious problems. Pasture land use is recommended and practiced where the usual agricultural productivity suffers due to drought condition, lack of surface or ground water irrigation and degraded terrain conditions.

There are also many surface water bodies which do not come under map able unit of the present figure (Fig. 3). Paleochannels, joints and thick weathered residuum of lower region (1) (Fig. 3) are potential sites for well development aiming at a consistent supply of water for livestock watering even during dry months. Water

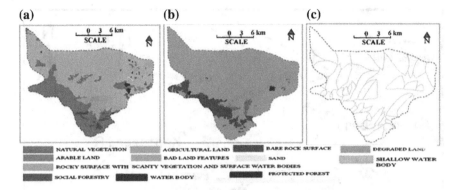

Fig. 2 **a** Land cover and land use map of the study area (post monsoon); **b** Land cover and land use map of the study area (pre monsoon); **c** Kangsabati river basin and sub basin map (after [13])

table lowers down up to 3 to 4 m from post monsoon to pre-monsoon season particularly in the weathered residuum. Most of the high stream flows are perennial in nature (Fig. 2c). Planar surfaces (3) with considerable spatial extent and deep weathered residuum present in the hilly region (2) (Fig. 3) offer good grazing potentiality and stock watering sites. Planar surfaces, inselberg or residual hill (4) and dissected plateau (Fig. 3) are the typical geomorphic features of an erosional landform. Paleochannels, joints and thick weathered residuum of lower region (1) (Fig. 3) are potential sites for well development aiming at a consistent supply of water for livestock even during dry months. Water table lowers down up to 3–4 m from post monsoon to pre-monsoon season particularly in the weathered residuum (Fig. 2a, b). Most of the high stream flows are perennial in nature. So, those stream courses like Saharjhor R, Bandu R and Kangsabati R, which are adjacent to the grazing ground, are suitable for stock-watering. Planar surfaces (3) with considerable spatial extent and deep weathered residuum present in the hilly region (2) (Fig. 3) offer good grazing potentiality and stock watering sites. Planar surfaces, inselberg or residual hill (4) and dissected plateau (Fig. 3) are the typical geomorphic features of an erosional landform.

3.1 Overlay Analysis

The database contains information in the form of maps that can be used to answer the user's problem are involved in overlay analysis. All those are necessary to establish a link between database and output that will provide that answer in the form of a map, tables or figures. In most geographical information system, it is assumed that the information in the database is present in the form of points, lines and areas and their associated attributes. The interpretation of two or more polygon nets by overlay is a special case of a much larger set of operations that can be used

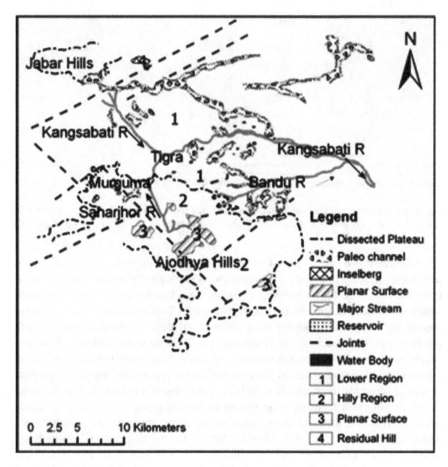

Fig. 3 Composite map depicts present and proposed potential stock-watering zones

to analyze area data. An overlay is a set of equally exclusive contiguous areas connected with a particular area. Each overlay is defined by a given attribute (Fig. 1c). Overlay is a concept in real world where different thematic layers in the digital map form are the intersections of two or more polygon nets. Previously polygon overlay in vector format was demanded enormous computing task but presently the commercial GIS software has solved this problem. The composite map has been produced through the overlaying process in GIS environment involving the thematic layers like slope, hydro geomorphology and lineament density [16] (Fig. 2c).

Since here the slope is high (>10°), the region is suitable for livestock grazing due to its agility and sure footedness. Whereas the area belonging to low to moderate slope suitable for grazing for both goat and cattle. Due to the presence of low to moderate lineament density and suitable hydro geomorphic features, livestock watering sites may be developed for the dry months. Potential livestock

(a)

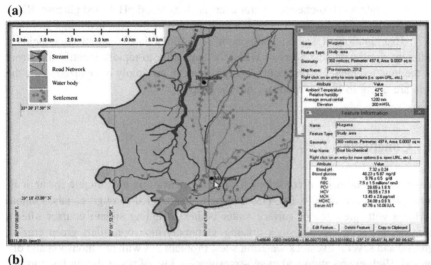

(b)

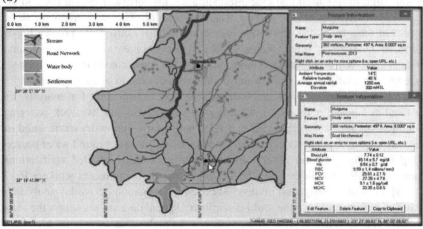

Fig. 4 a Non-spatial attributes including biochemical parameters of the pre-monsoon and **b** Non-spatial attributes including goat's biochemical parameters of the post-monsoon in the study area of Purulia; (after [15])

watering sites may be developed because of the presence of the alluvial fans and higher lineament density. The maps of Fig. 4a, b are clearly showing marked differences in biophysical and biochemical parameters between two seasons in Murguma, Purulia. Here we can compare the Fig. 4a with Fig. 4b, i.e. comparing bio-chemical parameters of goats under study of pre-monsoon season between the two present study areas of Purulia. It has been known that the purpose of a general system of animal health information has been to give information which subsequently provides better perceptive of the epidemiology of different disease and the continuous study of goat's bio-physical parameters such as rectal temperature, heart rate, pulse rate and bio-chemical parameters such as blood pH, blood glucose, RBC, PCV, MCV, MCH, MCHC values can give an idea of the goat heath in a real time [17]. Examination of spatial component of health related animal data yields a different significant advantage of GIS—the capability to rapidly identify the errors of the inputted data. Lost and out-of-range data can be simply recognized while the data are going to be incorporated into the map.

4 Conclusions

The present landmass of considerable spatial extension is not suitable for agricultural practice excepting a few places which are irrigated through canals. Plannar surfaces with presence of surface water bodies including stream courses offer a considerable scope for livestock grazing. Generally monsoon offers green grazing ground and several livestock watering sites. Problem lies with pre-monsoon and dry period. High evaporation and evapo-transpiration loss of water during hot and dry season leaves almost all the surface water bodies dry. Thematic layers on slope, lineament density, hydro geomorphology and drainage density are prepared. Post monsoon land cover offers many surface water storages along with perennial streams and their confluences. These locations are most convenient for livestock watering especially when these water bodies are adjacent to grazing ground.

While the thematic layers are overlayed in GIS environment for analysis, the resultant composite map shows suitable sites for grazing and livestock watering. Identification of new watering sites is needed particularly during dry months and for this reason several attempts have been made to locate water or ground water sources using integrated GIS and remote sensing techniques aided by filed verifications. These integrated techniques have been worked out by many recent workers [6, 7, 17, 18]. Present investigation involves successful application of overlay analysis in a GIS environment where the multi-disciplinary spatial data are set within the same geo-referencing scheme [19]. Thus, GIS can be sighted as a possible tool for a novel move towards of science [20], to maintain the animal health in terms of monitoring, observation as well as disease management policies.

References

1. Mishra, S.: Climate Change Adaptation in Arid Region of West Bengal, Climate Change Policy Paper III, WWF India (2012).
2. Mukherjee, P., Das, D.: A study on the development of basin and their hydro geomorphic features in and around Ajodhya plateau, eastern India, Proceedings of the International symposium on intermontane basins: Geology and resources(Chiang Mai, Thailand), AGID Publication, pp 409–417(1989).
3. GOI: Government of India: Waste land Map of West Bengal, Department of Land Resources. Ministry of Rural Development (2003).

4. NRSA: Integrated Mission for Sustainable Development, Technical Guideline, National Remote Sensing Agency, Department of Space, Govt. of India, Balanagar, Hyderabad (1996).
5. Vijith, H., Satheesh, R.: Evaluation of landuse pattern and geomorphology of parts of Western Ghats using IRS P6 LISS III Date 1E (1) Journal-AG, 88 pp 14–80 (2007).
6. Das, D.: An Investigation of Aquifer Recharge sites Identification using Spatial Technologies –In: Management of Aquifer recharge for Sustainability (Editor Peter Fox), Proceedings of the 6th International Symposium of Managed Aquifer Recharge (ISMAR-6), *ACACIA Publishing Incorporated*, Phoneix, Arizona, USA, pp 601–612. ISBN:0-9788283-9-9 (2007).
7. Krishnakmurthy, J., Venkatesh Kumar, N., Jayaraman, V., Manivel, M.: An approach to demarcate groundwater potential zones through remote sensing and geographical information system. *Int. J Remote Sensing.* 7; 1867–1884 (1996).
8. Malczewski, J.: GIS-based land-use suitability analysis: a critical overview Department of Geography, Progress in Planning 623–65 University of Western Ontario, London, Ont., Canada N6A 5C2 (2004).
9. Eadara, A., Harikrishna K.: Slope Studies of Vamsadhara River basin: A Quantitative Approach. *Int. J. Enginr Innovative Tech.* 3(1): 184–189 (2013).
10. Kumanan, C. J: Remote sensing revealed morphotectonic anomalies as a tool to neotectonicmapping- Experience south zxx India (2001).
11. Burrough, P. A.: Principles of GIS for Land Resource Assessment. Oxford Science Publication, Oxford (1990).
12. Cauvin, C., Escobar, F., Serradj, A.: Published by ISTE Ltd. And John Wiley & Sons, Inc. Great Britain & U.S.S. Vol. 1 (2010).
13. Roy, U., Das, D.: Pasture land use planning for rural and Non-arable land Management in a Chronically Drought prone area: a GIS Based approach In Resources and Development issues and concerns (Eds. Jana, N.C., Sivaramakrishnan, L.) Progressive Publishers, Kolkata, 152–159 (2013).
14. Bhatta, M., Das, D., Ghosh P.R.: The effect of ambient temperature on some biochemical profiles of Black Bengal goats (*Capra aegagrus hircus*) in two different agro-climatic zones in West Bengal, India. *IOSR J. Pharma. Biol. Sci.* 9 (4): 32–36 (2014).
15. Bhatta, M., Das, D., Ghosh, P.R.: A health GIS based approach to portray the influence of ambient temperature on goat health in two different agro-climatic zones in West Bengal, India, *Ind. J. Biol.* 2(1): 41–45 (2015).
16. Coles, E. M.: *Veterinary Clinical Pathology.* 3rd Ed., WB Saunders, Philadelphia (1980).
17. Ramaswamy, S.M., Anbazhagan, S.: Integrated terrain analysis in site selection for artificial recharge in Ayar basin, Tamil Nadu, India. Water Resource Journal ST/ESCAP/SER-C-/190. 43–48 (1996).
18. Dushaj, L. I., Salillari, V. S., Cenameri, M., Sallaku, F.: Application on GIS for land use planning: a case study in Central part of Albania, Research Journal of Agricultural Science, 41 (2) (2009).
19. Campbell, J. B.: *Introduction to Remote Sensing*, 2nd Ed. Guildford Press, New York (1996).
20. Burrough, P.A., McDonnell, R.A.: *Principles of Geographical Information Systems*, 1st ed., Oxford University Press Inc., New York, 35–57 (1998).

Wavelet-Based Image Compression Using SPIHT and Windowed Huffman Coding with Limited Distinct Symbol and It's Variant

Utpal Nandi and Jyotsna Kumar Mandal

Abstract A compression technique for image is proposed that is SPIHT algorithm based. Encoding of wavelet transformed quantized image is done using variants of Huffman coding. Comparisons are made among the binary un-coded SPIHT, SPIHT with Huffman coding, the proposed technique and its modified variant which show that the proposed techniques offer better PSNRs maintaining the same compression rates. But, the encoding time of the proposed techniques are slightly high.

Keywords Compression · Wavelet transform · Compression ratio · Encoding time · Decompression

1 Introduction

An effective lossy compression technique for image is wavelet-based image compression technique [1, 2]. The compression technique is based on multi-resolution analysis. Similar to the DCT-based JPEG [3], wavelet based technique considers an image as a collection of coefficient values where maximum values are close to zero. Thus, the image is represented with a few high coefficient values only. The technique is efficient over DCT-based JPEG since it analyzes the image without splitting into sub-images. The technique offers best compression ratio with better quality image. The technique has three main steps i.e. wavelet transform, quantization and encoding. The first and third steps are invertible. But, the second step is

U. Nandi (✉)
Department of Computer Science, Vidyasagar University,
Minapore 721102, West Bengal, India
e-mail: nandi.3utpal@gmail.com

J.K. Mandal
Department of Computer Science and Engineering, University of Kalyani,
Nadia 741235, West Bengal, India
e-mail: jkm.cse@gmail.com

© Springer Science+Business Media Singapore 2017
J.K. Mandal et al. (eds.), *Proceedings of the First International Conference on Intelligent Computing and Communication*, Advances in Intelligent Systems and Computing 458, DOI 10.1007/978-981-10-2035-3_44

435

not. Quantization reduces the precision of the floating point values of the transformed image. The wavelet transform generates huge zeros or near zero magnitudes. The encoding step is one of the important aspects of wavelet image compression. Any lossless technique can do the final encoding. The SPIHT algorithm [1] is a standard wavelet-based technique. In this paper, wavelet-based techniques using SPIHT with variants of Huffman coding are proposed. The brief discussion of SPIHT is done in Sect. 2. The variants of Huffman coding [4–6] are discussed in Sect. 3. Then, the discussions of proposed techniques are done in Sect. 4. The result for comparison between the SPIHT with Huffman and SPIHT with variants of Huffman coding are analyzed in Sect. 5. The conclusion is made in Sect. 6. The references are followed.

2 SPIHT Algorithm

Embedded Zero-tree Wavelet technique [2] is very simple and efficient. But, it is very slow. To make encoding and decoding faster and to obtain much better results, SPIHT [1] is introduced. It offers very high quality image and most widely used technique. It provides a wavelet-based technique standard. It has three main parts:

1. Partial ordering of wavelet transformed values by magnitude, with order transmission by subset partitioning that is used at the decoder end also
2. Refinement bit's ordered bit plane transmission
3. Across different scales, finding the self-similarity of wavelet transformed image.

The partial ordering is an output of comparison of wavelet transformed magnitudes to a set of thresholds $T_1, T_2, T_3 \dots T_n$ where $T_1 < T_2 < T_3 \dots < T_n$. Here, an element exceeding a specified threshold considers as significant. Otherwise, it is insignificant. It keeps LSP (List of Significant Pixels), LIP (List of Insignificant Pixels) and LIS (List of Insignificant Sets). The tree structure of the technique is depicted in Fig. 1. Total recovery of the image can be done by encoding all bits of the wavelet transform.

Fig. 1 Tree structure of SPIHT

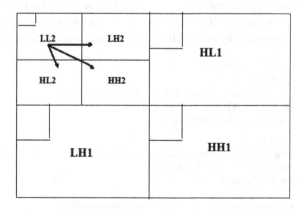

SPIHT deserves special attention because it provides high quality image and is optimized for progressive image transmission. It can be used for lossless compression also.

3 Variants of Huffman Coding

WHDS coding technique [4, 6] is a variant of Huffman coding [7] that uses a window buffer. The buffer keeps elements that are currently compressed. The total elements within the buffer are not restricted but type of elements by a specified value. The technique begins with a buffer without any elements and 0-node tree. Now each element is read one by one and encoded and inserted into the buffer. The tree is updated containing only elements of buffer. If the type of elements crosses the specified value, an element is deleted from buffer based on first in first out (FIFO) rule. The weight of the element is decremented and the tree is updated. The algorithm is given in Fig. 2. An effective variant of the WHDS technique is WHDSLRU [5, 6]. It maintains a window buffer to store recently processed symbols. Similar with WHDS technique, the restriction of window size is done by type of elements within the buffer. But, it exceeds a specified limit; least recently used symbol is removed from the window instead of oldest one. That is, the technique applied least recently uses (LRU) policy rather than FIFO. Initially, a 0-node Huffman tree is constructed and an empty window buffer is taken. A symbol from input file is read as new_smbl and coded using Huffman tree. The tree is updated by inserting new_smbl.

The new_smbl is kept in the buffer. If the type of element exceeds its specified limit, the least recently used symbol (LRU_smbl) is deleted from buffer and the tree is updated by removing the LRU_smbl. This is repeated until end of file.

Fig. 2 WHDS algorithm

Step 1: Initially window does not contain any element and Huffman tree has 0- node.

Step 2: Take elements one by one from file as new_smbl.

Step 3: Compress (or decompress) new_smbl and keep it in the buffer.

Step 4: Update the tree considering insertion of new_smbl.

Step 5: If the type of element cross the limit, delete the element (ols-smbl) based on FIFO from buffer and update the tree considering deletion of element ols-smbl.

Step 6: Repeat step 2 to 5 for all elements of file.

Step 7: Stop.

Fig. 3 SPIHT-WHDS
encoder

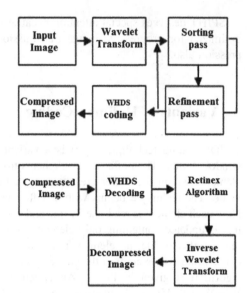

Fig. 4 SPIHT-WHDS
decoder

4 Proposed Technique

SPIHT is very fast and becomes more faster when omits entropy coding of the bit stream i.e. the slow arithmetic coding. But, if this entropy coding is omitted, the performance in term of compression rate degrades. The problem can be solved by using entropy coding techniques that are faster than arithmetic coding. The proposed technique uses WHDS as entropy coding and termed as SPIHT-WHDS. The SPIHT-WHDS encoder is shown in Fig. 3. First, the encoder obtains discrete wavelet transform of input image. The transformed image goes through the sorting pass, refine pass and produced binary un-coded data. Finally this data are further compressed by loss-less encoding technique WHDS and obtains final compressed image. The SPIHT-WHDS decoder is shown in Fig. 4. First, the compressed image is decoded by WHDS technique. Then, Retinex algorithm [8] is applied to the WHDS decoded image to improve the decompressed image quality. After that, inverse transform is applied to get final decompressed version. Then, a variant of the same is also proposed that replaces the entropy coding WHDS by WHDSLRU and termed as SPIHT-WHDSLRU since the performance of WHDSLRU is better than WHDS.

5 Results

For experimental purpose, a 200 × 300 gray scale image Lenna have been taken as shown in Fig. 5. The comparisons of encoding accuracy in terms of PSNR with increasing bit rate for Lenna image among the binary un-coded SPIHT

Fig. 5 Lenna image

Table 1 PSNR in dB

Bit rate (bpp)	SPIHT-binary un-coded	SPIHT-Huffman	SPIHT-WHDS	SPIHT-WHDSLRU
0.2	32.75	32.90	33.03	33.07
0.4	35.82	36.12	36.22	36.28
0.6	37.66	37.92	37.95	37.99
0.8	38.94	39.26	39.30	39.34
1.0	40.00	40.34	40.37	40.38

(SPIHT-binary un-coded) [1], SPIHT with Huffman coding (SPIHT-Huffman) [9], the proposed SPIHT-WHDS and its modified variant SPIHT-WHDSLRU are done as given in Table 1.

For bit rate = 0.2 bpp, the PSNR of the proposed SPIHT-WHDS is 33.03 that is better than the PSNRs 32.75, 32.90 offered respectively by existing SPIHT-Binary un-coded and SPIHT-Huffman. Again for the same bit rate, the proposed SPIHT-WHDSLRU offers more sophisticated result i.e. PSNR 33.07 than SPIHT-WHDS also. Similarly, for bit rate = 0.4, 0.6, 0.8, 1.0 also the performances offered by the proposed two are comparatively better in terms of reconstructed image quality measured by PSNR. And among these two proposed techniques, SPIHT-WHDSLRU offers better result. Then, the performance is also measured by another metric i.e. encoding time as given in Table 2. But, the encoding time of the proposed SPIHT-WHDS is 0.17 s for bit rate 0.2 bpp that is higher than existing SPIHT-Binary un-coded (0.06 s) and SPIHT-Huffman (0.12 s). Similarly for all other bit rates, the performances offered by the proposed two are slightly poor in terms of encoding time. And among these two proposed techniques, SPIHT-WHDS is faster.

Table 2 Encoding time in second

Bit rate (bpp)	SPIHT-binary un-coded	SPIHT-Huffman	SPIHT-WHDS	SPIHT-WHDSLRU
0.2	0.06	0.12	0.17	0.19
0.4	0.11	0.18	0.23	0.26
0.6	0.18	0.24	0.29	0.32
0.8	0.23	0.29	0.34	0.38
1.0	0.27	0.38	0.43	0.47

6 Conclusion

The proposed techniques try to make SPIHT faster without omitting entropy coding. Since the performance in term of compression rate degrades with omitting the entropy coding step. The techniques reduce the problem by using fast entropy coding (e.g. WHDS, WHDSLRU) of the bit stream instead of slow one (e.g. arithmetic coding). The performances offered by the proposed two SPIHT-WHDS and SPIHT-WHDSLRU are comparatively better in terms of reconstructed image quality measured by PSNR. And among these two proposed techniques; SPIHT-WHDSLRU offers better image quality. But, the proposed SPIHT-WHDS is slightly slow than existing SPIHT-Binary un-coded and SPIHT-Huffman. And among these two proposed techniques, SPIHT-WHDS is faster.

Acknowledgments The authors extend sincere thanks to the Department of Computer Science, Vidyasagar University, Paschim Medinipur and to the Department of Computer Science and Engineering, University of Kalyani Kolkata, West Bengal, India for using the facilities of infrastructure to develop the techniques.

References

1. Said, A., Pearlman, W. A.: A new, fast, and efficient image codec based on set partitioning in hierarchical trees. In: IEEE transaction on Circuits and Systems for Video Technology, Vol. 6, No. 3, pp. 243–250 (1996)
2. Shapiro, J. M.: Embedded image coding using zerotrees of wavelets coefficients. In: IEEE Trans. Signal Processing, vol. 41, pp. 3445–3462 (1993)
3. Wallace, Gregory K.,: The JPEG Still Picture Compression Standard. Communications of the ACM, Vol. 34, No. 4, pp 31–44 (1999)
4. Nandi, U., Mandal, J. K.: Windowed Huffman Coding with limited distinct symbols. In: 2nd International Conference on Computer, Communication, Control and Information Technology (C3IT-2012), vol. 4, pp. 589–594, Hooghly, India (2012)
5. Nandi, U., Mandal, J. K.: Windowed Huffman coding with limited distinct symbols by least recently used symbol removable. In: 3rd International conference on Front. of Intelligent Computing (FICTA-2014), pp. 489–494 (2014)
6. Nandi, U., Mandal, J. K.: Comparative study and analysis of Windowed Huffman Coding with limited distinct symbols and its variants. In: International Journal of Engineering Technology and Computer Research (IJETCR), Vol: 3, Issue 2, pp. 121–128 (2015)

7. Huffman, D.A.: A method for the construction of minimum-redundancy codes. In: Proceedings of the IRE, Vol. 40, No. 9, pp. 1098–1101 (1952)
8. Hao, M., Sun, X.: A modified Retinex Algorithm based on Wavelet Transformation. In: Second International Conference on Multi Media and Information Technology, pp. 306–309 (2010)
9. Mallaiah, A., Shabbir, S. K., Subhashini, T.: An Spiht Algorithm With Huffman Encoder For Image Compression And Quality Improvement Using Retinex Algorithm", International Journal Of Scientific & Technology Research, Vol. 1, Issue 5, pp. 45–49 (2012)

Extraction of Distinct Bifurcation Points from Retinal Fundus Images

Nilanjana Dutta Roy, Suchismita Goswami,
Sushmita Goswami, Sohini De and Arindam Biswas

Abstract With the immense acceptance and adoption of personal identification using biometrics, exigency of a reliable, faster and less expensive authentication process has arrived. In this novel work, we have attempted to exhibit some crucial features of human retina which are sufficient to build a secured biometric template in considerable amount of time, ignoring the other hazardous factors which may lead to even authentication failure. Experimental results prove that this proposed work would be able to identify the distinct bifurcation points in retinal fundus image, avoiding the crossovers consciously as they are not anatomically stable and could change their locations as an effect of some diseases. This work will reduce the complexity of any authentication algorithm drastically by concentrating only on existing bifurcations, ignoring tedious calculations on every junction points.

Keywords Personal identification · Biometrics · Distinct bifurcation · Crossover points

N.D. Roy (✉) · Suchismita Goswami · Sushmita Goswami · Sohini De
Department of Computer Science and Engineering,
Institute of Engineering and Management, Kolkata, India
e-mail: nilanjanaduttaroy@gmail.com

Suchismita Goswami
e-mail: suchismita.g24@gmail.com

Sushmita Goswami
e-mail: sushmita.g24@gmail.com

Sohini De
e-mail: sohini.de93@gmail.com

Arindam Biswas
Department of Information Technology, Indian Institute of Engineering
Science and Technology, Shibpur, Howrah, India
e-mail: barindam@gmail.com

© Springer Science+Business Media Singapore 2017　　　　　443
J.K. Mandal et al. (eds.), *Proceedings of the First International Conference
on Intelligent Computing and Communication*, Advances in Intelligent Systems
and Computing 458, DOI 10.1007/978-981-10-2035-3_45

1 Introduction

Retinal vasculature is an important feature for unique identification of an individual. The main characteristic features of the retinal vasculature are the bifurcation and crossover points. Normally, the blood vessels in human eye, which supplies blood and help in vision, have these two categories of the junction points on it. Literature show us that majority of the authors have successfully used these junction points including crossovers also as their unique feature for biometric authentication. Theoretically, it is feasible to identify every junction point in the fundus image which includes even the crossovers. Even though, the retinal structure is considered as a unique feature of an individual, yet changes may be observed in the locations of crossover points during ocular diseases and systemic cardiovascular disease [1]. It may vary according to the position of the high resolution fundus camera if the images are captured from a different angle, their location may change which further affects in the process of preparing templates. Blockage of vessel at crossover points, also known as arterial narrowing or nicking, is a symptom of hypertension. Moreover, arteriolar narrowing and abnormalities at the crossover points are caused by hypertension. Hence, conclusion can be drawn that crossover points are unstable when exposed to pathological conditions. The region where branching occurs from an existing vessel is known as a bifurcation point. Besides, bifurcation points have a greater distinctive power than the normal vessel segments [2, 6, 7]. Due to the instability of crossover points, the bifurcation points are often considered as the principal landmark in point matching image registration and authentication. The angle of bifurcation points during formation of branches are an important criteria for image registration. Hence, identification of bifurcation points is necessary for retinal image registration and successful biometric authentication.

2 Proposed Method

After a meticulous literature survey, we conclude with the fact that the motive of any personal identification process is to provide with extreme security by pointing out some of the crucial features in human retina, mostly the junction points which include both the bifurcations and crossovers for preparing biometric templates. Keeping the same motivation in mind, we have also taken this initiative to point out only bifurcation points as a relevant feature in human retina which would be competent enough to prepare the desired template accurately in spending minimum time and expenses.

The proposed method starts with the thinning process of any segmented fundus image taken from DRIVE [3] database. Initially, all the branch points(both crossover and bifurcation points) present on the image have been pointed out (refer Fig. 2c).

Then the normal crossover points are found from the image (refer Fig. 2d). Next, the algorithm proceeds by grouping all the closely connected bifurcation points and their corresponding angle calculation. In each point with acute angle and for any closely connected bifurcation points, difference between the angles of them are calculated. If the difference is very less (for example less than 10), we are marking the midpoint

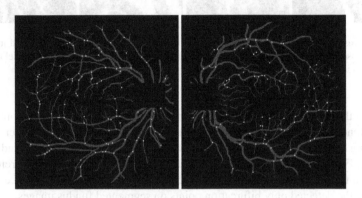

Fig. 1 Showing the detected bifurcation points on segmented fundus images

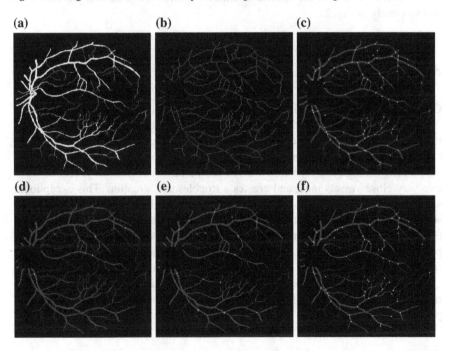

Fig. 2 Showing the different stages of the proposed algorithm **a** segmented image from DRIVE database **b** thinned image **c** all the junction points including both bifurcation and crossovers are found **d** normal crossover points are found **e** all the crossover points including normal crossovers and crossovers with arteriovenous nicking **f** distinct bifurcation points are found

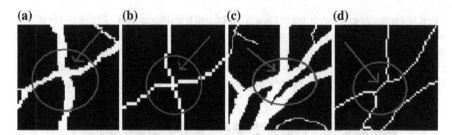

Fig. 3 Crossover point with arteriovenous nicking **a** normal crossovers and its thinned image **b** single crossover point appeared as two closed bifurcation points due to arteriovenous nicking and its thinned image

between these two closely connected bifurcation points as the crossover point with arteriovenous nicking (Fig. 4b, c). After tracing all the pairs, all the crossover points including normal crossover and crossover with arteriovenous nicking are found (refer Fig. 2e). Finally, all the crossover points from detected branch points are removed to obtain the distinct bifurcation points (refer Fig. 2f). So the result is successfully showing the detected only bifurcation points on segmented fundus images.

Please refer Fig. 1 for the desired result. The different stages of our algorithm are shown clearly in Fig. 2.

3 Algorithm

Crossover points are basically of two categories, one is the normal one and another is a false situation which is often detected as bifurcation point due to arteriovenous nicking occurs at the crossover points. Please refer the Fig. 3. A $3 * 3$ window is basically used here to scan image and to find out total number of neighbourhood white pixels surrounding each branch point. The term arteriovenous is the junction where two very close vessels meet and creates a troublesome situation. This exceptional situation often arises in the skeleton image and always there is a common probable scenario to identify this junction as two very close bifurcation points together. However, this process may lead to a false detection of real scenario as two bifurcation points also as shown the Fig. 4b, c where the two bifurcation points are really close in nature. With a hope to get a better solution here and to differentiate between true crossovers points from spurious results, another method is applied to detect geometric features of crossovers along with notion of two closely connected bifurcation points.

Algorithm 1 Distinct Bifurcation point

Ensure:

Consider segmented images from DRIVE database as its input

points ← detected branch points

bifurcation_points ← detected bifurcation points

diff ← difference between the angles at two close bifurcation points

Perform skeletonization on the segmented image

for each row in the image **do**

find all the junction or branch points

end for

for each detected branch points **do**

count the no of neighboring pixels including the chosen one by considering a small rectangular window surrounding it (The window is basically scanned to find out the total number of neighborhood white pixels surrounding the branch point)

if *count* == 5 **then**

mark the branch point as normal crossover and remove them from detected branch points

end if

end for

find the closely connected bifurcation points from the remaining detected branch points and group them individually

for each detected closely connected bifurcation pair **do**

find the angle at two close bifurcation points and mark as 1 and 2 (The angles are formed by any two vessels at the crossing are $a1$ and $a2$ shown in the following Fig. 4d)

if $\alpha1 <= 90$ and $\alpha2 <= 90$ **then**

$diff = abs(\alpha1 - \alpha2)$

else if *diff < threshold (approximately* 10) **then**

mark the midpoint of the two closely connected bifurcation points generated due to arteriovenous nicking as a single crossover that is represented by two closely connected bifurcation points in the thinned image, refer Fig. 4d

end if

end for

Then the result sets are found after the crossover points has been detected

(In the evaluation process, all crossovers within the closed region of optic disc, are excluded from evaluation. In addition, the very small tiny vessels may cause some errors while calculating the no of crossover points from a fundus image which are negligible)

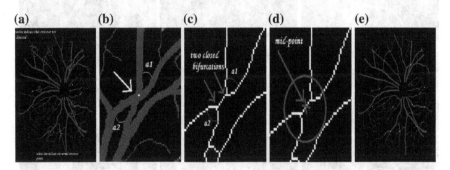

Fig. 4 Showing the different detected points on input images **a** Detected normal crossover points **b** Showing arteriovenous nicking **c** Showing arteriovenous nicking **d** Mid point of two closely connected bifurcation points detected as a single crossover point **e** All the crossover points detected through the above steps

4 Experimental Result

Due to inadequacy of the real data sets, all the experiments are essentially done on the segmented fundus images of a standard database called DRIVE [3]. The utmost outcome of this experiment is the detected prominent bifurcation points only which are very clearly and distinctly shown on the results, leaving a small portion of the Optic Disc region. On the contrary, the crossover points are also shown on the same images as a different view in order to substantiate the claim. Table 1 shows glimpses of results on a few images from DRIVE database and Table 3 claims the overall performance on DRIVE [3]. The performance evaluation is done based on number of bifurcation points which have been identified by the proposed method as true positives and the bifurcation points that are missed out and the number of false detection(spurious bifurcation points). The following formula helps to compute the performance of the proposed method (Fig. 5).

Table 1 Performance Evaluation on few images of DRIVE

	Image 12	Image 26	Image 19	Image 3
Actual bifurcations present	73	64	42	84
Detected bifurcation points	75	63	42	83
Correctly detected points	73	63	41	83
Bifurcation points not detected	0	1	1	1
Incorrect	2	0	0	0
Sensitivity	1	1	0.976	0.988
Specificity	0	0	0	0
Accuracy (%)	97.33	98.41	97.6	98.8

(a)   **(b)**   **(c)**

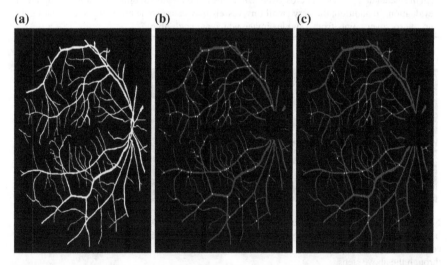

Fig. 5 Showing the different detected points on input images **a** segmented images from DRIVE database **b** detected distinct bifurcation points **c** reverse image showing distinct crossover points

Table 2 Formula to evaluate performance

Experimental result	Present	Absent
Positive	True positive (TP)	False positive (FP)
Negative	True negative (TN)	False negative (FN)

Table 3 Overall result on DRIVE database

Database	Sensitivity	Specificity	ROC	Accuracy (%)
DRIVE	0.98	0	0.9882	98.8

$$Performance = (\frac{(truly\ detected\ points - (spurious\ points + missed\ out\ points))}{ground\ truth\ points}) * 100.$$

5 Performance Evaluation

If the outcome from a prediction is P and the actual value is also P, then it is called a True Positive (TP); however if the actual value is N then it is said to be a False Positive (FP). Conversely, a True Negative (TN) has occurred when both the prediction outcome and the actual value are N, and False Negative (FN) is when the prediction outcome is N while the actual value is P (Tables 2 and 3).

$$Sensitivity = \frac{TP}{(TP + FN)} \tag{1}$$

$$Specificity = \frac{TN}{(TN + FN)} \tag{2}$$

$$Accuracy = \frac{(TP + TN)}{(P + N)} \tag{3}$$

6 Conclusion

Detection of only bifurcation points in a fundus image is a challenging task in true sense. The purpose of the work aims to avoid numerous unnecessary hazards of appraising crossover points as one of the crucial feature for authentication. So to focus on only bifurcation point which is anatomically stable and never changes [4, 5], is an important issue. We have tried to accomplish this work successfully and acquired a satisfactory result. But, due to inadequacy of real data, we have done the experiments in an available database named DRIVE [3].

References

1. U. T. V. Nguyen and A. Bhuiyan and L. A. F. Park and R. Kawasaki and T. Y. Wong and K. Ramamohanarao : Automatic Detection of Retinal Vascular Landmark Features for Colour Fundus Image Matching and Patient Longitudinal Study. In: 20th IEEE International Conference of Image Processing (ICIP), pp. 616–620, 2013.
2. A. Perez-Rovira, R. Cabido, E. Trucco, S.J. McKenna: Robust Efficient Registration via Bifurcations and Elongated Elements Applied to Retinal Fluorescein Angiogram Sequences. In: Medical Imaging, IEEE Transactions, Volume: 31, Issue: 1
3. The DRIVE database, Image sciences institute, university medical center utrecht, The Netherlands. http://www.isi.uu.nl/Research/Databases/DRIVE/, last accessed on 7th July, 2007
4. J. Hjek and M. Drahansk and R. Drozd : Extraction of Retina Features Based on Position of the Blood Vessel Bifurcation. In: Journal of Medical Research and Development (JMRD), pp. 55–58, volume 2, 2013
5. Wikipedia, http://www.en.wikipedia.org
6. Richard S, Snell, Michael A. Lemp, Clinical Anatomy of Eye, Wiley Publication, Print ISBN: 9780632043446, Online ISBN: 9781118690987, doi:10.1002/9781118690987
7. A. Edoardo and P. Roberto and G. Orazio and S. Francesco: Extraction of the Retinal Blood Vessels and Detection of the Bifurcation Points. In: International Journal of Computer Applications, pp. 29–34, volume 77, 2013

Automatic Measurement and Analysis of Vessel Width in Retinal Fundus Image

Suchismita Goswami, Sushmita Goswami and Sohini De

Abstract The unexpected changes in the width of retinal blood vessels can be recognized as one of the significant characters during the proliferation of several diseases. Clinical studies verify that width of retinal blood vessels may undergo changes during diabetes, hypertension, Retinopathy of prematurity and Proliferative Diabetic Retinopathy. Optical vessel swelling or Papilledema and optical neuritis are prominent ocular diseases that contribute to changes in the width. In this approach, width of retinal blood vessels has been determined to benefit the identification of such diseases. Every retinal fundus images exhibits several unique bifurcation points. The bifurcation points ensure stability and are not subject to any change even when prone to diseases. Identifying these points, our algorithm attempts to find the ratio of the width of a parent vessel with its corresponding child vessels at the bifurcation points. Experimental results prove that this ratio lies in the range of 1.1–1.8. Abrupt changes in the ratio ensures the probability of existence of ocular diseases.

Keywords Width · Optical vessel swelling · Parent vessel · Child vessel

1 Introduction

Human eye exhibits two categories of blood vessel, namely artery and vein. The oxygenated blood is carried by the arteries while the veins carry the deoxygenated blood. Despite their functional differences, structurally they both display several common features. They divide to form child nerves from a parent nerve and these points of division are termed as bifurcation points. Another structural feature is the crossover

Suchismita Goswami (✉) · Sushmita Goswami · Sohini De
Department of Computer Science and Engineering,
Institute of Engineering and Management, Kolkata, India
e-mail: suchismita.g24@gmail.com

Sushmita Goswami
e-mail: sushmita.g24@gmail.com

Sohini De
e-mail: sohinide93@gmail.com

© Springer Science+Business Media Singapore 2017   451
J.K. Mandal et al. (eds.), *Proceedings of the First International Conference on Intelligent Computing and Communication*, Advances in Intelligent Systems and Computing 458, DOI 10.1007/978-981-10-2035-3_46

point where one vessel crosses the other. However, literature review indicates that the crossover points are not anatomically stable and may deviate during ocular diseases and systemic cardiovascular diseases. Due to diseases like hypertension, diabetes and ocular surgery the crossover points may change their positions. Moreover, a pathological condition named arterial narrowing or nicking exists which indicates blockage of vessel at crossover points. It is caused by hypertension. Hence, our approach considers only the bifurcation points because they ensure stability. After identification of the bifurcation point, we attempt to measure the width of the parent nerve as well as the child nerves at these points. Theoretically it is not feasible to measure the width of every parent nerve. Moreover, as nerves move away from the optic disk their width tend to decrease. Hence, the nerves near the optic disk exhibit a larger width than the ones which are further away. As a result, manual attempt to measure the width of nerves can be a tedious task. Moreover, this may be utilized in diagnosing one of the severe retinal diseases, called Proliferative Diabetic Retinopathy. At the advanced stage of Proliferative Diabetic Retinopathy, lack of sufficient circulation of blood causes lack of oxygen in the retina. Growth of new fragile blood vessels are predominant in the retina as well as the vitreous, the transparent gel filling the interior of the eye. These abnormal vessels are prone to rupture and bleeding which may lead to hemorrhages in the retina. Further damage occurs when scar tissues develop in the retina. Clouding of the vision may also occur due to the fluid leak into the macula causing swelling of the macula, termed as macula edema. As the disease progresses, rise in the eye pressure is noted causing severe threat to vision. During Proliferative Diabetic Retinopathy the blood vessels become frail and the determination of width of the nerves will help us to identify any abrupt changes. This work will count and keep track of the number of normal vessels around the optic disc upto a certain distance. In this way, we may define a general statistics of the number count of blood vessels in a healthy human eye. It would be easier for any automated system to go in favor of PDR (Proliferative Diabetic Retinopathy) if any deformities found in the number and width around the optic disc abruptly.

2 Proposed Method

The algorithm starts with the segmented images from DRIVE [1] database as its input.

The segmented image is being thinned by using a standard morphological thinning operation of Matlab. Bwmorph function with the thinning operation works as follows: thinning is a morphological operation that is used to remove selected foreground pixels from binary images. It shrinks every object by reducing all lines to single pixel thickness without changing its structure and provides the binary image as output. Then all the branch points of the thinned image are calculated using another morphological function bwmorph along with the operation branchpoints [2]. Bwmorph function with the branchpoint operation works as follows:

$$M = \begin{bmatrix} 0 & 0 & 1 & 0 & 0 \\ 0 & 0 & 1 & 0 & 0 \\ 1 & 1 & 1 & 1 & 1 \\ 0 & 0 & 1 & 0 & 0 \\ 0 & 0 & 1 & 0 & 0 \end{bmatrix}$$

becomes

$$M = \begin{bmatrix} 0 & 0 & 0 & 0 & 0 \\ 0 & 0 & 0 & 0 & 0 \\ 0 & 0 & 1 & 0 & 0 \\ 0 & 0 & 0 & 0 & 0 \\ 0 & 0 & 0 & 0 & 0 \end{bmatrix}$$

The term branch points includes both bifurcation and crossover points.

As the crossover points are not anatomically stable and could change their location as an effect of some diseases, all the crossover points are removed from the previously detected branch points and the bifurcation points are successfully identified [3].

Then in order to find out width of the parent and child vessel near each bifurcation point, around every bifurcation point in the thinned image, a $3 * 3$ window is taken and whenever a white pixel is found, that pixel is being considered as the current pixel to find the next neighbourhood pixel. A $3 * 3$ window is taken because it is the optimum size to measure the connectivity around any branchpoint. A smaller window will fail to accommodate both parents and children, while a larger window will increase the search time. At the same time, the previously traversed pixel in the thinned image is removed to indicate that this pixel has been traversed. This process is repeated until a predefined no of pixels i.e., threshold value from the bifurcation point have been traversed for all the three branches. Result will give distinct point on the parent nerve and both the child nerves generated from the bifurcation point.

Anatomically, there are two possible scenarios of the vessels exist which are emanating out of the Optic Disc, i.e., either they tend to move along the horizontal direction, or they are prone to the vertical direction. So, to measure the width of any vessel at any point, careful steps has been taken about the directions and this could be done unerringly by comparing the number of white pixels along the vertical direction with the number of white pixel in the horizontal one. Then the minimum count of these two will be considered as the width of the present vessels at that point.

Start tracing the vessel horizontally by counting the number of white pixel until a transition takes place from white to black pixel. The same process is repeated for both the left and right directions from that point in the original segmented image, see Fig. 1a.

Then, vertical tracing is done again by counting number of white pixels until a transition takes place from white to black pixel. The process continues for both the upward and downward directions from that point in the original segmented image. Please refer Fig. 1b.

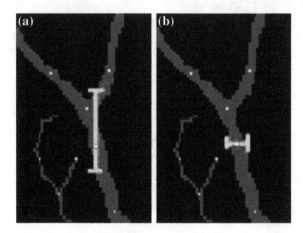

Fig. 1 Comparing this two count and as count along vertical direction is greater than the count along horizontal direction hence we can predict that tendency of the nerve is along the vertical direction. so the width of the nerve at that point is equal to the count along the horizontal direction **a** Counting the no of pixel while moving along the vertical direction i.e. upward and downward direction from the point **b** Counting the no of pixel While moving along the horizontal direction i.e. along *left* and *right* from the point

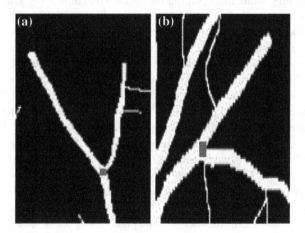

Fig. 2 The tendency of a vessel is to move along **a** vertical direction **b** horizontal direction

And then, the minimum count between horizontal and vertical directions is being selected as the width of the specified vessel (Fig. 2).

Finally, at each bifurcation point ratio of the width of the parent nerve to each child vessels are calculated (Fig. 3).

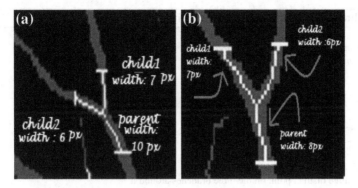

Fig. 3 Showing width calculated around two bifurcation points

2.1 Algorithm

3 Experimental Result

Due to inadequacy of the real data sets, all the experiments are essentially done on the segmented fundus images of a standard database called DRIVE [1]. The utmost outcome of this experiment is to measure the vessel width and to show a ration between parent and child vessels (Table 3). Tables 1 and 2 show the width measurement results from both DRIVE [1] and VARIA [4] database (Table 4).

Table 1 The width of individual major vessels around the optic disc from DRIVE database

Vessels	Image1	Image2	Image3	Image5	Image6
1	3	2	5	7	2
2	4	5	7	3	4
3	7	11	5	10	3
4	9	10	4	16	8
5	12	6	3	5	16
6	6	6	11	4	3
7	2	6	9	3	1
8	2	5	2	3	11
9	3	8	22	7	10
10	4	6	2	8	6
11	10	5	2	9	6
12	3	5	3	6	2

Algorithm 1 Width calculation for individual vessels

Ensure:
   $n_v \leftarrow$ *number of pixels in vertical direction*
   $n_h \leftarrow$ *number of pixels in horizontal direction*
   $n_v \leftarrow 0$
   $n_h \leftarrow 0$
   width $\leftarrow$ *total width of vessel*
   width $\leftarrow 0$
   parent $\leftarrow$ *width of the parent vessel*
   *child*1 $\leftarrow$ *width of one child vessel*
   *child*2 $\leftarrow$ *width of another child vessel*
   for each bifurcation point **do**
      mark three distinct point on the three branches around it
      for each of these points on the respective three branches **do**
         repeat
            move along the upward vertical direction
            $n_v \leftarrow n_v + 1$
            and move along the downward vertical direction also
            $n_v \leftarrow n_v + 1$
         until white to black transitions occur
         repeat
            move along the left horizontal direction
            $n_h \leftarrow n_h + 1$
            and move along the right horizontal direction
            $n_h \leftarrow n_h + 1$
         until white to black transitions occur
         if $n_v > n_h$ **then**
            width $: \leftarrow n_h$
         else
            width $: \leftarrow n_v$
         end if
         $n_h \leftarrow n_h + 1$
         $n_h \leftarrow n_h + 1$
      end for
      *calculate ratio*1 $: \leftarrow parent/child1$
      *calculate ratio*2 $: \leftarrow parent/child2$
   end for

The experimental results show around 97 % of the performance while performance is measured based on the calculation

$$Performance = (\frac{(truly\ detected\ points - (spurious\ points + missed\ out\ points))}{ground\ truth\ points}) * 100$$

Table 2 The width of individual major vessels around the optic disc from VARIA database

Vessels	Image1	Image3	Image6	Image9	Image14
1	8	3	4	7	5
2	11	5	10	11	15
3	12	2	6	5	7
4	8	5	3	12	5
5	14	2	9	13	13
6	2	12	8	9	9
7	1	4	6	12	7
8	1	3	12	15	3

Table 3 The Parent-child ratio at each bifurcation point from DRIVE database

Bif. points	Parent	Child1	Child2	Parent:child1	Parent:child2
350, 204	5	3	4	1.666667	1.25
350, 239	4	3	4	1.333333	1
352, 198	4	3	3	1.333333	1.333333
352, 210	4	3	4	1.333333	1
353, 137	5	3	3	1.666667	1.666667
354, 273	7	4	6	1.75	1.166667
355, 137	5	3	5	1.666667	1
355, 201	4	3	4	1.333333	1
392, 234	4	3	3	1.333333	1.333333

Table 4 A comparative study

Papers	Features	Time (in seconds)
Paper1 [5]	Laser techniques	10 approximately
Paper2 [6]	Utilizing electric field theory and graphical approach	9 approximately
Proposed	aforementioned algorithm	7 approximately

4 Complexity Analysis

Identification of Bifurcation points require linear amount of time for execution i.e., $O(n)$ where n denotes number of branch points remaining after removing all the crossover points. While calculating the width of the parent nerve and the child vessels, for each three distinct points in the above mentioned algorithm, we are moving in left and right direction from that point horizontally until a white to black transition occurs and similarly in vertical directions also. Both the cases, it consumes linear amount of time, i.e., the above algorithm results into a time complexity of

$O(n)$. Apart from this there would be obvious need of $O(n^2)$ amount of time while scanning the entire image for processing. And as there is no additional space or storage are required, we conclude that the space complexity is constant.

5 Conclusion

With the objective of Diagnosis of Proliferative Diabetic Retinopathy in its early stage by separating the newly generated fragile tiny vessels with the original ones and detection of any abnormality in the fundus image in mind, we have executed the proposed algorithm on 32 bit Matlab R2013a version on 32 bit Windows 7 running on Intel Core2 dual processor with 2 GB RAM. Experimental results show the width of individual major vessels around the Optic Disc and throughout the whole image. Additionally, from the experimental result we have found that the ratio of each parent to child vessels at every bifurcation point is within the range of 1.1–1.8. Also, for the images which are severely affected by the final stage of PDR, this method would be able to diagnose them effortlessly by keeping track width of the vessels around the optic disc. We hope this work would be worthwhile for the ophthalmologists also in analyzing the retinal images to diagnose PDR even at the initial stage.

References

1. The DRIVE database, Image sciences institute, university medical center utrecht, The Netherlands. http://www.isi.uu.nl/Research/Databases/DRIVE/, last accessed on 7th July, 2007
2. MatLab R2013a, http://in.mathworks.com/support/sysreq/sv-r2013a/
3. S. Saha, N. Dutta Roy, Automatic Detection of bifurcation points in retinal fundus images, International Journal of Latest Research in Science and Technology ISSN (Online): 2278–5299 Volume 2, Issue 2: Page No. 105–108, March–April (2013) http://www.mnkjournals.com/ijlrst.htm
4. VARIA database
5. Enrico Pellegrini, Gavin Robertson, Emanuele Trucco, Tom J. MacGillivray, Carmen Lupascu, Jano van Hemert, Michelle C. Williams, David E. Newby, Edwin JR van Beek, and Graeme Houston, "Blood vessel segmentation and width estimation in ultra-wide field scanning laser ophthalmoscopy", Biomedical Optic Express under Optical Society of America, Vol V, issue 12, 2014
6. Xiayu Xu, Joseph M. Reinhardt, Qiao Hu,Benjamin Bakall, Paul S. Tlucek, GeirBertelsen, "Retinal Vessel Width Measurement at Branchings Using an Improved Electric Field Theory-Based Graph Approach", November 2012 in Medical Imaging, SPIE, Volume 8314

Recognition of Handwritten *Indic* Script Numerals Using Mojette Transform

Pawan Kumar Singh, Supratim Das, Ram Sarkar and Mita Nasipuri

Abstract Handwritten Digit Recognition (HDR) has become one of the challenging areas of research in the field of document image processing during the last few decades. It has wide variety of applications including reading the amounts in cheque, mail sorting, reading aid for the blind and so on. In this paper, an attempt is made to recognize handwritten digits written in four different scripts *namely, Bangla, Devanagari, Arabic* and *Telugu* using Mojette transform. The Principal Component Analysis (PCA) is then applied for dimensionality reduction of the feature vector and also shortening the training time. Finally, a 48-element feature vector is tested on *CMATERdb3* handwritten digit databases using multiple classifiers and an average overall accuracy of 98.17 % is achieved using Multi Layer Perceptron (MLP) classifier.

Keywords Handwritten digit recognition · *Indic* scripts · Mojette transform · Principal component analysis · Multiple classifiers

1 Introduction

Handwritten digit recognition is the technique of identifying handwritten digits 0–9 (for a particular script) by the machine without human interaction [1]. The recognition of handwritten numerals is a difficult task due to the different handwriting

P.K. Singh (✉) · Supratim Das · Ram Sarkar · Mita Nasipuri
Department of Computer Science and Engineering, Jadavpur University,
Kolkata 700032, India
e-mail: pawansingh.ju@gmail.com

Supratim Das
e-mail: supratimdas21@gmail.com

Ram Sarkar
e-mail: raamsarkar@gmail.com

Mita Nasipuri
e-mail: mitanasipuri@gmail.com

© Springer Science+Business Media Singapore 2017   459
J.K. Mandal et al. (eds.), *Proceedings of the First International Conference
on Intelligent Computing and Communication*, Advances in Intelligent Systems
and Computing 458, DOI 10.1007/978-981-10-2035-3_47

qualities and styles those are subject to *inter-writer* and *intra-writer* variations. Due to this, building a generic recognizer that is capable of recognizing handwritten digits by infinitely large writers is almost impossible. Many recognition systems for various applications are proposed till date where higher recognition accuracy is always desired and achieved. Typically, the recognition systems are adapted to specific applications to achieve better performance. Digit recognition has applications in various fields, ranging from recognizing amounts from cheques to recognizing postal codes on postal envelopes. However, the extraction of the most informative features with high discriminatory power to improve the classification accuracy and reduction of its computation time still remain one of the most thrust areas.

From the literature review, it is observed that most of the good works [2–8] are done for digit recognition on *Roman* script, whereas relatively few works [9–12] have been reported for the digit recognition written in *Indic* scripts. This review also revealed that the methods described in the literature suffer from larger computational time mainly due to complexity in feature extraction and presence of various pre-processing stages, *viz*, size normalization, noise removal, skew correction, skeletonization, etc. In addition, the said recognition systems fail to meet the desired accuracy when exposed to different scripts. In this paper, we have investigated the effectiveness of feature extraction approach based on Mojette transform to capture discriminative features of handwritten digits written in four different scripts *namely, Bangla, Devanagari, Arabic* and *Telugu*.

2 Present Work

In the present work, a two-stage approach is proposed for estimating the feature vector towards the recognition of handwritten digits. In the first stage, a set of 190 features is extracted after the application of Mojette transformation on each handwritten numerals. In the second stage, PCA is applied for reduction of the dimensionality of feature vector. Finally, a 48-element feature vector is chosen for the classification of handwritten numerals. In the subsequent subsections, detail of work is described:

2.1 Mojette Transform

Mojette transformation [13] transforms an original two-dimensional image into a set of discrete one-dimensional projections. To our knowledge, the Mojette transform has not been applied before for recognizing the digit either in printed or handwritten form.

Mojette transform can be considered as a more general and strictly mathematically defined transformation for feature extraction from images which is equivalent to the so called *projection histograms features*. This transform, defined as a discrete version of Radon transform, can be expressed as a set of vectors whose each element (called a bin) is calculated as the sum of the value of pixels centered on a projection line defined by m and θ_i. Here θ_i is the projection angle and is defined as:

$$\theta_i = \tan^{-1}(q_i/p_i) \tag{1}$$

where, m is an integer that determines which of the various projection lines at the angle θ_i is being represented by the element. For each value of m and θ_i, the corresponding element of the Mojette transform vector [14] is given by

$$Proj_{p_i,q_i}(m) = \sum_{x=-\infty}^{\infty} \sum_{y=-\infty}^{\infty} f(x,y)\delta(m+q_iy-p_ix) \tag{2}$$

where, (x, y) defines the position of the pixel, $f(x, y)$ represents pixel value of the image, p_i and q_i define the projection angles, m determines the particular line of projection at the angle θ_i, and δ is the Kronecker delta function defined as below:

$$\delta(w) = 1 \; if \; w = 0$$
$$= 0 \; otherwise \tag{3}$$

For example, if $p_i = 1$, $q_i = 0$ and $m = 1$, the projection angle is $0°$ and the transform bins are the sums of pixel values in every row of the image; if $p_i = -1$, $q_i = 1$ and $m = 0$, the projection angle is $135°$ and the corresponding bins of the transform vector are the sums of pixel values on every diagonal of the image. Figure 1 illustrates the concept of the Mojette transform applied to images of 3×3 and 4×4 pixels respectively. The Mojette transform can reduce very significantly the size of the characteristic vector needed for adequate image representation, and thus the computational load for recognition. As the image size increases, the transform vector increases only linearly, while the number of pixels increases as the square of image dimension. By using the Katz formula, the number of bins B_i for the projection (p_i, q_i) for a digital image of size N × M is given as:

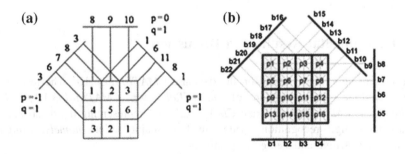

Fig. 1 Illustration of Mojette Transform for an image of size **a** 3×3 pixels and **b** 4×4 pixels

$$B_i = (M-1)|p_i| + (N-1)|q_i| + 1 \tag{4}$$

For the present work, each of the digit images are transformed at different angles of $\theta_i = 0°, 45°, 90°$ and $135°$. Given the size of the digit image as 32×32 pixels, the number of transform bins for each of the above mentioned projection angles are calculated as 32, 63, 32 and 63 respectively. Hence, a set of 190 features is extracted after the application of Mojette transformation which is much lesser than total number of image pixels.

2.2 Principal Component Analysis (PCA)

In the second stage, PCA is applied for feature dimensionality reduction. PCA [15] is a statistical method of converting a set of observations of possibly correlated variables by means of an orthogonal transformation into a set of values consisting of linearly uncorrelated variables, known as *principal components*. The number of *principal components* should be less than or equal to the number of original variables. The algorithm for PCA can be described as follows:

Step 1:	Compute the d-dimensional feature mean vector considering the whole feature dimension
Step 2:	Calculate the scatter matrix (or covariance matrix) considering the entire feature dimension
Step 3:	Compute the eigen vectors $(e_1, e_2, \ldots, e_d)$ along with their corresponding eigen values $(\lambda_1, \lambda_2, \ldots, \lambda_d)$
Step 4:	Sort the eigen vectors by decreasing eigen values and choose k eigen vectors with the largest eigen values to form a $d \times k$ dimensional matrix
Step 5:	The $d \times k$ eigen vector matrix is then used to transform the features onto the new feature subspace of dimension $k \times 1$

For our experiment, 190 features extracted using Mojette transform is transformed onto a 48-element feature vector which is chosen as the *principal component* for the current handwritten digit recognition purpose.

3 Experimental Results and Discussion

The present approach is tested on the database *CMATERdb3* [10]. The testing is presently done on four versions of *CMATERdb3*, viz., *CMATERdb3.1.1*, *CMA-TERdb3.2.1*, *CMATERdb3.3.1* and *CMATERdb3.4.1* representing the databases of handwritten digit recognition written in four major scripts *namely*, *Bangla*, *Devanagari*, *Arabic* and *Telugu* respectively.

Each digit images are first preprocessed using basic operations of skew corrections, and morphological filtering [16]. These images are then binarized using an adaptive global threshold value which is computed as the average of maximum and minimum intensities in that image. Gaussian filter [16] is applied to remove noisy pixels from the binarized digit images. The edges of the binarized digit images are then smoothen with the help of a well-known Canny Edge Detection algorithm [16]. Finally, each digit image is normalized to 32 × 32 pixels. A sample of handwritten digit images written in four scripts used for the current work is shown in Fig. 2. The said databases are *freely* available in our *CMATER* website (www. cmaterju.org) and at http://code.google.com/p/cmaterdb.

For *Devanagari*, *Arabic* and *Telugu* scripts, the datasets consist of 3000 digit samples for each script. The experimentations on these datasets are done by considering 2000 samples as training set and the remaining samples as test set where equal number of digits from each class is taken. For handwritten *Bangla* digits, a dataset of 4000 training samples are considered by choosing 400 samples for each of 10 digit-classes and the rest 2000 samples are selected for the testing purpose. The proposed approach is then applied on the preprocessed digit images of each script separately and evaluated using seven well-known classifiers *namely*, Naïve Bayes, Bayes Net, MLP, Support Vector Machine (SVM), Random Forest, Bagging and MultiClass Classifier. The graphical comparison of the identification accuracies of the said classifiers on the four databases is shown with the help of a bar chart in Fig. 3.

It is evident from Fig. 3 that MLP classifier achieves the highest recognition accuracies of 98.27, 97.5, 98.8 and 98.1 % on *CMATERdb*3.1.1, *CMATERdb*3.2.1, *CMATERdb*3.3.1 and *CMATERdb*3.4.1 respectively. The present work also calculates detailed error investigation (illustrating the average values for all the numerals of a particular script of MLP classifier) with respect to different well-known parameters *namely*, Kappa statistics, mean absolute error (MAE), root

Fig. 2 Sample handwritten images of numerals taken from *CMATERdb3* database written in: **a** *Bangla*, **b** *Devanagari*, **c** *Arabic*, and **d** *Telugu* scripts

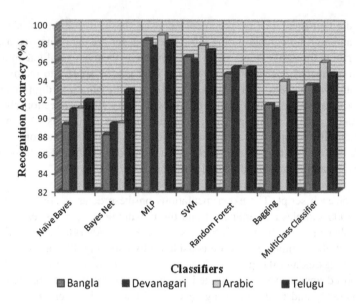

Fig. 3 Graph showing the recognition accuracies of the digit recognition technique using seven well-known classifiers on *Bangla*, *Devanagari*, *Arabic* and *Telugu* databases respectively

mean square error (RMSE), True Positive rate (TPR), False Positive rate (FPR), precision, recall, F-measure, and Area Under ROC (AUC) on all the four databases. Table 1 provides a statistical performance analysis of MLP classifier for numerals written in each of the aforementioned scripts.

Although the recognition accuracies achieved by the MLP classifier is quite convincing, but still the present technique suffers from some misclassifications. After observing the misclassified digits, it can be believed that the majority of misclassifications are seen in digits that are ambiguous either by noise, distortion, segmentation problem or peculiar writing style. Figure 4 shows some samples of misclassified digit images.

Table 1 Statistical performance measures of MLP classifier along with their respective means (styled in bold) achieved by the present technique for four handwritten script numerals

Scripts	Kappa statistics	MAE	RMSE	TPR	FP R	Precision	Recall	F-measure	AUC
Bangla	0.9807	0.0046	0.0559	0.983	0.002	0.983	0.983	0.983	0.996
Devanagari	0.9722	0.0064	0.0631	0.975	0.003	0.975	0.975	0.975	0.992
Arabic	0.9867	0.0039	0.0458	0.988	0.001	0.988	0.988	0.988	0.997
Telugu	0.9789	0.1601	0.2719	0.981	0.002	0.981	0.981	0.981	0.997
Mean	**0.9796**	**0.0437**	**0.1092**	**0.9817**	**0.002**	**0.9817**	**0.9817**	**0.9817**	**0.9955**

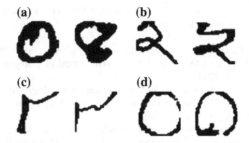

Fig. 4 Sample images of misclassified digits written in: **a** *Bangla* (the digit '0' misclassified as '4'), **b** *Devanagari* (the digit '2' misclassified as '5'), **c** *Arabic* (the digit '2' misclassified as '3'), and **d** *Telugu* (the digit '0' misclassified as '1') scripts respectively

4 Conclusion

Handwritten digit recognition has received substantial attention from the OCR research community during the last few decades. In this work, we have assessed the efficacy of Mojette Transform for handwritten digit recognition problem. This is done by capturing local discriminative features from handwritten digit recognition for attaining higher classification accuracy. The strength of the proposed method is assessed on the *CMATERdb3* handwritten digit database using multiple classifiers. Finally, MLP classifier is found to produce the highest recognition accuracies of 98.27, 97.5, 98.8 and 98.1 % on *Bangla, Devanagari, Arabic* and *Telugu* numeral scripts respectively. The attained accuracies confirm the suitability of Mojette transformation to handwritten digit recognition for scripts. Among the most important advantages of this feature extraction algorithm is that it does not require any normalization of digits where the most of the published works need digit normalization, which degrades the image quality. These features are also very simple to implement compared to other counterparts. The future scope of this work will be to append some structural features like concavity analysis which will help to remove some of confusions among similar digit classes. The present scheme will also be extended to more number of official *Indic* scripts which in turn would be a footstep towards a complete multi-script OCR system.

References

1. R. G. Mihalyi, *"Handwritten digit classification using support vector machines"*, Jacobs University, Bremen, Germany, 2011.
2. M. S. Akhtar, H. A. Qureshi, *"Handwritten Digit Recognition through Wavelet Decomposition and Wavelet Packet Decomposition"*, In; 8th International Conference on Digital Information Management (ICDIM), pp. 143–148, 2013.
3. Z. Dan, C. Xu, *"The Recognition of Handwritten Digits Based on BP Neural Network and the Implementation on Android"*, In: 3rd International Conference on Intelligent System Design and Engineering Applications, pp. 1498–1501, 2013.

4. U. R. Babu, Y. Venkateswarlu, A. K. Chintha, *"Handwritten Digit Recognition Using K-Nearest Neighbour Classifier"*, In: World Congress on Computing and Communication Technologies (WCCCT), pp. 60–65, 2014.
5. R. Ebrahimzadeh, M. Jampour, *"Efficient Handwritten Digit Recognition based on Histograms of Oriented Gradients and SVM"*, In: International Journal of Computer Applications, vol. 104, No. 9, pp. 10–13, 2014.
6. A. M. Gil, C. F. F. C. Filho, M. G. F. Costa, *"Handwritten Digit Recognition Using SVM Binary Classifiers and Unbalanced Decision Trees"*, In: Image Analysis and Recognition, Springer, LNCS 8814, pp. 246–255, 2014.
7. B. EL. Qacimy, M. A. Kerroum, A. Hammouch, *"Feature Extraction based on DCT for Handwritten Digit Recognition"*, In: International Journal of Computer Science, vol. 11, Issue 6, No. 2, pp. 27–33, 2014.
8. S. AL. Mansoori, *"Intelligent Handwritten Digit Recognition using Artificial Neural Network"*, In: International Journal of Engineering Research and Applications, vol. 5, Issue 5, pp. 46–51, 2015.
9. B.V. Dhandra, R.G. Benne, M. Hangarge, *"Kannada, telugu and devanagari handwritten numeral recognition with probabilistic neural network: a script independent approach"*, In: International Journal of Computer Applications 26, pp. 11–16, 2011.
10. N. Das, R. Sarkar, S. Basu, M. Kundu, M. Nasipuri, D. K. Basu, *"A genetic algorithm based region sampling for selection of local features in handwritten digit recognition application"*, In: Applied Soft Computing (ASC), 12(5), pp. 1592–1606, 2012.
11. N. Das, J. M. Reddy, R. Sarkar, S. Basu, M. Kundu, M. Nasipuri, D. K. Basu, *"A Statistical-topological feature combination for recognition of handwritten numerals"*, In: Applied Soft Computing (ASC), 12(8), pp. 2486–2495, 2012.
12. J. H. AlKhateeb, M. Alseid, *"DBN Based learning for Arabic Handwritten Digit Recognition Using DCT Features"*, In: 6th International Conference on Computer Science and Information Technology (CSIT), 2014.
13. J. P. Guedon, N. Normand, *"The Mojette Transform: The First Ten Years"*, In: Discrete Geometry for Computer Imagery, LNCS 3429, pp. 79–91, Poitier, France, 2005.
14. J. V´as´arhelyi, P. Serf`oz¨o, *"Analysis of Mojette Transform Implementation on Reconfigurable Hardware"*, In: Dagstuhl Seminar Proceedings 06141, Dinamycally Reconfigurable Architectures, 2006 Available: http://drops.dagstuhl.de/opus/volltexte/2006/746.
15. K. Pearson, *"On Lines and Planes of Closest Fit to Systems of Points in Space"*, In: Philosophical Magazine, 2(11), pp. 559–572, 1901.
16. R. C. Gonzalez, R. E. Woods, *"Digital Image Processing"*, vol. I. Prentice-Hall, India (1992).

A Survey of Prospects and Problems in Hindustani Classical Raga Identification Using Machine Learning Techniques

Sreeparna Banerjee

Abstract In this paper we present a survey of current research in Music Information Retrieval in North Indian Classical Music and describe all the characteristics of ragas used for classification. We then describe Bhatkhande's classification scheme and show how it can simplify the classification process of 100 ragas to 10 categories. We also discuss the issues that need to be addressed and the similarities and differences between Hindustani classical music and Western Classical music. Current research efforts on Raga identification are also described.

Keywords Pattern recognition of raga · Music information retrieval · Audio descriptors · Audio signal processing

1 Introduction

Music Information Retrieval (MIR) [1] is a relatively new field in Information Technology that has generated a lot of interest in the last two decades due to its applications in music generation, indexing, processing, recording, reproduction and music oriented services in multimedia. Extensive studies are being conducted in Western classical music [2–5] and notable efforts are also being made in Indian classical music [6–16]. There are two distinct traditions in Indian classical music, namely, Hindustani or North Indian music [6, 7, 9] and Carnatic or South Indian Music [8]. Although they are two distinct genres in Indian music, there are several similarities between them, the most important ones being the concepts of ragas and Talas. The initial work on MIR in Indian classical music has focused on raga identification using soft computing approaches like Support Vector Machine (SVM) [6], swara

Sreeparna Banerjee (✉)
Department of Natural Science and Computer Science and Engineering,
West Bengal University of Technology, Kolkata, India
e-mail: sreeparnab@hotmail.com

© Springer Science+Business Media Singapore 2017   467
J.K. Mandal et al. (eds.), *Proceedings of the First International Conference
on Intelligent Computing and Communication*, Advances in Intelligent Systems
and Computing 458, DOI 10.1007/978-981-10-2035-3_48

histograms [7], data mining algorithms [8] and exponential series analysis [9]. Hindustani Classical Music (HCM) has characteristics described below, some of which resemble Western Classical Music (WCM).

- A raga is essentially a melodic structure in HCM, which conforms to certain rules which dictate its usage of melodic notes (swaras) as well as govern the route of its ascent (arohan) and descent (abarohan) along a musical scale and also possess some preferred patterns or motifs (pakads) during its exposition. These ragas are meant to convey particular emotions. During the presentation of a raga, there are portions of the raga referred to as "gat" that are played using particular rhythms called the Ta ala and in a particular tempo for each of these rhythms.
- Like in Western music, Indian music is also tonal, where one tone, the tonic, provides an anchor around which other tones are organized. However, Western music expresses in tonality through harmony, whereas in Indian music the same is expressed through these musical forms, viz., the ragas. Furthermore, three factors, namely tonal hierarchy, harmonic functions of chords and modulation frequency [2] contribute to Western classical music (WCM). On the other hand, Indian music lacks both modulations and harmonic chord function.
- Tonal hierarchy does exist in Indian music with the melody being organized around the tonic referred to as Sa, which is a solfege of Do, and the fifth note, or sometimes the fourth note being the second most stable note. There is also a predominance of certain notes called the vadi swara followed by the samabadi swara which is generally a perfect fourth or fifth above the vadi swara. There are also anuvadi swaras which exist in the raga as well as vivadi swaras which are foreign to the ragas.
- The ragas must consist of at least five of the twelve notes (swaras) of the musical scale including the tonic and the fourth (Ma corresponding to the solfege Fa in the Western scale) or fifth (Pa corresponding to the solfege So). Indian music uses an additional mechanism called drone [2] which is a continuous sounding of the tonic and usually the fifth scale tone.
- Unlike its Western counterpart, where the musical scores of the composers have to be adhered to, Indian classical music allows ample scope for improvisation. In fact, the beauty and complexity of Indian classical music, be it Hindustani Classical Music (HCM) or Carnatic, is that, shown in Fig. 1, once the rules for raga composition are followed, the performer has complete freedom to compose and render the exposition.

However, this fact has made the job of identifying a raga extremely challenging as no two compositions, even by the same performer, are alike. Since ragas are meant to convey emotions, they have a certain character (devotional, erotic, bold, poignant, etc.,), have preferred times for rendition and can also be seasonal.

Fig. 1 Circular block
diagram of Thats

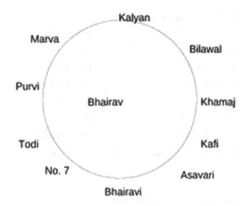

2 Characteristics of Hindustani Ragas

2.1 Characteristics

Hindustani or North Indian Classical Ragas can be composed from twelve notes or swards Any Raga must have at least five notes. A Raga with five notes (pentatonic) is called an Orabh raga, with six notes (hexatonic) is called Swarabh raga and all seven notes (septa tonic) is called complete or Sampurna. This characteristic of a Raga is called the jati.

Another important feature of a raga is its anga, which depicts whether the first half of the saptak (or octave) is more important in its rendition, in which case its is Poorvanga, or the second half, in which it is Uttaranga. Poorvanga ragas are generally played or sung during the latter part of the day whereas Uttarang ragas are generally sung in the first half of the day. Nyas swara is a stay note which has a prominent role in the ending of the raga.

Most of the melodies are composed in three octaves, namely, the mandra saptak (lower octave), madhya saptak (middle or most prominent octave) and tar saptak (higher octave). Ragas are built on scales consisting of seven tones. These scales are referred to as thats. The correspondence of the seven notes with the scale notes and their Western solfege syllables of the No major diatonic scale are given in Table 1. Keeping the tonic (Sa) and dominant (Pa) fixed, there other five notes, the remaining five that notes may be altered in one direction only. Re, Ga, Ha and Ni may be lowered by half a step while Ma may be raised by half a step. This gives rise to 32 possible seven note scales [2]. Since these pitch restrictions are absent in Carnatic music there are 72 possible scales known as Melakartha ragas.

The typical rendition of a raga consists of an alap which is free of metrics which invokes the mood of the raga followed by a gat based on a metric pattern. The gat can have different tempos which can conform to different tempos. An additional feature

Table 1 Scale notes/Swaras

Note	Indian note	Ratio to Sa (C)
C (Tonic)	Sa	1
D (Super tonic)	Re (shudh)	9/8
E (Mediant)	Ga (shudh)	5/4
F (Sub dominant)	Ma (shudh)	4/3
G (Dominant)	Pa	3/2
A (Submediant)	Dha (shudh)	5/3
B (Sub tonic/Leading note)	Ni (shudh)	15/8

Table 2 Scale notes/Swaras

That	Note	Vadi	Sambadi
Bilawal	C, D, E, F, G, A, B	A (Ni)	E (Ga)
Khamaj	C, D, E, F, G, A, Bb	E (Ga)	B (Ni)
Kafi	C, D, EbF, G, A, Bb	G (Pa)	C (Sa)
Asavari	CDEb FGAbBb	Ab (Dha)	Eb (Ga)
BhairavI	CDb Eb F GAb Bb	Ab (Dha)	Db (Re)
Bhairav	CD bEF # GAb B	Ab (Dha)	Db (Re)
Todi	CDb Eb F# GABb	A (Dha)	Eb (Ga)
Marwa	CDb EF # GAbB	Db (Re)	Ab (Dha)
Purvi	CDb EF # GAB	G (Pa)	Db
Kalyan	CDEF # GAB	E (Ga)	B (Ni)

of Hindustani Ragas, is the presence of Shrutis. Subtle nuances of a Raga can be expressed through minor variations (fixed in number for each of the notes) of a note (swara). These variations along with the twelve main notes give rise to 22 shrutis.

2.2 Bhatkhande's Classification

The most notable and most popular attempt at classification of ragas has been made by V.N. Bhatkhande. He had made the classification based on the swaras used by the ragas and called them thats. He claimed that the 32 thats mentioned above can be re binned into ten thats which have been named after the popular raga belonging to the that. The ten thats which is described in Fig. 1, their tones and their vadi and samavadi are listed in Table 2. According to Bhatkhande, no two versions of the same note (e.g. B and Bb) can occur during aroha or abaroha. Here b denotes komal (flat) and # denotes tibra (sharp).

3 Problems to Be Addressed in HCM

Some of the concerns that need to be addressed in order to develop suitable raga identification techniques for Music Information Retrieval (MIR) are mentioned in this section.

In addition to the seven notes given in Table 1, Hindustani Classical music ragas use ten micro tones in addition to five semitones corresponding to altered notes of Re, Ga, Ma, Dha and Ni giving rise to a total of 22 Shrutis.

The use of a drone instrument like the tanpura which provides the tonic or absolute frequency along with the fifth (Pa) or fourth (Ma) which provides a continuous pitch reference is an added consideration in HCM.

The presence of melodic motifs called bandish [17] in vocal music and asthayi in instrumental music are provide key phrases used for raga identification. HCM is homo-phonic in nature. Matching of timbre [18] of the vocal or instrumental performance with the accompanying instruments is an important aspect of music identification.

4 Comparison Between HCM and WCM

In both musical genres, the tonic plays a crucial role [19]. However, WCM is based on harmony, in which a group of notes called chords are played simultaneously [20]. HCM is based on melodic structures called ragas. These ragas can be rendered in several sub-genres like Dhrupad, Khayal, Thumri, etc., The micro tonal scale form based on a natural harmonic series reflects the mood of the raga.

The sub-genres of WCM include symphony orchestra and chamber music. Symphony is a musical composition with several movements, usually four, with each movement having a different tempo. Symphonies are usually orchestras with multiple instruments and sometimes, voices are included. Chamber music is composed of a small group of musical instruments.

HCM belongs to the chamber music category with one or two performers along with an accompanying background drone from the tanpura to provide fundamental frequency (pitch) reference and also an instrument to provide the tala or rhythm, which can be played with different layas (or tempo).

5 Music Information Retrieval(MIR)

As ragas form the backbone of HCM, MIR consists essentially of their classification, performed by first extracting low level features or audio descriptors [21] and subsequent identification using high level features [22], a process called audio data mining and use of metadata [23]. A typical low level search process could use LPI combined with a high level process like LSI [24].

After signal acquisition and signal thumb nailing to extract the most repetitive portion of the musical piece, which could provide a representative summary of the raga, feature extraction [25] is carried out. This feature extraction consists of low level features and high level features referred to as music content and music context, respectively [26].

The low level features [25] or music content are features that can be extracted directly from the audio signal can be classified as follows:

1 Time domain features like mean energy, zero crossings, number of silent frames.
2 Frequency domain based features like spectral centroid, roll-off, spectral flux, etc.,
3 Coefficient based features like MFCC (Mel frequency cepstral coefficients) LPC (Linear Predictive Coding), etc.,
4 Time-frequency domain features like pitch. This is the most widely used feature.

Loudness and timbre are low level features. Also sometimes referred to as the color or quality of sound. Timbre is related to three main properties of music signal. These are spectral envelop of shape and time variation of spectrum and time evolution of energy [27]. Audio descriptors are also a special set of low level features. High level features or musical context cannot be obtained directly from the audio signal but can be derived from it. These features include genre, melody, mood, rhythm and timbre in context of the melody and artist identification.

Some of the subfields of MIR [26] are feature extraction, similarity search, classification and applications like audio fingerprinting, which is a compact signature for content based audio retrieval. Feature extraction involves processes like timbre description, music transcription and melody extraction. Query based approaches fall under the similarity search category. Classification embraces emotion and mood recognition, genre classification, instrument classification, composer and performer identification among other techniques. Content based query and retrieval is another example of applications.

6 Related Work

The related work can be classified as work related to raga identification and works addressing concerns of Sect. 3.

6.1 Raga Identification

Since MIR in Indian Music is a relatively new field there is a small amount of literature available. One of the initial attempts was by Sahasrabuddhe and Upadhye [10] who used finite state automata to model a raga based on its swara constituent pattern. Pandey et al. [11] used the Hidden Markov Model (HMM) on swara sequences, treating them as word sequences based on swara alphabets in their Tansen raga recognition system. A Gaussian Mixture Model based on HMM using three features, namely, Chroma, MFCC (Mel Frequency Cepstral Coefficients), was used by Dighe et al. [12]. By combining these three features, they obtained a 62 dimensional feature vector. Pitch-class profile distribution with K-NN classifier and K-L divergence was used in [13]. Some of the research focused on a particular characteristic of a raga like [14] who compared the arohan and avarohan, [15] who studied vadi and samavadi and [16] who studied pakar.

6.2 Works Addressing Issues of Section 3

Attempts made to address the problems discussed in Sect. 3 are described below.

The harmonium is a keyboard system like the piano, but, unlike the piano where there are pauses between musical notes, the notes from the harmonium linger, thus producing continuous swaras. In [28], a novel method for raga recognition by utilizing the continuous sound producing property of the harmonium has been presented. At first, audio signals from the harmonium were segmented into separate frames for note detection purpose. Because the swaras are played continuously, proper onset and end-point detection is very critical for identification and this was done using two approaches, namely spectral flux determination and fundamental frequency estimation in the time-frequency domain analysis of each frame. Raga identification was treated as a template matching problem using a database of ragas to match the query audio signals with the prototypes in the database.

The homo-phonic aspect of HCM and timbre, whereas single melody line is accompanied by instruments like tanpura or harmonium was explored in [29]. The authors separated the singer from the accompanying instrument using a hybrid selection algorithm applied to six audio descriptors, namely, attack time, attack slope, zero crossing rate, roll-off, brightness, roughness and MFCC with 20 coefficients from the MIR toolbox of MATLAB. The k-means and a statistical classifier k nearest neighbors are used to identify and differentiate between human voice and instrument, although they have similar timbres. Since tonic is important in HCM, its identification should be the first step in automated raga identification, and this has been attempted by [29]. They used a four step process for music extraction. In the first step spectral peaks were isolated from the audio signal. Then the salience function determination was performed using the sum of weighted energies extracted at integer multiples (harmonics) of that frequency, followed by tonic candidate generation obtained from the pitch histogram of the entire audio signal and finally selecting the tonic as the highest peak. Classification was performed using the C4.5 decision tree of the Weka software.

Tuning of HCM was studied by [30] Stable fundamental frequency collection was done followed by construction of interval histogram. Their objective was to detect the twelve notes of the octave as well as further subdivisions of the octave to look for the presence of micro tones. Reference [24] extracts formants from audio files using

LPC for detection and then term document matrix which is decomposed using Single Vector Decomposition for Latent Semantic Indexing (LSI). The detected formant (query) is matched with the document.

Vocal expressions like glides (meend) and vibrato (andolan) can play an important role in raga identification. An example is the distinction between ragas Bhupali and Shudh Kalyan. An algorithm has been described by [27] to perform this operation. Firstly, the pitch curve was estimated and singing voice frames were identified. A pitch envelope was used to develop a canonical representation and, finally, templates were used for identification of expressions for creating a transcription of an audio signal.

7 Prospects

In summary it might be remarked that, although MIR in Hindustani Classical Music is in a nascent state, specially as compared with its Western counterpart, there has been a lot of progress in the past decade and, so, it is hoped that the future will witness remarkable progress in Raga identification, which is the mainstay of MIR in HCM.

References

1. Markus Schedule, Emilia Gmez, Julin Urbano, Music Information Retrieval: Recent Developments and Applications, Foundations and Trends R in Information Retrieval, Vol. 8, No. 2–3, pp. 127261, (2014).
2. Mary A. Castellano, J. J. Bharucha, Carol L. Krumhansl, Tonal Hierarchies in the Music of North India, Journal of Experimental Psychology: General, 113 (3), pp. 394–412 (1984).
3. G. Tzanetakis, G. Essl, P. Cook, Automatic Music Genre Classification of Audio Signals, Proceed. 2nd International Symposium of Music Information Retrieval, Bloomington, Indiana, USA, pp. 205–210, (2001).
4. Felix Weninger, Noam Amir, Ofer Amir, Irit Ronen, Florian Eyben, Bjorn Schuller, Robust feature extraction for automatic recognition of vibrato singing in recorded polyphonic music, Proceed. 37th International Conference on Acoustics, Speech and Signal Processing (ICASSP), Kyoto, Japan, pp. 85–88, (2012).
5. A. Ghias, J. Logan, D. Chamberlin and B.C. Smith, "Query by Humming- Musical Information Retrieval in an audio database" Proc. ACM Multimedia, pp 231–236 (1995).
6. Vijay Kumar, Harit Pandya, C.V. Jawahar, Identifying Ragas in Indian Music, Proceed. 22nd International Conference on Pattern Recognition (ICPR), pp. 767–772 (2014)
7. Pranay Dighe, Harish Karnick, Bhiksha Raj, Swara histogram based structural analysis and identification of Indian classical ragas, 14th. Conference of the International Society for Music Information Retrieval, Curitiba, Brazil, pp. 35–40, (2013).
8. K Oriya, R. Geetha Ramani, Shomona Gracia Jacob,Data Mining Techniques for Automatic recognition of Carnatic Raga Swaram notes, International Journal of Computer Applications, 52(10) pp. 4–10, (2012).

9. Soubhik Chakraborty, Saurabh Sarkar, Swarima Tewari and Mita Pal, An interesting application of simple exponential smoothing in music analysis, International Journal on Soft Computing, Artificial Intelligence and Applications (IJSCAI), 2(4), pp. 37–44, (2013).

10. H. Sahasrabuddhe and R Upadhye, "On the computational model of raag music of India", Workshop on AI and Music, European Conference on AI (1992)

11. G. Pandey, C. Mishra and P. Ipe, "Tansen: A system for automatic raga identification", Proc. Indian International Conference on Artificial Intelligence, pp. 1350–1363 (2003).

12. Pranay Dighe, Parul Agrawal, Harish Karnick, Siddartha Thota and Bhiksha Raj, Scale independent raga identification using chroma gram patterns and swara based features, IEEE International Conference on Multimedia and Expo Workshops, 2013.

13. Koduri Gopala-Krishna, Sankalp Gulati and Preeti Rao, A Survey of Raaga Recognition Techniques and Improvements to the State-of-the-Art, Sound and Music Computing, 2011.; OP. Chordia and A. Rae, Raag recognition using pitch-class and pitch-class dyad distributions, International Society for Music Information Retrieval, 2007.

14. S. Shetty abd K. Achary, Raaga Mining of Indian Music by Extracting Arohana-Avorahan patterns, International Journal of Recent trends in Engineering, 2009.

15. S. Shetty abd K. Achary,Raga Identification of Carnatic Music for Music Information Retrieval, International Journal of Recent trends in Engineering, 2009.

16. Vignesh Iswar, Shrey Dutta, Ashwin Bellur, Hema A. Murthy, Motif Spotting in an Alpana in Carnatic Music, International Society for Music Information Retrieval, 2013.

17. Joe Cheri Ross*, Vinutha T. P. and Preeti Rao, Detecting melodic motifs from audio for Hindustani classical music, 13th International Society for Music Information retrieval conference (ISMIR 2012) 2012.

18. Saurabh H. Deshmukh and S. G. Bhirud, North Indian Classical Musics Singer Identification by Timbre Recognition using MIR Toolbox, Intl. J. Computer Applications, 91(4) 1–4 (2014).

19. Justin Salamon, Sankalp Gulati and Xavier Serra, A multipitch approach to tonic identification in indian classical music, Proceed. International Society for Music Information Retrieval Conference 2012.

20. Parul Agarwal, Harish Karnick and Bhiksha Raj, A comparative study of indian and western music forms, Proceed. ISMIR 2013.

21. Saurabh H. Deshmukh, Divya P. Bajaj, S.G. Bhirud, Audio descriptive analysis of singer and musical instrument identification in north indian classical music, International Journal of Research in Engineering and Technology, 4(6), 505–508, June 2015.

22. Surendra Shetty, K K Achary and Sarika Hegde, Feature Extraction Computation and Automatic Raga Identification for Carnatic Ragas, nternational Journal of Information Processing, 7(2), 41–51, 2013

23. Preeti Rao, Audio metadata extraction: the case for Hindustani classical music, Proceed. of SPCOM 2012, IIScIISc Bangalore (2012)

24. P Kirthika, Rajan Chattamvelli, Frequency based audio feature extraction for raga based musical information retrieval Using LPC And LSI, Journal of Theoretical and Applied Information Technology, 69(3) 2014.

25. Balaji Thoshkahna, Algorithms for Music Information Retrieval, Master of Science Thesis, IISc Bangalore, April 2006.

26. Markus Schedl, Emilia Gomez and Julian Urbano, Foundations and trands in Information Retrieval, 8(2–3), 127–161 (2014)

27. Sai Sumanth, Miryala Kalika, Bali Ranjita, Bhagwan Monojit Choudhury, Automatically identifying vocal expressions for music transcription, Proceed. ISMIR 2013.

28. Rajshri Pendekar, S. P. Mahajan, Rasika Mujumdar, Pranjali Ganoo, Harmonium Raga Recognition, International Journal of Machine Learning and Computing, Vol. 3, No. 4, August 2013.

29. Justin Salamon, Sankalp Gulati and Xavier Serra, A multipitch approach to tonic identification in indian classical music,, Proceed. ISMIR Conference 2012.

30. Joan Serra , Gopala K. Koduri, Marius Miron and Xavier Serra, Assessing the tuning of sung indian classical musi, Proceed. ISMIR 2011

A Neuro Fuzzy Based Black Tea Classifying Technique Using Electronic Nose and Electronic Tongue

Sourav Mondal, Runu Banerjee(Roy), Bipan Tudu,
Rajib Bandyopadhyay and Nabarun Bhattacharyya

Abstract This paper presents a neuro-fuzzy classification technique using electronic nose, electronic tongue and the fused response of electronic nose and electronic tongue for the evaluation of black tea quality. In the tea industries an automated, neutral and low cost instrumental system to determine the overall tea quality is in great requirement. A general fuzzy rule based and neural network model can produces accurate predictions. But both models have some weakness. In this pursuit, Pseudo outer-product based fuzzy neural, a kind of fuzzy neural network classifying system has been attempted to classify tea grades. Results show that above model can classify in a better way compared to other models.

Keywords Electronic nose · Electronic tongue · Black tea analysis · Fuzzy c-means · Fuzzy neural network · POPFNN

1 Introduction

Tea is one of the most demanding beverages worldwide with an expanding market because of its certain flavor and aroma. Tea leaves processing techniques are one of major factor of determining the final marker price of tea. So before tea leaves are

Sourav Mondal (✉) · Runu Banerjee(Roy) · Bipan Tudu · Rajib Bandyopadhyay
Jadavpur University, Kolkata, India
e-mail: sourav.cemk@gmail.com

Runu Banerjee(Roy)
e-mail: runu_banerjee@iee.jusl.ac.in

Bipan Tudu
e-mail: bt@iee.jusl.ac.in

Rajib Bandyopadhyay
e-mail: rb@iee.jusl.ac.in

Nabarun Bhattacharyya
Centre for Development of Advanced Computing, Kolkata, India
e-mail: nabarun.bhattacharya@cdac.in

© Springer Science+Business Media Singapore 2017
J.K. Mandal et al. (eds.), *Proceedings of the First International Conference on Intelligent Computing and Communication*, Advances in Intelligent Systems and Computing 458, DOI 10.1007/978-981-10-2035-3_49

sold in the market, they need severe monitoring and analysis. In spite of favorable economic figures, tea industry follows orthodox method of tea testing by human panel known as 'Tea tasters'. The different varieties of tea are graded by tea taster based on color, taste and strength of the tea sample. However, this method is highly subjective with various human factors such as individual mental state, reduction of sensitivity due to infection, etc. Electronic nose and electronic tongue have the potential to classify tea samples in a more repetitive and accurate way.

Mimicking the sense of smell led to the development of Electronic Nose [1] and electronic tongue [2] has been developed for measures and compares taste. Previously authors have worked with fuzzy based classifiers [3]. In this work, FCM is used to get the significant information from the sensor responses as extracted features for individual sensory systems and those features are used to the next level classifier. These features are in the form of membership function vectors. The membership function vectors are fed into Pseudo Outer-Product based Fuzzy Neural Network (POPFNN) [4, 5] which analyze features and classify black tea according to their gradation.

2 Experimentation

Sensor responses of four different classes of tea samples obtained from electronic nose and electronic tongue are used for experimentation purpose, 12 samples for each class. Each sample consists of transient response from 5 sensors. Each transient response consists of 66 data points and 694 data points for electronic nose and electronic tongue, respectively.

An array of five MOS sensors [6] has been used to developed electronics nose and a three electrode potentiostat system with the working electrodes made up of five different noble metals are used to set up the electronic tongue system. A Platinum and an Ag/AgCl act as counter and reference electrode, respectively. Details description of sensor used in electronic nose and tongue are given in [7, 8].

3 Data Analysis

Data analysis from electronic nose and tongue is difficult because of innumerable compound present in the tea. These data can be analyzed by various pattern recognition models. In this work, POPFNN model is used with certain modifications as per our datasets along with other three classifying techniques. Among these four, POPFNN gives the best result compared to the fuzzy and Neural Network based model.

3.1 POPFNN Model Framework

Hybrid, POPFNN is a combination of neural networks and fuzzy systems. So it has the ability to eliminate the imperfections of the both techniques. POPFNN uses a self-organizing algorithm to learn and initialize the membership function of the input and output variables from a set of training data. The similar details of POPFNN used in this work are described in [9]. It is actually a Multi-Input Multi-Output (MIMO) system with a five-layer neural Network. As defuzzification layer is not required for our experiment so, a four layer POPFNN is shown in Fig. 1. Each layer in POPFNN performs a particular fuzzy operation. The number of neurons in the condition and the rule-base layers are defined in as:

$$l_2 = \sum_{i=1}^{l_1} J_i$$

$$l_3 = \prod_{i=1}^{l_1} J_i$$

where, J_i is the numeric labels for the ith input, l_1 is the number of inputs, l_2 is the number of neurons in the condition layer, and l_3 is the number of rules neurons,

An overall description of each layer of POPFNN is given as follows:

Input Layer: First layer is called input layer where sensor responses from nose and tongue are given to the condition layer. Sensor responses from nose and tongue are fuzzified using FCM at the second layer.

Net input: $f_i^I = n_i$
And Net output: $O_i^I = f_i^I$
Where: n_i = value of the ith input

Condition Layer: FCM is used in the first process to obtain the fuzzy information in the form of membership function vectors at the condition layer. The FCM [10]

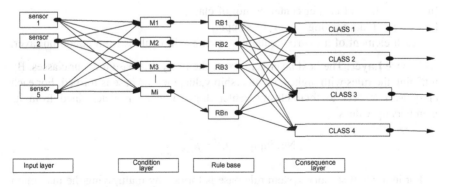

Fig. 1 Proposed POPFNN model

employs fuzzy partitioning such that a data point can belong to all groups with different membership values. This iterative algorithm is used to find the cluster centers by minimizing following objective functions in Eq. 1:

$$I_m = \sum_{i=1}^{F} \sum_{j=1}^{S} \mu_{ij}^m \left\| n_i - s_j \right\|^2, \ 1 \le m < \infty \tag{1}$$

where, m is the any real number greater than 1 which is weighting coefficient denoting the fuzziness of the cluster, μ_{ij} is the degree of membership of n_i in the cluster j, n_i is the i-th of d-dimensional measured data, s_j is the d-dimension cluster center, and $\left\| n_i - s_j \right\|$ denotes the Euclidian distance between n_i and s_j. Fuzzy partitioning is accomplished via an iterative optimization process of the objective function shown in above equation, with the update of membership μ_{ij} and the cluster center s_j by the following equations:

$$\mu_{ij} = \frac{1}{\displaystyle\sum_{k=1}^{S} \left(\frac{\left\| n_i - s_j \right\|}{\left\| n_i - s_k \right\|} \right)^{\frac{2}{m-1}}} \tag{2}$$

$$s_j = \frac{\displaystyle\sum_{i=1}^{F} \mu_{ij}^m \cdot n_i}{\displaystyle\sum_{i=1}^{F} \mu_{ij}^m} \tag{3}$$

where, F number of samples and S is the number of classes.

In the Fuzzy neural network as in Fig. 1, M1 to Mi is the membership values of classes of the electronic nose or electronic tongue respectively with respect to their cluster centers.

Net output: $O_k^{II} = \max\left(\mu_{[u \times v]_{[w \times x]}} \right) \forall [u \times v]_{[w \times x]}; \ 1 \ge u, v \ge 4$ and $u, v \in I$

where, $\mu_{[u \times v]_{[w \times x]}}$ = Membership value from FCM

[u × v] = [no. of cluster center × no. of class]

[w × x] = [data reading × no. of sample]

I = each element of a matrix containing membership values from condition layer.

Rule base Layer: The next layer is to create the fuzzy rule base for four classes. By retaining the rules with highest membership value, the rule base is formed. For each class with respect to their cluster center, highest value is determined from its membership values.

Net input: $f_k^{III} = \mu_{[u \times v]_{[w \times x]}}^{II}$

For individual sensory system rule base is formed by multiplying the maximum value of each element of every row of O_k^{III}. For each class with respect to their cluster center we get four rule bases. Rule bases from training are compared with unknown sample at Consequence Layer for classification purpose.

Four rule bases are formed using POP (Pseudo Outer Product) learning method during training:

RB_n = multiplying the maximum values of every rows of O_k^{III} matrix form rule base with respect to 4 cluster center.

where n = number of rule base and n varies from 1 to 4.

For combined sensory system we multiply two highest membership value of same class of nose and tongue.

Net Combined output: $O_{kC}^{III} = \max\left(\mu_{N[u \times v]_{[w \times x]}}^{II}\right) \times \max\left(\mu_{T[u \times v]_{[y \times z]}}^{II}\right)$;

and $u, v \in I$

where, $\mu_{N[u \times v]_{[w \times x]}}^{II}$ = output of the Condition Layer node that forms the antecedent conditions for the nose response.

$\mu_{T[u \times v]_{[y \times z]}}^{II}$ = output of the Condition Layer node that forms the antecedent conditions for the tongue response.

4 combined rule base is formed using same principle which is applied to form rule base for individual sensory system during training.

Consequence Layer: classification is done at Consequence Layer. Test sample should test simultaneously through both the sensor systems and for each individual sensor system rule bases are generated from the proposed model of Fig. 1 and fused rule base should compare with training model rule base with the help of Eq. (4).

Classification rate = (Degree of Testing − Degree of Training) / Degree of Training × 100

$$(4)$$

In our system, POPFNN has two fundamental modes. First is learning mode and second is classification mode. Black tea samples are used to train the POPFNN during learning mode. After training, membership function vectors (feature vectors) are extracted. These feature vectors are then used to initialize and adjust the parameter in POPFNN. Similarly during testing, feature vectors are extracted. Degrees are calculated from both training and testing data set by applying proper rule base technique. Procedure to classify samples is explained in Table 1.

3.2 Fuzzy Model Framework

For solving the problems based on pattern recognition Fuzzy framework described by Wang and Mendel [11] has been favored choice for the researchers. This fuzzy classifier model is a MIMO model framework which adopts the Takagi-Sugeno model where the antecedents are in the form of fuzzy and consequent is in crisp

Table 1 POPFNN algorithm

Step 1: Input (Input layer)
Sensor responses are given as input
Step 2: Minimize the objective function for S number of classes and F number of samples (Condition layer)
Step 2a: Randomly initialize the membership function
Step 2b: Calculate the cluster center
Step 2c: Updating the membership function with the help of cluster center
Step 2d: Evaluate the objective function
Step 3: Rule base is formed using POP learning method (Rule base layer)
Step 3a: Weight of the highest membership values is determined for each class with respect to their cluster center
Step 3b: Degree is calculated by multiplying the maximum values of every rows of O_k^{III} matrix
Step 4: Rule comparison and classification rate declaration (Consequence layer)
Step 4a: Degree of training and testing is obtained from sensor responses for both known and unknown tea samples by following the steps 1 to 3
Step 4b: Classification rate calculated by Eq. (4)

form. Wang Mendel method is applied on data from electronic nose and tongue for generating preliminary rules. Details of this work are described in [3].

3.3 Neural Network Framework with BP-MLP Topology

A three-layer BP-MLP model [12] with one input layer, one hidden layer and one output layer has been used in this work. Input layer has five nodes denotes transient response from five sensors. Only one hidden layer has been considered which has eight nodes and the number of output nodes is four as four different class of tea has been consider for correlation study.

3.4 Fuzzy Neural Network Framework (FNN)

This FNN model has four layers. First layer is called input layer which have five nodes. Second layer is used for generating membership function from sensors responses. Next layer represents BP-MLP neural network which is used for realizing fuzzy rules. Output layer is the last layer which gives the gradation of tea. Details of this work are described in [3].

Table 2 Comparative results of POPFNN classifier for different sensory systems

Trials	Electronic nose	Electronic tongue	Combined sensory system
1	83.69	87.82	95.44
2	86.34	85.42	94.80
3	88.42	83.24	95.77
4	74.33	88.75	97.05
5	82.38	80.60	98.05
6	89.50	85.71	93.14
7	88.34	92.32	96.41
8	90.12	89.04	95.51
9	82.67	74.78	92.18
10	81.16	83.70	91.94
Average	84.69	85.14	95.03

4 Results

Data analysis on 48 tea samples of four different class of tea of electronic nose and electronic tongue has been performed by fuzzy neural network analyses. Out of the total data set, 60 % of the data have been used for training, and the rest of 40 % were used as testing set. The summary of POPFNN results is given in Table 2.

Result shows that, for POPFNN classifier, the electronic tongue can classify better to a small degree than electronic nose. Average results from POPFNN are compared with other models of electronics nose, electronic tongue and combined sensory system which is shown in Table 3.

Study shows, our new improved fuzzy neural network model gives better assessment for tea quality analysis. Results also show that combined sensory response is quite improved than individual sensory system. So, this fast, accurate and effective fuzzy neural classifier has the ability to classify tea quality more precisely.

Table 3 Comparative results of classification rate between Fuzzy, neural network, FNN and POPFNN model for different sensory systems

Sensory system	Average classification rate (in %)			
	Fuzzy	Neural network	FNN	POPFNN
Electronic nose	74.60	75.14	76.52	84.69
Electronic tongue	79.41	77.34	81.61	85.14
Combination of electronic nose and electronic tongue	85.33	79.40	89.50	95.03

5   Conclusion

In this paper one fuzzy based model, one neural network model and two topologies of fuzzy neural networks based models are used to predict the tea quality. Four models are compared. It has been observed that fuzzy neural networks have ability to overcome the weakness of fuzzy and neural network model. But, this new POPFNN system outperformed all other models. Overall, the classifiers which are described here have the ability to assess the aroma, flavor and taste quality of black tea. This technique can be used for other similar applications also.

References

1. Keller, P.E.: Mimicking Biology: Applications of Cognitive Systems to Electronic Noses. Proceedings of the IEEE International Symposium on Intelligent Control/Intelligent Systems and Semiotics (1999).
2. Ivarsson, P., Holmin, S., Hojer, N. E., Krantz-Rulcker, C., Winquist, F.: Discrimination of Tea by means of a Voltammetric Electronic Tongue and Different Applied Waveforms. Sens. Actuators B, Chem., vol. 76, no. 1–3, pp. 449–454 (2001).
3. Banerjee(Roy), R., Modak, A., Mondal, S., Tudu, B., Bandyopadhyay, R., Bhattacharyya, N.: Fusion of Electronic Nose and Tongue Response Using Fuzzy Based Approach for Black Tea Classification. Procedia Technology, vol.10, pp. 615–622 (2013).
4. Zhou, R.W., Quek, C.: POPFNN: A Pseudo Outer-Product Based Fuzzy Neural Network. Neural Network 9 (9). 1569–1581 (1996).
5. Quek, C., Zhou, R.W.: The POP Learning Algorithms: Reducing Working Identifying Fuzzy Rules. Neural Network 14. 1431–1445 (2001).
6. Gas Sensors and Modules, http://www.figarosensor.com/gaslist.html.
7. Bhattacharyya, N., Bandyopadhyay, R., Bhuyan, M., Tudu, B., Ghosh, D., Jana, A.: Electronic Nose for Black Tea Classification and Correlation of Measurements with "Tea Taster" Marks. IEEE Trans. Inst. Meas. vol. 57, No. 7, (2008).
8. Palit, M., Tudu, B., Dutta, P. K., Dutta, A., Jana, A., Roy, J. K., Bhattacharyya, N., Bandyopadhyay, R., Chatterjee, A.: Classification of Black Tea Taste and Correlation With Tea Taster's Mark Using Voltammetric Electronic Tongue. IEEE Trans. Inst. Meas. vol. 59, pp 2230–2239 (2010).
9. Rong, L., Ping, W., Wenlei, H.: A Novel Method for Wine Analysis Based on Sensor Fusion Technique. Sensor and Actuators B66. 246–250 (2000).
10. Bezdek, J.C., Ehrlich, R., Full, W.: FCM: The Fuzzy C-Means Clustering Algorithm. Computers and Geosciences 10 (2–3), pp. 191–203 (1984).
11. Wang, L. X., Mendel, J. M.: IEEE Trans. Syst. Man Cybern., (1992).
12. Haykin, S.: Neural Networks-A Comprehensive Foundation (2nd ed.). Pearson Education, Asia, (2001).

Design of Novel Feature Vector for Recognition of Online Handwritten Bangla Basic Characters

Shibaprasad Sen, Ankan Bhattacharyya, Avik Das, Ram Sarkar
and Kaushik Roy

Abstract In the present work, a new feature vector has been designed towards recognition of handwritten online Bangla basic characters. At first, Center of Gravity (CG) of a particular character sample is determined. After that a circle enclosing the character sample is drawn whose radius is estimated as the distance of farthest data pixel from that CG. From this circular region, a 136-element feature vector is generated considering both the global as well as local information of the character sample. The feature set has been tested with several well-known classifiers on 10,000 isolated Bangla basic characters. Finally, Support Vector Machine (SVM) has produced 98.26 % recognition accuracy.

Keywords Online character recognition · CG-based circle · Global feature · Local feature

1 Introduction

Online Handwriting Recognition (OHR) is now becoming an upcoming area of research due to exponentially increasing popularity of devices like Take Note, iPad, Smartphones etc. Also individuals from major part of the society are becoming

Shibaprasad Sen (✉) · Ankan Bhattacharyya · Avik Das
Future Institute of Engineering and Management, Kolkata, India
e-mail: Shibubiet@gmail.com

Ankan Bhattacharyya
e-mail: ankan.bhattacharyya.94@gmail.com

Avik Das
e-mail: avikrik@gmail.com

Ram Sarkar
Jadavpur University, Kolkata, India
e-mail: raamsarkar@gmail.com

Kaushik Roy
West Bengal State University, Barasat, India
e-mail: kaushik.mrg@gmail.com

© Springer Science+Business Media Singapore 2017 485
J.K. Mandal et al. (eds.), *Proceedings of the First International Conference
on Intelligent Computing and Communication*, Advances in Intelligent Systems
and Computing 458, DOI 10.1007/978-981-10-2035-3_50

habituated to write information freely on those devices in their natural handwriting style. In these devices, written data are saved as online information. Not only this saves extra time but also it reduces the chances of mistyping that may happen while writing with a keyboard. Hence, researchers are showing interest about OHR. Though some good research works are available for Devanagari [1–4], and English [5–8] scripts but while, talking about the Bangla script, researchers has paid a little attention which is evident from the limited research materials available in the literature. Authors in [9] have prompted an approach for estimating the features in an unsupervised way, based on disparity space that embeds the local neighborhoods surrounded by the pen positions in the trajectory. In [10], Bhattacharya, N. et al. have reported segmentation as well as recognition techniques for cursive online Bangla texts. In this work, after passing through segmentation module, texts are broken into set of primitives. Such primitives may represent the character or parts from basic or compound character set. A method for recognition of those primitives has been devised then. A different technique based on combination of online and offline feature extraction procedure, have been applied for segmentation of hand-written online Bangla text is presented in [11]. Here, authors have considered directional feature for recognition of those segmented strokes. Another approach is mentioned in [12], where Roy, K. et al. have obtained elemental strokes from the character information and the sequential and then dynamic information at stroke level are collected from writing devices on the basis of pen movements. These information are served like feature values for their work. Bhattacharya et al. [13] have concentrated on annotation for unconstrained online Bangla handwriting samples and also have developed a substantial annotated dataset for those hand-writing specimens. A graphical user interface oriented semi-automatic path has been adopted towards annotation for character boundary levels. Authors in [14] have mentioned a challenging issue for unconstrained Bangla handwriting recog-nition i.e. large number of alphabet set of the Bangla script (around 300 different pattern considering the compound character set), which are complex enough in nature. In this paper authors have stated the fundamental observations for cursive Bangla handwriting on confined vocabulary set by combining Multi-Layer Per-ceptron (MLP) and Support Vector Machine (SVM). Authors in [15] have proposed a new approach for the recognition of cursively written Bangla word samples by describing feature estimation technique at sub stroke level and also proposed a Hidden Markov Model (HMM) model for recognition the same. Mondal et al. [16] have explored direction code histogram and point-float feature extraction approa-ches for Bangla alphabets. Soundness of the scheme was validated by HMM, Nearest Neighbor and MLP classifiers. According to authors in [17], features could be calculated from sequential as well as dynamic information which are found from the writing device. Produced feature values have been passed through quadratic classifier for recognition purpose. Sen et al. [18] have combined online (point based and structural features) and offline (quad tree based longest run and convex hull) feature extraction techniques towards handwritten Bangla characters recognition. Authors in [19] have discussed about a simple but effective feature extraction approach towards online handwritten basic Bangla characters called, distance based feature. Considering the number of speakers of Bangla languages, more research attention is required to develop the OHR system for Bangla script. In the current work, a new feature set is designed considering global as well as local information from CG-based circle generated over a character sample of the Bangla script.

2　Bangla Script

Bangla is the second most and fifth most popular language in India and in world respectively. This is also recognized as official language in Bangladesh. Bangla script is rooted from ancient Brahmi script. Writing pattern of Bangla has certain similarities with Dravidian languages, especially while considering the shape of vowels. But this script has more closeness with Aryan scripts, especially with Devanagari script. In Bangla script, every consonant serves as a syllable involving a built-in vowel written from left to right. Bangla alphabet set contains 39 consonants and 11 vowels. The concept of upper and lower case is absent in this script.

3　Database Preparation

A total of 200 specimens for individual Bangla character written by 100 distinct persons have been collected. The writers in this data collection drive have been selected from different sections of society considering their age groups, educational background, gender etc. 50 different character shapes belong to Bangla basic alphabet set, thereby form 10,000 total data samples. While collecting data, no constraint was imposed on the individuals without one thing; we requested that constituent strokes must be the part of basic stroke database for the script [12]. *Take note* device was used to store data. Representation of those data is described as a collection of pen points p_t, where t starts from 1 to M. p_t represents pen point having x coordinate x_t and y coordinate y_t with pen up or pen down information. Here, M describes total pen points for specimen character. During preprocessing stage firstly duplicate or repeated points have been removed from sample character as those were adding redundancy only. If p_i and p_k represents two successive pen positions, then ith point p_i is taken into account against kth point p_k if the Eq. (1) is satisfied.

$$x^2 + y^2 > m^2 \qquad (1)$$

where $x = x_i - x_k$ and $y = y_i - y_k$ and m yields the value 0 for all duplicate positions. After that characters have been normalized into 64-points without hampering the number of strokes and their structural patterns. The all these normalized patterns have been scaled to fit in a window of size 512×512.

4 Feature Extraction

4.1 CG-Based Circle

In this paper, we have mainly focused on combining both the global and local information of a character sample to represent it in the high-dimensional feature space. In both the cases, CG-based circle is drawn over a character sample and then different feature values are estimated from therein.

4.1.1 Global Information

In this feature extraction approach firstly, the CG of each character sample has been computed. Then the distance of farthest data pixel of the character sample from that CG is determined. This distance is then considered as radius of the circle enclosing the entire character sample. After forming the circle, we have computed the ratio of the radius of the circle and the distance between the CG of the character sample to the data point p_k, where $k = 1, 2, ..., 64$, as all character images are normalized into 64 points. In this way, we have estimated 64 ratios, and these values are used as feature for the recognition of said Bangla basic characters. Close observation revealed that these feature values have the power to explain the shape information of a character meticulously. Figure 1a, b show the formation of the CG-based circle and the computation of the one such ratio. Here G is the CG, K is the farthest data pixel from CG, R is the radius and p_1 is the arbitrary pixel point of that character. We have calculated the ratio of distances p_1G and R. The character is also rotated by 90° clockwise directions (see Fig. 1) and same procedure is repeated. As a result we have obtained 128 (i.e. 2×64) features for each character sample. The algorithm is described in Algorithm 1.

STEP I: BEGIN
STEP II: Calculate the CG of the character sample.
STEP III: Find the distance R from CG to farthest distant pixel from the CG.
STEP IV: Consider this distance R as radius and form the circle.
STEP V: For each pixel point, calculate the ratio of the distance from the pixel under consideration to CG and radius.
STEP VI: END

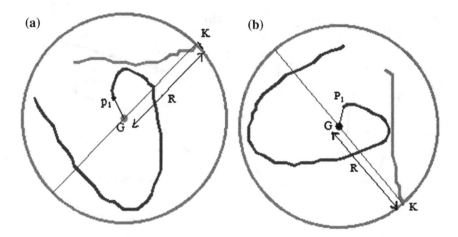

Fig. 1 CG-based circle generation enclosing the character ত for the estimating global feature at **a** 0° and **b** 90° rotations

Algorithm 1. Algorithm to compute global information from the CG-based circle.

4.1.2 Local Information

In this feature extraction technique, we have started with the CG-based circle estimated in the previous section. In Fig. 2a, b, we have described the situations for 0° and 90° clockwise rotational effect. Then we have divided the circular region into 4 sub-regions based on CG. We are basically interested about pixel distribution of the character sample in each sub-region. In Fig. 2a, b, yellow, red, green and blue colored portions describe the mass distribution of the character sample in each sub region. In each such region, we have calculated the ratio of the number of colored pixels to the background pixels. These ratios act as feature values in our current experiment. Algorithm 2 describes the steps used to calculate local feature.

STEP I: BEGIN
STEP II: Calculate CG of the character sample.
STEP III: Find the distance R from CG to the farthest data pixel point from the CG.
STEP IV: Consider this distance R as radius and form a circle.
STEP V: Circular region is divided into four sub-regions based on CG.
STEP VI: For each sub region, find the ratio of the number of data pixels to background pixels.
STEP VII: END

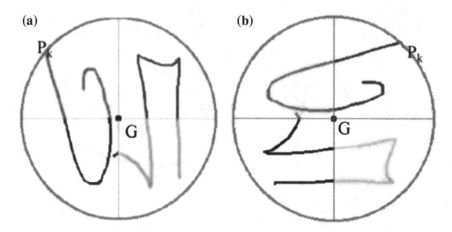

Fig. 2 Dissection of an image into four sub-regions to estimate local features at **a** 0° and **b** 90° (clockwise rotation)

Algorithm 2. Algorithm to compute the CG-based local feature.

From this procedure, we have computed 4 feature values considering 4 sub-regions which basically describe the mass distributions in the said sub-regions. The character sample is rotated by 90° in clockwise direction and Algorithm 2 is applied is similar fashion. Hence, we have designed 8 (i.e. 2 * 4) feature values from local information.

Therefore, combining both global and local information, each character sample is represented by a 136-element feature set (i.e. 128 + 8). The strength of global, local and combined feature sets while recognizing the character set of Bangla script are tested separately.

5 Result and Discussion

It is already mentioned that there are 50 different character shapes in Bangla script and 200 samples for each character are collected for the current experiment. Standard classifiers like SVM, Sequential Minimal Optimization (SMO), MLP, NaiveBayes, BayesNet, and Simple Logistic are used. As described in previous section that, a total of 136 (128 global and 8 local) features are calculated for the recognition purpose. We have then applied the approach called Principal Component Analysis (PCA) to estimate the most discriminating feature vector. In this way, 18, 14 and 5 features are selected when we consider the combined, global and local features respectively. Then we have applied 5-fold-cross validation scheme on total dataset. Success rates of different classifiers are enlisted in Table 1 for all these situations. Figure 3 graphically represents the behavior of different classifiers for these 3 cases. In Table 1, gray cells represent the highest recognition accuracy. The

Table 1 Accuracies of different classifiers in recognizing online handwritten Bangla characters

Classifiers	Success rate (%)		
	Global feature	Local feature	Combined feature
SVM	96.56	50.68	98.26
SMO	88.71	40.55	94.23
MLP	86.50	47.68	91.50
NaiveBayes	84.30	41.42	89.59
BayesNet	82.89	37.40	88.24
Simple logistic	80.32	39.02	90.00

recognition rate for SVM is 98.26 % (maximum) when the combined feature vector is applied. C-SVC type SVM has been selected from LibSVM library and kernel is accessing radial basic function; exp $(-gamma * |u-v|^2)$ having eps = 0.001 and gamma = 0.0. In Fig. 3 blue, brown and green lines represent the results obtained by said classifiers for global, local and combined features respectively.

It is evident from Fig. 3 that the green line is always above the blue and brown lines, which indicates the performances of all the above mentioned classifiers are better for the combined feature than considering only global/local features. This is because the combined feature vector rightly estimates the shape information of the specified character. Though the result is promising, still some error cases are observed. We have listed (see Table 2) few some cases where misclassification among each other is major. Apart from this list, we have also found some other

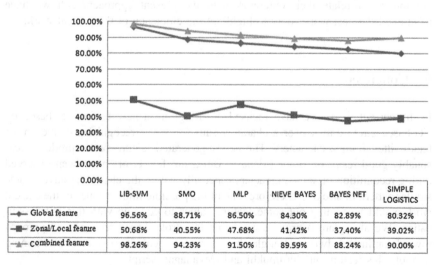

	LIB-SVM	SMO	MLP	NIEVE BAYES	BAYES NET	SIMPLE LOGISTICS
Global feature	96.56%	88.71%	86.50%	84.30%	82.89%	80.32%
Zonal/Local feature	50.68%	40.55%	47.68%	41.42%	37.40%	39.02%
combined feature	98.26%	94.23%	91.50%	89.59%	88.24%	90.00%

Fig. 3 Graphical representation of the success rates produced by different classifiers towards the recognition of online handwritten Bangla basic characters

Table 2 Few misclassified characters observed by the present technique (tested with SVM)

Original character along with sample count	Misclassified as
ম(200)	স(8)
চ(200)	ঢ(7)
ঋ(200)	ঝ(5)

misclassified characters but as they are very few in numbers, so we have not listed the same. After looking into samples minutely we can say that cursive nature of Bangla script and strong structural similarities between character samples are the reasons for the misclassification.

Authors in [19] fragmented the sample alphabets into few numbers of segments for the dataset used in current experiment and then calculated the distances among all pixel points from each other which were considered as feature values. Authors achieved maximum success rate of 98.20 % which is marginally less than the success rate achieved by the present technique (i.e. 98.26 %). Distance based feature tries to describe the shape structure whereas feature set, applied in the present work, not only keeps track the structural information of the characters but also considers the pixel distribution of the same in different zones of the character which makes the actual difference in achieving better recognition accuracy. The work mentioned in [12] extracted point based as well as structural feature towards recognition and achieved 92.9 % success rate. Sen et al. [18] combined online and offline information for online isolated Bangla character recognition with 83.92 % accuracy. As these two approaches were tested on separate datasets hence it is not possible to correlate their outcomes with the present approach, still we have mentioned the past works only to highlight the efficiency of the present work.

6 Conclusion

In the current work, we have focused on designing a new feature set based on CG-based circle drawn over a character sample for the recognition of the online handwritten Bangla characters. The newly generated feature vector produces reasonably good result on limited dataset. But again few errors have been observed during recognition of some characters especially where the characters have almost similar pattern of strokes. Therefore, one of the possible future scopes of the current work would be finding out some feature values to cope up with this misclassification. We also plan to provide a scheme for stroke based recognition approach towards online handwritten Bangla character/word recognition. We are also aiming to apply this feature on Gurumukhi and Devanagari script.

References

1. Kubatur, S., Sid-Ahmed, M., Ahmadi, M.: A neural network approach to online Devanagari handwritten character recognition. In: international conference on High Performance Computing and Simulation, 2012, DOI:10.1109/HPCSim.2012.6266913.
2. Swethalakshmi, H., Sekhar, C.C., Chakravarthy, V.S.: Spatiostructural features for recognition of online handwritten characters in Devanagari and Tamil scripts. In: Proceedings of International Conference on Artificial Neural Networks, 2007, vol. 2, pp. 230–239.
3. Swethalakshmi, H., Jayaraman, A., Chakravarthy, V.S., Sekhar, C.C: On-line handwritten character recognition for Devanagari and Telugu scripts using support vector machines. In: Proceedings of International Workshop on Frontiers in Handwriting Recognition, 2006, pp. 367–372.
4. joshi, N., Sita, G., Ramakrishnan, A.G., Deepu, V.: Machine recognition of online handwritten Devanagari characters. In: Proceedings of International Conference on Document Analysis and Recognition, 2005, pp. 1156–1160.
5. Tappert, C.C., Suen, C.Y., Wakahara, T.: The state of online handwriting recognition. In: IEEE Trans. on Pattern Analysis and Machine Intelligence, 1990, pp. 787–807.
6. Agarwal, S., Kumar, V.: Online Character Recognition. In: Proceedings of the Third International Conference on Information Technology and Applications, 2005, pp. 698–703.
7. Zafar, M.F., Mohammad, D., Anwar, M.M.: Recognition of Online Isolated Handwritten Characters by Back propagation Neural Nets Using Sub-Character Primitive Features. In: IEEE 9th International Conference INMIC, 2006, pp. 157–162.
8. Vescovo, G.D., Rizzi, A.: Online Handwriting Recognition by the Symbolic Histograms Approach. In: IEEE International Conference on Granular Computing, 2007, pp. 686–690.
9. Frinken, V., Bhattacharya, N., Pal, U.: Design of unsupervised Feature Extraction system for on-Line handwriting Recognition. In: 11th IAPR International Workshop on Document Analysis and Systems, 2014, pp. 355–359.
10. Bhattacharya, N., Pal, U., Kimura, F.: A System for Bangla Online Handwritten Text. In: International Conference on Document Analysis and Recognition, 2013, pp. 1335–1339.
11. Bhattacharya, N., Pal, U.: Strokes Segmentation and Recognition from Bangla Online Handwritten Text. In: Frontiers in Handwriting Recognition, 2012, pp. 740–745.
12. Roy, K., Bandhopadhyay, A., Mondal, R.: Stroke-Database Design for Online Handwriting Recognition in Bangla. In: International Journal of Modern Engineering Research, 2012, pp. 2534–2540.
13. Bhattacharya, U., Banerjee, R., Baral, S., Dey, R., Parui, S.K.: A semi automatic annotation scheme for Bangla online mixed cursive handwriting samples. In: International Conference on Frontiers in Handwriting Recognition, 2012, pp. 680–685.
14. Mohiuddin, Sk., Bhattacharya, U., Parui, S.K.: Unconstrained Bangla Online Handwriting Recognition based on MLP and SVM. In: Proceedings of joint workshop on Multilingual OCR and Analytics for noisy unstructured Text Data, DOI:10.1145/2034617. 2034635, 2011.
15. Fink, G.A., Vajda, S., Bhattacharya, U., Parui, S., Chaudhuri, B.B.: Online Bangla Word Recognition Using Sub-Stroke Level Features and Hidden Markov Models. In: International Conference on Frontiers in Handwriting Recognition, 2010, pp. 393–398.
16. Mondal, T., Bhattacharya, U., Parui, S.K., Das, K., Mandalapu, D.: On-line Handwriting Recognition of Indian Scripts - the first Benchmark. In: 12th International Conference on Frontiers in Handwriting Recognition, 2010, pp. 200–205.
17. Roy, K., Sharma, N., Pal, U.: Online Bangla Handwriting Recognition System. In: International Conference on Advances in Pattern Recognition, 2006, pp. 117–122.

18. Sen, S.P., Paul, S.S., Sarkar, R., Roy, K., Das, N.: Analysis of different classifiers for Online
    Bangla Character Recognition by Combining both Online and Offline Information. In: 2nd
    International Doctoral Symposium on applied computation and security Systems, 2015.
19. Sen, S.P., Sarkar, R., Roy, K.: A Simple and Effective Technique for Online Handwritten
    Bangla Character Recognition. In: 4th International Conference on Frontiers in Intelligent
    Computing: Theory and Application, 2015.

An Eigencharacter Technique for Offline-Tamil Handwritten Character Recognition

R.N. Ashlin Deepa and R. Rajeswara Rao

Abstract Accuracy in handwritten character recognition system is a challenge in the area of pattern recognition because of a variety of writing styles. Eigenface is a method that has been widely used in face recognition systems. This method is proposed in the field of handwritten character recognition, in this paper. Here, Eigencharacters are created from a 2-D training set of images and weight vectors are generated. These weight vectors are used as feature vectors for classification. The classification is performed using Euclidean Distance, k-NN and SVM classifiers. Experimental results proved that the proposed Eigencharacter method using Euclidean distance produced good classification accuracy.

Keywords Pattern recognition · Handwritten character recognition · Eigencharacter · Weightvector · Classification

1 Introduction

Handwritten character recognition system is a branch of pattern recognition with lot of challenges. Many character recognition systems have been developed to increase the accuracy in the classification of character images. Off-line handwritten character recognition involves the automatic conversion of scanned document image into a static representation of character image [1]. In on-line handwritten character recognition [2], the coordinates of pen-tip movements are stored as one-dimensional vector, and are converted into feature vector for recognition. Selection of feature extraction method is probably one of the most important

R.N. Ashlin Deepa (✉)
Gokaraju Rangaraju Institute of Engineering and Technology, Bachupally,
Kukatpally, Hyderabad 500090, Telangana, India
e-mail: deepa.ashlin@gmail.com

R. Rajeswara Rao
University College of Engineering, Vizianagaram, Andhra Pradesh, India
e-mail: raob4u@yahoo.com

© Springer Science+Business Media Singapore 2017   495
J.K. Mandal et al. (eds.), *Proceedings of the First International Conference on Intelligent Computing and Communication*, Advances in Intelligent Systems and Computing 458, DOI 10.1007/978-981-10-2035-3_51

characteristics for achieving high performance [3]. Feature extraction phase is followed by recognition phase.

Eigenface is the method which has been frequently used for the facial recognition system [4]. Apart from the traditional approaches, where object geometry is used for recognition, eigenvalues and eigenvectors for a collection of images were generated that determines the eigenspace [5]. The input image is projected to this eigenspace and is recognized. In many pattern recognition problems, Eigenface method is used. In [6], a set of two-dimensional training images are used to generate and vectorize eigenfaces, then the data set undergoes Singular Value Decomposition (SVD). Then the associated eigenvectors having higher values with the most vital principal components are retained, and thus constructs the "eigenfaces". Eigenvectors are also used in clustering, an important technique in machine analysis [7]. The eigenvectors of affinity/similarity matrix is derived from the dataset and is used for clustering the data using a special clustering algorithm. They also have analyzed the eigenspace characteristics that, in a data affinity matrix, not every eigenvector is relevant and informative for clustering. Thus the selection of eigenvectors is very critical because irrelevant/uninformative eigenvectors may lead to poor clustering.

In [8], the generalized eigenvectors of the class is used for feature extraction and these features are invariant to linear transformations of the input. Then the subset of the generalized eigenvectors are selected from each class. The eigenvalues far from 1 shows highly discriminative features. The top few eigenvectors are selected as top eigenspaces that are cheaper to compute than bottom eigenspaces. The feature extraction using eigenvectors needs to be analyzed as the true class-conditional second moment matrices have shown low effective rank [9]. General dimensionality reduction approach, PCA uses eigenvectors to select best features [10]. There are many applications where eigenvalues and eigenvectors are used as a significant part and are greatly used in dimensionality reduction.

2 Motivation

From the above discussions it is known that eigenvectors play an important role in object recognition and feature extraction. In facial recognition, eigenfaces are widely used, and produces very good accuracy. The method can show a better performance in other pattern recognition areas too. In this paper, we would like to examine how eigencharacters can be analyzed in aiding handwritten character recognition. In Eigencharacters method, few variables can represent largest variances in image characteristics.

In this paper, Sect. 3 describes the methodology, Sect. 4 describes Experimentation and Analysis of the result, Sect. 5 describes the Conclusion and References.

3 Methodology

In this research, a fast, simple and more accurate computational model of character recognition system is developed. When an input character image is represented as a vector or a pattern, the system must determine to which character class the input character image belongs to. The eigencharacters technique, utilizes PCA and relies on information theory. The principal components used in eigencharacters are eigenvectors of the covariance of the matrix of character images, while each character is a point in n space where n is the number of pixels in each character image.

The pertinent information of a character image needs to be extracted and kept as feature vector and is compared against an existing database of character images. Most variation in the images of different classes in the dataset is depicted by the largest eigenvalues and corresponding eigenvectors. These eigenvectors describe the features that characterize the variation between the character images, and each character image can be represented as a linear combination of all the eigenvectors.

Based on the above idea, the procedure used in this work is given in Algorithm1.

3.1 Algorithm1

Suppose that the database consists of M images. Each image is represented by i × j matrix. The algorithm follows the steps in [1].

Step1: Transform the M images into column vectors of length n, where $n = I \times j$ is shown in Fig. 1a. Now concatenate the columns of M_1 to transform Image in column vector Γ_1 as Fig. 1b.

Step2: After transforming all images of training dataset, we will get the following set I:

$$I = \{\Gamma_1, \Gamma_2, \Gamma_3, \dots \dots \Gamma_m\} \tag{1}$$

Step3: Generate mean image ψ from each Γ_m where $1 \leq m \leq M$. Ψ is computed as follows.

$$\psi = \frac{1}{M} \sum_{i=1}^{M} \Gamma i \tag{2}$$

Fig. 1 **a** Column vector of Images. **b** Concatenated column vector

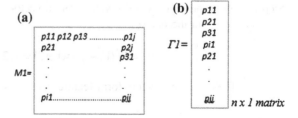

Step4:   Calculate the difference of each Γi, $i \in M$, $1 \leq i \leq M$, and Ψ the mean-subtracted value.

$$\Phi i = \Gamma_i - \psi \tag{3}$$

where $i = 1, 2, 3,\ldots\ldots M$ and a set of matrix is obtained which is represented as A.

$$A = [\Phi_1, \Phi_2, \ldots, \Phi_M]n \times M \text{ matrix} \tag{4}$$

Now this collection of data form the normalized training set which defines the character images i.e. n space for character images. In order to reduce the dimensionality while preserving the variance (information of interest), we need to construct a covariance matrix C for each Φi where $1 \leq i \leq M$.

$$C = \frac{1}{M} \sum_{i=1}^{M} \Phi i \Phi i', \text{ an } n \times n \text{ matrix} \tag{5}$$

By applying transformations, the covariance matrix obtained C will be equal to the following.

$$C = AA' \text{ where } C \text{ is } n \times n \text{ matrix}, A \text{ is } n \times M \text{ matrix and } A' \text{ is } M \times n \text{ matrix} \tag{6}$$

Step 5:   Find eigenvector matrix and corresponding eigenvalues from the covariance matrix C and sort the eigenvectors by highest eigenvalues. These eigenvectors give Eigencharacters E_i of the training set images.

Step 6:   From the eigencharacters generated in the above step, the projection matrix or weight matrix P is generated for each image in training set as given below.

$$P(\varphi i, Ej) = \Phi i \cdot Ej, 1 \leq i \leq M, 1 \leq j \leq M \tag{7}$$

Step 7:   If linear combinations of each of the Eigencharacters are used, it is possible to roughly reconstruct the database images. It is possible using the following formula. Let R_i be the reconstructed image.

$$Ri = \sum_{n=1}^{M} \Phi i En \text{ where } i = 1, 2, 3 \ldots \ldots M \tag{8}$$

Step 8:   From the Eigencharacters, the weight vector is generated for each training image in the dataset

$$Wi = Ri'(\Gamma i - \psi) \text{ where } i = 1, 2, 3, \ldots \ldots M \tag{9}$$

These weight vectors form feature vector FV = {$W1, W2\ldots\ldots WM$}.

Step 9: When an input character image is received, it is normalized and is subtracted from the mean image generated in Step 3. Then the weight of the subtracted input image with each of the Eigencharacters is calculated and this forms the feature vector of the input image F. Use Euclidean distance to find the minimum distance between F and the weight vector generated in Step 8. The index i of W_i, having minimum distance with the weight of input character image is used for determining the class of the input.

4 Experimentation

In this work, the dataset used contains handwritten Tamil characters with 200 character images. Out of these, 150 character images are considered for training and 50 images are considered for testing. The dataset is formed using an optical scanner which digitizes the images with an appropriate resolution of 300 to 1000 dots per inch. Each character image is preprocessed (noise removal, binarization, edge detection and thinning). The preprocessed images in the training dataset is given to the *Algorithm1* as discussed in Sect. 3.1. The input character images are also preprocessed and kept in the database as test dataset.

The implementation of the algorithm is done in MATLAB. For the purpose of displaying the output in this paper, only 5 classes are considered, each with 5 character images and thus the total number of images in the training set becomes 25. To reduce the error due to lighting conditions, each character image in the training set is normalized. To normalize the character image the standard deviation and mean of each image was normalized to mean standard deviation and mean. The value of mean standard deviation and mean used in the experiment was 80 and 100 respectively. Let mean value of pixels in the image is m and the standard deviation of the pixels in the image is std, then the normalized image is calculated by,

$$\text{The normalized image} = (\text{image} - \text{m}) * 80/std + 100$$

The mean image is calculated from all the training dataset images. Figure 2 gives the training set character images and mean image. Each normalized image in the training dataset is subtracted from mean character image and set A is generated with 25 subtracted images A = $\{\Phi_1, \Phi_2, ..., \Phi_{25}\}$. Table 1 gives the Eigen values and the corresponding Eigencharacters.

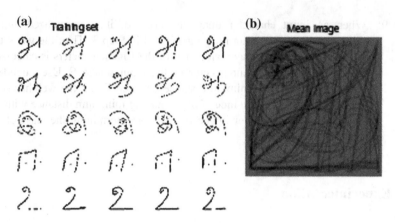

(a) Tahhgset **(b)** Mean Image

Fig. 2 **a** Training dataset of Tamil alphabets (Row 1: 'AH' Row 2: 'AAH' Row 3: 'E' Row 4: 'EE' Row 5: 'U'. **b** Mean Image

Using the Eigencharacters E_i created, the weight vector W is generated $W = \{W_1, W_2....W_{25}\}$. When an input character image is received, it is normalized and the subtracted image is generated by subtracting from the mean image of the training set. Now, weight vector Wt is calculated with respect to all the 25 Eigencharacters as shown in the Table 2. The input image and reconstructed image from the weight vector is shown in Fig. 3. Then the normalized Euclidean distance *Dist(i)* is calculated between W_t and W_i where i = 1, 2, 3,.... 25.

Finally for classification, index of the training character image, with *min(Dist)* is recovered that recognizes the class of the input character image. The input shown in Fig. 3 belongs to class 2. The training set is organized such that images from 1 to 5 belong to class 1, images from 6 to 10 belong to class 2, images from 11 to 15 belong to class 3, images from 16 to 20 belong to class 4 and images from 21 to 25 belong to class 5. The Table 2 shows the weight vector generated by $\Phi 8$ i.e. the subtracted image8 in the training set, with each of the 25 Eigencharacters. The Euclidean distance between the weight of the input image W_t and weight of each of the training image is calculated as Dist(i). It is shown that the min(Dist) = *41240.6865660360*, with Rank = 1, which belongs to the image8 in the training dataset. The image8 belongs to the class 2. Thus the input character image is classified to the respective class.

Figure 4 gives the pictorial representation of Table 2. The classification rate can be calculated using the following formula. The accuracy for 10 classes is shown in Table 3.

Table 1 Eigencharacters and eigenvalues of the training set character images

Eigencharacter	Eigenvalue	Eigencharacter	Eigenvalue	Eigencharacter	Eigenvalue
1	8.953212814759302E10	10	7.1058830929822008E8	19	5.8235199547438038E8
2	8.2205967867370484E9	11	6.9046870590556748E8	20	5.5839138640070668E8
3	8.2106222897745876E9	12	6.7414551077844428E8	21	5.5250704846077338E8
4	8.0236141874051117E8	13	6.6437072715663418E8	22	5.3026522607281488E8
5	8.3955156058822668E8	14	6.3856208264623218E8	23	5.1134797793458818E8
6	8.1423795835493368E8	15	6.2429941171446788E8	24	4.5773781616715 9E8
7	7.6165828078442168E8	16	5.7419040174466818E8	25	4.4614542017572 5E8
8	7.3496255623668048E8	17	5.9908034167784338E8		
9	7.3178141823426728E8	18	5.8448179678960698E8		

Table 2 Classification of input character image showing minimum distance as Rank

S. no	W8 = weight of 8th training image w.r.t. 25 eigencharacters $W8 = R8'(\Gamma 8 - \psi)$	Wt = Weight of input image w.r.t. 25 eigen characters	Normalized Euclidean distance Dist(i) = norm (Wi) − norm(Wt) where i = 1–25 character images in trainingset	Rank (minimum Euclidean distance)
1	33583.3149617131	−804.278306281754	44195.2881748533	25
2	2210.43477073423	863.740097060812	43783.1846247561	20
3	3216.43213839671	1845.04556342332	44022.9183009512	23
4	−7916.74392124914	−1872.89200478926	43499.2105940432	15
5	−5119.94459893754	−1154.45006355428	43796.3717405141	21
6	−1296.73574542512	−463.986841693624	41702.9897436268	2
7	42.0746184583586	964.392016654172	42908.9518274501	3
8	13657.1502168239	2822.62676901556	*41240.6865660360*	*1*
9	5701.03819335059	1260.59448027981	43154.9748310210	4
10	9340.37483843989	390.888574974326	43182.8693932593	7
11	−2259.98565138188	−1040.30940490628	43567.5151620835	16
12	−761.982805273231	−140.398221891933	43317.3089457852	13
13	−11137.1780585846	−429.374277778779	43309.7738765573	12
14	7020.09844000009	254.550424231885	43223.6166132643	10
15	−956.529857167489	−527.393485257671	43185.6418512871	8
16	2224.16586204357	−193.733459561314	43180.4347696610	6
17	−1557.03693941740	−250.259633659507	43467.5299547714	14
18	−5657.03429323119	52.7627361719196	43765.6360010378	19
19	3242.09525705100	−393.794943673880	43732.1225637856	18
20	845.962433989053	202.412629502224	43212.1646380453	9
21	3394.71174132402	−191.921195061879	43302.2782953869	11
22	1788.61060546031	−568.156187918784	44103.0426861925	24
23	3103.01344372280	458.806552731795	43175.0213654764	5
24	−2349.01756412409	−995.435155153510	43694.6788159210	17
25	−534.983176570472	−815.189386340303	43929.2970243021	22

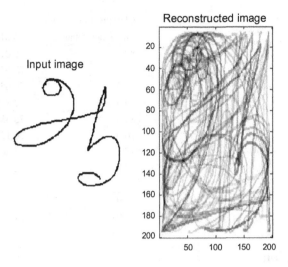

Input image

Reconstructed image

Fig. 3 Input image (Tamil alphabet 'AAH') and reconstructed image

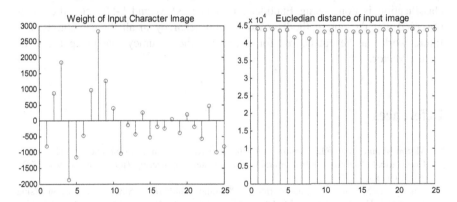

Fig. 4 Classification based on minimum Euclidean distance. (*Left*) The weight of the input vector w.r.t training set. (*Right*) The minimum Euclidean distance 4.1241e + 04 w.r.t 8th image in the training set which belong to class 2 (Tamil alphabet 'AAH')

$$Accuracy = \frac{total\ number\ of\ correctly\ classified\ objects}{total\ number\ of\ classified\ objects} * 100$$

The experimentation is done with 150 character images of 10 classes for training set and 50 for testing set.

Table 3 Classification accuracy for the 10 class labels. Featurevector: weight vector using eigencharacters

Class label	Classification accuracy %		
	Euclidean distance	SVM	k-NN
1	80	75.01	67.01
2	79	77.63	68.47
3	78.25	76.05	70.74
4	79.5	72.8	72.00
5	80.6	75.43	69.16
6	76.03	76.12	70.3
7	77.25	74.26	62.05
8	79.2	72.2	70.73
9	81	74.57	69.00
10	78.9	70.32	66.05
Average	78.97	74.44	68.55

5   Conclusion

In this study, we have used Eigencharacters to generate feature vector for hand-written character images. The features are extracted and used for classification. The classification is done using Euclidean Distance, SVM and k-NN classifier. It is observed that the our method provides good accuracy in all three methods. But future work can be considered for improving the accuracy rate using the same feature vector.

References

1. Meenu M and Jyothi RL, "Handwritten Character Recognition: A Comprehensive Review on Geometrical Analysis", IOSR Journal of Computer Engineering (IOSR -JCE), Vol 17(2), 2015, Page(s): 83–88.
2. Thierry A, Sanparith M and Patrick G, "Online Handwritten Shape Recognition Using Segmental Hidden Markov Models", IEEE Trans. On Pattern Analysis And Machine Intelligence, Vol 29(2), 2007, Page(s): 205–217.
3. Ashlin Deepa RN and Rajeswara Rao R, "Feature Extraction Techniques for Recognition of Malayalam Handwritten Characters: Review", International Journal of Advanced Trends in Computer Science and Engineering, Vol. 3(1), 2014, Page(s):481–485.
4. Huang, Kevin, "Principal Component Analysis in the Eigenface Technique for Facial Recognition". Senior Theses, Trinity College, Hartford, CT 2012.
5. Ovidiu Ghita and Paul F Whalen, "Object recognition using eigenvectors", *Proc. SPIE* 3208, Intelligent Robots and Computer Vision XVI: Algorithms, Techniques, Active Vision, and Materials Handling, 85 (September 26, 1997); doi:10.1117/12.290331.
6. Peter Kajenski, "Handwritten Character Recognition in Ancient Manuscripts", STANFORD CS 229 PROJECT, December 14, 2012.
7. Tao Xiang and Shaogang Gong, "Spectral clustering with eigenvector selection", Pattern Recognition 41, 2008, Page(s): 1012–1029.

8. Nikos Karampatziakis and Paul Mineiro, "Discriminative Features via Generalized Eigenvectors", Proceedings of the 31 st International Conference on Machine Learning, Beijing, China, W&CP Vol 32, 2014. JMLR.

9. Bunea, F. and Xiao, L. On the sample covariance matrix estimator of reduced effective rank population matrices, with applications to fPCA. ArXiv e-prints, December 2012.

10. Manal Abdullah, Majda Wazzan and Sahar Bo-saeed, "Optimizing Face Recognition Using PCA", International Journal of Artificial Intelligence & Applications (IJAIA), Vol. 3(2), March 2012.

Text and Non-text Separation in Handwritten Document Images Using Local Binary Pattern Operator

Showmik Bhowmik, Ram Sarkar and Mita Nasipuri

Abstract Development of an automated system for handwritten document analysis is being considered as an important research topic since last few decades. Digitized documents, either handwritten or printed, contain a mixture of text and non-text elements which need to be separated for designing a document layout analyzer or even an Optical Character Recognizer. In this paper, a technique is described to separate the text objects from the non-text objects present in a handwritten document image. For this purpose, a Rotation Invariant Local Binary Pattern (RILBP) based texture feature is used to represent the said components, at the feature space. Finally, the classification is carried out using an Artificial Neural Network based classifier called, Multi-layer Perceptron (MLP). The system provides an impressive result on a database comprising of 100 handwritten document images.

Keywords Text/Non-text separation · Rotation invariant local binary pattern · Handwritten document images

1 Introduction

A significant number of researches are going on to develop an efficient document analysis system to transit from a paper-based document environment to its electronic counterpart due to its several advantages. Such system provides a means for converting the information (either text or non-text) present in a document to a

Showmik Bhowmik (✉)
Department of Computer Science & Engineering, Dumkal Institute of Engineering and Technology, Dumkal, India
e-mail: showmik.cse@gmail.com

Ram Sarkar · Mita Nasipuri
Department of Computer Science & Engineering, Jadavpur University, Kolkata, India
e-mail: raamsarkar@gmail.com

Mita Nasipuri
e-mail: mitanasipuri@gmail.com

© Springer Science+Business Media Singapore 2017   507
J.K. Mandal et al. (eds.), *Proceedings of the First International Conference on Intelligent Computing and Communication*, Advances in Intelligent Systems and Computing 458, DOI 10.1007/978-981-10-2035-3_52

digital form, suitable for interactive graphics and text editing. The primary goal of a document image analysis system is to interpret the text and graphics (or non-text data), appear in a digitized document separately to carry out the subsequent processing. The separated text can be fed to an OCR system for later retrieval and a required set of processing can be carried out on the non-text data to make them usable in their digitized form.

Varied type of components present in document images makes the text/non-text separation a challenging task. Generally, two categories of documents are there namely, Structured documents and Unstructured documents. Among them, printed ones are considered as the structured documents. Text data and non-text data present in a printed document have uniform style, which makes the separation task a bit easier, compared to the handwritten documents. Handwritten documents are basically the unstructured documents, which contain components of varied shapes and styles. In the proposed work, a method is reported to isolate the text components from non-text ones, appeared in a handwritten document image using textured based feature.

2 Related Work

Segmentation of a page into text and non-text objects, is a necessary preprocessing stage before the elements are fed to a suitable OCR engine. Okun, et al. presented a method in [1] for document zone classification using connected component and run-length information. In [2], Roy et al. discussed a set of improvements (like overlapping region handling, color separation) to text/graphics discrimination task to make it suitable for Map. Bukhari et al. [3] offered a segmentation schema involving connected component classification based on shape and context information as a feature vector. In [4], Sarkar, et al. described a technique for text/non-text separation using Spiral Run Length Smearing Algorithm (SRLSA), which is basically the modified version of traditional Run Length Smoothing Algorithm (RLSA). In [5], Zirari, et al. reported a method for text and graphics classification using graph based modeling and structure analysis. In [6], Delaye et al. presented a well-explained review of contextual information modeling for the categorization of text and non-text strokes in online handwritten documents using conditional random field.

3 Motivation

Due to the non-uniformity regarding the shapes and sizes of the components, the discrimination of text and graphics components in a handwritten document image is more complex than a printed document image. Generally, the OCR system is designed to process the document images having only text data. Therefore, the separation of these two types of components needs to be done primarily. Beside this, sometimes the texts need to be separated and extracted from natural images or from the videos to identify street signs, license plates of the vehicles etc., to build up an image and video indexing and retrieval system [7]. These can also be useful in various applications like automation of navigation system, traffic control and other systems. In this work, a method is constructed to isolate the text components from the non-text components (like table, figures etc.) in handwritten document images.

4 Present Work

In this work, at first, the connected components are identified from the gray level handwritten document images and then the histogram of the RILBP is computed for each component to represent them in the feature space. The final classification is carried out using MLP to recognize the components as text or non-text (see Fig. 1).

4.1 Component Extraction

In this work, the LBP based texture feature is estimated on each component present in the handwritten documents, which are extracted from the gray level images. To extract the components from the gray level document images, the 8-way CCL algorithm [8, 9] is applied.

4.2 Feature Extraction

In the current work, feature extraction is carried out on each of the gray scale components. The text and non-text components reside in an image may have different texture and as the RILBP proved itself well in texture classification, it is used in this work to separate text and graphics components.

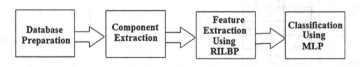

Fig. 1 Diagrammatic representation of the present work

Rotation Invariant Local Binary Pattern (RILBP)

Ojala et al. proposed Gray scale and RILBP in [10] for the purpose of rotation invariant texture sorting. The simplicity and efficiency of this operator magnetized the attention of researchers from various domains of image processing. It has achieved impressive results in texture classification [11], face recognition [12], and shape localization etc.

For a pixel in any gray scale image, LBP code of the same is computed by estimating the gray level difference with its neighborhoods.

Let, N represents the total number of circularly symmetric neighbors present on a circle of radius R, centered at the pixel under consideration. If the coordinate of the pixel under consideration is (0, 0), the coordinates of the pth neighbor pixel can be computed as

$$(R * cos(2\pi p/N), R * sin(2\pi p/N))$$
$$\text{where, } p = 0, 2, \ldots\ldots\ldots, N - 1. \tag{1}$$

Now, the gray scale invariant LBP code for the pixel under consideration can be computed as follows,

$$LBP_R^N = \sum_{p=0}^{N-1} s(GL_p - GL_c)2^p \tag{2}$$

$$s(x) = \begin{cases} 0, & x < 0 \\ 1, & x \geq 0 \end{cases} \tag{3}$$

Here, GL_c is taken as gray level intensity of the center pixel and GL_p is the gray level intensity of the pth neighbor (see Fig. 2).

The LBP_R^N operator yields 2^N distinct values, representing the 2^N different binary patterns which could be generated by the N pixels lie in the neighboring set. In case, the image is rotated, then the gray values GL_N would respectively move along the perimeter of the circle approximated around GL_0. If we rotate a specific binary pattern (except all 0's or all 1's), it will result in different LBP_R^N values. The outcome of rotation is detached by defining the LBP_R^N as follows,

$$LBP_{N,R}^{ri} = \min\{ROR(LBP_R^N, i) | i = 1, 2, \ldots, N - 1\} \tag{4}$$

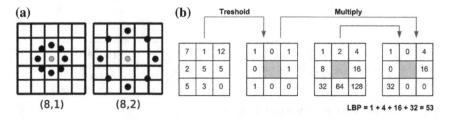

Fig. 2 **a** Illustration of the circularly symmetric neighbor set and **b** LBP code computation

Here, ROR(x, i) does a circular bit-wise i number of right shifts on any N-bit number x.

'U' value of any pattern represents the number of spatial transition (0 to 1 changeover) in the pattern. This can formally be defined as,

$$U\left(LBP_R^N\right) = \left|s(GL_{N-1} - GL_c) - s(GL_0 - GL_c)\right| + \sum_{p=1}^{N} \left|s(GL_p - GL_c) - s(GL_{p-1} - GL_c)\right|$$

(5)

The 'Uniform' LBP pattern represents the pattern with limited transitions $(U\left(LBP_R^N\right) \leq 2)$ in the circular binary presentation. The improvement of rotation invariance of the LBP is achieved by considering 'Uniform' patterns as follows,

$$LBP_{N,R}^{riu2} = \begin{cases} \sum_{p=0}^{N-1} s(GL_p - GL_c) & \text{if } U\left(LBP_R^N\right) \leq 2 \\ N+1 & \text{otherwise} \end{cases}$$

(6)

Superscript riu2 imitates the application of rotation invariant 'uniform' patterns having U value of at most 2.

After the computation of LBP code for all the pixels in the image, a histogram of the generated LBP code is built. Suppose the size of the image is RxW. The histogram can be estimated as,

$$Hist(l) = \sum_{i=1}^{R} \sum_{j=1}^{W} f(LBP_{N,R}^{riu2}(i,j), l)$$

(7)

where, l $\in$ [1, k].

Here, k represents the maximum LBP pattern value.

In the present work, during the computation of RILBP ($LBP_{N,R}^{riu2}$) for each pixel in the identified connected component, 8 circularly symmetric neighbors are considered (N = 8) with radius R = 1, which produce 10 feature values (10 components of the calculated histogram) for each connected component.

5 Experimental Results

The current technique is evaluated on a dataset of 100 handwritten document pages. These documents are collected from various sources such as classroom notes, homework and assignments etc. of the students of different Schools/Colleges/Universities of West Bengal, India. The contents of the documents are topics from various subjects. After collecting the data, we have scanned the same at 300 dpi and the scanned images are stored as 24 bit color images in BMP format. The database consists of 34,439 text and graphics components in total. For the evaluation task the entire database is grouped into 3 sets namely, Training, Validation and Test sets. 70 % of the total components are used to build the Training set and the remaining components are equally divided into two sets to

Table 1 Distribution of text components and non-text elements in Training, Validation and Test sets

Training set			Validation set			Test set		
Total	Text	Non-text	Total	Text	Non-text	Total	Text	Non-text
24107	23045	1062	5166	4935	231	5166	4931	236

build Validation and Test sets. The distribution of text components and as well as non-text components in the said sets is given in Table 1.

For the classification task a neural network based classifier, called MLP, with one hidden layer is applied in the present work. The classifier is trained with the learning rate of 0.3. First, the performance of the classifier is observed by varying the number of hidden layer neurons and it is found that the classifier performed well when the number of neurons in the hidden layer is set to 10. Then the experiment is carried out with 10 hidden layer neurons for different number of iterations to check the performance. Figure 3a shows the overall accuracy of the system by varying the number of neurons in the hidden layer whereas Fig. 3b displays the outcome of the model at various iterations. The recognition accuracies achieved by the classifier are 90.7, 91.3, 90.6 and 90.8 % in training, validation, testing and overall evolution process respectively (see Table 2).

Figure 4 shows the confusion matrices for training, validation and test sets respectively. In these confusion matrices '1' indicates 'class 1' means 'Text' and '2' indicates 'class 2' means "Non-text or Graphics". In Fig. 5 the corresponding Receiver Operating Characteristics (ROC) curves are given.

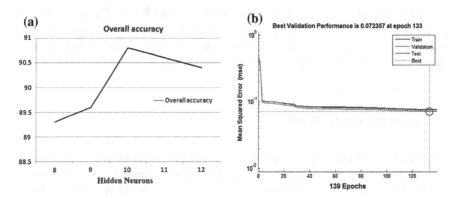

Fig. 3 Overall accuracy of the proposed text and non-text separation technique for different number of hidden neurons, **b** System performance for various epochs of MLP

Table 2 Accuracy achieved by the proposed system

Training set (%)	Validation set (%)	Test set (%)	Overall evolution (%)
90.7	91.3	90.6	90.8

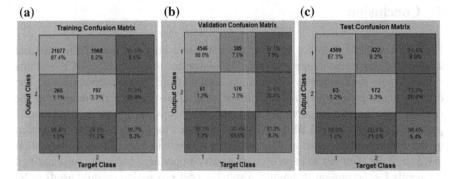

Fig. 4 Confusion matrices for **a** Training, **b** Validation and **c** Test sets respectively

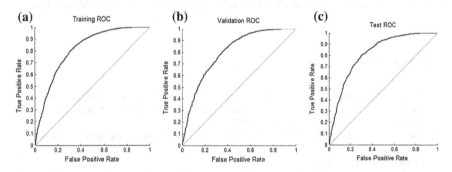

Fig. 5 **a** ROC curve for **a** Training, **b** Validation and **c** Test sets respectively

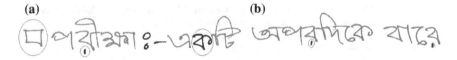

Fig. 6 Some wrongly classified components

6 Error Case Analysis

In most of the cases the proposed system performs reasonably well but there are also cases in which system fails to identify the component's true class. Figure 6 shows some error cases, where encircled components are wrongly classified. In this figure 'green' is used to show the text and 'red' is for non-text components.

7 Conclusion

Isolating text from graphics present in a document image is one of the essential steps of the document image analysis. The reason for this is that the OCR systems cannot handle the document if it contains both text and graphics components. The discrimination process is quite difficult in case of handwritten documents as the components do not have any specific shape/size/layout. In this work, a method is presented, which uses RILBP based texture features for text/non-text classification of the components present in handwritten document images. Though the system performs quite well but to cope up with small components which are misclassified frequently by the present technique, a suitable post-processing module needs to be added. Another frequently occurring issue, component overlapping, is not considered here, which is also required to be addressed in future. With minor modification, the proposed scheme could be suitably applied to printed documents too.

References

1. Okun, O., Dœrmann, D., & Pietikainen, M. (1999). *Page segmentation and zone classification: the state of the art* (No. LAMP-TR-036). OULU UNIV (FINLAND) DEPT OF ELECTRICAL ENGINEERING.
2. Roy, P. P., Lladó, J., & Pal, U. (2007, March). Text/graphics separation in color maps. In *Computing: Theory and Applications, 2007. ICCTA'07. International Conference on* (pp. 545–551). IEEE.
3. Bukhari, S. S., Azawi, A., Ali, M. I., Shafait, F., & Breuel, T. M. (2010, June). Document image segmentation using discriminative learning over connected components. In *Proceedings of the 9th IAPR International Workshop on Document Analysis Systems* (pp. 183–190). ACM.
4. Sarkar, R., Moulik, S., Das, N., Basu, S., Nasipuri, M., & Kundu, M. (2011, November). Suppression of non-text components in handwritten document images. In *Image Information Processing (ICIIP), 2011 International Conference on* (pp. 1–7). IEEE.
5. Zirari, F., Ennaji, A., Nicolas, S., & Mammass, D. (2013, May). A simple text/graphic separation method for document image segmentation. In *Computer Systems and Applications (AICCSA), 2013 ACS International Conference on* (pp. 1–4). IEEE.
6. Delaye, A., & Liu, C. L. (2014). Contextual text/non-text stroke classification in online handwritten notes with conditional random fields. *Pattern Recognition, 47*(3), 959–968.
7. Chen, D., Bourlard, H., & Thiran, J. P. (2001). Text identification in complex background using SVM. In *Computer Vision and Pattern Recognition, 2001. CVPR 2001. Proceedings of the 2001 IEEE Computer Society Conference on* (Vol. 2, pp. II-621). IEEE.
8. Yapa, R. D., & Harada, K. (2008). Connected component labeling algorithms for gray-scale images and evaluation of performance using digital mammograms. *International Journal of Computer Science and Network Security, 8*(6), 33–41.
9. Park, J. M., Looney, C. G., & Chen, H. C. (2000, March). Fast connected component labeling algorithm using a divide and conquer technique. In *Computers and Their Applications* (pp. 373–376).
10. Ojala, T., Pietikäinen, M., & Mäenpää, T. (2002). Multiresolution gray-scale and rotation invariant texture classification with local binary patterns. *Pattern Analysis and Machine Intelligence, IEEE Transactions on, 24*(7), 971–987.

11. Ojala, T., Mäenpää, T., Pietikainen, M., Viertola, J., Kyllönen, J., & Huovinen, S. (2002). Outex-new framework for empirical evaluation of texture analysis algorithms. In *Pattern Recognition, 2002. Proceedings. 16th International Conference on* (Vol. 1, pp. 701–706). IEEE.

12. Ahonen, T., Hadid, A., & Pietikainen, M. (2006). Face description with local binary patterns: Application to face recognition. *Pattern Analysis and Machine Intelligence, IEEE Transactions on, 28*(12), 2037–2041.

Page-to-Word Extraction from Unconstrained Handwritten Document Images

Pawan Kumar Singh, Sagnik Pal Chowdhury, Shubham Sinha, Sungmin Eum and Ram Sarkar

Abstract Extraction of words directly from handwritten document images is still a challenging problem in the development of a complete Optical Character Recognition (OCR) system. In this paper, a robust word extraction scheme is reported. Firstly, applying Harris corner point detection algorithm, key points are generated from the document images which are then clustered using well-known DBSCAN technique. Finally, the boundary of the text words present in the document images are estimated based on the convex hull drawn for each of the clustered key points. The proposed technique is tested on randomly selected 50 images from *CMATERdb*1database and the success rate is found to be 90.48 % which is equivalent to the state-of-the-art.

Keywords Word extraction · Handwritten documents · Harris corner point · DBSCAN clustering · Convex hull · *Cmaterdb*1.1 · *Bangla* script

P.K. Singh (✉) · Ram Sarkar
Department of Computer Science and Engineering, Jadavpur University, Kolkata, India
e-mail: pawansingh.ju@gmail.com

Ram Sarkar
e-mail: raamsarkar@gmail.com

S.P. Chowdhury · Shubham Sinha
Department of Computer Science and Technology, Indian Institute
of Engineering Science and Technology, Shibpur, Howrah 711103, India
e-mail: sagnik.pc@gmail.com

Shubham Sinha
e-mail: bitan1994@gmail.com

Sungmin Eum
University of Maryland Institute for Advanced Computer Studies,
College Park, USA
e-mail: cloud9min@gmail.com

© Springer Science+Business Media Singapore 2017
J.K. Mandal et al. (eds.), *Proceedings of the First International Conference on Intelligent Computing and Communication*, Advances in Intelligent Systems and Computing 458, DOI 10.1007/978-981-10-2035-3_53

517

1   Introduction

Page segmentation is an important preprocessing step for any Document Image Analyzer (DIA). It is a mechanism to segment documents into meaningful entities like paragraphs, text lines, words, and finally separating each character constituting the word images. Therefore, extraction of words is one of the key modules of the DIA before the character set is to be processed by the OCR system. Hence, isolating individual words accurately would make significant contribution towards developing suitable OCR system. Evidently, better correctness of segmentation result yields lesser errors in recognition. But, due to inconsistency of handwriting styles from varied writers belonging to different backgrounds, segmenting the handwritten word images accurately is always an exigent task [1].

In the OCR research fraternity, it is a common methodology to perform the text line extraction which immediately follows the word extraction procedure. Wide variation in inter-line gaps, or even multi-level skewness of the text-lines make the process more challenging. Furthermore, if the words are touched in the successive text lines, this problem becomes more unsolvable as the entire text lines would be segmented erroneously [2]. The said problems are very much common in the unconstrained handwritten text documents. Similarly, when words are extracted from each text lines, this error multiplies due to added approximations in the word extraction process. This elevates the chances of a particular text word not being extracted properly. But, if it is possible to develop some methodology to extract the word directly from the document images then the possibilities of error proportionally decreases. This concept inspires us to build up a direct page-to-word segmentation procedure for unconstrained handwritten documents.

From the literature survey performed in [2], it can be seen most of the works addressing on segmentation of handwritten text line, word or character have been focused on either *Roman*, *Chinese*, *Japanese* or *Arabic* scripts. Whereas a limited amount of work described in the state-of-the-art [3] has been done for unconstrained handwritten *Bangla* documents. Sarkar et al. [4] used Spiral Run Length Smearing Algorithm (SRLSA) for word segmentation. The algorithm first segmented the document page into text lines and then smears the neighboring data pixels of the connected components to get the word boundaries which merges the neighboring components and thus word components in a particular text line is extracted. Saha et al. [5] proposed a Hough transform based methodology for text line as well as word segmentation from digitized images. Ryu et al. [6] considered word segmentation of *English* and *Bangla* scripts as a labeling problem. Each gap in a line was labeled either as inter-word or intra-word. A normalized super pixel representation method was first presented that extracted a set of candidate gaps in each text line. The assignment problem was considered as a binary quadratic problem as a result of which pairwise relations as well as local properties could be considered. In [7], initially the contour of the words present in a given text line were detected and then a threshold was chosen based on Median White-Run Length (MWR) and Average White Run-Length (AWR) present in the given text line.

After that the word components were extracted from the text lines based on the contour and the previously chosen threshold value. At last, these words were represented in bounded boxes. A few more works [8–11] for word extraction was already done for other *Indic* scripts like *Oriya*, *Devanagari*, *Kannada* and *Tamil* handwritten documents. It can be observed from the literature study that for the above mentioned word extraction approaches, the text lines are first considered and then words were extracted from them. But, till date there is no work available which addresses the direct page-to-word segmentation for unconstrained *Bangla* script.

2 Proposed Work

The proposed work is a two-stage approach to extract the words directly from the handwritten document images written in *Bangla* script. In the first stage, the key points are estimated using Harris corner point detection algorithm. In the second stage, the estimated key points are grouped into clusters using DBSCAN clustering algorithm. Finally, a convex hull is formed considering the corner points belonging to a particular cluster which ultimately helps to determine the required word boundary. Both the stages are described in detail in the following subsections:

2.1 Harris Corner Point Detection

A point having two dominant and distinct directions of an edge in a local neighborhood of that point is defined as a corner. The Harris corner detector [12] is a widely used point detector in the domain of image processing. This is so because of its strong invariance to rotation, scale, illumination variation and noise present in the image. The Harris corner detector is ideally based on the neighboring auto-correlation function of an image. This auto-correlation function measures the local changes of an image with patches which move a small amount in various directions. It is based on the Moravec Operator [13] which is used to compare the error between shifted patches with the original image using the sum of squared differences. The key to Harris detector is the variation of intensity within a sliding window $w(x, y)$ (with displacement u and v in the x- and y-directions respectively) written as:

$$E(u, v) = \sum_{x, y} w(x, y)[I(x+u, y+v) - I(x, y)]^2 \tag{1}$$

where, $w(x, y)$ is the window weighting function at position (x, y), $I(x, y)$ is the intensity at (x, y) and $I(x+u, y+v)$ is the intensity at the moved window $(x+u, y+v)$. The aim is to maximize $E(u, v)$. Applying Taylor's expansion and some arithmetic operations, Eq. (1) can be written as follows:

$$E(u,v) \approx \sum_{x,y} w(x,y) \left(u^2 I_x^2 + 2uv I_x I_y + v^2 I_y^2 \right)$$

$$= (u,v) \left(\sum_{x,y} w(x,y) \begin{bmatrix} I_x^2 & I_x I_y \\ I_x I_y & I_y^2 \end{bmatrix} \right) \begin{pmatrix} u \\ v \end{pmatrix} = (u,v) M \begin{pmatrix} u \\ v \end{pmatrix} \tag{2}$$

Here, (I_x, I_y) is the gradient at position (x,y). Matrix M is known as structure tensor of a pixel which is actually an unbiased estimate of the covariance matrix of the gradients for the pixels within the window. It actually provides a characterization of information of all pixels within the window. Then, the eigen values λ_1 and λ_2 of the matrix M are calculated to determine if window corresponds to a corner or not.

- If $|R|$ is small, that is, $\lambda_1 \approx 0$ and $\lambda_2 \approx 0$ then the pixel (x,y) has no features of interest and the region is flat.
- If $R < 0$, that is, $\lambda_1 \approx 0$ and λ_2 has large positive value, then the region is an edge.
- If R is large, that is, λ_1 and λ_2 have large positive values, then the region is a corner.

In order to avoid computing the eigen values which are computationally expensive, in original Harris detector, the authors suggested using the following score R defined by:

$$R = \lambda_1 \lambda_2 - k(\lambda_1 + \lambda_2)^2 = \det(M) - k(trace(M))^2 \tag{3}$$

Figure 1 shows the execution of Harris corner points on a portion of the document page written in *Bangla* script. Harris corner detector is used due to its robustness to textured images and it detects only strong shifted corner positions such as those with an "L" shaped or "T" junction. This technique is also computationally inexpensive.

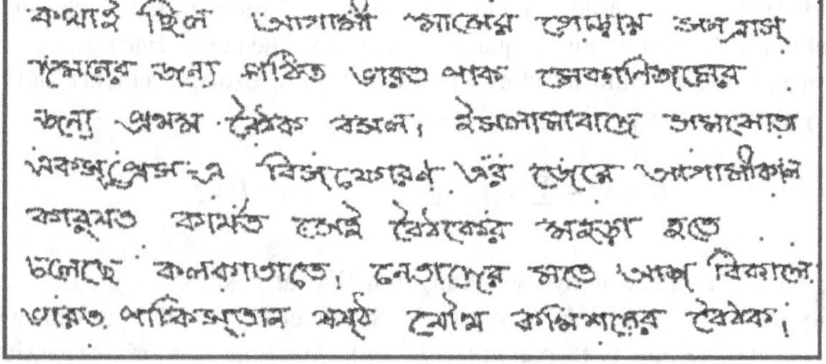

Fig. 1 Detection of Harris corner points for a portion of a sample document written in *Bangla* script

2.2 DBSCAN Clustering

The feature points generated from Harris corner point detection are passed on to the Density-Based Spatial Clustering of Applications with Noise (DBSCAN) algorithm [14], where the feature points are grouped based on minimum number of points to form a cluster and minimum distance of a feature from another feature. For a given set of points, it forms a cluster considering the points which are close to each other, marking as outlier, points that lie alone in low-density regions. DBSCAN requires two parameters: the distance up to which points are to be checked i.e., eps (ε) as well as the least number of points needed to form a dense region (*minPts*). Neighborhood of this point up to distance ε is retrieved, and if there are sufficient number of points, a starting cluster is formed. We have determined the values of ε and *minPts* on trial-and-error basis while executing the DBSCAN algorithm. The advantage of using DBSCAN algorithm is that it is not required to know the number of clusters in the dataset a priori, as opposed to the other popular clustering algorithms such as *k*-means. Moreover, it considers the noise present in the data and it is also robust to outliers.

2.3 Extraction of Word Images

After realizing all the points belonging to each cluster, a convex hull is drawn considering all the key points belonging to a particular cluster to approximate the word boundary. Convex hull [15] is defined by a set of points which forms the outermost boundary of the said point. The set of points used to draw convex hull is used to define the boundary and its centroid is calculated. Centering this centroid, the region of the convex hull is extrapolated by 20 % to accommodate the outer part(s) of the word on which the Harris corner points are not detected. Figure 2 shows the circumscribing convex hull drawn over the cluster of key points to extract the word images for a piece of a document image written in *Bangla* script.

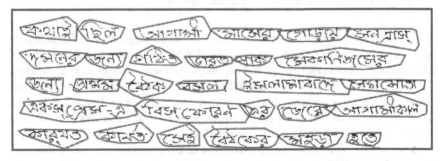

Fig. 2 A portion of handwritten *Bangla* document illustrating the convex hull drawn based on the set of points attained after DBSCAN clustering algorithm for extracting the word images

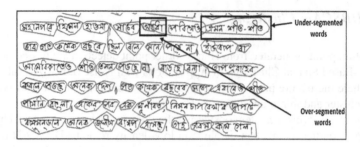

Fig. 3 Illustration of over-segmented and under-segmented word images

2.4 Removal of Over-Segmentation and Under-Segmentation Errors

A simple post-processing technique has been applied to cope up with the two major error cases: over-segmentation and under-segmentation of the words. The possible causes of these errors are either wrong detection of Harris corner points or improper clustering of the corner points around the word images. If a single word component is erroneously broken down into two or more parts (two or more words are recognized as a single word), then it is considered as over-segmentation (under-segmentation) error. To combine over-segmented components, the spatial distance between two neighbouring convex hulls are measured to verify their closeness and the two convex hulls are merged if they are close enough. For under-segmented errors, the vertical histogram of the word image is considered and the minima valley is calculated which considers the gap in between two or more consecutive words. This gap is taken into consideration to separate the word images. Figure 3 shows the scenarios of over-segmentation and under-segmentation. Though this technique is found to be useful, but there still exists a lot of over-segmentation and under-segmentation errors.

3   Experimental Results and Discussion

For the experimental evaluation of the present methodology, a set of 50 document pages are randomly selected from *CMATERdb*1 handwritten databases [16]. *CMATERdb*1.1.1 contains 100 document images written solely in *Bangla*, whereas *CMATERdb*1.2.1 contains document images written in *Bangla* mingled with some *English* words. For manual evaluation of the accuracy of word extraction technique, we have considered the errors produced due to under- and over-segmented word images. In both the cases, such extracted words are also treated as wrongly extracted words. The performance evaluation of the present technique is shown in Table 1. Figure 4 illustrates the present word extraction technique on a sample document page.

Table 1 Performance evaluation of the present word extraction technique on *CMATERdb*1 handwritten databases

Database	*CMATERdb*1
Number of document pages	50
Actual number of words present (T)	6761
Number of words extracted experimentally	7472
Number of over-segmented words (O)	361
Number of under-segmented words (U)	350
Success rate $[(T-(O+U)*100)/T]$	**90.48 %**

Fig. 4 Illustration of successful word extraction from a sample handwritten *Bangla* document page

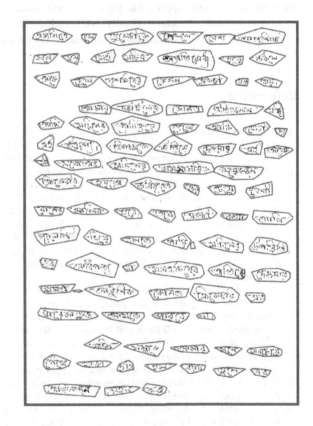

4 Conclusion

In the proposed work, we have designed a direct page-to-word segmentation algorithm for unconstrained *Bangla* handwritten document images. In general, researcher in this domain first identifies the text lines and the extracted text lines are fed to the word extraction module. Considering the typical complexities of the unconstrained handwritings, we have decided to avoid the text line extraction module which otherwise would have generated unwanted errors, cumulative in nature; in due course which would lessen the recognition accuracy of the word extraction module. Experimental result on the *CMATERdb*1 handwriting databases has shown that the proposed technique yields the state-of-the-art performance. Another advantage of this technique is that it could handle the skewness present the in the text line to some extent as we have extracted the words directly from the document images. Though the results are encouraging. The proposed technique still suffers from over- and under-segmentation issues even after we tried to rectify these errors to some extent. Appropriate post-processing modules need to be developed to cope up with these situations. Finally, we could say that with minor modifications, this technique could be successfully applied to other *Indic* scripts documents too.

References

1. R. Sarkar, S. Malakar, N. Das, S. Basu, M. Kundu, M. Nasipuri, "Word *Extraction and Character Segmentation* from *Text Lines of Unconstrained Handwritten Bangla Document Images*", In: Journal of Intelligent Systems, vol. 2, Issue 3, pp. 227–260, 2011.
2. R. G. Casey, E. Lecolinet, "*A survey of methods and strategies in character segmentation*", In: IEEE Transactions on Pattern Analysis and Machine Intelligence, vol.18, pp. 690–706,1996.
3. C. J. Kumar, G. Singh, R. Rani, R. Dhir, "*Handwritten Segmentation in Bangla script: A Review of Offline Techniques*", In: International Journal of Advanced Research in Computer Science and Software Engineering, vol. 3, Issue 1, pp. 135–140, 2013.
4. R. Sarkar, S. Moulik, N. Das, S. Basu, M. Nasipuri, D. K. Basu, "*Word extraction from unconstrained handwritten Bangla document images using Spiral Run Length Smearing Algorithm*", In: Proc. of 5th Indian International Conference on Artificial Intelligence (IICAI), pp. 71–90, 2011.
5. S. Saha, S. Basu, M. Nasipuri, D. K. Basu, "*A Hough Transform based Technique for Text Segmentation*", In: Journal of Computing, vol. 2, Issue 2, pp. 134–141, 2010.
6. J.Ryu, H. Koo, N.Ik. Cho, "*Word Segmentation Method for Handwritten Documents based on Structured Learning*" In: IEEE Signal Processing Letters, vol. 22, no. 8, pp. 1161–1165, 2015.
7. F. Kurniawan, A. R. Khan, D. Mohamad, "*Contour vs Non-Contour based Word Segmentation from Handwritten Text Lines: an experimental analysis*", In: International Journal of Digital Content Technology and its Applications, vol. 3(2) pp. 127–131, 2009.
8. N. Tripathy, U. Pal, "*Handwriting segmentation of unconstrained Oriya text*", In: Proc. of 9th IEEE International Workshop on Frontiers in Handwriting Recognition, pp. 306–311, 2004.
9. A. S. Ramteke, M. E. Rane, "*Offline Handwritten Devanagari Script Segmentation*", In: International Journal of Scientific & Engineering Research, vol. 1, Issue 4, pp. 142–145, 2012.
10. H. R. Mamatha, K. Srikantamurthy, "*Morphological Operations and Projection Profiles based Segmentation of Handwritten Kannada Document*", In: International Journal of Applied Information Systems (IJAIS), vol. 4, No. 5, pp. 13–19, 2012.
11. S. Pannirselvam, S. Ponmani, "*A Novel Hybrid Model for Tamil Handwritten Character Segmentation*", In: International Journal of Scientific & Engineering Research, vol. 5, Issue 11, pp. 271–275, 2014.
12. C. Harris, M. Stephens, "*A combined corner and edge detector*", In: Alvey vision Conference, vol. 15, 1988.

13. H. Moravec. *"Obstacle Avoidance and Navigation in the Real World by a Seeing Robot Rover"*, In: Tech Report CMU-RI-TR-3 Carnegie-Mellon University, Robotics Institute, 1980.
14. M. Ester, H. P. Kriegel, J. Sander, X. Xu, *"A density-based algorithm for discovering clusters in large spatial databases with noise"*, In: Proc. of 2nd International Conference on Knowledge Discovery and Data Mining, vol. 96, pp. 226–231, 1996.
15. R. C. Gonzalez, R. E. Woods, *"Digital Image Processing"*, vol. I. Prentice-Hall, India (1992).
16. R. Sarkar, N. Das, S. Basu, M. Kundu, M. Nasipuri, D. K. Basu, *"CMATERdb1: a database of unconstrained handwritten Bangla and Bangla–English mixed script document image"*, In: International Journal of Document Analysis and Recognition, vol. 15, pp. 71–83, 2012.

A Computer Vision Framework for Detecting Dominant Points on Contour of Image-Object Through Thick-Edge Polygonal Approximation

Sourav Saha, Saptarshi Roy, Prasenjit Dey, Soumya Pal, Tamal Chakraborty and Priya Ranjan Sinha Mahapatra

Abstract This paper presents a computer vision framework for detecting *dominant boundary-points* on an object's contour through polygonal approximation of the shape without loss of its significant visual-interpretation. The proposed framework attempts to approximate a polygonal representation of the contour with each polygonal-side having a meaningful thickness to handle noisy curvatures with irregular bumps. The vertices of the polygon are extracted through a novel recursive strategy. The merit of such a scheme depends on how closely it can represent the shape with minimal number of vertices as dominant points without losing its inherent visual characteristics. As per our observation, the proposed framework seems to perform reasonably well in approximating the shape of an object with a small number of dominant points on the contour.

Keywords Shape representation · Polygonal approximation · Regression line

This work has been supported by DST Purse Scheme, University of Kalyani

Sourav Saha (✉) · Saptarshi Roy · Prasenjit Dey · Soumya Pal · Tamal Chakraborty
Institute of Engineering & Management, Kolkata, India
e-mail: souravsaha1977@gmail.com

Saptarshi Roy
e-mail: saps.roy2010@gmail.com

Prasenjit Dey
e-mail: prsnjt002@gmail.com

Soumya Pal
e-mail: soumya1432010@gmail.com

Tamal Chakraborty
e-mail: tamalc@gmail.com

P.R.S. Mahapatra
Department of Computer Science & Engineering, University of Kalyani, Kalyani, India
e-mail: priya_cskly@yahoo.co.in

© Springer Science+Business Media Singapore 2017   527
J.K. Mandal et al. (eds.), *Proceedings of the First International Conference on Intelligent Computing and Communication*, Advances in Intelligent Systems and Computing 458, DOI 10.1007/978-981-10-2035-3_54

1 Introduction

Computer Vision researchers have always been interested to extract key descriptive characteristics from an object's contour for effective shape modeling. Detection of *dominant contour points* for extracting meaningful segments is a critical aspect of automated shape analysis as it helps in representing contours efficiently with reduced redundancy. Some of the notable efforts put in by the researchers so far towards extracting dominant points are discussed below.

1.1 Related Work

Dominant points are commonly identified as the points with local maximum curvatures on the contour. There are many algorithms developed to detect dominant points. Most of them can be classified into two categories: (1) Polygonal approximation approaches and (2) Corner detection approaches. We shall confine our discussion to polygonal approximation approach as the proposed method is primarily developed based on it. The most popular polygonal approximation algorithm proposed by Ramer [1] recursively splits the curve into two smaller pieces at the point with maximum deviation from the line segment joining two end points and a threshold value for the maximum deviation is preset to terminate the recursive process. For sequentially partitioning a curve, Wall and Danielsson [3] developed a method by finding the point at which the deviance of area per unit length surpasses a stipulated value. Kankanhalli [5] and Masood et al. [7] offered an iterative mechanism which starts with a set of probable dominant points and then deletes most redundant dominant points at each iterative step. The detection of an optimally approximated polygon using a dynamic programming strategy was demonstrated by Dunham [4]. Huang and Sun [6] proposed a genetic algorithm to find the polygonal approximation. Kolesnikov used *ISE*-bound estimation to construct shortest path as polygonal approximation in a feasibility graph comprising of contour points as its nodes [8]. Most of these polygonal approximation schemes perform fairly well on smooth contours but fail to treat bumpy contours satisfactorily.

In this paper, we propose a computer vision framework for detecting dominant boundary-points on an object's contour through thick-sided polygonal approximation. The proposed framework attempts to find a polygonal representation of the contour with each polygonal-side having a thickness. The thickness is sensibly chosen to handle contour curvatures with irregular bumps. The experimental results show that the method is effective in detecting dominant points. In Sect. 2, we will illustrate the basic principle of the proposed framework. The experimental results are discussed in Sect. 3. Some concluding remarks are made in Sect. 4.

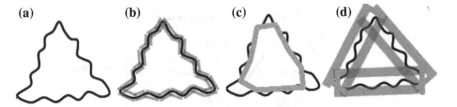

Fig. 1 **a** Object. **b** and **c** Traditional polygonal approximation, **d** Proposed scheme

2 Proposed Framework

This section discusses our proposed framework in detail. As mentioned earlier the proposed scheme for detecting dominant boundary points is based on polygonal approximation of closed curve. It is observed that most of the traditional polygonal approximation schemes fail to address the presence of bumpy irregularities along the contour effectively as they tend to produce misleading polygonal representation either with too many sides (Fig. 1b) or with too few sides (Fig. 1c). Our proposed framework attempts to handle such noisy curve using a thick-edge polygonal approximation (Fig. 1d). Figure 1 intuitively demonstrates the suitability of the proposed strategy. The proposed algorithm takes a sequence of boundary points as input which is obtained through tracing contour of an object by adopting Moore's strategy [2, 10].

2.1 Thick-Edge Polygonal Approximation of the Contour Using Polyline

In geometry, a closed *polyline* is a connected series of line segments which can be used to represent a polygon wherein its last segment connects the first segment. Over the years, various *polyline* approximation techniques have been widely explored for representing a curve in order to generalize its shape. However, it has not been studied extensively from computer-vision perspective. In this paper, we have focused on computer-vision perspective in developing a strategy to approximate the contour of an object with *polyline*. The following section illustrates the basic principle of our proposed algorithm.

Basic Principle of the Proposed Scheme with Illustration Here, we illustrate basic principle of the proposed algorithm with reference to Fig. 2. In this example, we have approximated the curve $\overset{\frown}{P_1 P_2}$ by a *polyline* with user specified thickness W_{TH}. The following procedural steps describe the working principle to perform the intended task.

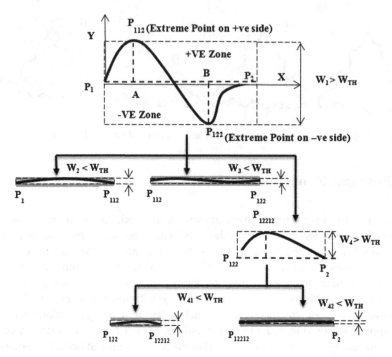

Fig. 2 Polyline fitting to a digital curve

Basic Principle of the Proposed Polyline Approximation

Step 1:   Determine the regression-line based on the points of the curve $\widehat{P_1P_2}$

Step 2:   Partition all the points of the curve $\widehat{P_1P_2}$ into two sets based on whether they are lying on positive or negative side of the regression line $\overrightarrow{P_1X}$.

Step 3:   Find extreme points on both sides. In our example, P_{112} is the extreme point on positive side and P_{122} is the extreme point on negative side of the regression line $\overrightarrow{P_1X}$.

Step 4:   Determine thickness of the rectangular strip bounding the curve $\widehat{P_1P_2}$ as $W_1 = P_{112}\overline{A} + P_{122}\overline{B}$ where A and B are projections of P_{112} and P_{122} respectively on the regression line $\overrightarrow{P_1X}$.

Step 5:   If $W_1 > W_{TH}$ then split the curve $\widehat{P_1P_2}$ into three possible curve segments namely $\widehat{P_1P_{112}}$, $\widehat{P_{112}P_{122}}$, $\widehat{P_{122}P_2}$ and repeat from step 1 for each segment.

Step 6:   Else if $W_1 \leqslant W_{TH}$ then treat two end points of the input curve as consecutive vertices of a *polyline* i.e. as dominant points.

2.2 Proposed Algorithm

The recursive algorithm developed for approximating a closed digital curve with thick-polyline is formally presented in algorithmic format as *doThickPolyLineApprox*. At each recursive step, it determines the regression-line based on the points of the input curve segment. Points on both sides of the regression-line are considered separately in order to find out two extreme points as discussed in previous section. In the algorithm presented here, *extremePosSidePt* and *extremeNegSidePt* refer to the extreme points on positive side and negative side of the regression line respectively. The thickness of the enclosing rectangle of the input curve can be computed as the sum of individual distances of the extreme points from the regression-line. As long as the thickness of the bounding rectangle of the input curve remains larger than a user specified polyline-thickness, it continues to split the input curve around extreme points to generate multiple segments and makes recursive call individually with each of these decomposed segments as input parameter for *polyline* approximation. Otherwise two end-points of the input curve are added to the *PolyLineVertexList* without making any further recursive call as they can represent consecutive vertices of *polyline* with desired thickness. The average computational complexity of the algorithm can be worked out as $O(N \ln N)$ based on the recurrence relation:
$T(N) = N + aT\left(\frac{N}{a}\right)$ where $0 < a < 3$ and $T(1) = O(1)$.

3 Experimental Results and Analysis

Evaluation of performance is a crucial issue for the proposed framework, mainly due to the subjectivity of human vision based judgment. The merit of such a scheme depends on how closely a polygon with a small vertex-set maximally captures visually significant area of the object. The most widely used criteria, for estimating effectiveness of the scheme examines—(a) *Area Difference Rate* that considers area-difference of the approximating polygon with original image object and (b) *Contour Compression Rate* that considers difference between the number of polygonal vertices and the number of object-boundary points. We have combined both of these frequently used measures in order to evaluate the performance of our framework quantitatively. The areas of the polygon and also of the object are computed in terms of number of pixels lying inside the closed contour of their representations. The mathemetically formulated expression used to measure the merit of our scheme is also presented here. It is understandably framed to ensure that (a) larger contour-compression-rate coupled with smaller area difference rate must produce higher merit-score (b) smaller contour-compression-rate coupled with larger area difference rate must produce lower merit-score (c) contour-compression-rate must be larger than area difference rate in order to produce positive merit-score. The performance of the proposed framework has been evaluated by conducting experiments on the MPEG-7 test dataset [9]. The obtained result presented in Table 1 shows that the pro-

---

Algorithm 1: DOTHICKPOLYLINEAPPROX

Input: Global *PointList*: A list of points representing the curve;
startPt and *endPt* referring two end-points of the curve-segment currently
considered for approximation;
polyLineThickness as thickness of the *polyline*
Output: Global *PolyLineVertexList*: List of dominant points

1 Determine the *regressionLine* based on the list of points representing input curve
2 *extremePosSidePt* ← *findExtremePositiveSidePoint(startPt, endPt, regressionLine)*
3 *extremeNegSidePt* ← *findExtremeNegativeSidePoint(startPt, endPt, regressionLine)*
4 *d1* ← *findDistance(extremePosSidePt, regressionLine)*
5 *d2* ← *findDistance(extremeNegSidePt, regressionLine)*
   /* compute thickness of the rectangular strip enclosing the
      input curve   */
6 *thickness* ← *d1 + d2*
7 **if** *polyLineThickness* ⩾ *thickness* **then**
     | /* add two end-points of the input curve to List of
     | dominant points: PolyLineVertexList   */
8 | *addToPolyLineVertexList(startPt, endPt)*
     /* else make recursive call   */
9 **else if** *only extremePosSidePt exists* **then**
10 | *doThickPolyLineApprox(startPt, extremePosSidePt)*
11 | *doThickPolyLineApprox(nextToExtremePosSidePt, endPt)*
12 **else if** *only extremeNegSidePt exists* **then**
13 | *doThickPolyLineApprox(startPt, extremeNegSidePt)*
14 | *doThickPolyLineApprox(nextToExtremeNegSidePt, endPt)*
15 **else if** *both extremePosSidePt AND extremeNegSidePt exists* **then**
16 | *pt1* ← *extremePt appearing first in PointList representing input curve*
17 | *pt2* ← *extremePt appearing second in PointList representing input curve*
18 | *doThickPolyLineApprox(startPt, pt1)*
19 | *doThickPolyLineApprox(nextToPt1, pt2)*
20 | *doThickPolyLineApprox(pt2, endPt)*
21 **return** *PolyLineVertexList*

---

posed scheme performs reasonably well as compared to popular Ramer's method [1]. However as per our observation, the performance of our scheme is also sensitive to the presumed edge-thickness value of the approximating polygon which needs to be chosen sensibly to obtain better result. On selection of larger edge-thickness value, the proposed scheme produces an approximating polygon having small number of vertices whereas choice of smaller edge-thickness value leads to generation of a polygon having large number of vertices. Extensive experimentation is therefore required in order to formulate the preferred edge-thickness value instead of choosing it intuitively for obtaining optimal performance.

$$\Delta area = ABS(PixelCountInsideObject - PixelCountInsidePolygon)$$

$$AreaDifferenceRate = \frac{\Delta area}{PixelCountInsideObject} \tag{1}$$

$$\Delta contour = ABS(NumberOfContourPoints - NumberOfPolygonVertices)$$

Table 1 Result: thick-edge polygonal approximation (edge-thickness = 8 pixel)

Data	Polygon	Polygon	Merit	Merit
Sample	(Ramer)	(Proposed)	(Ramer)	(Proposed)
Bat			89	98
Bird			87	97
Butterfly			88	95

(continued)

Table 1 (continued)

Data Sample	Polygon (Ramer)	Polygon (Proposed)	Merit (Ramer)	Merit (Proposed)
Frog			85	96
Personal-car			89	98

$$ContourCompressionRate = \frac{\Delta contour}{NumberOfContourPoints} \qquad (2)$$

$$Merit = (ContourCompressionRate - AreaDifferenceRate) \times 100 \qquad (3)$$

4 Conclusion

This paper presents a relatively new idea for dominant point detection on an object's contour. Extensive investigations and analysis of various stages are vital for an accurate assessment of the limitations, and requirements of the employed framework. As per our observation, the proposed framework seems to perform reasonably well in approximating the shape of an object by producing a polygon with its sides having sensibly chosen thickness and such an idea leads to provisioning a suitable strategy to effectively handle noisy bumps along the contour.

References

1. Ramer, U. : An iterative procedure for the polygonal approximation of plane curves, Com-puter Graphics and Image Processing 1 (1972) 244–256.
2. Pavlidis, T.: Algorithms for Graphics and Image Processing, Computer Science Press, Rockville, MD, (1982) 143–186.
3. Wall, K., Danielsson, P.E. : A fast sequential method for polygonal approximation of digitized curves. Computer Vision, Graphics, and Image Processing 28 (1984) 220–227.
4. Dunham, J.G. : Optimum uniform piecewise linear approximation of planar curves. IEEE Transactions of Pattern Analysis and Machine Intelligence 8 (1986) 67–75.
5. Kankanhalli, M.S. : An adaptive dominant point detection algorithm for digital curves. Pattern Recognition Letters 14 (1993) 385–390.
6. Huang, S.C., Sun, Y.N.: Polygonal approximation using genetic algorithms. Pattern Recognition 32 (1999) 1409–1420.
7. Masood, A., Haq, A.S.: A Novel Approach to polygonal Approximation of Digital Curves. Journal of Visual Communication & Image Representation (2007) 264–274.
8. Kolesnikov, A.: ISE-bounded polygonal approximation of digital curves, Pattern Recognition Letters 33 (2012) 1329–1337.
9. MPEG-7 Shape Matching, http://www.dabi.temple.edu/~shape/MPEG7/dataset.html
10. Saha, S., Basak, J., Mahapatra, P. R. S. : A Hierarchical Convex Polygonal Decomposition Framework for Automated Shape Retrieval. Proceedings of Second International Conference INDIA 2015, Volume 339, Series of Advances in Intelligent Systems and Computing, Springer India (2015) 783–792.

Kuan Modified Anisotropic Diffusion Approach for Speckle Filtering

Abhishek Tripathi, Vikrant Bhateja and Aditi Sharma

Abstract Synthetic Aperture Radar (SAR) is a coherent imagery tool used for extracting information in astronomy and meteorology. But these images are generally corrupted with a granular noise called speckle, making it difficult for extracting information. In this paper, the Anisotropic Diffusion (AD) filter is modified by incorporating the use of the Kuan filter for speckle removal. In the proposed work, the image is segmented into two regions based on the value of conduction function; the homogeneous region is processed using the Kuan filter and the other non-homogeneous region is processed with anisotropic diffusion. This modified AD filter provides better detection of weak edges and effective reconstruction of structural content with enhanced image restoration features. Further, based on the simulation results and via image quality metrics analysis; the proposed work is claimed better.

Keywords Speckle suppression · Image restoration · Adaptive and diffusion algorithms

1 Introduction

SAR is an acquisition device that actively results in radiation and also captures the backscattered signals from a small portion of the resolution cell. The received signal is complex, as output from the in-phase and quadrature channels can be viewed as the incoherent sum of several backscattered waves [1, 2]. SAR data is utilized in

Abhishek Tripathi (✉) · Vikrant Bhateja · Aditi Sharma
Department of Electronic and Communication Engineering, Shri Ramswaroop
Memorial Group of Professional College (SRMGPC), Lucknow 227105, U.P, India
e-mail: abhishek1.srmcem@gmail.com

Vikrant Bhateja
e-mail: bhateja.vikrant@gmail.com

Aditi Sharma
e-mail: aditiii065@gmail.com

© Springer Science+Business Media Singapore 2017
J.K. Mandal et al. (eds.), *Proceedings of the First International Conference on Intelligent Computing and Communication*, Advances in Intelligent Systems and Computing 458, DOI 10.1007/978-981-10-2035-3_55

various environment monitoring applications like mining, oceanography, sea and ice monitoring, oil pollution monitoring, etc. [3–5]. However, these images are generally contaminated with a highly unordered and chaotic noise called speckle (granular noise). The object when illuminated by coherent source and its surface structure is approximately of the order of the wavelength of the incident radiation, the reflected wave consists of contributions from various independent scattering areas [6]. In SAR systems, the difference between the true mean values of the image pixels and the measured values which degrades the visual quality of images is generally referred to as speckle noise [7–11]. The speckle in SAR images hinders the ability to locate and extract information and other fine details, thereby area discrimination of both textural and radiometric aspects become less efficient [12]. Therefore, reduction of speckle and reconstruction of image features is the fundamental step in pre-processing SAR data. Restoration algorithms for speckled SAR images have been broadly classified based on Minimum Mean Squared Error (MMSE), Linear Minimum Mean Squared Error (LMMSE), Bayesian, Non-Bayesian approaches, etc. [13, 14]. The MMSE and LMMSE criterion minimizes the mean squared error that corresponds to shrinkage of the noisy coefficients by a factor inversely related to its Signal to Noise Ratio (SNR). However, in the lowest levels of wavelet decomposition, it does not respect the Gaussian assumption and hence is inferior to Bayesian despeckling approaches. These Bayesian approaches (Lee Filter, Frost Filter and Kuan filter [14]) are based on local statistics where the centre pixel intensity is computed inside the moving window using the average intensity values. Further, the aforesaid local statistics filters [14] were extended by A. Lopes et al. [12] which filtered the image by segmenting it into three regions-homogeneous regions, non-homogeneous regions and isolated points. Bayesian approaches provide satisfactory speckle reduction but smoothens the image which leads to edge distortion and blurring. The Non-Bayesian approaches employing AD filtering [13] use the diffusion method that removes speckle from an image by modifying it via the Partial Differential Equations (PDE) [15–20]. It involves the use of a continuous and monotonically decreasing conduction function that classifies the image into heterogeneous and homogeneous regions while the use of gradient factor helps in discrimination of true and false edges. This helps in edge preservation and restoration of image content that are important for interpretation or data extraction. However, the conventional AD is computationally complex and does not prove satisfactory for images corrupted with speckle, which is a multiplicative noise [21, 22]. It is found that the conventional AD performs well for images which are corrupted by additive noise while image that contains speckle, the AD enhances it rather than reducing the corruption [3]. Recently, various improved works have been reported in Bayesian approach by Tripathi et al. [16], Misra et al. [17], Dong et al. [23] and Ching-Ta et al. [24] but these filters display good results for other noise models like impulse or salt pepper noise and hence blur the speckle affected images. Also, amended works in Non-Bayesian approach are described in Bhateja et al. [18–22] and Ashutosh et al. [25] give improved filtering but these results in higher complexity and large window sizes leading to blurring of images. Therefore, the proposed work combines the two approaches Bayesian and

Non-Bayesian approaches to attain positive traits of both approaches under a single despeckling algorithm. The following paper is structured as follows: Sect. 2 describes the proposed methodology, Sect. 3 presents the simulation results and discussions, Sect. 4 concludes the work.

2 Proposed Methodology

2.1 Background

The diffusion function of the Perona and Malik AD filter (PMAD) uses the gradient operator for the purpose of preserving edges. PMAD uses the fundamental concept to modify the conductivity in the nonlinear diffusion equation:

$$\frac{\partial}{\partial t}I(x,y,t) = \nabla.(c(x,y,t)\nabla I) \tag{1}$$

where $I(x, y, t)$ is an image, t is the time scale and $c(x, y, t)$ is the monotonically decreasing function of the image gradient which is defined as the conduction function given in Eq. (2).

$$c(x,y,t) = \frac{1}{1 + \left(\frac{||\nabla I||}{k}\right)^2} \tag{2}$$

where k is referred to as diffusion constant. If $c(x, y, t)$ is equal to 1, linear isotropic diffusion is achieved while when $c(x, y, t)$ tends to 0, non linear anisotropic diffusion takes place [2]. Further the transformed pixel is estimated within a spatial sub window of size 3×3 using the local gradient, which is calculated using nearest-neighbor differences as:

$$\begin{aligned}
\nabla_N I_{i,j} &= I_{i-1,j} - I_{i,j} \\
\nabla_S I_{i,j} &= I_{i+1,j} - I_{i,j} \\
\nabla_E I_{i,j} &= I_{i,j+1} - I_{i,j} \\
\nabla_W I_{i,j} &= I_{i,j-1} - I_{i,j}
\end{aligned} \tag{3}$$

Subscripts N, S, E, and W (North, South, East, and West) describe the direction of computation of the local gradients. The equation for the transformed pixel is given by:

$$I_t(i,j) = I(i,j) + \lambda(\nabla_E I.c_E + \nabla_W I.c_W + \nabla_N I.c_N + \nabla_S I.c_S) \tag{4}$$

The gradient operator of the PMAD diffusion constant helps in the detection of sharp edges but it poses constraints in case of non-sharp edges. Hence, it leads to unnecessary blurring and distorts the high frequency structure of the image. Also, the conventional AD performs well for images which are corrupted by additive noise while enhance the corruption of speckle affected images [3]. Thus, the proposed method enables efficient speckle reduction and edge preservation for better data extraction. The aforesaid shortcomings of conventional AD will be addressed in the Kuan Modified AD approach in Sect. 2.2.

2.2 Kuan Modified AD Filter

In the proposed work, the hybridization of Bayesian and Non-Bayesian approaches is achieved so as to acquire the positive traits of both the approaches. The operational procedure of the proposed approach begins by initializing a spatial window (w) of size 3 × 3 over the noisy image. The process continues with the determination of statistical parameters within the window which include: standard deviation (Sd), mean (m), AD parameters like gradient and conduction function using Eqs. (2) and (3). The standard deviation characterizes edge and textural information while the mean depicts the gross structure of an image. The calculated value of the conduction function using Eq. (2) helps in segmenting the image into the homogeneous and the non-homogeneous regions. If the value of the conduction function $c(x, y, t)$ tends to 1, it is regarded as the homogeneous region and the image is processed using Kuan filter given by Eq. (5)

$$It(i,j) = cp(i,j)wf + m(1 - wf) \tag{5}$$

The Kuan filter estimates the center pixel (cp) in the filter window by using the MMSE criteria. It approximates the Lee filter for calculating the signal estimate from the local standard deviation (Sd) and mean (m) with better accuracy.

The various non-stationary image statistical parameters needed for the Kuan algorithm are: Effective No. of Looks (ENL) which depicts the amount of speckle reduction in the homogeneous area and Noise Variation of Coefficient (C_u) computed globally for the entire image. Other local parameters being Image Variation of Coefficient (C_i) which defines the speckle content, Threshold (C_{max}) factor that discriminates the true and false edges and the Weighting Function (w_f) required for calculating the transformed pixel. The aforesaid parameters are mathematically defined under Table 1 [26]. While, if the value of the conduction function $c(x, y, t)$ is 0, the region is referred to as non-homogeneous and the image is processed normally using the PMAD filter with step size 0.05. Here the transformed pixel is calculated using Eqs. (3) and (4). This filter depends on coefficient of variation and conduction function to reduce speckle and also preserve radiometric information. Along with speckle suppression and preservation of detailed areas, it also lowers the unnecessary computational complexity. Further, the results and discussion section of this paper proclaims the proposed work better.

Table 1 Parameters required for Kuan algorithm

Effective no. of looks	$L = \frac{\mu^2}{\sigma^2}$
Noise variation of coefficient	$C_u = \sqrt{\frac{1}{L}}$
Image variation of coefficient	$C_i = \frac{Sd}{m}$
Threshold	$C_{max} = \sqrt{1 + \frac{2}{L}}$
Weighting function	$w_f = \dfrac{1 - \left(\frac{C_u}{C_i}\right)^2}{1 + C_u^2}$

3 Results and Discussions

The proposed work is validated using various test images taken from SAR systems of NASA [27]. The experimental procedure initiates with the normalization of the image in context to bring down the pixel intensity between the ranges 0–1. The work has been simulated on different variance levels ranging from 0–0.05. At first the noisy SAR image (I) is given as input and its mean (μ) and standard deviation (σ) are computed for the further processing. The noisy image is then processed by a filtering kernel (w) of size 3 × 3 generally. Next, various parameters like standard deviation (σ_w) mean (m), gradients (G_i) and conduction coefficient (c) are computed for the respective kernel or window. Now if the value of the conduction function tends to 1, the image is processed using the Kuan filter while if the value tends to 0, the gradients and conduction function in four directions have been calculated using the conventional AD algorithm. Further, the transformed pixel is calculated where the value of λ is experimentally chosen and the step size is taken as 0.05. This process repeats till the spatial window reaches the last pixel and lastly the denoised image is displayed. The results have been simulated on two test images—1 and 2 for various speckle variances. The obtained filtered results for test image-1 are shown in Fig. 1 with low, medium and high speckle variance. It can be observed from Fig. 1 that the image structural features are well preserved along fine details without any penalty imposed by the edges. In addition, the suppression of speckle is achieved at single iteration of the proposed filter and even at higher speckle variances, *PSNR* and *SSIM* are well maintained. The performance of the proposed approach on test image 1 has been benchmarked in Fig. 1 and is compared with other conventional filters [14] in Tables 2 and 3. These tables depict the efficiency of the proposed approach in terms of above mentioned quality metrics (*PSNR* & *SSIM*). The overall drop in the *PSNR* value is approximately 3 dB with variation of noise from 0.001–0.05 and the high values of *SSIM* affirms the confinement of structural content throughout the image. The second set of images (test image-2) is carried out for the visual discrimination of the proposed work with other filters. Image in Fig. 2c seems to be blurred while images of Fig. 2b, d, e are left out with some residual speckle. The proposed algorithm thereby overcomes the above

(a)   (b)   (c)

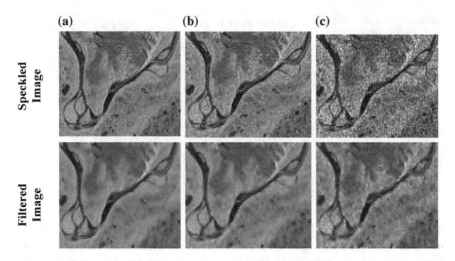

Fig. 1 Simulation Results for Proposed Speckle Suppression Methodology on test image 1at **a** low (0.001), **b** medium (0.01) and **c** high (0.05) speckle intensities

Table 2 PSNR (in dB) values of various speckle suppression techniques on image 1

Noise variance	Lee filter [14]	Frost filter [14]	Kuan filter [14]	PMAD filter [13]	Proposed method
0.001	16.4325	18.0007	16.3805	20.8371	22.5751
0.005	16.1500	17.9729	16.6016	20.3941	21.8525
0.03	14.5814	17.7815	15.5439	17.4601	20.6210
0.05	13.4367	17.6623	15.4674	16.2614	19.7581

Table 3 SSIM values of various speckle suppression techniques on image 1

Noise variance	Lee filter [14]	Frost filter [14]	Kuan filter [14]	PMAD filter [13]	Proposed method
0.001	0.7344	0.7780	0.7332	0.9027	0.9719
0.005	0.7242	0.7771	0.7403	0.8939	0.9683
0.03	0.6728	0.7711	0.7402	0.8251	0.9510
0.05	0.6452	0.7693	0.7386	0.7919	0.9367

limitations and gives better despeckling results as observed from Figs. 1 and 2. This modified filter yields better detection of weak edges and effective reconstruction of structural content. The resulted denoised image shows improved filtering with better enhanced features and preserved edges.

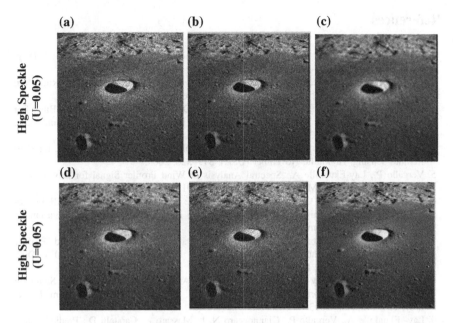

Fig. 2 Comparison of results of speckle suppression from various filtering techniques. **a** Noisy SAR image, Results using **b** Lee filter [14], **c** Frost filter [14], **d** Kuan filter [14], **e** PMAD filter [13], **f** Proposed methodology

4   Conclusion

SAR images possess non-uniform landscape and consist of sharp edges which pose difficulties in filtering and extraction of fine details. The local statistics speckle suppression filters although less computationally complex, blur the images along with speckle reduction. These conventional algorithms generally lead to over filtering of the edges and it becomes more prominent at higher speckle variances. While on the other side, AD filter performs well for images containing additive noise while enhances the corruption of speckled images. The proposed work hybridizes the Bayesian and Non-Bayesian approaches and adequately averages homogenous regions with better preservation of edges and textural information. Also, blurring is diminished leading to finer restoration of image details. Simulation results give the visual and objective performance evaluation for the proposed work for both the test images. Hence, the proposed work is claimed better and can be further effectively applied for environment monitoring.

References

1. Chan Y. K., Koo V. C.: An Introduction to Synthetic Aperture Radar (SAR). In: J. Prog. Electromag. Res. B 2 27–60, 2008.
2. Lay- Ekuakille A., Pellicano D., Dellisanti C., Tralli F.: SAR Aided Method for Rural Soil Evaluation. In: SPIE2002 Remote Sensing, Crete (Greece), 103–112, 2002.
3. Griffo G., Piper L., Lay-Ekuakille A., Pellicano D., De Franchis E.: Modelling A Buoy For Sea Pollution Monitoring Using Fiber Optics Sensors. In: 4th Imeko Tc19 Symposium, Lecce, Italy, 182–186, June 2013.
4. Lay Ekuakille A., Scarano A.V.: Progressive Deconvolution of Laser Radar Signals. In: SPIE Remote Sensing, Honolulu, November (USA), 319–326, 2004.
5. Vergallo P., Lay-Ekuakille A.: Spectral Analysis of Wind Profiler Signal for Environment Monitoring. In: IEEE I2MTC, Graz, Austria, May 2012.
6. Bhateja V., Tripathi A., Gupta A., Lay-Ekuakille A.: Speckle Suppression in SAR Images Employing Modified Anisotropic Diffusion Filtering in Wavelet Domain for Environment Monitoring. In: Measurement, 74, 216–246, 2015.
7. Lay-Ekuakille A., Trotta A.: On the Missing Data Problem in Rass Wind Profiler Measurements: An Algorithm based on Functional Differential Equations. In: ERAD02, Delft, Holland, 212–216, December 2002.
8. Pelillo V., Piper L., Lay- Ekuakille A., Arnesano A., Lanzolla A., Andria G.: Sampling Optimization for Monitoring Contaminated Soils. In: 4th Imeko TC19 Symposium, Leece, Italy, 110–113, June 2013.
9. Lay- Ekuakille A., Vergallo P., Giannoccaro N. I., Massaro a., Caratelli D.: Prediction and Validation of Outcomes from Air Monitoring Sensors and Networks of Sensors. In V International Conference on Sensing Technology, ICST2011, Nov 28th- Dec 1st 2011, Palmerston North, New Zealand, 73–78, 2011.
10. Lay Ekuakille A., Tralli F., Tropeano M.: 2000, Land Modification Measurements Using ERS-2 Satellite Images. XVI IMEKO World Congress, Vienna, Austria, 25–28, September 2000.
11. Lay-Ekuakille A.: Atmospheric Remote Sensing: Temperature Sensing: Temperature Data Accuracy Evaluation of Rass-Wind Profiler System. In: XVII IMEKO World Congress, June 2003, Dubrovnik, Croatia, 2086–2089, 2003.
12. Lopes A., Touzi R., Nezry E.: Adaptive Speckle Filter and Scene Heterogeneity. In: IEE Transactions on Geoscience and Remote Sensing, 28(6), 992–1000, 1990.
13. Perona P., Malik J.: Scale Space and Edge Detection using Anisotropic Diffusion. In: IEEE Transactions on Pattern Analysis and Machine Intelligence, 12 (7), 629–639, 1990.
14. Argenti F., Lapini A., Bianchi T., Alparone L.: A Tutorial on Synthetic Aperture Radar Images. In: IEEE Geoscience and Remote Sensing Magazine, 1 (3), 6–35, October 2013.
15. Singh S., Jain A., Bhateja V.: A Comparative Evaluation of Various Despeckling Algorithms for Medical Images. In: Proc. Of (ACMICPS) CUBE International Information Technology Conference and Exhibition, 32–37, Pune, India, 2012.
16. Bhateja V., Tripathi A., Gupta A.: An Improved Local Statistics Filter for Denoising of SAR Images. In: Proc. of (Springer) 2nd International Symposium on Intelligent Informatics (ISI'13), 235, 23–29, Mysore, India, 2013.
17. Bhateja V., Misra M., Urooj S., Lay-Ekuakille A.: Bilateral Despeckling Filter in homogeneity Domain for Breast Ultrasound Images. In: Proc. 3rd (IEEE) International Conference on Advamce in Computing. Communication and Informatics (ICACCI-2014), 1027–1032, Greater Noida (U.P.), India, 2015.
18. Bhateja V., Singh G., Srivastava A., Singh J.: Despeckling of Ultrasound Images using Non-Linear Conductance Function. In: Proc. (IEEE) International Conference of Signal Processing and Integrated Networks (SPIN-2014), 722–726, Noida (U.P.), India, 2014.

19. Srivastava A., Bhateja V., Tiwari H.,: Modified Anisotropic Diffusion Filtering Algorithm for MRI. In: Proc. (IEEE) 2nd International Conference on Computing for Sustainable Global Development (INDIACom-2015), 1885–1890, New Delhi, India, 2015.
20. Bhateja V., Tripathi A., Gupta A., Lay-Ekuakille, A.: Speckle Suppression in SAR images Employing Modified Anisotropic Diffusion Filtering in Wavelet Domain for Environment Monitoring. In: Elsevier Measurement Journal, 74, 246–254, 2015.
21. Bhateja V., Singh G., Srivastava A., Singh J.: Speckle Reduction in Ultrasound Images using an Improved Conductance Function based on Anisotropic Diffusion. In: Proc. (IEEE) 2014 International Conference on Computing for Sustainable Global Development, 619–624, 2014.
22. Bhateja V., Singh G., Srivastava A.: A Novel Weighted Diffusion Filtering Approach for Speckle Suppression in Ultrasound Images. In: Proc. of (Springer) International Conference on Frontiers in Intelligent Computing Theory and Application (FICTA 2013), 247, 459–466, Bhubaneswar, India, 2013.
23. Dong Y., Xu S.: A New Directional Weighted Median Filter for Removal of Random-Valued Impulse Noise. In: IEEE Signal Processing Letters, 14(3), 193–196, 2007.
24. Ching-Ta L., Tzu-Chun C.: Denosing of Salt and Pepper Noise Corrupted Image using Modified Directional Weighted Median Filter. In: Pattern Recognition Letters, 33 (10), 1287–1295, 2012.
25. Ashutosh P., Agarwal S., Chand S.: A New and Efficient Method for Removal of High Density Salt and Pepper Noise Through Cascade Decision based Filtering Algorithm. In: Procedia Technology, 6, 108–117, 2012.
26. Sharma A., Bhateja V., Tripathi A.: An Improved Kuan Algorithm for Despeckling SAR Images. In: Information Systems Design and Intelligent Application, 434, 663–672, 2016.
27. National Aeronautics and Space Administration, http://history.nasa.gov/ap11fj/photos/41-p. htm.

Part VII
Security and Cryptography

Dual-Image Based Reversible Data Hiding Scheme Through Pixel Value Differencing with Exploiting Modification Direction

Jana Biswapati, Giri Debasis and Mondal Shyamal Kumar

Abstract A dual image based reversible data hiding (RDH) method using pixel value difference (PVD) and exploiting modification direction (EMD) are introduced in this paper. First enlarge the cover image size and then select a pair of pixel for data embedding. We embed 4 bits secret message through pixel value difference and 3 bits secret message using embedding function of exploiting modification direction within a pixel pair. We obtain two modified pairs of pixel which contain 7 bits secret message. We then distribute these pixel pairs among dual stego images depending on a shared secret key (K). The recipient successfully obtain secret message and retrieve original image using same shared secret key. We compared our proposed method with other existing techniques and find out averagely good results with respective to payload.

Keywords Pixel value difference · Dual image · Reversible data hiding · Image interpolation · Exploiting modification direction

Jana Biswapati (✉)
Department of Computer Science, Vidyasagar University, Midnapore 721102,
West Bengal, India
e-mail: biswapati.jana@mail.vidyasagar.ac.in

Giri Debasis
Department of Computer Science and Engineering, Haldia Institute of Technology,
Haldia 721657, West Bengal, India
e-mail: debasis_giri@hotmail.com

M.S. Kumar
Department of Applied Mathematics with Oceanology and Computer Programming,
Vidyasagar University, Midnapore 721102, West Bengal, India
e-mail: shyamal_260180@yahoo.com

© Springer Science+Business Media Singapore 2017   549
J.K. Mandal et al. (eds.), *Proceedings of the First International Conference
on Intelligent Computing and Communication*, Advances in Intelligent Systems
and Computing 458, DOI 10.1007/978-981-10-2035-3_56

1 Introduction and Related Work

Data hiding is the way of hidden data communication through cover media. Data embedding method based on the least significant bits (LSB) substitute is presented by Turner [1]. The LSB substitution techniques can achieve an acceptable visual quality when it embed three bits into three LSBs. To enhance the visual quality, Mielikainen [2] proposed a data embedding scheme called LSB matching revisited. Wu and Tsai [3] suggested a hiding scheme using PVD which can embed a good amount of secret data. After that Wang et al. [4] presented a message concealing technique using PVD and modulus function which enhance quality and increases embedding capacity. Zhang and Wang [5] improve Mielikainen's scheme. Kieu and Chang [8] modify extraction function which is proposed by Zhang and Wang's scheme. Recently, researchers are developing data hiding schemes to enhance embedding capacity and minimize the distortion when embed large data. Jung and Yoo [6] proposed interpolation based data hiding scheme. Lee and Huang [7] improved data hiding capacity using interpolation. Shen and Huang [9] presented the scheme using PVD and improving EMD but the scheme was not reversible. Qin et. al. [10] proposed EMD as a reversible scheme. To improve data hiding in terms of the payload using pair of pixel, we propose dual-image based data hiding scheme which achieve reversibility through PVD and improve EMD and provide hiding capacities up to 1.75 bpp with good visual quality.

1.1 Motivation

Here, we introduce an innovative data hiding method using PVD and EMD to achieve high embedding capacity.

- The motivation of this method is to improve the payload and accomplish reversibility. Data hiding through PVD was irreversible. EMD has been used with PVD to get reversibility.
- Another motivation is to enhance stego quality with out compromising embedding capacity. To do this, we modify the embedding function f() using two pixel pair (C_a, C_b) and (P_a, P_b) and again we modify pixel pair (E_a', E_b') by adding or subtracting 4 before distribute among dual image.
- Another motivation is to enhance security in data hiding. We distribute modified pixel pair among dual stego images, stego-1 (S1) and stego-2 (S2) depending on a shared secret key bit stream to enhance security.

Organization of this paper is mentioned here. Section 2 outlines embedding and extraction scheme. Section 3 present experimental results with comparisons. Finally, Sect. 4 present the conclusion.

2 Proposed Scheme

A new RDH method through PVD and modified EMD on interpolated dual image has been proposed. We achieve high data embedding capacity with reversibility. The detail data embedding and data extraction procedure are described here.

2.1 Data Embedding Stage

The cover image I of hight H and width W are taken. First interpolate the original image I and generate cover image C of size $(2H \times 2W)$ using the following equation.

$$
C(i,j) = \begin{cases}
I(p,q) \\
\quad \{\text{where, } p = 1 \ldots H, q = 1 \ldots W, \\
\quad\quad i = 1,3,\ldots (2H-1), j = 1,3,\ldots (2W-1)\} \\
(C(i,j-1) + C(i,j+1))/2 \\
\quad \{\text{if } i \bmod 2 \neq 0, j \bmod 2 = 0, | i = 1 \ldots (2H-1), \\
\quad\quad j = 1 \ldots (2W-1)\} \\
(C(i-1,j) + C(i+1,j))/2 \\
\quad \{\text{if } i \bmod 2 = 0, j \bmod 2 \neq 0, | i = 1 \ldots (2H-1), \\
\quad\quad j = 1 \ldots (2W-1)\} \\
(C(i-1,j-1) + C(i-1,j+1) + C(i+1,j-1) + \\
\quad\quad C(i+1,j+1))/4 \\
\quad \{\text{if } i \bmod 2 = 0, j \bmod 2 = 0, | i = 1 \ldots (2H-1), \\
\quad\quad j = 1 \ldots (2W-1)\}
\end{cases}
\tag{1}
$$

Now, consider the secret message M which are divided into stream of 7 bits that is $M = \sum\{M_i | i = 1, 2, \ldots, M/7\}$. Each 7 bits are again divided into two parts, where 4 bits are embedded through PVD method and 3 bits are embedded using modified embedding function of EMD method. To hide 4 bits secret data, we select pixel pair (C_a, C_b) from the image C. Then compute the difference d as follows, $d = |C_a - C_b|$. The difference value $|d|$ belongs to the range from 0 to 255. In our scheme, we use the same range table R with n contiguous subrange R_b, $\{R_b | b = 1, 2, \ldots, n\}$ which are used in both embedding and extraction stage. Each subrange R_b has a lower and a upper bound, namely l_b and u_b respectively and $R_b \in [l_b, u_b]$. The width w_b of each subrange R_b is obtained by $w_b = u_b - l_b + 1$. Here, each subrange R_b has same width w_b that is 16. The number of embedded bits are decided by the range table, where the difference $|d|$ is mapped into R. The number of embedded bits t are calculated by $t = \lfloor \log_2(w_b) \rfloor$.

Since, the length of each subrange of range table is 16, that means 4 bits secret data are embedded within each pixel pair. We select 4 bits from secret message M_i and convert it into decimal value v. To embed secret message v, we compute new difference $|d'|$ using $d' = v + l_b$. Now, we calculate following parameters m, c and f as $m = d' - d$, $c = \lceil m/2 \rceil$ and $f = \lfloor m/2 \rfloor$. We get updated pixel pair (P_a, P_b) by modifying the pixel pair (C_a, C_b) either by adding or subtracting parameters c and f by which 4 bits secret data are embedded. If $(C_a > C_b)$, then the modified pixel pair (P_a, P_b) is obtained using $P_a = C_a + f$ and $P_b = C_b - c$ else $P_a = C_a - c$ and $P_b = C_b + f$. After that we apply our improver modification direction function to embed next 3 bits of secret messages within two pixel pairs

Fig. 1 Block diagram of message embedding stage

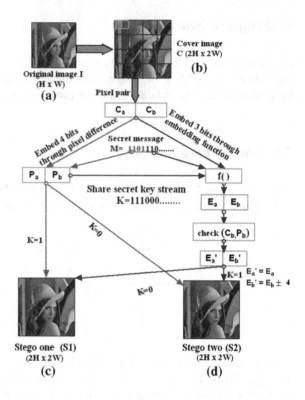

(C_a, C_b) and (P_a, P_b). We select 3 bits secret message from M_i and convert into decimal value v_1. Now, we propose modified function $f()$ using two pixel pair $(C_a, C_b, \text{and } P_a, P_b)$ by the formula $f(C_a, C_b, P_a, P_b) = (C_a \times 1 + C_b \times 2 + P_a \times 3 + P_b \times 4) \bmod 9$. If $f(C_a, C_b, P_a, P_b) = v_1$, then the new pixel pair (E_a, E_b) is same as the pixel pair (C_a, C_b) that is $E_a = C_a$ and $E_b = C_b$. If they are not equal, then calculate $f_1()$ such that $f_1() = v_1$. When we use $f_1()$ function, we will store the pixel value C_a into the new pixel E_a to achieve reversibility. The $f_1()$ function is calculated using the pixels value (C_a, C_b, P_a, P_b) as $f_1() = [1 \times C_a + 2 \times (C_b - (x \times sign(P_b - C_b))) + P_a \times 3 + P_b \times 4] \bmod 9$. Now, we found the value of x such that $f_1()$ value is equal to v_1. Where, x is an integer, $x \in \{1, 2, \ldots, 5\}$ and $sign()$ return 1 or -1 depend on the value of $(P_b - C_b)$. So the new modified stego pixel pair (E_a, E_b) is calculated as follows. $E_a = C_a; E_b = (C_b - (x \times sign(P_b - C_b)))$. We introduce another modification in this section of our propose scheme to enhance better image quality on pixel pair (E_a, E_b). The modified pixel pair (E_a', E_b') is computed as

$$(E_a', E_b') = \begin{cases} (E_a, E_b + 4) & \text{if } P_b > C_b \\ (E_a, E_b - 4) & \text{if } P_b \leq C_b \end{cases} \qquad (2)$$

Finally, we get two stego pixel pair (P_a, P_b) and (E_a', E_b'). To enhance the security, we distribute stego pixel pair among dual image using a share key stream K. For $K = 1$, the pixel pair (P_a, P_b) is stored within the stego -1 and the pixel pair (E_a', E_b') is stored within the stego-2 image. If the K is 0 then the pixel pair (P_a, P_b) is stored within the stego-2 and the pixel pair (E_a', E_b') is stored within the stego-1. The block diagram of embedding stage is depicted in Fig. 1. The original image I of size $(H \times W)$ is shown in Fig. 1a. Figure 1b shows the cover image C of size $(2H \times 2W)$. Figure 1c and d are shown as the two stego image $S1$ and $S2$ respectively of size $(2H \times 2W)$.

2.2 Extraction Procedure

During secret message extraction, we first check the shared secret key bit K. If K is 1, then we select pixel pair (E_a', E_b') from the stego image S2 else we select pixel pair (E_a', E_b') from the stego image S1. Then we recover the original image I from pixel pair (E_a', E_b'). To do this, we first recover pixel (I_a) of the original image I, check the pixel pair (E_a', E_b') which belongs to the odd row of S1 or S2. If the pixel

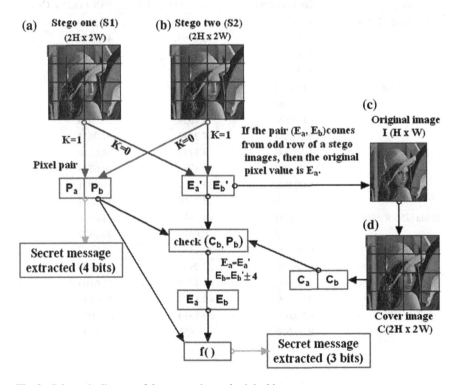

Fig. 2 Schematic diagram of data extraction and original image recovery stage

pair retrieved from odd row, then the original pixel value (I_a) is obtained by $I_a = E'_a$. After retrieval the original image I, we generate the cover image C using Eq. (1). To retrieve the secret message from dual image we first select pixel pair from the stego images S1 $(H \times W)$and S2 $(H \times W)$ in raster scan order and apply pixel value difference or modified exploited modification direction method depending on the share secret key stream K. If the K is 1 then we retrieve pixel pair(P_a, P_b) from the stego image S1 and the pixel pair (E'_a, E'_b) from the stego image S2. If the key is 0 then the pixel pair(P_a, P_b) is retrieved from the stego image S2 and the pixel pair (E'_a, E'_b) is retrieved from the stego image S1. Also we select pixel pair (C_a, C_b) from the cover image C in raster scan order. Then we apply pixel value differencing method on the pixel pair (P_a, P_b) to extract secret data. Compute the difference d' as $d' = |P_a - P_b|$. The bits length t are computed from this pair using the range table R where the difference d' is mapped as $t = \lfloor \log_2(w_b) \rfloor$, where $w_b = (u_b - l_b + 1)$. Here, l_b and u_b be the lower bound and upper bound of each subrange R_b respectively. The secret message v are extracted as $v = d' - l_b$. We convert v into binary form of t bits. To get the pixel pair (E_a, E_b) from the modified pixel pair (E'_a, E'_b) we follow

Table 1 Data embedding capacity of proposed method with PSNR

Input image	Data (in bits)	PSNR (S1 vs. C)	PSNR (S2 vs. C)
Cameraman (256 × 256)	1,60,000	44.8730	44.6110
	4,00,000	40.8789	40.6836
	6,00,000	39.1323	38.9067
	9,16,656	37.2853	37.0608
House (256 × 256)	1,60,000	44.8721	44.6118
	4,00,000	40.8752	40.6821
	6,00,000	39.1334	38.9054
	9,16,656	37.2841	37.0611
F16 (256 × 256)	1,60,000	44.8733	44.6128
	4,00,000	40.8746	40.6773
	6,00,000	39.1364	38.9071
	9,16,656	37.2818	37.0621
Zelda (256 × 256)	1,60,000	44.8814	44.6053
	4,00,000	40.8817	40.6877
	6,00,000	39.1332	38.9097
	9,16,656	37.2951	37.0705
Babbon (256 × 256)	1,60,000	44.8719	44.6109
	4,00,000	40.8774	40.6146
	6,00,000	39.1379	38.9053
	9,16,656	37.2855	37.0634

$$(E_a, E_b) = \begin{cases} (E'_a, E'_b - 4) & \text{if } P_b > C_b \\ (E'_a, E'_b + 4) & \text{if } P_b \leq C_b \end{cases} \tag{3}$$

Then we retrieve secret data v_1 from the pixels (E_a, E_b, P_a, P_b) using the function f() as

$$f(E_a, E_b, P_a, P_b) = (E_a \times 1 + E_b \times 2 + P_a \times 3 + P_b \times 4) \, mod \, 9 \tag{4}$$

We get $v_1 = f(E_a, E_b, P_a, P_b)$ and convert v_1 into 3 bits binary form. The schematic diagram of data extraction procedure and recovering cover image is depicted as in Fig. 2, where Fig. 2a and b are the two stego image S1($2H \times 2W$) and S2($2H \times 2W$) respectively.

Table 2 Comparison between dual image based existing methods and our proposed method

Methods	Measure	Images					
		Lena	Peppers	Boat	Goldhill	Zelda	Baboon
Lee et al. (2009)	PSNR (S1)	51.14	51.14	51.14	51.14	51.14	51.14
	PSNR (S2)	54.16	54.17	54.16	54.16	54.17	54.14
	PSNR (Avg.)	52.65	52.66	52.65	52.65	52.66	52.64
	Capacity(bpp)	0.75	0.75	0.75	0.75	0.75	0.75
Lee and Huang (2013)	PSNR (S1)	49.76	49.75	49.76	49.77	49.77	49.77
	PSNR (S2)	49.56	49.56	49.57	49.57	49.58	49.56
	PSNR (Avg.)	49.66	49.66	49.67	49.67	49.68	49.77
	Capacity(bpp)	1.07	1.07	1.07	1.07	1.07	1.07
Chang et al. (2013)	PSNR (S1)	39.89	39.94	39.89	39.9	39.89	39.91
	PSNR (S2)	39.89	39.94	39.89	39.9	39.89	39.91
	PSNR (Avg.)	39.89	39.94	39.89	39.9	39.89	39.91
	Capacity(bpp)	1.53	1.52	1.53	1.53	1.53	1.53
Qin et al. (2014)	PSNR (S1)	52.11	51.25	51.11	52.11	52.06	52.04
	PSNR (S2)	41.34	41.52	41.57	41.34	41.57	41.56
	PSNR (Avg.)	46.72	46.39	46.84	46.72	46.82	46.80
	Capacity(bpp)	1.16	1.16	1.16	1.16	1.16	1.16
Lu et al. (2015) (k = 5)	PSNR(1)	30.67	32.86	30.32	34.06	35.86	30.42
	PSNR(2)	37.18	36.89	36.91	36.76	36.90	37.50
	PSNR(Avg.)	33.92	34.87	33.61	35.41	35.38	33.96
	Capacity(bpp)	1.25	1.25	1.25	1.25	1.25	1.25
Proposed Scheme	PSNR (S1)	37.28	37.28	37.28	37.28	37.29	37.28
	PSNR (S2)	37.06	37.06	37.06	37.06	37.07	37.06
	PSNR (Avg.)	37.17	37.17	37.17	37.17	37.18	37.17
	Capacity(bpp)	1.75	1.75	1.75	1.75	1.75	1.75

3 Experimental Results

The experiment has been performed using the standard image of size (256×256). Table 1 shows the PSNR of stego-1 (S1) and stego-2 (S2). The PSNR of S1 and S2 deviates 44 dB to 37 dB when data bit changes from 1,60,000 to 9,16,656 bits. The payload is computed as follows.

$$B = \frac{2H \times W \times r}{2H \times 2W \times s}, \tag{5}$$

where $H = 256$ and $W = 256$, r is the bits to be embedded and s is the number of stego images. The payload $B = 1.75$ bpp.

Table 2 presents the comparison of our scheme with other dual images based existing methods. From Table 2, we observed that PSNR of the suggested scheme is higher than 37 dB when payload is 9,16,656 bits, which is lower than Qin et al. [10], Lu et al. [16], Chang et al.'s [11, 12] and Lee et al.'s [14, 15] and Chang et al. [13] schemes. The embedding capacity of this scheme is 1.75 bpp. We observed that our scheme gain good results than existing state-of-the art methods. Our proposed method is better with respect to embedding capacity but the PSNR is slightly dropped.

4 Conclusion

In this approach, a shared secret key K is used which guarantees security. Without K, receiver can not retrieve secret data. Data hiding through PVD was not reversible. The reversibility has been accomplish in our approach through dual image. Our scheme gains payload 1.75 bpp, that is better than recently developed dual image based scheme. The proposed scheme still maintains good PSNR which is greater than to 37 dB when we embed maximum secret data that is 9,16,656 bits. The embedding capacity and security of our proposed scheme is better than other existing scheme.

References

1. Turner, L. F., Digital data security system, Patent IPN, WO 89/08915, (1989).
2. Mielikainen. J., LSB matching revisited. IEEE Sgnal processing Letters, 13(5), pp. 285–287, (2006).
3. D. Wu, W. Tsai, A steganographic method for images by pixel-value differencing. Pattern Recognition Letters, vol. 24, pp. 1613–1626, (2003).
4. Chung-Ming Wang, Nan-I Wu, Chwei-Shyong Tsai, Min-Shiang Hwang, A high quality steganographic method with pixel-value differencing and modulus function, The Journal of System and Software, (2007).
5. X. Zhang, S. Wang, Efficient steganographic embedding by exploiting modification direction, IEEE Communications Letters, vol. 10, no. 113, pp. 781–783, (2006).

6. K. Jung, K. Yoo, Data hiding method using image interpolation, Comput. Stand. Interfaces vol. 31, pp. 465–470, (2009).
7. C. Lee, Y. Huang, An efficient image interpolation increasing payload in reversible data hiding, Expert Syst. Appl. vol. 39, pp. 6712–6719, (2012).
8. Kieu TD, Chang CC. A steganographic scheme by fully exploiting modification directions, Expert Syst Appl, vol. 38, pp. 10648–10657, (2011).
9. Shu-Yuan Shen and Li-Hong Huang, A data hiding scheme using pixel value differencing and improving exploiting modification directions, computers and security, vol. 48, pp. 131–141, (2015).
10. C. Qin, C.C. Chang, and T.J. Hsu, Reversible Data Hiding Scheme Based on Exploiting Modification Direction with Two Steganographic Images, Multimedia Tools and Applications, pp. 1–12, (2014).
11. C.C. Chang, T.D. Kieu, and Y.C. Chou, Reversible Data Hiding Scheme Using Two Steganographic Images, Proceedings of IEEE Region 10 International Conference (TENCON), pp. 1–4, (2007).
12. C.C. Chang, Y.C. Chou, and T.D. Kieu, Information Hiding in Dual Images with Reversibility, Proceedings of Third International Conference on Multimedia and Ubiquitous Engineering, pp. 45–152, (2009).
13. C.C. Chang, T.C. Lu, G. Horng, Y.H. Huang, and Y.M. Hsu, A High Payload Data Embedding Scheme Using Dual Stego-images with Reversibility, Proceedings of Third International Conference on Information, Communications and Signal Processing, pp. 1–5, (2013).
14. C.F. Lee and Y.L. Huang, Reversible Data Hiding Scheme Based on Dual Stegano-Images Using Orientation Combinations, Telecommunication Systems, vol. 52, no. 4, pp. 2237–2247, (2013).
15. C.F. Lee, K.H. Wang, C.C. Chang, and Y.L. Huang, A Reversible Data Hiding Scheme Based on Dual Steganographic Images, Proceedings of the Third International Conference on Ubiquitous Information Management and Communication, pp. 228–237, (2009).
16. T.C. Lu, C.Y. Tseng, and J.H. Wu, Dual Imaging-based Reversible Hiding Technique Using LSB Matching, Signal Processing, vol. 108, pp. 77–89, (2015).
17. J. Fridrich, J. Goljan, R. Du, Invertible authentication, in: Proceedings of the SPIE, Security and Watermarking of Multimedia Contents, vol. 4314, SanJose, CA, pp. 197–208, January (2001).

Cryptanalysis of an Asymmetric Image Cryptosystem Based on Synchronized Unified Chaotic System and CNN

Musheer Ahmad, Faiyaz Ahmad and Syed Ashar Javed

Abstract Cheng et al. proposed an asymmetric image encryption scheme which is based on adaptively synchronised chaotic cellular neural network and unified chaotic systems in [Communication in Nonlinear Science and Numerical Simulation 18(10) 2825–2837 2013]. The cryptosystem was asymmetric in nature and the synchronization error converged quickly to zero. Numerical simulations for performance evaluation included synchronization effectiveness, cryptosystem robustness and statistical analyses like key space, key sensitivity and NPCR/UACI analyses, all with effective results. But, this paper demonstrates the cryptanalysis of Cheng et al. cryptosystem by exploiting inherent deficiencies of encryption algorithm like low robustness and poor plain-image sensitivity. It is done by mounting the proposed cryptographic CPA or KPA attack which leads to successful retrieval of original plaintext image. The simulated cryptanalysis shows that Cheng et al. cryptosystem is not suitable for practical utility in image security applications.

Keywords Image encryption · Cryptanalysis · Security · Unified chaotic system · Synchronization

1 Introduction

The massive growth in the usage of multimedia imagery on the internet and otherwise has led to an increased need for secure transmission of images. Image transmission over the network corresponds to a wide variety of applications in

Musheer Ahmad (✉) · Faiyaz Ahmad · S.A. Javed
Department of Computer Engineering, Faculty of Engineering and Technology,
Jamia Millia Islamia, New Delhi 110025, India
e-mail: musheer.cse@gmail.com

Faiyaz Ahmad
e-mail: ahmad.faiyaz@gmail.com

S.A. Javed
e-mail: asharjaved7@gmail.com

© Springer Science+Business Media Singapore 2017   559
J.K. Mandal et al. (eds.), *Proceedings of the First International Conference
on Intelligent Computing and Communication*, Advances in Intelligent Systems
and Computing 458, DOI 10.1007/978-981-10-2035-3_57

domains like defense, medicine, business, banking and many more. Any breach in confidentiality of images by unauthorized agents can adversely affect the functioning of these applications [1]. With the parallel advancement of the malicious attacks, improved techniques to secure the image transmission have become imperative. Over the years, numerous cryptographic techniques for multimedia data have been proposed. At the same time, cryptanalysis or the breaking of these cryptosystems without getting access to the secret key has helped to identify the security flaws in those encryption systems and methods [2–9]. With the highly prevalent use of images nowadays, and countless encryption algorithms to secure them, cryptanalysis has become an important field. Numerous standardized algorithms for the cryptanalysis of symmetric keys exist [4, 5]. As far as asymmetric keys are concerned, these cryptosystems rely on hard mathematical problems as a basis for their security. Therefore, the cryptanalysis of such ciphers boils down to an attack to solve these mathematical problems. Another aspect of asymmetric encryption is that its cryptanalysis has the opportunity to make use of the information gained from the public key. For the purpose of security investigations of ciphers, Shannon's maxim is considered as an appropriate assumption which simply states: *the enemy knows the system except the secret key*. Keeping that in mind, some of the classical methods for cryptanalysis are as follows [1]:

1. Cipher-text only attack: In this attack, the attacker assumes to have the access of ciphertext which can be utilised to predict the plaintext due the inherent statistical bias in any sensible message.
2. Known-plaintext attack (KPA): In this case, the attacker supposes to have access to part of plaintext for a particular ciphertext which can be used for deciphering the complete message.
3. Chosen-plaintext attack (CPA): Here, the attacker is having a temporary access to encryption mechanism without the key which can be in turn used for finding the secret key.
4. Chosen-ciphertext attack (CCA): Here, the attacker holds temporary access to decryption mechanism which can lead to the breach of the secret key.

In 2013, Cheng et al. gave an image encryption procedure which is established on synchronizing a unified chaos (UCS) and cellular neural network (CNN), which is asymmetric in nature [10]. The first system used is the unified chaotic system which is a combination of different systems, like chaotic Lorenz system, chaotic Lu system, and the chaotic Chen system. The unified system is adaptively synchronized with cellular neural networks using drive-response architecture with systems 1 and 2 respectively. The error trajectory of the defined architecture asymptotically converges to zero. The image cryptosystem is developed with a pair of asymmetric keys and a parameter update law for synchronization of the parameters of the two systems. The cipher image is obtained by the conventional process of bit-wise EXOR of the diffused pixels. The ciphered pixels are converted to a decimal set and the set is reordered to form the final cipher image. In terms of performance, various simulations are used to scrutinize the algorithm by the authors. Even though the simulated security metrics show good results, on careful observation, certain inadequacies like the dependency of the decimal values on the keystream render the algorithm broken for practical utility. In this paper, our contribution includes careful security probe of Cheng et al. cryptosystem to find underlying deficiencies. And, by exploiting the deficiencies of Cheng et al. cryptosystem, the proposed cryptographic CPA and KPA attacks are formulated to completely retrieve original image with no prior knowledge of secret key.

2 Cheng et al. Encryption Algorithm

The image encryption algorithm used by Cheng et al. uses two different systems one for drive system and other for response system which produce asymmetric keys [10]. The first system is actually combination of chaotic systems, viz. the Chen, Lu, and Lorenz chaotic system, which are united to form a unified chaotic system. The UCS system is described below:

$$dx_1/dt = (25\alpha + 10)(x_2 - x_1)$$
$$dx_2/dt = (28 - 35\alpha)(x_1 - x_1x_3 + (29\alpha - 1)x_2)$$
$$dx_3/dt = x_1x_2 - (8 + \alpha)x_3/3$$

where $x(t) = [x_1\ x_2\ x_3]_T$ are system variables at time t, and α is system parameter, the unified system exhibits chaotic character for $\alpha \in (0, 1)$. The above system behaves as the general chaotic Chen system, when $\alpha \in (0.8, 1]$, and is called chaotic Lorenz system, when $\alpha \in [0, 0.8)$. The chaotic Chen and Lorenz systems appear alike, but they are topologically dissimilar. The Lü system exists in above UCS system for $\alpha = 0.8$. The Lu system fills the void exist between two systems. The governing equation of chaotic CNN is described as:

$$\dot{x}_i(t) = -c_ix_i(t) + \sum_{j=1}^{n} a_{ij}f_j(x_j(t)) + \sum_{j=1}^{n} b_{ij}f_j(x_j(t - \tau_j))$$

where $i = 1, 2, ..., n$, n is count of neurons in CNN, x_i is the system variable for ith neuron, $c_ix_i(t)$ is a suitable bounding function. The part $f_i(x) = (|x + 1| - |x - 1|)/2$ describes the response of the neurons to one another. The matrix $A = (a_{ij})_{n \times n}$ defines the weights of connected neuron in the CNN, while the matrix $B = (b_{ij})_{n \times n}$ decides the weights of neuron, with delay τ_j, in CNN network.

For synchronizing UCS system with CNN, the authors used system parameter α, the update law for parameter, with the given control inputs, $u(t) = [u_1\ u_2\ u_3]_T$ is given as follows:

$$\hat{a} = 25e_1(x_1 - x_2) + e_2(35x_1 - 29x_2) + e_3x_3/3$$

The dynamics of synchronization error asymptotically converges to zero. For a detail description of the synchronization mechanism, the readers are encouraged to go through Ref. [10]. The cryptosystem is then prepared according to the following algorithm.

Encrypt-Cheng()

1. Arrange pixels of the input image having size M $\times$ N to get set $S = \{S_1, S_2, \dots, S_{M \times N}\}$.
2. Initial states of the synchronization for CNN and uncertain unified chaotic system are respectively $[z_1(0)\ z_2(0)\ z_3(0)]_T$ and $[x_1(0)\ x_2(0)\ x_3(0)]_T$. Achieve synchronization at time t_s.
3. Use $k_e = \{x_1(0), x_2(0), x_3(0), a, t_c\}$ as encryption *key* tuple.
4. After t_c, the unified chaotic system is iteratively executed for another M $\times$ N/3 times to get encryption keys k_e:

   4.1 *for* k = 1 to M $\times$ N/3

   for i = 1 to 3

   $$T_{3 \times (k-1) + i} = abs(x_{ik}) - floor(abs(x_{ik}))$$

   next i

   next k

   4.2 $T = \{T_1, T_2, \dots, T_{3 \times (k-1)+i}, \dots, T_{M \times N}\}$, $k = 1, 2, \dots, M \times N/3$, $i = 1, 2, 3$.

   $$K = round\left(mod\left(T \times 10^{12}, 256\right)\right)$$

   4.3 Encryption keys is $k_e = de2bi(K)$

5. Evaluate $C_{3 \times (k-1)+i} = de2bi(S_3 \times (k-1) + i)$ XOR k_e
6. The cipher set C is converted to a decimal set D as $D_{3 \times (k-1)+i} = bi2de(D_{3 \times (k-1)+i})$.

Schematic block diagram of Cheng et al. image cryptosystem is depicted in Fig. 1.

3 Cryptanalysis of Cheng et al. Cryptosystem

A cryptosystem's security level should ideally be determined only by the quality of the secret key. Kerckhoffs's principle (reformulated or independently stated by Claude Shannon) states that the secrecy of the key is the only undisclosed entity in cryptography and the rest is all in public knowledge [8]. The attacker, practically

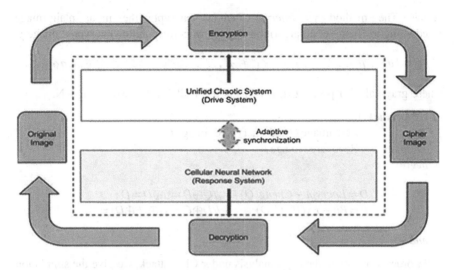

Fig. 1 Cheng et al. asymmetric image cryptosystem

too, can compromise the design of the encryption or decryption mechanism through any of the classic attacks and subsequently break the system. This implies that for any cryptosystem, security should rely on the system itself and not the secrecy of the system. Security through obscurity is a poor technique for controlling the vulnerabilities of the system. Given this, a mediocre key quality coupled with the classic attacks can give access to the whole system.

The security strength of Cheng et al. encryption algorithm depends on the secrecy of initial values assigned to secret key. The design of algorithm unveils that if, anyhow, attacker knows the generated encryption keystream k_e, he/she can retrieve the plain image from unauthorizedly received encrypted image from communication channel. Thus, the encryption keystream k_e is equivalent secret key of cryptosystem. Therefore, instead of trying to know the actual secret key values, an attacker may design method to find the keystream k_e. This can be achieved by exploiting the following inherent flaws of algorithm under probe.

- The encryption keystream k_e for plain-image remain unaltered when dissimilar plain-images are ciphered. The generation of keystream is independent to pending plain-image information.
- The algorithm has high key sensitivity which makes the brute-force attack infeasible. However, it doesn't have plain-image sensitivity.

Assume that the attacker has gained temporary access to Cheng et al. cryptosystem and encrypted image C which is to be decoded. Assume P be holding original data image to be retrieved from captured cipher image C.

Cryptographic chosen-plaintext (CPA) attack needs specially designed image to reveal keystream k_e. A zero image Q containing all pixels with zero gray values is designed for the purpose. The revelation of P from C under CPA attack is provided below. The method $y = Encrypt\text{-}Cheng(x)$ encrypts the input plain-image x according to Cheng et al. algorithm and return corresponding encrypted image y.

$$Q = \{q_{1,1}, q_{1,2}, \cdots\cdots, q_{1,N}, q_{2,1}, q_{2,2}, \cdots\cdots, q_{2,N}, \cdots\cdots, q_{M,N-1}, q_{M,N}\}$$

where gray-value of pixel at (i, j) is $q_{i,j} = 0$ for all $i = 1 \sim M, j = 1 \sim N$.

CPA-attack()

    Input   :   Zero image Q and encrypted image C

    Output :   Recovered image P

begin

$$D = Encrypt - Cheng(Q) \qquad //Q \oplus D = 0 \oplus D = D$$
$$P = Bit - ExOR(C, D) \qquad //D \oplus C = D \oplus (P \oplus D) = P$$

end

In order to illustrate the cryptanalysis under CPA attack, we give the simulation results in Fig. 2.

The known-plaintext (KPA) attack doesn't needs any specially designed image. Instead, it takes the access of one plain image and its accompanying ciphered image and analyzes them to reveal useful plain-image information. Let the attacker has an access to original image P_1 and its ciphered image C_1. The revelation of P from C under KPA attack is as:

KPA-attack()

    Input   :   plain $-$ image P_1 and its encrypted image C_1

    Output :   Recovered image P

begin

$$D = Bit - ExOR(P_1, C_1) \qquad //P_1 \oplus C_1 = P_1 \oplus (P_1 \oplus D) = D$$
$$P = Bit - ExOR(D, C) \qquad //C \oplus D = (P \oplus D) \oplus D = P$$

end

The simulation results of cryptanalysis under KPA attack are portrayed through Fig. 3.

(a)   **(b)**   **(c)**   **(d)**

Fig. 2 Cryptanalysis under chosen-plaintext attack. **a** Q, **b** D, **c** Received **C**, **d** Recovered P

Fig. 3 Cryptanalysis under known-plaintext attack. **a** P_1, **b** C_1, **c** $P_1 \oplus C_1 = D$, **d** Received C, **e** Recovered P

Plain image sensitivity is essential for preventing cryptographic attacks. Any minor refitting in plaintext image should be uncorrelated to the consequent change in the cipher image. In fact, even minor changes in the plain image should bring about highly incongruous deviations in the cipher image. But, the algorithm has poor sensitivity of keystream k_e generated for each image during the process. To illustrate the degree of effect of this loophole, we consider two similar plain images P1 and P2 which differ only by a single pixel. These images are encrypted using the Cheng et al. algorithm and named C1 and C2 respectively. The keystream generated in the two processes are key dependent which is kept same. The similar keystreams generated result in similar cipher images C1 and C2 which only differ by a single pixel. The difference of images C1 and C2 produces a zero image (a black image). This lapse in the Cheng et al. algorithm is simulated below. This shows that the cryptosystem under study fails to exhibit Shannon's properties of confusion and diffusion for a strong cryptosystem (Fig. 4).

Plain-image_Sensitivity()
Input : P_1 and P_2 be images having just one differing pixel
Output : Conflict between C_1 and C_2 as image J

(a)   (b)   (c)

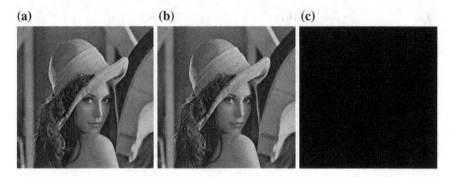

Fig. 4 Simulation of poor plain-image sensitivity. **a** P_1, **b** P_2, **c** J

begin

$$C_1 = Encrypt - Cheng(P_1)$$
$$C_2 = Encrypt - Cheng(P_2)$$
$$J = Bit - ExOR(C_1, C_2) \qquad //since\ x \oplus x = 0$$

end

4   Conclusion

Here, we proposed to break an asymmetric image encryption algorithm recently presented by Cheng et al. which adaptively synchronizes two chaotic systems namely cellular neural network and uncertain unified chaotic system to produce the secret encryption keystream. The algorithm used keystream which fluctuated with the change in the keystream and but don't with the minute change in the pending plain image information. This made the system prone to successful cryptanalytic attacks. This paper also demonstrated the lack of system sensitivity in relation to the plain image through simulated attacks which successfully retrieved the plain image. The analysis and subsequent cryptanalysis of the Cheng et al. algorithm shows its unsuitability for practical image encryption applications.

References

1. Schneier, B.: Applied Cryptography: Protocols Algorithms and Source Code in C, Wiley, New York (1996)
2. Çokal, C., Solak, E.: Cryptanalysis of a chaos-based image encryption algorithm. Physics Letters A 373(15), 1357–1360 (2009)
3. Rhouma, R., Solak, E., Belghith, S.: Cryptanalysis of a new substitution-diffusion based image cipher. Communication in Nonlinear Science and Numerical Simulation 15(7), 1887–1892 (2010)

4. Li, C., Lo, K.T.: Optimal quantitative cryptanalysis of permutation-only multimedia ciphers against plaintext attacks. Signal processing 91(4), 949–954 (2011)
5. Bard, G.V.: Algebraic Cryptanalysis. Springer, Berlin (2009)
6. Ahmad, M.: Cryptanalysis of chaos based secure satellite imagery cryptosystem. In: Contemporary Computing. Springer Berlin Heidelberg, 81–91 (2011)
7. Sharma, P.K., Ahmad, M., Khan, P.M.: Cryptanalysis of image encryption algorithm based on pixel shuffling and chaotic S-box transformation. In: Security in Computing and Communication. Springer-Verlag Berlin Heidelberg, CCIS 467, 173–181 (2014)
8. Ahmad, M., Ahmad, F.: Cryptanalysis of Image Encryption Based on Permutation-Substitution Using Chaotic Map and Latin Square Image Cipher. In: Proceedings of the 3rd International Conference on Frontiers of Intelligent Computing: Theory and Applications. Springer International Publishing Switzerland, AISC 327, 481–488 (2015)
9. Ahmad, M., Khan, I.R., Alam, S.,: Cryptanalysis of Image Encryption Algorithm Based on Fractional-Order Lorenz-Like Chaotic System. In: Emerging ICT for Bridging the Future. Springer International Publishing Switzerland, AISC 338, 381–388 (2015)
10. Cheng, C-J., Cheng, C-B.: An asymmetric image cryptosystem based on the adaptive synchronization of an uncertain unified chaotic system and a cellular neural network. Communications in Nonlinear Science and Numerical Simulation 18(10), 2825–2837 (2013)

Chaotic Map Based Image Encryption in Spatial Domain: A Brief Survey

Monjul Saikia and Bikash Baruah

Abstract Image encryption is a process of conversion of original image into an unintelligible form, which is not understood by anyone except authorized parties after performing decryption operation on it with the help of a secret key. Image encryption is a challenging task due to its large amount of data and correlation among pixels restricts using traditional encryption algorithm. So, to obtain an efficient and robustness against security violation during image transmission chaotic based image encryption techniques are proven to be more suitable. These techniques are considered more effective due to low computational power, high sensitivity to initial conditions. In this paper, we discuss about chaos and their properties and we give a general overview of chaotic map based image encryption scheme and various phases in the process.

Keywords Chaotic map · Image encryption · Confusion · Diffusion

1 Introduction

Operation on spatial domain in an image is a technique that are based on direct manipulation of pixels on it. In cryptography, image encryption in spatial domain is the process of encrypting the image pixel values directly in such a way that only authorized parties can retrieve the original information and can read it. Among the two broad categories of cryptography namely Symmetric-key cryptography and Asymmetric-key cryptography, typically symmetric cryptography is widely used in case of image encryption.

Monjul Saikia (✉) · Bikash Baruah
Department of Computer Science and Engineering, North Eastern Regional Institute
of Science and Technology, Nirjuli, Arunachal Pradesh, India
e-mail: monjuls@gmail.com

Bikash Baruah
e-mail: baruahkikash9@gmail.com

© Springer Science+Business Media Singapore 2017 569
J.K. Mandal et al. (eds.), *Proceedings of the First International Conference on Intelligent Computing and Communication*, Advances in Intelligent Systems and Computing 458, DOI 10.1007/978-981-10-2035-3_58

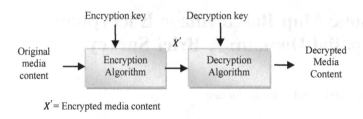

X' = Encrypted media content

Fig. 1 General encryption and decryption model

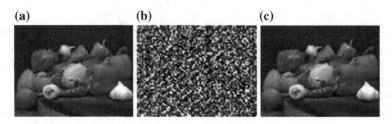

Fig. 2 a Original image, **b** Encrypted image, **c** Decrypted image

The concept of multimedia encryption was introduced in early 1980s [1] and became a hot research topic since then. The practice of implementation can be classified into three broad categories namely raw data encryption, compressed data encryption, and partial encryption. In initial stage of multimedia encryption a very few encoding techniques were proposed which is mostly for raw data with the help of pixel scrambling or permutation. The method of pixel permutation on the image or video is done so that resulting data is unintelligible. Algorithms like space filling curves [1] and Eurocrypto [2] which confuses the relation between adjacent image pixels or video pixels were used in early days. These methods changes the relation between adjacent pixels values, therefore the subsequent compression operations are not applicable further.

In recent years, with the development of Internet technology, multimedia applications [3–5] are seemed to be more real-time application demanded and security is becoming main concern [6]. Partial encryption is method of encrypting only parts of the sensitive multimedia data that are significant in human perception which improves encryption efficiency. Thus, the real-time requirement can be met keeping the file format unchanged and it benefits the bandwidth requirements in transmission process.

Figure 1 describes general flow diagram of image encryption process. Here original image is encrypted using the key and then encrypted image is obtained. This encrypted image and the key is then send to destination. In the receiver side Encrypted Image is then decrypted using the same key and the original image is retrieved. Figure 2, shows an example of the original image, encrypted image and decrypted image.

2 Types of Encryption on Spatial Domain

Encryption on spatial domain determines the operations on each pixel values. i.e., operation is done in pixel directly. Complete Encryption as well as Partial Encryption can be performed on spatial domain. In complete Encryption technique the whole image values are considered and encrypt each pixel where as in Partial Encryption part of an image is considered for encryption. It may be of two types: region based encryption and bit plane encryption.

In region based encryption firstly the region or the sensitive area which have to be encrypted is selected using some detection mechanism and only these selected coordinates are encrypted using different encryption technique.

In bit-plane encryption single original image is divided into number of bit planes. For a 256 gray level image length of each pixel value is 8-bit. So, 256 gray level images can be converted into 8 different bit planes from MSB (Most Significant Bit) to LSB (Least Significant Bit). Among all the bit planes only some of the bit planes are being encrypted. Now combination of all encrypted and non encrypted 8-bit planes will give the encrypted image.

3 Chaotic Map

In mathematics the theory of Chaos [7] is the area of study that discusses the nonlinear things that are effectively impossible to predict or control and are highly sensitive to the initial conditions. The important characteristics of a Chaos are as follows:

- it must be sensitive to initial conditions
- it must be topologically mixing
- it must have dense periodic orbits

It can be said that any function that fulfils the above mentioned behavior are called a chaotic function. The map or the graph obtained by plotting the values which is again found by infinite iteration of that function is called Chaotic Map for that function. In the recent years, researchers have developed many Chaotic Maps [7] by studying different real life incidents or events. These maps are widely used for different encryption techniques specially for image encryption. The first challenging task in chaotic map based image encryption is the selection of one or more map(s) which will be suitable according to the designers' encryption objectives. A comparative performance analysis of chaotic based encryption on colour image encryption and its cryptographic requirements were discused in [8].

Here we are going to discuss some of the chaotic maps. Usually, for more security, 2D maps are extended to 3D maps before using it in encryption process. Some of the maps and its 3D extensions are given below:

A: *Arnold cat map*

The well-known Cat map is a two-dimensional invertible chaotic map introduced by Arnold and Avez [9, 10]. The mathematical formula is:

Fig. 3 Geometrical
explanation of Arnold cat map

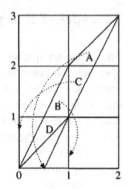

$$\begin{bmatrix} x_{n+1} \\ y_{n+1} \end{bmatrix} = \begin{bmatrix} 1 & 1 \\ 1 & 2 \end{bmatrix} \begin{bmatrix} x_n \\ y_n \end{bmatrix} \bmod n = A \begin{bmatrix} x_n \\ y_n \end{bmatrix} \bmod n \tag{1}$$

where n is the dimension of the image (Fig. 3).

This 2D cat map can be written for $N \times N$ matrix is,

$$\begin{bmatrix} x_{n+1} \\ y_{n+1} \end{bmatrix} = \begin{bmatrix} 1 & a \\ b & ab+1 \end{bmatrix} \begin{bmatrix} x_n \\ y_n \end{bmatrix} \bmod N = A \begin{bmatrix} x_n \\ y_n \end{bmatrix} \bmod N \tag{2}$$

The extension of 2D cat map to 3D [10, 11] is,

$$\begin{bmatrix} x_{n+1} \\ y_{n+1} \\ z_{n+1} \end{bmatrix} = A \begin{bmatrix} x_n \\ y_n \\ z_n \end{bmatrix} \bmod N \tag{3}$$

$$A = \begin{bmatrix} 1 + a_x a_z b_y & a_z & a_y + a_x a_z + a_x a_y a_z b_y \\ b_z + a_x b_y + a_x a_z b_y & a_z b_z + 1 & a_y a_z + a_x a_y a_z b_y b_z + a_x a_z b_z + a_x a_y b_y + a_x \\ a_x b_x b_y + b_y & b_x & a_x a_y b_x b_y + a_x b_x + a_y b_y + 1 \end{bmatrix}$$

B: *Chen's Chaotic Map*

Chen's 3D chaotic map [1] can be expressed

$$\begin{aligned} \dot{x} &= a(y-x) \\ \dot{y} &= (c-a)x - xz + cy \\ \dot{z} &= xy - bz \end{aligned} \tag{4}$$

C: *Logistic Map*

Logistic mapping [6, 10–12] was originally proposed by P. Verhulst in 1845, but
has become widely known through the work by R. May. It is the simplest among all
the chaotic maps (Fig. 4).

Fig. 4 Chen's Chaotic Map
with a = 35, b = 3 and
c = 28

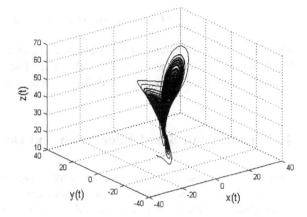

Fig. 5 Chaotic Logistic Map

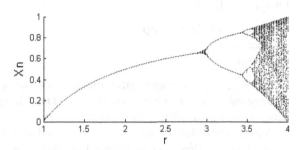

Fig. 6 Chaotic Logistic Map
sequences

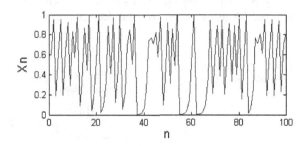

$$x_{n+1} = rx_n(1 - x_n) \tag{5}$$

Here, r is the constant and x_n is the state value (Fig. 5).

Example: Chaotic sequence generated by a Logistic Map with values of
$r = 3.9999$ and initial value of $x_n = 0.60232$ is given in Fig. 6.

D: *Chebyshev map*

Chebyshev Map [9, 13] is also one type of Chaotic Map which is trigonometric
function with Chebyshev polynomial. The equation is,

$$x_{n+1} = T_k(x_n) = \cos(k \cdot \cos^{-1} x_n), \quad x_n \in [-1, 1] \tag{6}$$

where k and $x(n)$ are parameter and state value, respectively. Choosing a value of $k \in [2, \infty)$, the system shows chaotic behaviour. The initial value of x_n and the parameter k are used as the key for diffusion module in later stage.

4 Chaotic Map Based Encryption and Decryption Model

Though there are several Chaotic Map based Image Encryption Techniques are available, but the whole process or techniques can be generalized into three different phases. These three phases are followed in all Chaotic Map based Image Encryption. These are:

- Selection of chaotic maps
- Confusion
- Diffusion

For designing these three steps different schemes are used, depending on the requirements and the objectives of individuals design, i.e., level of security, necessity of key sensitivity, fastness etc.

A generalized flow diagram of Chaotic Map based image Encryption process is given in Fig. 7. Here initially different chaotic maps and its behavior have to be analyzed. Depending on the users requirement one or more map(s) have to be selected for using encryption and decryption.

Fig. 7 Chaotic map based encryption model

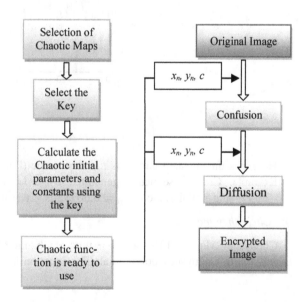

The next step is to select a secret key to be used as initial condition for the chaotic map with effective length considering brute force attack. Then to determine initial parameters and constants of the chaotic maps so that it gives proper chaotic behavior.

After that the selected chaotic maps are ready to use in encryption process which produces x_n, y_n in every iteration.

In Confusion stage chaotic mapping functions are applied to the original image to rearrange the pixel values. This process may be repeated several times to give more confusion to the output image. In case of partial image encryption [14, 15], separation of the bit planes and performing confusion to the bit values within the planes are done in this step.

The next and final step is the diffusion stage to get the complete encrypted image. In this stage, each pixel values of the output generated by confusion stage is XORed with the chaotic function values. This step can also be repeated for multiple times to achieve the higher security level.

Finally Encrypted Image is obtained to transmit to the receiver side. Similarly a generalized decryption process of chaotic map based image encryption is given in Fig. 8.

In decryption model the process is repeated in reverse order. But, the main difference is that here, everything is predefined, i.e., chaotic maps, key, diffusion technique and confusion technique. The key have to be considered same as used in encryption model. Then to calculate the initial states and constants same process has to be followed. Once the chaotic map is ready to use, first diffusion step has to done on encrypted image in reverse order. The output of the diffusion step will be the input of confusion step. Finally after completion of confusion step, final decrypted image can be retrieved.

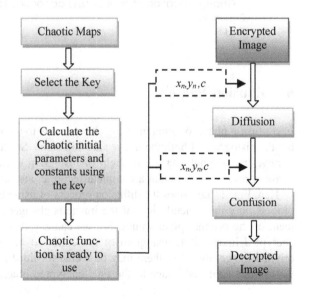

Fig. 8 Chaotic map based decryption process

Fig. 9 Example of confusion using Arnold Map **a** Original image **b** 1st Iteration

5 Confusion Phases

In confusion phase, pixel positions of the original image are interchanged using some chaotic maps or mapping functions. Usually, the pixel values are not changed in this phase unless bit-plane confusion is not done. Some of the diffusion techniques used so far in chaotic map based encryption process are discussed below:

Example 1 The 3D Arnolds Cat Map is given by Eq. 2. Here $n = 256$ is dimension of the image (jump.png). Here $\begin{bmatrix} x \\ y \end{bmatrix}$ represents the original pixel positions and $\begin{bmatrix} x' \\ y' \end{bmatrix}$ represents the new pixel positions after the Arnolds Cat Map transformation (Fig. 9).

> *Algorithm: Arnold Confusion*
> *Start*
> *While there exist pixels*
>   *Change location of pixel at (x,y) according to transformation.*
> *End loop*
> *End*

6 Diffusion Phase

In confusion phase, original image is changed due to the interchange of pixel values but the histogram of both the images are still same. So, it may easier for attacker to retrieve the original image with help of histogram equalization. Hence, diffusion phase is necessary to obtain a complete secure cipher image (Fig. 10).

Here Fig. 11 describes the diffusion process of using chaotic maps. In confusion process only the general view of the image is changed, because of the rearrangement of the original pixel values. So, we can say that confused image is also an encrypted image. But, histogram of the confused image remains same with the original image, therefore there may be a clue for third party to attack the image if we use this confused image as the final encrypted image.

Fig. 10 Confusion of a
4 × 4 matrix

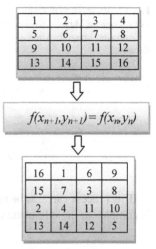

Fig. 11 Diffusion of a 4 × 4
matrix

37	34	34	40
30	39	40	38
31	32	33	37
31	37	31	37

⇩

$$f'(i,j) = x_n \oplus f(i,j)$$

⇩

165	88	197	180
128	251	230	181
49	29	195	34
98	15	229	147

Chaotic sequence
10000000
01111010
11100111
10011100
10011110
11011100
11001110
10010011
00101110
00111101
11100010
00000111
01111101
00101010
11111010
10110110

f(i,j): pixel values

The main objective of diffusion is to change the image pixel values so that histogram of the original image and encrypted image will be completely different.

Example 2 With help of a Logistic map as in Eq. 5 chaotic sequences are generated and performed XOR operation on the pixels of confusion image (Fig. 12).

Algorithm: Diffusion
Start
  While there exist pixels
  $f'(i,j) = x_n \otimes f(i,j)$, where x_n is chaotic sequence
  End loop
End

Fig. 12 Example of diffusion using Logistic Map **a** Confusion image **b** After diffusion

(a) **(b)**

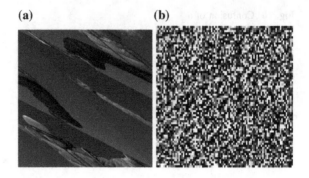

This process can also be repeated for more than once. Once the diffusion process is completed, the output image is our final encrypted image.

7 Conclusion

In this paper we were discussed chaotic map based image encryption scheme in spatial domain. Due to the important property of chaos, that is the chaotic systems are extremely sensitive to the initial conditions, makes chaotic encryption techniques more suitable for image encryption process. These chaotic maps can be used in image confusion and diffusion processes. In this paper we were trying to elaborate the different phases of chaotic based image encryption with the help of examples and illustrate that with the help of proper selection of chaotic maps and initial conditions to it, the essential security requirements in an image can be achieved.

References

1. Y. Matias and A. Shamir. 1987. A video scrambling technique based on space filling curves. Proceedings on Advances in Cryptology-CRYPTO'87, Lecture Notes in Computer Science, vol. 293, 398–417
2. CENELEC (European Committee for Electro technical Standardization). 1992(December). Access control system for the MAC/packet family: EUROCRYPT. European Standard EN 50094. Brussels: CENELEC
3. Shiguo Lian "Multimedia Content Encryption: Techniques and Application", CRC Press, ISBN 987-1-4200-6527-5, pp. 43–85
4. M. Saikia, S.J. Bora and Md. A. Hussain "A Review on Applications of Multimedia Encryption" in ISBN: 987-81-8487-088-6 in national conference on Network Security- issues, challenges and Techniques, at Tezpur University
5. K. Sakthidasan, B. V. Santhosh Krishna, "A New Chaotic Algorithm for Image Encryption and Decryption of Digital Color Images", International Journal of Information and Education Technology, vol. 1, no. 2, June 2011

6. N.K. Pareeka, Vinod Patidar, K.K. Sud, "Image encryption using chaotic logistic map", Image and Vision Computing 24 (2006) 926–934
7. https://en.wikipedia.org/wiki/List_of_chaotic_maps
8. K S Tamilkodi, (Mrs) N Rama, "A comprehensive survey on performance analysis of chaotic colour image encryption algorithms based on its cryptographic requirements", International Journal of Information Technology, Control and Automation (IJITCA), vol.5, no.1/2, April 2015
9. https://en.wikipedia.org/wiki/Chebyshev_polynomials
10. Pawan N. Khade, Prof. Manish Narnaware, "3D Chaotic Functions for Image Encryption", IJCSI International Journal of Computer Science Issues, vol. 9, Issue 3, no 1, May 2012ISSN (Online): 1694-0814
11. Guanrong Chen, Yaobin Mao, Charles K. Chui, "A symmetric image encryption scheme based on 3D chaotic cat maps," Chaos, Solutions and Fractals, vol. 21, no. 3, pp. 749–761, 2004.
12. Mayank Mishra, Prashant Singh, Chinmay Garg, "A New Algorithm of Encryption and Decryption of Images Using Chaotic Mapping", International Journal of Information & Computation Technology. ISSN 0974-2239, vol. 4, no. 7 (2014), pp. 741–746
13. Xin Ma, Chong Fu, Wei-min Lei, Shuo Li, "A Novel Chaos-based Image Encryption Scheme with an Improved Permutation Process", International Journal of Advancements in Computing Technology, vol. 3, no. 5, June 2011
14. Nitumoni Hazarika, Monjul Saikia, "A Novel Partial Image Encryption Using Chaotic Logistic Map", 2014 International Conference on Signal Processing & Integrated Networks (SPIN), IEEE, 2014
15. Monjul Saikia, Nitumoni Hazarika, Margaret Kathing "Partial Image Encryption using Peter De Jong Chaotic Map based Bit-Plane Permutation and it's Performance Analysis" published in ACEEE Fifth International Conference on Recent Trends in Information, Telecommunication and Computing ITC 2014 on Mar 21st at Chandigarh, India ISBN: 978-94-91587-21-3 Search DL ID: 02.ITC.2014.5.5 pp. 1–10 URL: http://searchdl.org/index.php/conference/view/751

A Novel Approach to E-Voting Using Multi-bit Steganography

Soura Dutta, Xavier Das, Ritam Ganguly and Imon Mukherjee

Abstract In our paper, we have proposed a new mechanism of E-voting using two layers of security using steganographic technique. The basic idea conveyed in our paper is simple, but novel. We have dealt with the Personal Identification Number (PIN) and the fingerprint for establishing uniqueness among the individual voters, in order to make a vote count. The techniques used here are the Least Significant Bit (LSB) embedding and the Minimal Impact Decimal Digit Embedding (MIDDE). If the steps are followed backwards, we retrieve the PIN and the fingerprint, which is impossible without prior knowledge of the embedding used. It has also been seen that our algorithm is foolproof against statistical attacks and malicious attempts of recovery.

Keywords Information security · E-voting · Steganography · Decimal digit embedding · Fingerprint hiding

1 Introduction

"Steganography" derives its name from the Greek word '*steganos*' which stands for 'covered' and '*graphein*' which stands for 'writing'. When we speak of steganography, we basically refer to the act of hiding certain information within some other kind of information, which is probably of a different or same type of medium.

Soura Dutta · Xavier Das · Ritam Ganguly · Imon Mukherjee (✉)
Departement of Computer Science & Engineering,
St. Thomas' College of Engineering & Technology, Kolkata 700023, India
e-mail: duttasoura@gmail.com

Xavier Das
e-mail: xav1994das@gmail.com

Ritam Ganguly
e-mail: ganguly.ritam@gmail.com

Imon Mukherjee
e-mail: mukherjee.imon@gmail.com

© Springer Science+Business Media Singapore 2017   581
J.K. Mandal et al. (eds.), *Proceedings of the First International Conference on Intelligent Computing and Communication*, Advances in Intelligent Systems and Computing 458, DOI 10.1007/978-981-10-2035-3_59

In the domain of e-voting using steganography, Chordia et al. [1] have mentioned that both the biometric information and the password are necessary, where they have used the fingerprint as the cover image. In the work of Nikita et al. [7], cryptography has been used to minimize the risk attacks, in addition to steganography, which adds extra layer of security. Lokhande et al. [6] have proposed that the cover image and the secret image should be composed into a single image on the basis of a key so that any further attack would require the complete knowledge of the key and the image templates.

We use steganography to operate upon the fingerprint, passport image and the PIN of the voter. To strengthen the security, the fingerprint with PIN is also embedded in the passport image, thereby requiring a two phase decryption in order to retrieve the PIN back. However, this technique lies in the frequency domain and going for a greater amount of change in a single pixel value of the cover can account for a significant rise in distortion which is efficiently managed in our algorithm. Our algorithm is randomly placing the PIN in the fingerprint image to protect it against statistical attacks. In this entire process, the goal remains that the fingerprint and PIN should not be tampered or even detected because they are the unique credentials that this mechanism would require in order to make a vote count.

The following notations and abbreviations, as shown in Table 1, have been used throughout the chapter.

Table 1 Notations and abbreviations used

$FP^{(F)}$:	The grayscale fingerprint in which the PIN to be embedded	$I^{(R)}_{x,y}$:	The intensity of red component of each pixel of cover image
$FP^{(S)}$:	The fingerprint that is taken from user		
$P^{(REQ)}$:	The PIN requested to be entered by the user	$I^{(G)}_{x,y}$:	The intensity of green component of each pixel of cover image
$C_{x,y}$:	The generated stego image		
B:	The binary equivalent of the $P^{(REQ)}$	$I^{(B)}_{x,y}$:	The intensity of blue component of each pixel of cover image
N_R:	A chosen random row number		
$S^{(1)}$:	The $FP^{(S)}$ embedded with $P^{(REQ)}$	$\delta[i]_{x,y}$:	Difference of last digit of $I_{x,y}$ of ith component and ith digit of $I^*_{x,y}$
$H^{(1)}$:	The height of hidden fingerprint		
$W^{(1)}$:	The width of hidden fingerprint		
H:	The height of stego image	$I^*_{x,y}$:	The intensity of each pixel of secret fingerprint image
W:	The width of stego image		
T:	Information Template		

2 Proposed Method

By this method, we make the voting process safe and secure. This procedure is divided into two modules which are *Registration* and *Login*.

The registration module allows the user to register for the distance-voting mechanism using Algorithm 3. The system auto-generates a PIN and the user's fingerprint is taken as input, which is read in grayscale. The user also inputs his passport image. N_R is generated and is embedded into the first row and the PIN is embedded into N_R+1th row. Thus, $FP^{(F)}$ is produced. $FP^{(F)}$ is embedded in passport image using Algorithm 1 (MIDDE), in the user end, as a result $C_{x,y}$ is generated and $C_{x,y}$ is transferred to the system server.

The login module allows the user to log in to cast his vote using Algorithm 4. The user inputs $P^{(REQ)}$ and $FP^{(S)}$. $C_{x,y}$ is retrieved from the database and using Algorithm 2 we decode $C_{x,y}$ to get $FP^{(F)}$. From $FP^{(F)}$, N_R is retrieved and in the N_R+1th row of $FP^{(S)}$, the $P^{(REQ)}$ is further embedded (using LSB matching) into $FP^{(S)}$ and $S^{(1)}$ is produced, which is matched with $FP^{(F)}$. Only in the case where $FP^{(F)}$ matches with $S^{(1)}$, the user is permitted to vote. The voting information is embedded into the lower half of $FP^{(S)}$, as given by the following template T. The template consists of the voter's name, the candidate's name, the candidate's party and the date.

The space that we allocate for T is 150 bytes, which is to be hidden in the lower section of the fingerprint image. Having embedded this into the $FP^{(S)}$, we embed $FP^{(S)}$ into the passport image using the MIDDE and send it back to the database.

The process can be well understood from Fig. 1.

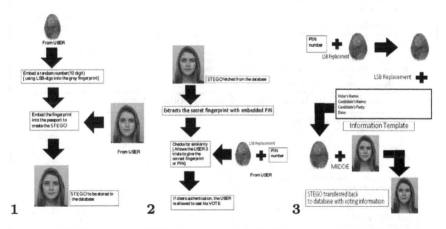

Fig. 1 *Left* (1): Registration, *Middle* (2): Login, *Right* (3): Information template (T) embedding module

Input: One cover image and secret image to hide
Output: Stego image

1   Read the intensity value ($I_{x,y}$) of the pixels of the cover image;
2   Read the intensity value ($I^*_{x,y}$) of the pixels of the fingerprint image;
3   Extract each digit of $I^*_{x,y}$ and store in array $k^*_{x,y}[i], i \in \{1,2,3\}$;
4   Extract the last digit of $I^{(R)}_{x,y} I^{(G)}_{x,y} I^{(B)}_{x,y}$ store in $k[i]_{x,y}, i \in \{1,2,3\}$;
5   Find difference vector $\delta[i]_{x,y} \leftarrow k[i]_{x,y} - k^*_{x,y}[i]$;
6   **for** *every pixel of R, G, B* **do**
7   **if** $\delta[i]_{x,y} > 5$ **then**
8 | SET $I_{x,y} \leftarrow I_{x,y} + (10 - \delta[i]_{x,y})$;
     end
9   **else if** $\delta[i]_{x,y} < -5$ **then**
10 | SET $I_{x,y} \leftarrow I_{x,y} - (10 - |\delta[i]_{x,y}|)$;
     end
11   **else**
12 | SET $I_{x,y} \leftarrow I_{x,y} - \delta[i]_{x,y}$;
     end
   end
13   Repeat the above steps until the secret image is embedded uniformly in the cover image;
14   The stego image containing the fingerprint is generated by this algorithm;

Algorithm 1: MIDDE Embedding Algorithm.

Input: The stego image
Output: The secret fingerprint image

1   **for** *x=1 to h_1* **do**
2   **for** *y=1 to w_1* **do**
3   Read intensity value of stego image $C_{x,y}$;
4   Extract last digit of $C^{(R)}_{x,y} C^{(G)}_{x,y} C^{(B)}_{x,y}$ store in $k[i]_{x,y}, i \in \{1,2,3\}$ vector respectively;
5   **for** *i=1 to 3* **do**
6 | SET $I^*_{x,y} \leftarrow I^*_{x,y} \times 10 + k[i]_{x,y}$;
       end
7   SET $y \leftarrow y + \lfloor w/w_1 \rfloor$;
     end
8   SET $x \leftarrow x + \lfloor h/h_1 \rfloor$;
   end
9   The secret fingerprint image is generated by this algorithm;

Algorithm 2: MIDDE Extraction Algorithm.

Input: Fingerprint ($FP^{(F)}$), a passport image
Output: A stego image $C_{x,y}$

1   A ten digit PIN is auto-generated, displayed and stored in an array ;
2   A row is chosen randomly and that row number (N_R) is embedded into the first row of the fingerprint image by LSB matching;
3   LSB of the pixels of N_R+1th row of the fingerprint ($FP^{(F)}$) are matched with B;
4   The fingerprint obtained is further embedded in the passport image, using Algorithm 1;
5   To maintain uniform distribution of the fingerprint image in the passport image, the row and column spacing taken appropriately;
6   The embedded passport image is saved as stego in the database;

Algorithm 3: Algorithm of Registration Module.

Input: Stego image $C_{x,y}$, PIN ($P^{(REQ)}$), fingerprint ($FP^{(S)}$)
Output: Authentication to cast vote and secure transmission

1   Extract $FP^{(F)}$ from the stego image using Algorithm 2;
2   Extract N_R from first row of $FP^{(F)}$;
3   Read $FP^{(S)}$ and $P^{(REQ)}$;
4   N_R is embedded in first row of $FP^{(S)}$ by LSB matching;
5   Select pixels of row N_R+1;
6   The PIN $P^{(REQ)}$ is embedded into $FP^{(S)}$ by matching the LSBs;
7    $S^{(1)}$ is now generated. $S^{(1)}$, should now be same as that of $FP^{(F)}$;
8   Match $S^{(1)}$ with $FP^{(F)}$;
9   **if** *match is a success* **then**
10   Allow to cast vote;
11   Embed T (filled with voting information) in lower half of the $FP^{(S)}$ using LSB Embedding;
12   Re-embed $FP^{(S)}$ in the passport image using Algorithm 1;
13   Send the embedded stego image to the database;
   end
14   **else**
15 | Go to Step 3 for two more chances;
   end

Algorithm 4: Algorithm of Login Module.

The equations used in Algorithm 1 are described below.

$$\delta[i]_{x,y} = k[i]_{x,y} - k^*_{x,y}[i],$$ (1)

where $i \in \{1, 2, 3\}$ as per component of cover and digit number of secret pixel.

$$C_{x,y} = \begin{cases} I_{x,y} - (10 - |\delta[i]_{x,y}|), & \text{if } \delta[i]_{x,y} < [-5]; \\ I_{x,y} - \delta[i]_{x,y}, & \text{if } \delta[i]_{x,y} \in [-5, 5]; \\ I_{x,y} + (10 - \delta[i]_{x,y}), & \text{otherwise.} \end{cases}$$ (2)

Our algorithm is based on following two assumptions.

Assumption 1: The size of cover must be larger than the fingerprint.
Assumption 2: Both the sender and receiver knows the dimensions of hidden finger-print image.

This MIDDE Algorithm 1 concentrates on hiding decimal digits directly instead of converting it to its binary form, since only digits from 128 to 255 require all eight bits, 0 to 127 may require any number of bits between one to seven. So, changing eight bits for a digit like 16, results in three redundant bit change. So, if we store each decimal digit directly, we can accommodate a digit in two to three LSBs only. First, from Eq. (1), we find the difference between digit to be hidden $\{k^*_{x,y}[i] : i \in \{1, 2, 3\}\}$ and the least significant digit of cover pixel $\{k[i]_{x,y} : i \in \{1, 2, 3\}\}$. As described in Eq. (2), the alteration of Least Significant Digit is done by adding or subtracting the difference or 10's complement of the difference from the cover pixel. So in cases of digits 0–3 and 7–9, the alteration affects only least significant two bits of original binary eight bits of cover pixel, but in case of digits 4 (or 6) and 5, the alteration expands up to three LSBs.

In Fig. 2, we have taken a particular pixel whose intensity for the red component, the green component and the blue component respectively are 235, 136 and 231. We take the last digits of every component and subtract every digit of 189 from those

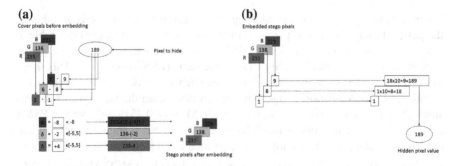

Fig. 2 a MIDDE embedding, b MIDDE decoding

Fig. 3 Fingerprint
embedding (*1st row*) and
stego image embedding (*2nd
row*) of a sample passport
image

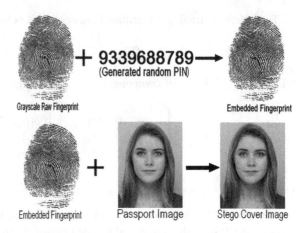

digits. The subtraction is performed in the following manner, i.e., subtracting 9 from 1, 8 from 6, 1 from 5. Accordingly, by our Eq. (2), we get the pixels 229, 138 and 231. Likewise, in the extraction procedure in Fig. 2, we take the digit 9 from 229, 8 from 128 and 1 from 231 and get the digit back which results to 189.

3 Experimental Observations and Contrast with Existing Algorithms

In this section, we will like to concentrate on the other Biometric E-Voting Algorithms available and also focus on analysing the fidelity of our stego images, therefore embedding accuracy of Algorithm 1.

3.1 Image Fidelity Analysis

Let us consider that the fingerprint image's dimension be $h \times w$ pixels and that of the passport image be $H \times W$ pixels. If their dimensions are equal or comparable to some extent, the Mean Square Error (MSE) becomes very high whereas the Signal to Noise Ratio (SNR) and Peak Signal to Noise Ratio (PSNR) decreases. This signifies that the total error in the image is high and hence, not desired. To avoid such a situation, we use a passport image that is much larger than the fingerprint's dimensions. Performing the tests, we get the average MSE as 0.2519 and average PSNR as 54.7517. The MSE value is much lesser and the PSNR is higher which is quite optimal. The comparison of sample passport image and stego image can be seen in Fig. 3. The visual perceptibility can be measured with the help of MSE, SNR and PSNR. We find low MSE and high PSNR value as shown in Table 2, that proves in spite of

Table 2 Table to show accuracy attributes after double layer embedding

Sample image	Size	MSE	SNR (dB)	PSNR (dB)
Passport1.bmp	1544 × 1113	0.0947	49.8131	58.3658
Passport2.bmp	600 × 600	0.3475	44.3124	52.7216
Passport3.bmp	800 × 599	0.3136	45.8158	53.1677

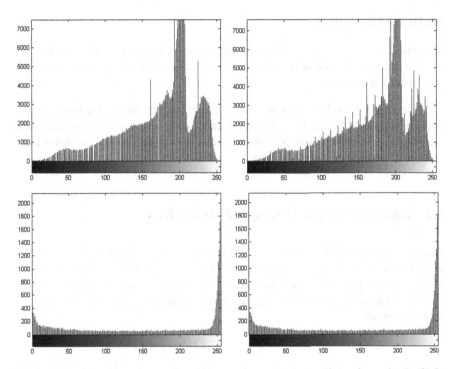

Fig. 4 Histograms of the red component of the sample passport image (*1st*) and stego image (*2nd*), raw grayscale fingerprint (*3rd*) and embedded grayscale fingerprint (*4th*)

this double layer embedding, we get pretty satisfying PSNR, MSE values as ultimate embedding result.

Histogram Analysis

Definition 1 A histogram is a graphical representation of the distribution of numerical data. It is an estimate of the probability distribution of a continuous variable (quantitative variable).

Without loss of generality, we select red component of sample and stego for comparison in Fig. 4. We see that the histograms between the passport image and the stego image are almost the same without any imperceptible visual difference. Any change in characteristic is hardly detectable.

BP0 BP0 BP1 BP1 BP2 BP2 BP3 BP3 BP4 BP4 BP5 BP5 BP6 BP6 BP7 BP7
Cover Stego Cover Stego Cover Stego Cover Stego Cover Stego Cover Stego Cover Stego Cover Stego

Fig. 5 Comparison of bitplane slices of sample passport image and stego image

Bit Plane Slice Analysis

Definition 2 A bit plane of a digital discrete signal (such as image or sound) is a set of bits corresponding to a given bit position in each of the binary numbers representing the signal.

For example, for 8-bit data representation there are 8 bit planes: the first bit plane contains the set of the most significant bit, and the 8th contains the least significant bit. We divide the sample passport image and stego in eight bit slices and comparing them pairwise in Fig. 5 in this section.

3.2 Robustness Against Sample Pair Analysis

Dumitrescu et al. [2, 3] mention that the test is valid only for natural images. We embed PIN into the fingerprint. Fingerprint is not considered as natural image. Hence, our method withstands the test. So our method is not breakable by Sample Pair Analysis. In other words, our distance e-voting mechanism is secured from statistical attacks.

3.3 Robustness Against Correlation Analysis

Normalalized Correlation Coefficient (Say S) can be calculated as:

$$S = \frac{\sum_{x,y=0}^{H \times W - 1}(I_{x,y} \times C_{x,y})}{\sum_{i,j=0}^{H \times W - 1} I_{x,y}^2} \tag{3}$$

In Table 3, let X be the cover image and Y be the stego image.

The sample pair test also reveals that the embedding accuracy is very high, although it decreases for lower-size images. In the work of Kahdum [5], the calculation of Normal Correlation Metric was done and it was found that the correlation metric was always greater than one in every test case. But, in our paper the correlation tests of the cover images against the stego images are less than one, although extremely close to one, showing very high accuracy. The results are as follows (Table 4),

Table 3 Attack test results

Set (X, Y)	Size (in pixels)	Correlation of X against X	Correlation of X against Y	Sample Pair Test
Set 1	(1113 × 1544)	1.0000	0.9999	0.0104
Set 2	(599 × 800)	1.0000	0.9999	0.0312
Set 3	(600 × 600)	1.0000	0.9994	0.2714

3.4 Performance Against Existing Techniques

Unlike the state-of-the-art works [6–8], where the fingerprints are not matched during login and registration, but here we attempt to do so. Hence, they cannot be properly termed as a 'Biometric' Technique of distance voting, because they only use fingerprints as cover images of Personal Identification Numbers or Secret Keys as stated in technique [1]. But our method aims at introducing the matching between two fingerprints (as grayscale images) provided at two different times, before allowing the voter to cast his vote. Fingerprints are matched not in their stand-alone forms, but in a state where the PIN is embedded into the fingerprint. This paper also presents dual layers of security powered by LSB matching technique and MIDDE Algorithm 1, which reduces the probability of success of different attacks. Also, randomizing the row number, in which the PIN shall be stored in, is an added advantage since it does not follow any pattern. In the work of Rura et al. [9], it is seen that they have used ballot papers and have implemented the concept of visual cryptography. However, this works only for black and white ballot papers. Secondly, it is not possible for the ballot paper to be digitally recognized. Ballot papers will require manual comprehension to be understood. Contrary to above, in our approach, the use of LSB and MIDDE algorithms help in the decryption of the voting details. In addition to the work of [4], we have added a new method of conveying the voting information securely in embedded form without taking any extra space with the fingerprint image that was embedded in the cover.

Table 4 Normalized correlation metric value

Image sets	Normalized correlation metric value
Cover image described in [5]	1
Stego image described in [5]	1.02883

4 Conclusion

In this paper, we see that this approach to e-voting using both LSB and Multi-bit steganography adds two layers of security, therefore being a novel approach. This paper is resistant to statistical and visual attacks like chi-square and LSB-Enhancement because of the randomness in the position of the PIN, which is previously encrypted. The PSNR values haves also proved to be quite good and our MIDDE algorithm has stood the test against sample-pair analysis.

References

1. Chordia, P., Chavan, P., Patil, B., More, P.: On Vote Secured Online Voting, International Journal of Innovative Research in Computer and Communication Engineering, vol. 1(9), pp. 2117–2120 (2013)
2. Dumitrescu, S., Wu, X., Memon, N.: Steganalysis of Random LSB Embedding in Continuous-tone Images, IEEE ICIP, vol. 3, pp. 641–644, New York, USA (2002)
3. Dumitrescu, S., Wu, X., Wang, Z.: Detection of LSB Steganography via Sample Pair Analysis, IEEE Transactions on Signal Processing, vol. 51(7), pp. 1995–2007 (2003)
4. Dutta, S., Das, X., Ganguly, R., Mukherjee, I., and Bhattacharjee, S., "A Novel Distant E-Voting Mechanism Using Dual Layer Security", In Proceedings of International Conference on Telecommunication Technology & Management (ICTTM- 2015), IIT-Delhi, India, (2015), ISBN: 987-0-9926800-5-3
5. Kahdum, A. I.: Stegananlysis using Image Quality Metrics (Normalized Correlation Metric), Al - Faith Journal, no. 32, pp. 44–51 (2008)
6. Lokhande, S.: E-Voting through Biometrics and Cryptography - Steganography Technique with conjunction of GSM Modem, Emerging Trends in Computer Science and Information Technology (ETCSIT2012), pp. 38–42 (2012)
7. Nikita, M., Patil, C., Chavan, S., Raut, S. Y.: Secure Online Voting System Proposed By Biometrics And Steganography International Journal of Emerging Technology and Advanced Engineering, vol. 3(5), pp. 213–217 (2013)
8. Swaminathan, B., Dinesh, J. C. D.: Highly Secure Online Voting System with Multi Security using Biometric and Steganography, International Journal of Advanced Scientific Research and Technology, vol. 2(2), pp. 195–203 (2012)
9. Rura, L., Issac, B., Haldar, M. K.: Secure Electronic Voting System based on Image Steganography, IEEE Conference on Open Systems (ICOS2011), Langkawi, Malaysia, pp. 80–85 (2011)
10. http://www.pixelsquid.com
11. http://www.canstockphoto.com/images-photos/fingerprint

Comparative Analysis of Classification Techniques in Network Based Intrusion Detection Systems

Sunil Kumar Gautam and Hari Om

Abstract An Intrusion Detection System (IDS) monitors the system events and examines the log files in order to detect the security problem. In this paper, we analyze the classification algorithms, especially Entropy based classification, Naïve classifier, and J48 using KDD-CUP'99 dataset to detect the different types of attacks. The KDD-Cup'99 dataset is a standard dataset for analysing these type of classification techniques. In KDD-CUP'99 dataset, each instance corresponds to either attack or normal connection. The KDD-Cup'99 dataset contains mainly four types of attack, namely, DOS, U2R, R2L, Probe and these four types of attacks also have subcategories attacks. In this paper, we carry out simulations on the KDD-Cup'99 dataset for all four types of attacks and their subcategories. The back, land, Neptune, pod, smurf, teardrop belong to DoS; the rootkit, Perl, loadmodule, buffer-overflow belong to U2R; the FTP-write, spy, phf, guess-passwd, imap, warezclient, warezmaster, multihop belong to R2L, and the Ipsweep, nmap, portsweep, satan belong to the probe. The simulation results show that the entropy based classification algorithm gives high detection rate and accuracy for normal instances over the J48 and Naïve Bayes classifiers.

Keywords Intrusion detection system · Naïve Bayes · J48 · KDD-CUP'99 dataset

S.K. Gautam (✉) · Hari Om
Department of Computer Science & Engineering, Indian School of Mines,
Dhanbad, India
e-mail: gautamsunil.cmri@gmail.com

Hari Om
e-mail: hariom4india@gmail.com

© Springer Science+Business Media Singapore 2017
J.K. Mandal et al. (eds.), *Proceedings of the First International Conference
on Intelligent Computing and Communication*, Advances in Intelligent Systems
and Computing 458, DOI 10.1007/978-981-10-2035-3_60

1 Introduction

Intrusion Detection Systems have accomplished to find a real time intruder in network systems and provide Confidentiality, Integrity, Availability (CIA) model for data security [1]. The Internet is a platform for many users to gather and exchange the information among them. As a result, several applications have been developed in recent years for user's flexibility. An IDS provides security to information from malicious activities in a network by removing the unauthenticated data that contains malicious programs. These systems monitor the activities of a network and determine whether these activities are malicious or legitimate. The malicious programs (attacks) are classified into following five different categories: Denial of Service (DoS), User to Root (U2R), Remote to Local (R2L), Information Gathering (Probe), and Normal. An IDS detects the intrusion by using two types of methodologies i.e., either misuse detection or anomaly detection. The former searches for specific patterns or sequences of programs and user behaviours that match the well-known intrusion scenario. In contrast, the latter methodology detects the predefined programs and also novel programs which are randomly generated by attackers. Therefore, the anomaly detection methodology is more useful for intruder detection [2]. Due to rapid growth of data on the network, the detection systems need data dimensionality reduction techniques or features selection algorithms for large dataset such as KDD-CUP'99 dataset. Feature selection algorithms or reduction techniques minimize the dimensionality of a dataset by eliminating the redundant features [3]. In this paper, we apply classification techniques, namely Naïve Bayes, J48 [4], and Entropy based algorithm [5] on the KDD-CUP'99 dataset. This study shows that the entropy based algorithm is more efficient for detecting the attacks in network based intrusion detection system. The remainder of the paper is organized as follows. Section 2 discusses various intrusion detection systems and Sect. 3 describes related works. Section 4 presents research methodologies and simulation results are described in Sect. 5. Finally, in Sect. 6 we draw the conclusion.

2 Intrusion Detection System

An Intrusion Detection System (IDS) provides insurance of CIA features and also examines all processes, analyzes the events in a network. It detects suspicious activities and sends a message about the security problems to the network administrator. The IDSs are mainly divided in two categories, namely Host based Intrusion Detection System (HIDS) and Network based Intrusion Detection System (NIDS). An HIDS is capable of detecting intrusions on a single system only. The advantage of this system is that it can analyze the end-to-end encrypted communication activities. An NIDS analyses the entire network data packets as well as it can be applied in a single host environment. However, it cannot analyze a wireless protocol and also detect attacks in the encrypted messages [6].

In a large dataset, detection of malicious activities is a challenging task because the large dataset contains lot of redundant data, for instance the KDD'99, a huge dataset, contains lot of duplicate data. Therefore, to remove the redundant data various classification techniques such as statistical decision theory, artificial neural network, support vector machines, Naïve Bayes classifier, and J48 are used. These classification techniques are able to differentiate the normal and abnormal (intrusions) connections. This can be restricted to consider a two-class problem, focussing on distinguishing attacks or normal patterns. The main aim is to produce a classifier that works correctly on the unseen examples (test dataset). The KDD (Knowledge Discovery and Data Mining) CUP'99 dataset was derived from the DARPA in 1998. The KDD dataset contains 5 million instances with 41 features out of which 9 are nominal features and 31 are continuous features. In this paper, we take 10 % of the training data from the original dataset that contains 494,021 instances and the test dataset contains 331,029 instances. The KDD dataset contains 22 types of attacks, which are classified into the following four categories [7]:

- Denial of Service (DoS) attacks: In DoS, an attacker wants to keep the system resources too busy by sending continuous requests to the server. Due to attacker's bogus requests, the legitimate user can't access the machine.
- User-to-Root (U2R) attacks: In U2R, an attacker initiates with access to a normal user account on the system and is able to exploit vulnerability to gain root access to the system.
- Remote to Local attack (R2L): In R2L, an attacker sends the packets to a machine over a network that exploits the machines vulnerability to illegally gain local access as a user.
- Probe attack: In probe, an attacker scans a network to gather information or find known vulnerabilities.

The attacks have 22 subcategories consisting of the given attacks. The back, land, neptune, pod, smurf, teardrop belong to DoS; the rootkit, perl, loadmodule, buffer-overflow belong to U2R; the ftp-write, spy, phf, guess-passwd, imap, warezclient, warezmaster, multihop belong to R2L; and the Ipsweep, nmap, portsweep, satan belong to Probe.

3 Related Work

Feature selection is a vital issue in selecting appropriate features from a dataset and discarding the irrelevant ones. The objective of a features selection technique is to find a subset of features that describes the given problem with minimum degradation of performance. The feature selection algorithms have following advantages [8]:

- Relevant selection algorithm easily describes the dataset behaviour, which helps to visualize the data processing.
- The feature selection algorithm reduces the dataset size, which often reduces the computing cost.
- Simplicity that helps in using simpler models and gaining speed.

Hence, feature selection is an important aspect for developing a subset by removing the irrelevant features from an original dataset. Researchers have developed many features selection algorithms such as support vector machine based techniques, correlation based techniques, chi-square test based techniques [9]. Liu et al. propose a binary particle swarm optimization with support vector machine for reducing the redundant features as well as computational cost. This technique eliminates noisy features and reduces the time interval during the selection of relevant features [10]. Feature selection is a changeling task when the dataset's features are unlabelled; thus requiring a new mechanism to solve this problem. The semi supervised learning provides wrapper type. The genetic algorithm based semi-feature selection methods change the unlabeled samples to a primary labelled training datasets with the help of a classifier. This methodology is applicable in both wrapper type (based on backward feature elimination) selection algorithm and traditional filter type (select features based on discriminant criteria) feature selection algorithms [11]. The correlation based feature selection algorithm can also be used for feature selection with classifier-independent function. Inter-correlation between features is also useful in selecting features [12]. Kumar et al. have used this algorithm reducing the image features and the features of hand shape and palm texture [13]. Liu et al. have performed dimension regression by using weighted chi-squared test integrated with sliced inverse regression [14]. Enormous size datasets contain several duplicates or irrelevant features that consume the time and space in analysis process affecting the results accuracy. Erick et al. have proposed the filter and traditional approaches for feature selection for diverse applications such as remote sensing data. Features can be ranked individually by using filtering methods based on the squared statistics and the feature subsets can be selected based on their ranking [15]. The feature selection algorithms use two approaches, i.e. traditional wrapper and filter for selecting the particular instances. Zhu et al. have proposed a hybrid methodology for feature classification on a memetic framework that improves the classification performance and accelerates the search in identifying important feature subsets [16]. Julia et al. have discussed an unsupervised feature selection method to reduce the higher dimensions of the dataset that contains a number of features [17].

4 Research Methodology

4.1 Entropy Based Classification

Gautam et al. [5] have discussed an algorithm based on entropy, which is the average amount of information contained in the dataset. The entropy characterizes uncertainty about the source of information. This algorithm classifies different types of attacks. The rate of entropy is less than when the class distribution is pure, i.e. it belongs to single class. The rate of entropy is larger when the class distribution is impure i.e. class distribution belongs to many classes. In this paper, we take two types of datasets, namely training and testing datasets. The entropy based classification algorithm is given below:

input: Training dataset form KDD dataset of size $n \times m$.
Testing dataset also from KDD dataset of size $n \times m$

$$n = number\ of\ attack's\ records$$
$$m = number\ of\ features\ selected.$$

output: Number of records correctly classified.
Number of records incorrectly classified.
Efficiency of algorithm using confusion matrix.

4.2 J48 Algorithm

J48 algorithm, developed by Ross Quinlan, is an extension of the Iterative Dichtomiser (ID3) algorithm that is used for classification of class labels. The J48 algorithm is more advance than ID3 due to additional features such as accounting for missing values, decision tree pruning, continuous attribute value ranges, derivation of rules, etc. It creates a decision tree by divide-and-conquer paradigm to split a root node into a subset of two partitions till the leaf node (target node) occurs in the tree. It creates the rules for target instances and given the decision tree structure of the dataset. The verification process in J48 is done by C4.5 algorithm. The C4.5 is implemented on Weka tools, freely available on http://www.cs.waikato. ac.nz/ml/weka. The Waikato Environment for Knowledge Analysis (Weka) is a data mining tool that provides a number of options associated with tree pruning. This tool requires a Java virtual machine (JVM) for executing J48 classes. The big dataset is split into more than one class, which are represented by decision tree and J48 algorithm does not contain any code for building this decision tree [4, 18–20].

4.3   Naïve Bayes (NB)

The Naïve Bayes classifier is widely used for classification problem in data mining and machine learning. It is more popular due to its easiness and detection rate of the correct instances. The Naïve Bayes classifier removes the data redundancy and improves the classification accuracy. This classifier is more popular for real time problems due to its feature that are given below [21].

- It is not sensitive to irrelevant features.
- Independence of features is assumed.
- It is optimal if the features are really independent.
- It has low storage requirement.

The Naïve Bayes classier is applied in KDD cup'99 dataset in network based intrusion for classifying network attacks. The Naïve Bayes has some drawback also such as it is not suitable for high dimensional data due the underflow and overfitting problem. Chandra et al. have solved this problem and improved the accuracy for high dimensional dataset [22, 23]. The Naïve Bayes classification is not applicable for miscellaneous dataset. This type of dataset contains both categorical and numeric datasets. Hsu et al. have extended the Naïve Bayes method to solve this issue by statistical theory when handling numeric instances [24]. In this paper, we implement the Naïve Bayes classification by Weka tools, thus we get corrected classified instances.

5   Experimental Results

In this paper, we have done a comparative study between the Naïve Bayes (NB) Classifier, J48 algorithm, and Entropy based algorithm for DoS, U2R, R2L, and Probe. The experiments have carried out on Knowledge Discovery and Data Mining (KDD)'99 cup dataset, a real time dataset provided by Defense Advanced Research Projects Agency (DARAPA), United States. It contains all information about the network connection, such as protocol type, connection duration, and login types, etc. In these experiments, we have taken 10 % of the KDD dataset. The KDD datasets categorized in 22 sub-subcategories attacks. We have done our simulations using MatLab and Weka tools. The WEKA tool is an open source platform inde-pendent software which is freely available and contains several in-built datamining and machine learning algorithms. We have used the WEKA as API in MatLab and converted the dataset file into Attribute-Relation File Format (.arff), thus refining the dataset. In next steps, we have calculated the classification accuracy in terms of confusion matrix. A confusion matrix (see Table 1) contains all information about the actual class and predicated class and classification has been done by the clas-sification system. Further, in confusion matrix, the TP (true positive) and TN (true

Table 1 Confusion matrix

Actual class	Predicated class	
	C	NC
C	TN	FP
NC	FN	TP

Here, *C* Anomaly Class, *NC* Normal Class
TN True Negative, *FP* False Positive
FN False Negative, *TP* True Positive
Accuracy = (TN + TP)/(TN + TP + FN + FP), Precision (P) = TP/(TP + FP)

Table 2 Detection Rate (%) in DoS attacks

Attacks	Naive Bayes	J48 algorithm	Entropy based algorithm
Back	74.76	91.72	94.85
Land	92.68	93.12	94.12
Neptune	80.43	87.81	92.47
Pod	77.79	83.45	87.96
Smurf	91.82	92.34	94.59
Teardrop	87.62	90.15	92.18

negative) combination is known as correctly classified instances; the FP (false positive) and FN (false negative) addition is known as incorrectly classified.

Here, we have done a comparative study among the Naïve Bayes (NB) Classifier, J48 algorithm, and our existing Entropy based algorithm for all attacks. Table 2 contains the detection accuracy rate for all Denial of Service attacks that are shown in Fig. 1 using bar graph. Tables 3, 4 and 5 contain the detection rates of U2R, R2L and Probe and these attacks, which have been shown in Figs. 2, 3 and 4, respectively.

Fig. 1 Denial of service attacks

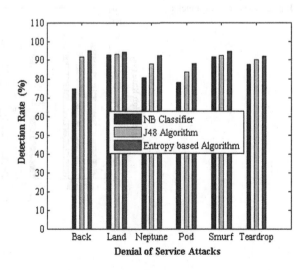

Table 3 Detection Rate (%) in U2R attacks

Attacks	Naive Bayes	J48 algorithm	Entropy based algorithm
RootKit	87.21	90.75	92.65
Perl	90.41	94.25	95.49
Load Module	92.78	93.78	94.15
Buffer Overflow	94.56	95.12	95.87

Table 4 Detection Rate (%) in R2L attacks

Attacks	Naive Bayes	J48 algorithm	Entropy based algorithm
Ftp_Write	92.85	93.78	95.78
Spy	94.58	95.77	96.15
Phf	92.78	94.65	92.16
Guess_pwd	85.19	93.78	94.18
Imp	90.12	92.43	95.29
Warezclient	78.12	88.94	91.63
Warenmaster	92.56	93.75	91.63
Multihop	89.12	91.98	93.88

Table 5 Detection Rate (%) in probe attacks

Attacks	Naive Bayes	J48 algorithm	Entropy based algorithm
Ipsweep	86.52	90.54	93.85
Nmap	88.34	91.26	94.56
Portsweep	90.44	93.65	92.89
Satan	79.85	92.78	95.12

Fig. 2 User to root attacks

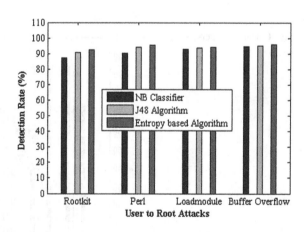

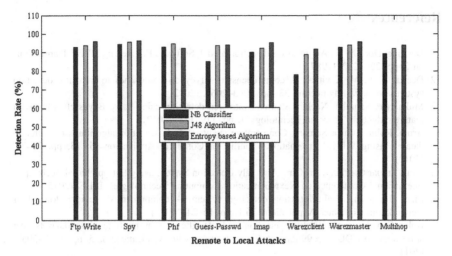

Fig. 3 Remote to local attacks

Fig. 4 Probe attacks

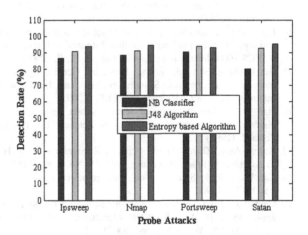

6   Conclusions

This paper has discussed the detection rate of all individual attacks, i.e. its focus has been on the DoS, U2R, R2L and Probe subcategories attacks. Experimentally, the entropy based algorithm performs significantly over the Naïve Bayes classifier and J48 algorithm. Under R2L subcategories, the Phf attack detection rate values are comparable to that of the J48 algorithm and under Probe subcategories the Ports-weep attack detection rate values are comparable to that of the J48 algorithm. In future, we will use some more datasets to validate the entropy based algorithm and also improve the results for the Phf and Portsweep attacks.

References

1. Denning, Dorothy E.: An intrusion-detection model. Software Engineering, IEEE Transactions on 2, pp. 222–232. (1987)
2. Deepa, A. J., V. Kavitha.: A comprehensive survey on approaches to intrusion detection system. Procedia Engineering 38, pp. 2063–2069. (2012)
3. Mukherjee, Saurabh, Neelam Sharma.: Intrusion detection using naive Bayes classifier with feature reduction. Procedia Technology 4, pp. 119–128. (2012)
4. Arora, Rohit, Suman Suman.: Comparative analysis of classification algorithms on different datasets using WEKA. International Journal of Computer Applications 54.13, pp. 21–25, (2012)
5. Sunil Kumar Gautam, Hari Om.: Anomaly Detection System using Entropy based Technique. International Conference on Next Generation Computing Technologies, IEEE, (2015)
6. Liao, Hung-Jen, et al.: Intrusion detection system: A comprehensive review. Journal of Network and Computer Applications 36.1, pp. 16–24. (2013)
7. Bolon-Canedo et al.: Feature selection and classification in multiple class datasets: An application to KDD Cup 99 dataset. Expert Systems with Applications 38.5, pp. 5947–5957. (2011)
8. Guyon et al.: Eds. Feature extraction foundations and applications. Vol. 207. Springer, (2008)
9. Singh, Rajdeep et al.: Analysis of Feature Selection Techniques for Network Traffic Dataset. Machine Intelligence and Research Advancement (ICMIRA), 2013 International Conference on. IEEE, (2013)
10. Liu, Weili, Dexian Zhang.: Feature subset selection based on improved discrete particle swarm and support vector machine algorithm. Information Engineering and Computer Science, 2009. ICIECS 2009. International Conference on. IEEE, (2009)
11. Bu, Hualong, Shangzhi Zheng, Jing Xia.: Genetic algorithm based Semi-feature selection method. Bioinformatics, Systems Biology and Intelligent Computing, 2009. IJCBS'09. International Joint Conference on. IEEE, (2009)
12. Hall M.A.: Correlation-based feature subset selection for machine learning. Doctorate dissertation, Department of Computer Science, University of Waikato, Hamilton, New Zealand, (1999)
13. Kumar Ajay, Zhang David.: Personal Recognition Using Hand Shape and Texture. IEEE Transaction on Image Processing, Vol. 15, No. 8, pp. 2454–2461, (2006)
14. Liu, Huan, Rudy Setiono.: Chi2: Feature selection and discretization of numeric attributes. Tai. IEEE, (1995)
15. Cantú-Paz Erick, Newsam Shawn, Kamath Chandrika.: Feature selection in scientific applications. 10th ACM SIGKDD international conference on Knowledge discovery and data mining, Seattle, WA, USA, pp. 788–793, (2004)
16. Zexuan Zhu.: Wrapper–Filter Feature Selection Algorithm Using a Memetic Framework, IEEE Transactions on Systems, Man, and Cybernetics—Part b: Cybernetics, Vol. 37, No. 1, pp. 70–76,(2007)
17. Handl Julia, Knowles Joshua.: Feature Subset Selection in Unsupervised Learning via Multi objective Optimization. International Journal of Computational Intelligence Research, Vol. 2, No. 3, pp. 217–238, (2006)
18. Chandolikar, Mrs NS, V. D. Nandavadekar.: Efficient algorithm for intrusion attack classification by analyzing KDD Cup 99. Wireless and Optical Communications Networks (WOCN). 2012 Ninth International Conference on. IEEE, (2012)
19. Kaur, Gaganjot, and Amit Chhabra.: Improved J48 Classification Algorithm for the Prediction of Diabetes. International Journal of Computer Applications 98.22 (2014)
20. Gupta, D. L., A. K. Malviya, Satyendra Singh.: Performance analysis of classification tree learning algorithms. IJCA) International Journal of Computer Applications 55.6 (2012)
21. Hsu, Chung-Chian, Yan-Ping Huang, Keng-Wei Chang.: "Extended Naive Bayes classifier for mixed data." Expert Systems with Applications 35.3, pp. 1080–1083, (2008)

22. Farid, Dewan Md, et al.: "Hybrid decision tree and naive Bayes classifiers for multi-class classification tasks." Expert Systems with Applications 41.4, pp. 1937–1946, (2014)
23. Chandra, B., Manish Gupta.: Robust approach for estimating probabilities in Naïve–Bayes Classifier for gene expression data. Expert Systems with Applications 38.3, pp. 1293–1298, (2011)
24. Baron, Grzegorz.: "Influence of Data Discretization on Efficiency of Bayesian Classifier for Authorship Attribution." Procedia Computer Science 35, pp. 1112–1121, (2011)

A New and Resilient Image Encryption Technique Based on Pixel Manipulation, Value Transformation and Visual Transformation Utilizing Single–Level Haar Wavelet Transform

Arindrajit Seal, Shouvik Chakraborty and Kalyani Mali

Abstract Lossless image cryptography is always preferred over lossy image cryptography. In this approach the authors have proposed a very resilient and novel image encryption/decryption algorithm. Initially the image is first converted to frequency components and the encryption is performed on sub-bands and the encrypting algorithm is found to be very strong, reliable and strong. The encryption algorithm involves pixel breakup into two parts and reversing parts of the pixel. The results show a deviation of pixel between the images present in the original and encrypted domains. The decryption algorithm is exactly the encryption algorithm in reverse. The proposed algorithm is evaluated by standard measures and it is seen to be attack-resistant to well-known attacks.

Keywords Cryptography · Lossless · Single-level Haar transform · Image encode

1 Introduction

Cryptography is a means of concealing important information so that it is not seen by intruders. A technique in which important information is hidden and kept a secret while it is passed over unsecured network or communication path comprises the basic principle of cryptography. Image encryption is different to that of text encryption based cryptography. Image encryption are mainly of two types namely

Arindrajit Seal (✉) · Shouvik Chakraborty · Kalyani Mali
CSE Department, University of Kalyani, Kalyani, India
e-mail: arindrajit.seal@gmail.com

Shouvik Chakraborty
e-mail: shouvikchakraborty51@gmail.com

Kalyani Mali
e-mail: kalyanimali1992@gmail.com

© Springer Science+Business Media Singapore 2017 603
J.K. Mandal et al. (eds.), *Proceedings of the First International Conference on Intelligent Computing and Communication*, Advances in Intelligent Systems and Computing 458, DOI 10.1007/978-981-10-2035-3_61

lossless and lossy. Lossy images are images which are not the exact replica of the original images. If the losses are marginal then there is no problem in accepting lossy images but generally lossless encryption is much more preferred. Various image encoding or encryption methods can be broadly grouped into 3 major categories:-permutation of positions, transformation of values and visual transformation. The proposed algorithm uses all the three categories. The proposed algorithm uses Discrete Wavelet Transform (DWT) with only a single-level of decomposition. The original image is got back by using the Inverse Wavelet Transformation.

2 Related Works

There is a major difference between an image and a text. Images are very large. Another difference lies in the fact that whenever a compression technique is used on text there is rarely any loss. For this reason both lossy and lossless encryption is paid heed to for images. There are two main forms of image encryption namely Spatial Domain Cryptography and Frequency Domain Cryptography. Our proposed algorithm uses two of these features contiguously.

The paper proceeds as follows. Sections 2.1 and 2.2 introduces the concepts of spatial domain and Frequency domain Cryptography. Section 3 introduces the Haar Wavelet Transform. Section 4.1 highlights the Proposed Algorithm. Section 4.2 shows us the Decryption Algorithm. Section 5 deals with Results and Discussion and Sect. 6 deals with conclusion.

2.1 Spatial Domain Cryptography

Many schemes in the spatial domain have been proposed. Maniccam and Bourbakis [1] proposed a lossless approach of performing compression losslessly. Bhatnagar and Wu [2] proposed a selective method which is dependent on curves of filling of spaces, pixels of particular locations and/or chaotic maps with nonlinear property and decomposition using singular value. Yen and Guo [3] presented an image encoding or encryption method which is dependent on a sequence that is binary and which is developed from a chaotic approach. Hou [4] has used the properties of human vision for decrypting encrypted images. In his work, the author has given three techniques for visual encryption of a grayscale image and also color images dependent on halftone method. Abdalla and Yahya [5] has stated a shuffle encryption method performing nonlinear byte replacement. The method executes a shuffling operation which is partially based on the input data and uses the key which is given by the user. Chen et al. [6] had proposed a technique to encode a colour image using the transform termed as ARNOLD as well as the method of interference. In their experiment, a colour image is divided into three separate channels (Red, Green, and Blue), and every channel is then encoded or encrypted to two unique, random phase masks.

2.2 Frequency Domain Cryptography

Researchers have attempted to use both image or block based transform methods. In the first case, the entire image is changed and in the second case the image pixels are partitioned into blocks of same sizes and then each block is transformed one after the other. Tedmori and Al-Najdawi [7] had presented a lossless technique in the frequency transform domain. In their work, the encoding algorithm uses the discrete cosine transform (DCT) to convert any given image into its frequency components. The sub-bands are involved in encryption by using a weighting factor and various swaps.

3 Haar Wavelet

The modus operandi of image transform is to lessen the correlation in the transformed image so that it cannot be deciphered. The DWT is a wavelet transform (such as Daubechies wavelets, Haar wavelet etc.) where discrete analysis of wavelets occur. The main advantage of DWT over block-based is temporal resolution where spatial and frequency information is achieved. In this paper we have used the Haar transform. It was proposed by a Hungarian mathematician Alfred Haar in 1910. It is very useful in real life cases of signal and image compressions in computer engineering. The transform is derived from the Haar matrix. A fine example of a 4 × 4 Haar transformation matrix is given in Fig. 1.

The Haar wavelet analogy or transform can be seen as a sampling process where rows of the matrix behave as fine resolution samples.

The main function of a DWT when used in a 2-Dimensional discrete image having N × N dimensions is: each and every row of a 2D image is made to pass through both low-pass and a high-pass filters (Lx and Hx) and the resulting matrix of each filter is sampled by 2nd factor to get L and H. L is the main image which is passed through a low-pass filter and also sampled in the (horizontal) x-direction and H is the detailed image which is passed through a high-pass filter and also sampled in the x direction. Then we perform the same action on each and every column just as we performed on rows as given in Fig. 2. We get four bands namely LL, LH, HL and HH. All of the above bands are part of the frequency domain and the interesting thing is that after applying Inverse Haar Transform we can get back the exact original image as before. A live example of the Haar Transform of 1st level is given in Fig. 3.

Fig. 1 The 4 × 4 Haar transformation matrix

1	1	1	1
1	1	-1	-1
$\sqrt{2}$	$-\sqrt{2}$	0	0
0	0	$\sqrt{2}$	$-\sqrt{2}$

Fig. 2 The Haar wavelet
decomposition

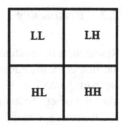

Fig. 3 Example of applying
Haar wavelet (original image
of Lena—*Left* and image
produced by Haar wavelet
transform—*Right*)

4 Proposed Algorithms

4.1 Encoding Algorithm

The encryption is done using single-level Haar Wavelet Transform of the actual
image. The actual image is broken down into sub-bands by making it pass through
DWT (Discrete Wavelet Transform). The sub-bands are LL1, LH1, HL1, and HH1.
The algorithm is summarized in a flowchart as shown below in Fig. 4.

The algorithm starts at the second stage when every pixel value is converted into
its inverted binary 8 bit form. The inverted binary sequence is divided into two parts
and these are inverted separately. Two parts are concatenated and converted into its
decimal value and the original pixel is replaced with this value. The decimal value
of the reverse of the inverted binary value is taken and substituted for every pixel of
LL1. The next step involves inverting or reversing the sign of the other frequencies
by multiplying the details matrices LH1, HL1, and HH1 by a value of (−1) as
follows: LH1 - > LH1 × − 1, HL1 - > HL1 × − 1, HH1 - > HH1 × − 1.

The reason behind performing the sign reverse operation is that the value of the
sinusoid corresponds to the subtraction of the darkest and brightest portions of
the image. A value less than zero signifies contrast-reversal or inversion. Swapping
the values of LH1 with HH1 and vice versa also swapping the values of HL1 with
LL1 and vice versa is done. The resultant LL1 matrix is further vertically divided

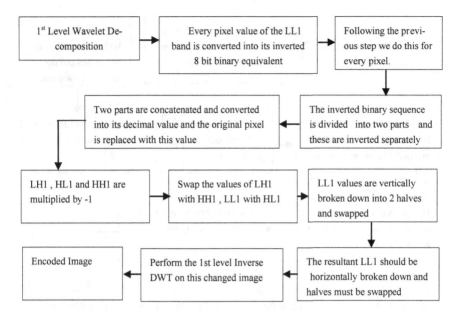

Fig. 4 The encryption algorithm

into two halves, the left and the right halves are swapped. The resultant LL1 matrix is again divided horizontally into two halves the top and the bottom halves are swapped. Perform the single-level inverse discrete transform on the image and we get the final encoded image.

4.2 Decryption Algorithm

Decryption is just the reverse of the encryption method. The decryption algorithm begins with first performing a single-level Haar wavelet transform of the encoded image. The Fig. 5 shows the block diagram of the decryption algorithm.

First we need to break LL1 into two halves upper and lower. Swap the two halves horizontally. After that the image is broken vertically and the two halves are swapped. Next we swap the values of LH1 with HH1 and LL1 with HL1. The next step is to invert the sign of the other frequencies by multiplying the details coefficient matrices LH1, HL1, and HH1 by a value of (-1) as follows: LH1- > LH1 $\times -1$, HL1- > HL1 $\times -1$, HH1- > HH1 $\times -1$. After that every pixel value of the LL1 band is converted into its inverted 8 bit binary equivalent. The inverted binary sequence is divided into two parts and these are inverted separately. The next step comprises the entire binary sequence being inverted for every pixel. The decimal value of the reverse of the inverted binary

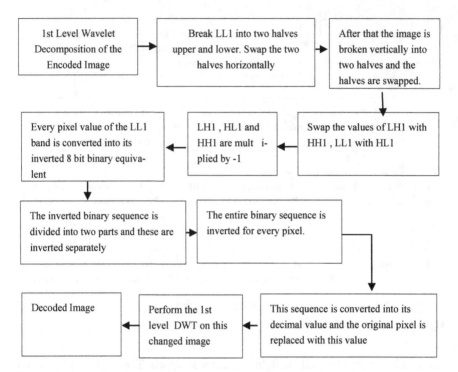

Fig. 5 The decryption algorithm

value is taken and substituted for every pixel of LL1. Then we perform the 1st level inverse DWT on the matrix to get the final decoded image.

5   Experimental Results and Analysis

The ultimate objective of the image encryption procedures is to produce an encrypted image which is impossible to comprehend. Image quality will worsen that is certain. The proposed method worsens the image quality during encryption but at the end we get back the actual image after decrypting it. In the later sections, different parameters and results are shown.

5.1   Correlation Coefficients

Correlation coefficients are tested in three directions namely horizontal, diagonal and vertical directions. It is provided in the table down below.

Table 1 provides the comparison of approaches.

Table 1 Original and encrypted image comparison

Encryption method	Test image	Horizontal		Vertical		Diagonal	
		Original	Encrypted	Original	Encrypted	Original	Encrypted
Proposed	Lena	0.946	0.0054	0.973	0.0318	0.921	0.0038
	Lake	0.958	0.0024	0.958	0.0038	0.929	0.0013
Tedmori and Najdawi [7]	Lena	0.987	0.0035	0.936	0.0025	0.927	0.0115
	Lake	0.976	-0.0750	0.904	−0.0039	0.912	0.0181

Table 2 Comparison of PSNR values

Test image	Proposed		Tedmori and Najdawi [7]	
	O-D	O-E	O-D	O-E
Lena	Undefined	0.0067	Undefined	0.0027
Lake	Undefined	0.0061	Undefined	0.0033

Table 3 Comparison of NPCR and UACI values

Test image	Proposed		Tedmori and Najdawi [5]	
	NPCR %	UACI %	NPCR %	UACI %
Lena	99.842	38.720	99.961	39.971
Lake	99.775	41.703	99.953	41.871

5.2 PSNR

PSNR is a short form for Peak Signal to Noise Ratio. It is a known parameter. The PSNR results in an undefined value only if the approach is lossless.

Table 2 provides the difference of approaches.

5.3 NPCR and UACI

The effect of changing just one pixel in the actual image and then getting a completely uncorrelated encoded or encrypted image, two famous measures are employed: Number of Pixels Change Rate (NPCR) and Unified Average Changing Intensity (UACI).

Table 3 provides the comparison of our methods with standard methods.

Figure 6 shows few of the results of the proposed algorithms. They are given in order to do a comparison.

Fig. 6 From top to bottom, standard images "Lena", "Cameraman" and "Hestain". *Right* to *left*, the figures show the original, encoded and decoded outputs

6   Conclusion

There are various image encryption algorithms proposed in this field of image cryptography. Our algorithm is a lossless image encryption algorithm. The decrypted image is equal to the actual image which shows the algorithm is lossless. Standard tests show the usefulness of the algorithm against benchmark algorithms. Our proposed algorithm proves to be a resilient, robust and highly secure algorithm in the field of image cryptography.

References

1. S. Maniccam, N. Bourbakis, Lossless image compression and encryption using SCAN, Elsevier J. Pattern Recognit., 34(6) (2001) 1229–1245.
2. G. Bhatnagar, Q.M. Jonathan Wu., Selective image encryption based on pixels of interest and singular value decomposition, Elsevier Digital Signal Process., 22(4) (2012) 648–663.
3. J. Yen, J. Guo, A new mirror-like image encryption algorithm and its VLSI architecture, J. Pattern Recognit. Image Anal., 10(2) (2000) 236–247.
4. Y. Hou, Visual cryptography for color images, Elsevier J. Pattern Recognit., 36(1), (2003) 1619–1629.
5. A. Yahya, A. Abdalla, A Shuffle Image-Encryption Algorithm, J. Comput. Sci., 4(12) (2008). 999–1002.
6. W. Chen, C. Quan, C.J. Tay, Optical color image encryption based on Arnold transform and interference method, Elsevier Optics Commun., 282(18) (2009) 3680–3685.
7. S. Tedmori, N. Al-Najdawi, Lossless Image Cryptography Algorithm Based on Discrete Cosine Transform, Int. Arab J. Inf. Technol., 9(5) (2012) 471–478.

A Scheme for QR Code Based Smart Door Locks Security System Using an ARM Computer

Suprakash Mukherjee and Subhendu Mondal

Abstract This paper deals with a new approach to implement QR codes in door locks security. In this work an advanced security system is presented using Unique QR Identification (or UQID) code, which is specially designed to be used in door locks. The UQID system presented here is a new methodology implemented to provide security services to hotel rooms along with better hospitality to guests. A guest books a room online on the hotel website and immediately after booking a room a QR code is emailed to the guest. The guest can save this QR code in his phone/smart watch/tablet or any other device with a display. When visits the hotel he simply holds this QR code against the door computer which authenticates whether the right QR code has been presented by the guest and unlocks the door or keeps the door locked accordingly. The QR code sent to the guest at the time of registering for the hotel room is the QR code which is indeed the key to access the room for the guest.

Keywords QR code · Door locks · ARM computer · Hotel security

1 Introduction

An electronic lock is usually connected to an access control system and works by means of an electronic current. The use of smart phones have most likely increased popularity of smart locks which have begun to be used more commonly in residential areas. Additionally, smart locks are gaining momentum in co-working spaces and offices. Electronic lock systems implementing, passwords, security tokens, biometrics etc. offer a variety of means of authentication; many of which are

Suprakash Mukherjee (✉)
AEIE Department, Heritage Institute of Technology, Kolkata, India
e-mail: bubai.suprakash007@gmail.com

Subhendu Mondal
MCA Department, Meghnad Saha Institute of Technology, Kolkata, India
e-mail: subhendu.mondal93@gmail.com

© Springer Science+Business Media Singapore 2017 613
J.K. Mandal et al. (eds.), *Proceedings of the First International Conference
on Intelligent Computing and Communication*, Advances in Intelligent Systems
and Computing 458, DOI 10.1007/978-981-10-2035-3_62

non exhaustive. Perhaps the most widely popular door lock is the RFID based swipe card access control which is very popular and is used widely in hotel doors.

The proposed system presented here uses implicit web based authentication because of which the guest does not need to carry any extra tangible key in his/her pocket since the system can unlock information get key to the room is in the phone of the guest in the form of an image. The UQID code based entry system is unique as it does away with traditional keys and swipe cards. This system so developed uses a simple web interface where in approved users are added to a MySQL database along with their hash encrypted representative user specific string of the generated UQID code. This UQID code so generated is emailed to the respective user's address using a simple MTP (Mail Transfer Protocol). The end user client used here in this scheme is an ARM based Linux computer that has capability to scan UQID codes via a system connected CMOS imaging sensor.

The end user or the guest needs to perform some simple tasks in order to use this system. A guest registers online on the hotel website to book a room in a particular date range. On successful completion of room booking the guest receives an email with a QR code attachment in the form of an image file. The guest at the time of his/her arrival at the hotel uses the same QR code as the key to enter the respective room after check in. Since this QR code is an image file the guest can simply download it into his phone/smart watch/tablet or any other display device. The guest needs to just hold this QR code to be scanned by a CMOS imaging sensor connected to the ARM computer at the door which has the capability to scan and authenticate the QR code and unlock the door upon its validation.

The system uses MD-5 cryptographic hash encryption during the validation of the QR code. Details about the implementation of this have been described under the Sect. 2.2 of this paper.

2 Schematics

The UQID system mainly consists of the two system components for the proper implementation of the system, namely the hotel website and the client ARM computer at the door. The real time implementation of the system used is a Raspberry Pi B+ embedded Linux ARM computer connected to a Logitech USB standard webcam (CMOS imaging sensor) and the demo hotel website which has been built on PHP and it uses MySQL database.

2.1 Algorithm and Functional Description

The UQID system involves a series of processes that are carried out when a guest books a room on the hotel website. The complete technical functionality of the system has been described. When a guest books a room successfully on the hotel website a 16 bit alphanumeric unique ID/random number is generated on the hotel web server. This randomly generated number undergoes two processes. Firstly, it's representative QR code is generated and emailed to the guest to the registered email. Secondly, the unique ID and valid timeframe information is transmitted to the hotel onsite server behind firewall. A MD5-1way cryptographic hash of the unique ID is generated by using the PHP md5() built in function. This hash is saved in the hotel database. Note that neither the unique ID nor the QR code is saved anywhere in the database after. Any traces of these are immediately deleted from the server after the QR code is mailed and the hash is generated. Only the hash is saved in the database. This hash is then sent to the client door ARM computer. The QR code is created by using the PHP qrlib [1, 2] which is an open source library downloaded from sourceforge.net. The following figures describe the process of QR code generation using the qrlib library in PHP on the hotel web server (Figs. 1, 2, 3 and 4).

When the guest wants to access his hotel room he presents his QR code he received at the time of booking to the CMOS imaging sensor connected to the ARM computer. The QR code is scanned and decoded into its original unique ID using ZBar. ZBar is an open source barcode and 2D barcode (QR code) reading software package. After the QR code is decoded it's respective MD5 cryptographic hash is generated by using "hashlib" which is a built in python library. This hash generated at the door computer is matched with the hash received from the hotel web server. If the hashes match the door disengages for 10 s before it engages back again. If the hashes do not match the door remains engaged and the information is recorded.

The MD5-hash encryption and its properties are discussed in detail under Sect. 2.2 of the paper.

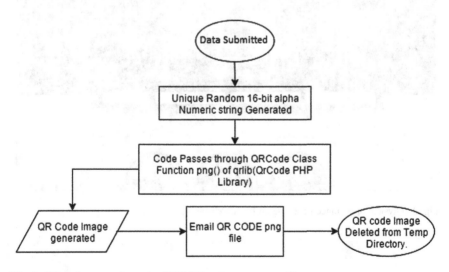

Fig. 1 QR code generation using PHP QR code open source library

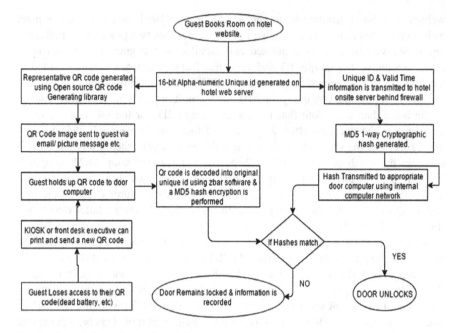

Fig. 2 The functional description of the system

Fig. 3 Physical appearance of the demo hotel website

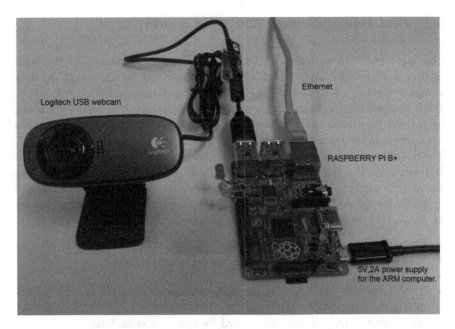

Fig. 4 Physical appearance of the door computer and webcam

2.2 Software Security

The system as described above implements MD5 cryptographic hash encryption. The MD5 message-digest algorithm is a widely used cryptographic hash function producing a 128-bit (16-byte) hash value, typically expressed in text format as a 32 digit hexadecimal number. MD5 is commonly used to verify data integrity and has also been utilized in a wide variety of cryptographic applications.

A cryptographic hash function is a hash function which is considered practically impossible to invert, that is, to recreate the input data from its hash value alone [3]. MD5-1 way cryptographic hash encryption is a one way process (i.e., a hash can be created from an id but not the other way round) so even if the security of the database is compromised and a hacker is able to retrieve the hashes from the website, one cannot create a new copy of the QR code to gain access to the room since neither any QR code or any unique id is saved in the database. Rather only the cryptographic hash is saved, which as explained earlier is a one way process, hence the unique ID can never be traced from a hash.

Moreover the system uses an ARM computer that runs embedded Linux. The primary difference between Linux and many other popular contemporary operating systems is that the Linux kernel and other components are free and open-source software. Due to its low cost and ease of customization, Linux is often used in embedded systems. Linux is also an enough secure and stable platform that enhances security of the system here.

2.3   Real Time Implementation of the System

The real time implementation of the proposed system comprised of the hotel website built on PHP and it uses MySQL database. On the client end the door computer used is a Raspberry Pi B+ ARM computer running Linux. A Logitech C310 standard USB webcam is connected to the Raspberry Pi. To show the status of whether the electric door lock is engaged or disengaged two Leds have been used, one indicating that the door is locked while other indicates that it is unlocked. The Raspberry Pi is connected to the internet via the Ethernet port through which it is linked to the hotel web server. Apart from these an open source software package called ZBar has been implemented in the UQID system that is used for reading and decoding the QR codes. ZBar [4] has various features that enhances the efficiency of the UQID system greatly. Some of the features of ZBar are listed below [4].

- Cross platform-Windows, Linux, Unix, iOS, embedded etc.
- Real time scanning and high speed.
- Small memory footprint.
- Small code size—the EAN decoder and core scanner represent under 1 K lines of C code.
- Not limited to images also can be used for real time video feed.
- No floating point operations.
- Works fine for embedded applications using inexpensive hardware and processors.

The figures above show the physical outlook of the system in the real time simulation process (Figs. 3 and 4).

3   Analysis of Efficacy of the System

3.1   Advantages

3.1.1   Uniqueness of UQID System Over RFID/Swipe Card Based Systems

There are various reasons that make this system more efficient and better in many aspects with respect to the conventional swipe card based door lock systems which are mostly popular with hotel room security. A few features of the UQID system which gives it an edge over the other conventional systems are pointed out below:

- The guest need not carry any extra weight in his/her pocket since the key to the room is in the phone of the guest in the form of an image. Whereas in swipe card access a card is needed to be carried, this adds to an extra weight and occupies extra room in the wallet of the guest.

- If there are multiple guests (a group of friends/family) sharing the same room then the QR code can be easily shared among them so that all of them can independently gain access to the room. In case of swipe card access it would not be possible without having multiple cards for each room.
- In case a guest loses the QR code he can simply download it again from the email he received or request a new QR code from the hotel website instead of bothering the hotel staff with such small issues. Requesting a new QR code on the website would immediately email him a new generated QR code while invalidating the previous one.
- Implementing the UQID system also gives an impressive response time. The response time (time gap between presenting the QR code and unlocking the door) of the system has been experimentally found out to be less than 1 s.
- The system is completely automated and networked. So information regarding when some person has entered the room or left is saved in the database which can be easily fetched if required under any issue.
- This system is compatible with all pre-existing electronic locks. So upgrading the hotel security system is cost-friendly and easy. Any pre-existing lock can be easily hooked up to the system and the system is ready for operation.
- One of the most important factor that gives the UQID system that it has an estimated cheap installation cost and maintenance cost. Also this system doesn't require additional cost incurred in RFID based systems due to the physical keys.
- This system can also work on cheap hardware like cheaper camera to read the QR codes. Also QR code reading is much faster for an optical device than reading bar codes.
- The system also uses MD5-1 way cryptographic has encryption which also lets the system achieve good security. The security provided in the system is better explained in the Sect. 2.2 of the paper.
- If the guest has a smart watch he can more easily use the QR code with the help of that to access the room. Also if the guest has no device at all he can simply print it from the kiosk or the front desk and that would also work equally well.
- The maintenance cost is also quiet less and can be further reduced by making the hardware minimalistic and modular.

3.1.2 Uniqueness of UQID System Over Other QR-Code/Bar Code Based Security Systems

There are many variants of door security solutions implementing QR codes in the system that are available in the market. But the way things have been implemented so far in this technology has various drawbacks due to which QR code door security facilities aren't so popular. Most QR code implementing door security facilities implement the technique the other way round i.e., instead of the user having the QR code, the QR code is on the door front and the user through his/her smart phone, tablet, other QR code reading devices has to scan the QR code and when the scanning is done by the registered gadget for the door then the lock disengages.

- Response time for such systems is more since there are a number of steps on the guest end that he has to do to access the room. Also delays in the server side and slow internet connection on the guest's device are problems often faced in such systems. Whereas the response time (gap between presenting the QR code and disengaging the lock) in the UQID system is less than 1 s.
- User's gadget failure and breakdown would lead to many hassles involved in both the users and service providers end. On the other hand in the UQID system even if there is a gadget failure the user can still print out the QR code from the front desk or kiosk and use it as a temporary key to access the room.
- Maintenance cost of such systems involves greater budget and more expert personals. The UQID system is easy to install and use and also has low maintenance budget.
- Such systems require firstly internet on the scanning device or some other connectivity to correctly send the server the QR code information from the registered gadget to disengage the lock. The UQID system implements QR codes the other way round. The user has a QR code instead. So no need for high end gadgets and internet required by the guest.
- Moreover an app on the smart gadget is required with a good camera to read the QR code, which means the user needs to spend on a better gadget to use the system. UQID system just requires a decent sized display device that can display the QR code to unlock the door.
- These kinds of system not only involve greater expenditure on the side of the guest but also expenses are more on the system installer. The UQID system is cheaper and user friendly on both ends.
- Bar codes require more expensive readers. UQID system can run on a budget camera as used in the project here due to use of QR codes.

3.2 Limitations of the System

Like every technology this system also has a limitation i.e., the QR code can easily be copied and thus reach unwanted people. So it would be advisory to the guest/user to keep the QR code safe in their gadgets and see to it that no unwanted person is able to access the QR code in their device or simply steal it by copying. But on the up side the existing electronic door locks are even more vulnerable to attack by hackers than the UQID.

4 Discussions

Security systems have a wide range of applications. The UQID system can not only bring better hospitality solutions in hotels but also can be put to a number of uses. Home automation is one of the upcoming technologies where smarter systems are being developed. QR code authentication as a home security solution can prove to be one of the user friendly and cost effective technologies to the latter. QR code authentication gets rid of conventional keys like swipe card/tokens and gives a much more portable solution to this. Also unlike systems that implement biometrics the UQID system is much cheaper facilitating good software security as well. Using smarter door security has also increasingly risen in the past decade in offices as well. Most of the offices in the world now use smarter systems in door security and thus this calls an urgent need to provide a cost effective as well as secure solution along with easy usability on the customer's end. Since smart phones are getting increasingly popular it is becoming a general demand of the consumers to give them the freedom to do all tasks they perform in their day to day life through their smart phone. The UQID system thus solves various problems faced in the hospitality industry at a very appealing cost. This system provides complete automation in security with lesser maintenance and system monitoring requirements. On an overall note the UQID system provides decent mechanical and software security solutions at a very effective cost and minimalistic design with better hospitality to guest and users using the system.

References

1. http://phpqrcode.sourceforge.net/.
2. https://sourceforge.net/projects/phpqrcode/.
3. Schneier, Bruce. "Cryptanalysis of MD5 and SHA: Time for a New Standard". *Computerworld.* Retrieved 15 October 2014.
4. http://zbar.sourceforge.net/.

Image Steganography Using BitPlane Complexity Segmentation and Hessenberg QR Method

Barnali Gupta Banik and Samir Kumar Bandyopadhyay

Abstract Image Steganography is an immemorial technique of data hiding behind an Image known as vessel image, camouflaging the covert image from the outside world. In this paper a novel algorithm of image steganography has been proposed where two techniques are used—(i) Bit Plane Complexity Segmentation (BPCS) analysis and (ii) QR Decomposition of linear algebra to choose the region where the full secret message is embedded without exposing its existence.

Keywords Bit plane complexity segmentation · Image steganography · QR decomposition

1 Introduction

There are different techniques available for image steganography. However Bit Plane Complexity Segmentation Analysis is comparatively new and less explored till date. As image can be defined as a matrix of pixel values, thus it is useful to operate techniques of linear algebra over image. After rigorous study it has been found that QR decomposition technique of linear algebra is never been applied in image steganography. In this paper the same technique is applied along with BPCS analysis for image steganography which shows very good result at the end.

B. Gupta Banik (✉)
Department of Computer Science & Engineering, St' Thomas College
of Engineering & Technology, Kolkata, India
e-mail: gupta.barnali@gmail.com

S.K. Bandyopadhyay
Lincoln University, Lincoln, Malaysia
e-mail: skb1@vsnl.com

© Springer Science+Business Media Singapore 2017   623
J.K. Mandal et al. (eds.), *Proceedings of the First International Conference on Intelligent Computing and Communication*, Advances in Intelligent Systems and Computing 458, DOI 10.1007/978-981-10-2035-3_63

2 Literature Survey

One of the important factors of Image Steganography technique is capacity of secret message. Traditional techniques of LSB substitutions use only least significant bit (LSB) or sometimes multiple LSB to embed secret data [1]. But repeated testing has shown that such substitution can carried out till 5th least significant bit at maximum. After 5th bit substitution, secret message starts revealing its existence which is against the objective of steganography [2]. An experiment has shown that in case of 24 bit true color image, capacity of embedding secret image is 1/8th of the total size. Other popular image steganography method to embed secret data in transform domain can be carried out either by using Discrete Cosine Transformation (DCT) or by Discrete Wavelet Transformation (DWT) techniques. In DCT, the mid-frequency band is explored to embed secret data by comparing nearly equivalent coefficient values. In DWT, low frequency components holds actual image data, hence high frequency components can be used to embed secret message data [3]. After analyzing aforementioned traditional techniques it can be stated that the capacity of the secret message doesn't go beyond 10 % of the cover image. In [4], authors have explored the concept that human cannot perceive any change in shape information in a complex binary pattern. That entire section can be replaced with secret message data in BPCS steganography, thus using this technique capacity of secret data insertion may increase up to 50 % original vessel data.

A. Bit Plane Complexity Segmentation

Bit-Plane Complexity Segmentation (BPCS) Steganography was developed by Kawaguchi and Eason in [4, 5]. The first step of BPCS is conversion of normal image to Canonical Gray Code (CGC) from Pure Binary Code (PBC). All natural images are coded with PBC. However it has major drawbacks of 'Hamming Cliff' which signifies two numerical nearby values may have their bit representations with larger hamming distance [6]. An example can be drawn using two integers 7 and 8 are represented with PBC with 4 bits, it comes like—0111 and 1000. Here hamming distance is 4 which is quite high indicating a small change in pixel values may reflect much in output. To overcome this problem it is better to use Gray code which ensures hamming distance among two successive numbers is always 1. PBC provides much better region for secret data embedding in BPCS. But due to Hamming Cliff problem, CGC is preferred over PBC in BPCS [7]. Binary code can be easily converted to gray code by using the following formula:

$$g = g_1 g_k$$
$$g_k = b_{k-1} \oplus b_k \tag{1}$$

Table 1 Table for comparing Hamming distance between PBC and CGC

Decimal	Binary	Hamming distance	Gray	Hamming distance
1	0001	–	0001	1
2	0010	2	0011	1
3	0011	1	0010	1
4	0100	3	0110	1
5	0101	1	0111	1
6	0110	2	0101	1
7	0111	1	0100	1
8	1000	4	1100	1

Fig. 1 Bit plane slicing of an image

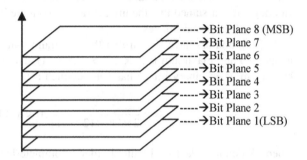

where $b_k = b_1, b_2, ...b_n$, b_1 is the most significant representation and $\oplus$ represents XOR operation Table 1.

The next step of BPCS is to decompose the image into set of bit planes, which is also known as bit plane slicing. An image is accumulation of pixels. Suppose 1 pixel can be represented by 8 bits—then it can be imagined that the image can be sliced into 8 bit planes [5] where plane 1 contains all least significant bit (LSB), plane 2 contains all 2nd least significant bits, and likewise plane 8 contains all most significant bits (MSB) as shown in Fig. 1.

If an image I is comprised of n bit pixel, those can be disintegrated to a series on n binary images. For gray-scale image, it would be

$$I = (I_1, \ I_2, \ I_3, \ \ldots \ldots I_n)$$

If I is a color image then,

$$I = (I_{R1}, I_{R2}, \ \ldots . I_{Rn}, I_{G1}, I_{G2}, .., I_{Gn}, \ \ldots I_{B1}, I_{B2}, .., I_{Bn}) \tag{2}$$

where I_{R1}, I_{G1}, I_{B1} is most significant bit (MSB) plane and I_{Rn}, I_{Gn}, I_{Bn} is least significant bit (LSB) plane. The complexity of each bit plane increases from MSB to LSB monotonically [8].

Third step of this method is to divide each bit plane into 8 × 8 consecutive and non-overlapping blocks followed by calculating complexity of each block. If the complexity of an image block is larger than the threshold (typically 0.3) then that block is regarded as noise like region. Secret data can be inserted with highest precision in these noise like regions which are the most complex part of the vessel image and hence very much suitable for data embedding.

Image complexity has been initially defined by the American mathematician George David Birkhoff as number of elements the image consists of [9]. In [4] Kawaguchi and Eason used black-and-white border's length of an image as a parameter to formulate image complexity. The image is treated as complex when border is lengthy, else it can be considered as simple. The black-and-white border's entire length is same with total count of differing color by the rows and columns of an image. It is assumed that the image frame hassquare 2 m × 2 m pixels where m is 8 to 12 for normal images.

In [10], authors have shown that the minimum number of color changes in an image is 0 and maximum is $2 \times 2^m \times (2^m - 1)$. The Image Complexity denoted by α of m × m binary image has been defined as:

$$\alpha = \frac{k}{2 \times 2^m \times (2^m - 1)}, 0 \leq \alpha \leq 1 \tag{3}$$

where, k denotes black-and-white border's complete length of the given image.

In BPCS method, image is segmented without any information about its content. Sometimes it may happen that a bit plane is in between of noisy and informative region. In that case black-and-white border complexity α may depict the block as complex and embedding data there may reveal its existence at the end. To cope with such issue, in [10]—another 2 complexity measures have been suggested which are run-length irregularity and border noisiness. If a bit plane has significant run-length irregularity as well as large border noisiness, it can be treated as complex one.

B. QR Decomposition of Upper Hessenberg Matrices

In linear algebra, QR decomposition of a matrix A into a product of Q and R can be shown as—

$$A = QR \tag{4}$$

where Q is orthogonal matrix satisfies a condition of $Q^T Q = I$, where Q^T is the transform of Q, I is Identity Matrix and R is upper triangular matrix.

By reversing the order of the product of Q and R, the below equation can be derived

$$RQ = Q*AQ \tag{5}$$

where Q * AQ is the similarity transformation of A. This generates a sequence of matrices A_k starting with $A_0 = A$ and given by:

$$A_k = R_k Q_k \tag{6}$$

where Q_k and R_k represents QR factorization of A_{k-1} such that $A_{k-1} = Q_k R_k$

This decomposition is used to solve linear systems of equations [11]. Generally Gram Schmidt method is applied for basic QR decomposition [12, 13], the complexity of which is $O(n^3)$. This method is a bit slow and expensive as it needs more repeated steps to reach convergence. In such case time complexity can increase up to $O(n^4)$ which implies scope for further improvement [14]. The desired improvement can be achieved in this algorithm by introducing Hessenberg matrix, the structure of which is given below:

$$H = \begin{bmatrix} u & \cdots & \cdots & \cdots & u \\ u & \ddots & & & \vdots \\ 0 & \ddots & \ddots & & \vdots \\ \vdots & \ddots & \ddots & \ddots & u \\ 0 & \cdots & 0 & u & u \end{bmatrix} \tag{7}$$

A matrix $H \in \mathbb{C}^{n \times n}$ is called Hessenberg Matrix if its elements below the lower off diagonal are zeros, $h_{i,j} = 0$ when $i > j + 1$. The matrix H is called unreduced Hessenberg matrix if $h_{i,j+1} \neq 0 \forall i = 1, 2, .., n - 1$.

To reduce the time complexity of decomposition technique down to $O(n^2)$, the following two step approach has been considered:

STEP I: Compute Hessenberg Matrix H
It is beneficial to use Householder reflector to reduce Hessenberg Matrix [15]. A matrix $P \in \mathbb{C}^{n \times n}$ of form $P = I - 2uu^*$ where $u \in \mathbb{C}$ and $\|u\| = 1$ is called a Household reflector. Household reflector has some favorable properties like:

(a) It is always hermitian $P = P^*$
(b) It is always orthogonal since, $P^2 = I$
(c) It satisfies $Px = \alpha e_1$ where x is vector, $\alpha: = \rho \|x\|$, $\rho = \pm 1$, e_1 is the unit vector.

These properties lead to formula given below:

$$u = \frac{x - \alpha e_1}{\|x - \alpha e_1\|} = \frac{z}{\|z\|}, \text{ where } z: = x - \alpha e_1 = \begin{bmatrix} x_1 - \rho \|x\| \\ x_2 \\ \vdots \\ x_n \end{bmatrix} \tag{8}$$

The algorithm for computing Hessenberg Matrix is given below:

Algorithm 1 for computing Hessenberg Matrix.

INPUT: A matrix $A \in \mathbb{C}^{n \times n}$

METHOD: for j=1, 2... n-2 do

      Compute U_j where using eq. 8 where $x^T = [a_{j+1,j}, \ldots, a_{n,j}]$

Compute $P_j A$: $A_{j+1:n, j:n} := A_{j+1:n, j:n} - 2u_j(u_j^* A_{j+1:n, j:n})$

      Compute $P_j A P_j^*$: $A_{1:n, j+1:n} := A_{1:n, j+1:n} - 2(A_{1:n, j+1:n})u_j^*$

end;

OUTPUT: A Hessenberg matrix H of A.

STEP II: Apply Basic QR method to Matrix H

In linear algebra, 'Givens rotation' is defined as a rotation in a plane spanned by two coordinate axes and named by its inventor Wallace Givens. The QR decomposition can be applied to a Hessenberg matrix H using Givens rotation, which can be represented by the following matrix:

$$G(i,j,\theta) = \begin{bmatrix} 1 & \cdots & 0 & \cdots & 0 & \cdots & 0 \\ \vdots & \ddots & \vdots & & \vdots & & \vdots \\ 0 & \cdots & k & \cdots & -l & \cdots & 0 \\ \vdots & & \vdots & \ddots & \vdots & & \vdots \\ 0 & \cdots & l & \cdots & k & \cdots & 0 \\ \vdots & & \vdots & & \vdots & \ddots & \vdots \\ 0 & \cdots & 0 & \cdots & 0 & \cdots & 1 \end{bmatrix} \tag{9}$$

where $\theta = [k, l]$, $k = \cos\theta$ and $l = \sin\theta$ appear at the junction of ith row and jth column.

As stated in [16], QR decomposition computes $Q^T A = R$. Let $[k, l]$ = givens (a, b) which calculates k and l such that,

$$\begin{bmatrix} k & -l \\ l & k \end{bmatrix}^T \begin{bmatrix} a \\ b \end{bmatrix} = \begin{bmatrix} r \\ 0 \end{bmatrix}, \text{ where } r = \sqrt{a^2 + b^2}$$

Now, let G (i, j, k, l)^T is givens rotation matrix which revolves ith and jth elements of a vector v clockwise by an angle θ so that $\cos\theta = k$ and $\sin\theta = l$, such that if $v_i = a$ and $v_j = b$, then the updated vector will be $u = G(i, j, k, l)^T v$, where $u_i = r = \sqrt{a^2 + b^2}$ and $u_j = 0$.

The QR factorization of m × n matrix H is evaluated as follows:
Algorithm 2 for computing QR factorization.

```
Q = I;
R = H;
    For p = 1 : n do
        For q = m : -1 : p + 1 do
            [k, l] = givens (r_{q-1,p}, r_{ip});
            R = G (q, p, k, l)^T R;
            Q = QG (q,p, k, l);

    End;
End;
```

3 Proposed Method

In this research article, it is proposing to analyze the cover image first using Bit plane Complexity Segmentation to choose the region for secret image embedding without revealing its existence. In the next step by using QR decomposition the secret message embedding has been performed.

The steps for embedding algorithm are as below:

STEP-1: Convert the cover image from Pure Binary Coding to Canonical Gray Coding.

STEP-2: Segment the image into several bit planes.

STEP-3: Calculate complexity of bit planes by Eq. (3)

STEP-4: Choose the noisy bit plane whose $\alpha > 0.3$

STEP-5: After selection of noisy bit plane prepare this as a matrix.

STEP-6: Apply algorithm 1 to get Hessenberg Matrix of the selected matrix.

STEP-7: Apply algorithm 2 to decompose the matrix into Q and R part

STEP-8: Divide the secret image into series of blocks

STEP-9: Embed the secret image at the Q part by two steps:

    (a) Calculate pseudo random number p

    (b) To encrypt message use, $C = Q + p*S$ where C is stego bit plane and S is the secret message

STEP-10: Calculate M = C x R;
STEP-11: Reconstruct M as stego image.

Retrieving secret message from stego image can be done by the extraction algorithm:

STEP-1: Retrieve the stego bit plane from the Stego Image by calculating complexity of bit planes.
STEP-2: Get the secret information by compare the stego bit plane with original bit plane like $Q_s = Q_c - Q_o/p$ where Q_s is the Q part of secret image, Q_c is the Q part of stego bit plane, Q_o is the Q part of original bit plane.
STEP-3: Calculate $N = Q_s \times R$
STEP-4: Prepare N as secret image.

4 Results and Analysis

The result of the proposed embedding algorithm has been shown in the Fig. 2. First the cover image is broken into 8 bitplanes followed by each bit plane's complexity has been calculated. Depending on the bitplane complexity one bitplane is chosen as cover media where the secret message is embedded resulting stego bit plane. Finally stego image is constructed using this stego bit plane along with rest 7 bitplanes. There is no such perceptual difference found between original cover image and the stego image.

Using the extraction algorithm, the secret image can be successfully recovered from the stego image. The Peak Signal to Noise Ratio (PSNR) for original cover image and stego image has been calculated shown in Table 2, which is quite high and impressive value. Also the quality measurement has been done using SSIM (Structural SIMilarity) Index. SSIM for stego and original cover image is close to 1 which implies that there are no structural differences between these images. Here lies the successfulness of this algorithm showing very high quality stego image although it contains the secret message.

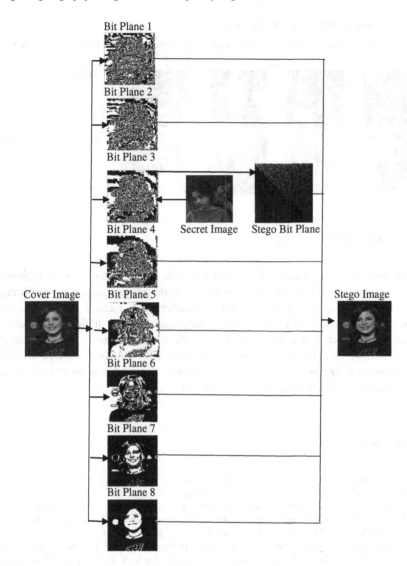

Fig. 2 Implementation of proposed embedding algorithm

Table 2 Table for PSNR and SSIM values

Cover Image	Secret Image	Stego Image	Recovered Image	PSNR (dB)	SSIM
				44.16	0.9733
				51.46	0.9985

5 Conclusion

There are different techniques already in place for image steganography. However in this article, a novel approach to embed the secret image within the cover image has been proposed using BPCS and Hessenberg QR method which was not explored before. The main objective of steganography are hiding data without revealing existence of it as well as recovering the secret message without damaging the content—both of these objectives are maintained here. As both PSNR and SSIM Index shows very good result, thus the proposed approach can be considered as successful for implementing image steganography.

References

1. Neil F. Johnson, SushilJajodia, George Mason University, "Exploring Steganography: Seeing the Unseen", Computing Practices, 1998 IEEE pg. 26–34.
2. Barnali Gupta Banik and Samir Kumar Bandyopadhyay, "An Image Steganography Method On Edge Detection Using Multiple Lsb Modification Technique", Journal of Basic and Applied Research International, International Knowledge Press, 9(2): 75–80, 2015.
3. Barnali Gupta Banik, Samir Kumar Bandyopadhyay, "A DWT Method for Image Steganography", International Journal of Advanced Research in Computer Science and Software Engineering, Volume 3, Issue 6, June 2013 ISSN: 2277 128X.
4. EijjiKawagauchi and Richard O. Eason, "Principle and Application of BPCS-Steganography" in Proc. SPIE, vol. 3529, 1998, pp. 464–473.
5. EijjiKawagauch and Richard O. Eason, KIT Steganography Research Group (Kyushu Institute of Technology,) http://datahide.org/BPCSe/index.html.
6. Nazmul Siddique, HojjatAdeli, "Computational Intelligence: Synergies of Fuzzy Logic, Neural Networks and Evolutionary Computing", Wiley Publications.
7. Shrikant S. Khaire, Dr. Sanjay L. Nalbalwar, "Review: Steganography – Bit Plane Complexity Segmentation (BPCS) Technique", International Journal of Engineering Science and Technology Vol. 2(9), 2010, 4860–4868.
8. "Bit Plane". PC Magazine. Retrieved 2007-05-02.

9. SCHA R., BOD R. "ComputationeleEsthetica", Informatie en Informatiebeleid 11, 1 (1993), 54–63.
10. MichiharuNiimi, Hideki Noda and EijiKawaguch, "An image embedding in image by a complexity based region segmentation method", IEEE Image Processing, 1997. Proceedings., International Conference on (Volume:3).
11. Weisstein, Eric W. "QR Decomposition." From MathWorld—A Wolfram Web Resource. http://mathworld.wolfram.com/QRDecomposition.html.
12. Raymond Puzio, Keenan Kidwell. "Proof of Gram-Schmidt orthogonalization algorithm" (version 8). PlanetMath.org.
13. Harvey Mudd College Math Tutorial on the Gram-Schmidt algorithm.
14. Lecture notes "QR algorithm"- Elias Jarlebring - Autumn 2014.
15. Per-OlofPersson, MIT OpenCourseWare on Introduction to Numerical Methods, "Hessenberg/Tridiagonal Reduction" October 2006.
16. Che-Rung Lee, Lecture Notes 6: Givens rotation, December 22, 2011.
17. Jeng-Shyang Pan, Hsiang-Cheh Huang, Laxmi c. Jain, Wai-Chi Fang, "Intelligent Multimedia Data Hiding: New Directions", Springer.

A Novel Scheme for Analyzing Confusion Characteristics of Block Ciphers

Dipanjan Bhowmik, Avijit Datta and Sharad Sinha

Abstract In this paper, a scheme aimed at analyzing the confusion characteristics of block ciphers has been proposed. The scheme analyzes the S-Boxes which are the source of confusion in block cipher. The test results obtained from the application of the proposed scheme on DES S-Boxes as well as on sole AES S-Box has been listed in the paper. The proposed scheme subsequently could very well be a part of a good test suit aimed at comprehensively analyzing the cryptographic strength of block ciphers.

Keywords Block cipher · Confusion · S-Boxes

1 Introduction

According to Shannon [9], the cryptographic strength of a block cipher is primarily defined by two aspects, namely *confusion* and *diffusion*. On one hand, diffusion analyzes the degree of randomness in the ciphertext given a plaintext, on the other hand, confusion measures the degree of non-linearity in the ciphertext for a given plaintext.

Linearity can be understood using the following analogy. Suppose an arbitrary input with property X is taken and put into a magic box. If the output property Y of the magic box can be guessed with some degree of confidence, then the magic box is said to be somewhat linear.

Dipanjan Bhowmik · Sharad Sinha (✉)
Department of Computer Science and Application, University of North Bengal, Siliguri, India
e-mail: ssinha.nbu@gmail.com

Dipanjan Bhowmik
e-mail: howzat.dipanjan@gmail.com

Avijit Datta
Department of Computer Application, Siliguri Institute of Technology, Siliguri, India
e-mail: avijit.go2avi@gmail.com

© Springer Science+Business Media Singapore 2017   635
J.K. Mandal et al. (eds.), *Proceedings of the First International Conference on Intelligent Computing and Communication*, Advances in Intelligent Systems and Computing 458, DOI 10.1007/978-981-10-2035-3_64

In case of block ciphers, the section that is responsible for non-linearity or confusion are the Substitution Boxes or the S-Boxes. So, when it is said that the confusion characteristic of a block cipher is to be analyzed, it actually means analyzing the underlying S-Boxes.

Till date many properties have been laid down which an S-Box should posses. In general, the characteristics that must be met by an S-Box [1, 6, 8] are:

- **Bijection**: When the S-Box is $n \times n$ bits, the mapping between input vector and output vector must be injective (one-one) and surjective (onto). However, this property does not hold true if the S-Box is $m \times n$ bits where $m \neq n$.
- **Balance**: It requires that each and every Boolean vector responsible for the S-Box must have equal number of 0s and 1s.
- **Non-linearity**: It means that S-Box should not be a linear mapping between the input and the output. If an S-Box is somewhat linear, then attacks such as Linear Cryptanalysis due to Matsui [5] exist which will exploit the linearity of the S-Box to break the cipher. If an S-Box is designed in a way that it achieves high degree of non-linearity, then it will give a bad approximation when approximated with linear Boolean Functions, which will make the cryptosystem difficult to break.
- **Strict Avalanche Criterion**: The Strict Avalanche Criterion (SAC) requires that if any input bit is changed, then each output bit must change with a probability of one-half. In other words, it can be stated that if an input bit changes there should be significant, as well as random, changes in the output vector. Though this is essentially referred to as the diffusion characteristic of the cipher, a good S-Box also plays an important role in achieving this property. An S-Box with good cryptographic strength should have 50 % dependency on each input bit.
- **Bit Independence Criterion**: The Bit Relationship Criterion, also referred to as the **Correlation Immunity**, requires that each output bit should act independently from each other. That is, there should not be any dependence of one output bit on any other output bit.

2 Proposed Scheme

The testing algorithm proposed in this paper primarily addresses the last two properties mentioned above, namely the Strict Avalanche Criterion (SAC) and Bit Independence Criterion (BIC). It binds both the properties into a single test. The proposed scheme takes as input an $m \times n$ S-Box and produces two upper triangular matrices, namely *BRT* matrix and *BRT_Total* matrix. Figure 1 shows a schematic diagram of an m × n S-Box [7].

The algorithm goes through all possible m-bit input vectors and from each of the input vectors, it generates the n-bit output vector. After obtaining the original output

Fig. 1 S-Box

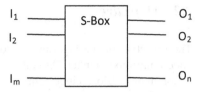

vector, each of the input bits is flipped (one at a time) and this produces a different output vector, which is then bitwise XORed with the original output vector to generate a row of the SAC matrix [2]. Thus, for each original input vector a corresponding SAC matrix of order $m \times n$ is generated. Each possible pair of columns are then compared to generate the upper triangular matrix for the particular input. This process is then repeated for all possible input vectors and ultimately the upper triangular matrix *BRT_Total* is generated, which gives an overall perspective of The Strict Avalanche Criterion and The Bit Independence Criterion.

2.1 Algorithm

Input: an $m \times n$ S-Box.

Method:

1. For each possible input P (m-bits)
   a. Obtain the output C (n-bits) corresponding to the given m-bit input.
   b. For each of the m bits in the original input
      i. Flip the ith bit
      ii. Use the modified input to obtain the output from the S box
      iii. Store it as the ith vector in the *OUT* matrix.
      iv. Bitwise XOR the output obtained in the above step with the original output C and store the same as the ith entry of the *SAC* matrix.
   c. Compare each column (say i) of the *SAC* matrix with every other Column (say j), count the number of matches and store the count as the jth entry of the ith row of the *BRT* matrix.
   d. Add the count obtained in the previous step to the $(I,j)^{th}$ entry of the BRT-Total matrix.

Output:

1. *BRT-Total* matrix that gives an overall perspective.
2. *BRT* matrices that is concerned with a particular input vector.

2.2 Objective

The objective of the algorithm is to analyze the confusion characteristics of the underlying block cipher. As Substitution boxes (S-Boxes) are the source of confusion in most Block ciphers, the algorithm ultimately analyzes the security aspects of the S-Boxes. The characteristics of the underlying S-Boxes, which are tested using this test are:

1. *Bit Independence Criterion*
2. *Strict Avalanche Criterion.*

3 Experimental Results

Two of the most widely documented block ciphers were put to test, the results are presented in Sects. 3.1 and 3.2.

3.1 Experimental Results for DES S-Boxes

The S-Boxes of Data Encryption System (DES) [3] were put to test using the proposed approach. The results obtained are depicted in Table 1.

Moreover, the detailed output corresponding to input $X = (56)_{10}$ are as follows:
X = 56, Y = 3
—THe OUTput matrix corresponding to X = 56—
1 0 1 0   10
1 0 1 0   10
0 1 0 1   5
1 1 1 1   15

Table 1 Experimental results for DES S-Boxes

S-Box#	Expected mean	Observed mean	Variance	Standard deviation	Coefficient of variance
1	192	168.6667	73.06667	8.547904	5.067927
2	192	172.0000	64.0000	8.000000	4.651163
3	192	182.6667	36.26667	6.022181	3.296815
4	192	162.6667	324.2667	18.00741	11.07013
5	192	172.6667	79.46667	8.914408	5.162784
6	192	176.6667	124.2667	11.1475	6.309903
7	192	183.3333	130.6667	11.43095	6.235065
8	192	172.6667	41.06667	6.408328	3.711387

1 1 0 1 13
0 1 0 1 5

—THe SAC matrix corresponding to X = 56—

1 0 0 1
1 0 0 1
0 1 1 0
1 1 0 0
1 1 1 0
0 1 1 0

Bit Relationship Factor between Col 0 and Col 1 = 1
Bit Relationship Factor between Col 0 and Col 2 = 0
Bit Relationship Factor between Col 0 and Col 3 = 4
Bit Relationship Factor between Col 1 and col 2 = 5
Bit Relationship Factor between Col 1 and Col 3 = 1
Bit Relationship Factor between Col 2 and Col 3 = 2

Expected mean	3
Observed mean	2.166667
Variance	3.766667
S.D	1.94079
Coefficient of variance	89.57493

3.2 Experimental Results for AES S-Box

The lone S-Box of Advanced Encryption Standard (AES) [4] was also put to test using the proposed approach, the results of which are as follows:

Expected mean	1024
Observed mean	1144
Variance	1430.519
S.D	37.8222
Coefficient of variance	3.306136

The detailed results corresponding to input $X = (250)_{10}$ are obtained as follows:
X = 250, Y = 45

—THe OUTput matrix corresponding to X = 250—

0 0 0 0 1 1 1 1 15
1 0 0 1 1 0 1 1 155
1 0 1 1 1 0 1 1 187

```
1 0 0 0 1 0 0 1   137
0 0 1 0 1 1 0 1   45
0 1 0 1 0 1 1 1   87
1 1 1 1 0 1 0 0   244
1 1 0 1 1 0 1 0   218
```

—THe SAC matrix corresponding to X = 250—

```
0 0 1 0 0 0 1 0
1 0 1 1 0 1 1 0
1 0 0 1 0 1 1 0
1 0 1 0 0 1 0 0
0 0 0 0 0 0 0 0
0 1 1 1 1 0 1 0
1 1 0 1 1 0 0 1
1 1 1 1 0 1 1 1
```

Bit Relationship Factor between Col 0 and Col 1 = 3
Bit Relationship Factor between Col 0 and Col 2 = 4
Bit Relationship Factor between Col 0 and Col 3 = 6
Bit Relationship Factor between Col 0 and Bol 4 = 5
Bit Relationship Factor between Col 0 and Col 5 = 3
Bit Relationship Factor between Col 0 and Col 6 = 7
Bit Relationship Factor between Col 0 and Col 7 = 5
Bit Relationship Factor between Col 1 and Col 2 = 5
Bit Relationship Factor between Col 1 and Col 3 = 3
Bit Relationship Factor between Col 1 and Col 4 = 6
Bit Relationship Factor between Col 1 and Col 5 = 6
Bit Relationship Factor between Col 1 and Col 6 = 4
Bit Relationship Factor between Col 1 and Col 7 = 4
Bit Relationship Factor between Col 2 and Col 3 = 2
Bit Relationship Factor between Col 2 and Col 4 = 5
Bit Relationship Factor between Col 2 and Col 5 = 5
Bit Relationship Factor between Col 2 and Col 6 = 3
Bit Relationship Factor between Col 2 and Col 7 = 7
Bit Relationship Factor between Col 3 and Col 4 = 5
Bit Relationship Factor between Col 3 and Col 5 = 3
Bit Relationship Factor between Col 3 and Col 6 = 7
Bit Relationship Factor between Col 3 and Col 7 = 3
Bit Relationship Factor between Col 4 and Col 5 = 4
Bit Relationship Factor between Col 4 and Col 6 = 6
Bit Relationship Factor between Col 4 and Col 7 = 6
Bit Relationship Factor between Col 5 and Col 6 = 4
Bit Relationship Factor between Col 5 and Col 7 = 4
Bit Relationship Factor between Col 6 and Col 7 = 4

Expected mean	4
Observed mean	4.607143
Variance	1.951058
S.D	1.396803
Coefficient of variance	30.3182

3.3 Discussion

From these results, it is clear that DES S-Boxes vary a great deal with respect to Bit Independence Criterion as well as Strict Avalanche Criterion, which is certainly not ideal. On the other hand, the sole AES S-Box shows relatively better statistical properties with respect to Bit Independence Criterion and Strict Avalanche Criterion. In other words, AES S-Box doesn't vary as much as the DES S-Boxes.

4 Conclusion

After obtaining the results of the test on each and every S-Box of DES as well as the single AES S-Box, coefficient of variance has been computed. Coefficient of variance is defined as a statistical measure of the dispersion of data points in a data series around the mean. It is calculated as follows:

$$\text{Coefficient of Variance} = \text{Std. Deviation}/\text{Mean}$$

Coefficient of variance represents the ratio of the standard deviation to the mean, and it is a useful statistic for comparing the degree of variation from one data series to another, even if the means are drastically different from one another.

Figure 2 depicts the Standard Deviation and Coefficient of Variance of each of the 8 DES S-Boxes as well as the AES S-Box.

As evident from the plot, the coefficient of variance in case of DES S-Boxes varies in the range of (3, 12). The results closer to the lower end of the spectrum indicate that those S-Boxes perform well with respect to the test but results obtained at the higher end of the spectrum indicate that those S-Boxes do not perform well with respect to the underlying test. In contrast the results obtained from the sole S-Box of AES yields coefficient of variance close to 3 indicating the S-Box performs well with respect to the underlying test.

As discussed in Sect. 1, the two fundamental aspects of a block cipher are confusion and diffusion. Though there are tests to analyze the cryptographic strength of the block cipher as a whole as well as to analyze the diffusion characteristic of the underlying cipher but there are hardly any tests aimed at precisely analyzing the confusion characteristics of the underlying cipher. In such a situation,

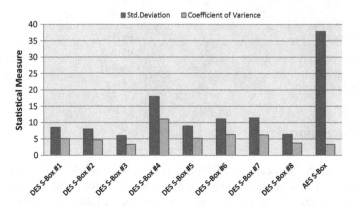

Fig. 2 Performance of AES versus DES

the proposed test could very well serve as a tool to analyze the confusion characteristic of the underlying block cipher and may be included as a part of a comprehensive test suit for analyzing the cryptographic strength of block ciphers.

References

1. Adams, C and Tavares, S, 1990, "The structural design of cryptographically good s-boxes", Journal of cryptology, 27–41.
2. Bhowmik, D, Datta, A, & Sinha, S, 2015, "Measuring Diffusion Charactarestic of Block Ciphers: The Bit Relationship Test", National Conference on Computational Technologies-2015, International Journal of Computer Sciences and Engineering (IJCSE), vol-3, Special Issue -1, 76–80.
3. Coppersmith, D, 1994, "The Data Encryption Standard and its Strength Against Attacks", IBM Journal of Research and Development, 38(3) 243.
4. Daemen, J, and Rijmen, V (March 9, 2003). "AES Proposal: Rijndael" (PDF). National Institute of Standards and Technology. p. 1.
5. Matsui, M, and Yamagishi, A, "A new method for known plaintext attack of FEAL cipher". *Advances in Cryptology - EUROCRYPT1992.*
6. Mister, S and Adams, C, 1996, "Practical S-Box Design", Workshop on Statistical Areas in Cryptology, SAC 1996, Workshop Record.
7. Paar, C. and Pelzl, J., 2010. "Understanding Cryptography", Berlin: Springer-Verleg.
8. Rodwald, P and Mroczkowski, P, 2006, "How to create" good "s-boxes?", 1st International Conference for Young Researchers in Computer Science, Control, Electrical Engineering and Telecommunications ICYR 2006, Zielona *Góra, Poland, 18–20.*
9. Shannon, C,1949, "Communication Theory of Security Systems", Bell Systems Technical Journal, vol-28.

Part VIII
Wireless Sensor Network

Source-Initiated Routing to Support Large-Scale Pseudo-Linear ZigBee Networks

Uttam Kumar Roy

Abstract IEEE 802.15.4 compatible instruments are resource-limited; thereby cannot run conventional resource-hungry routing algorithms. Even though, ZigBee supplied a simple routing algorithm for tree networks, it cannot handle long linear networks typically set up in mines, roadside, agricultural field, mountains etc. his paper proposes a light-weight routing logic based on source routing that supports arbitrary long networks. The proposed algorithm is especially suitable for networks that look almost linear. The algorithm leverages the properties that a pseudo-linear network has limited branching. Experimental results show that this flexible mechanism exhibits excellent packet delivery performance.

Keywords Zigbee · Pseudo-linear tree networks · Source routing

1 Introduction

There are many situations where network topologies are often linear or almost linear. Examples include networks formed in mines, roadside (Fig. 1), agricultural field, mountains etc. To have cheap solutions, scientists are trying to use the novel IEEE 802.15.4/ZigBee [1] technology. As it encourages low cost devices; thereby resource-limited, ZigBee [1] provides a light-weight and table-free routing scheme as well as an addressing scheme. However, it does not support long and asymmetric networks.

In [2–9], we already have proposed several potential solutions to overcome the problems. In this paper, we have proposed a table-free, light-weight routing algorithm based on source routing that supports arbitrary long networks. The proposed algorithm is especially suitable for pseudo-linear networks where nodes typically have one or a few children nodes. The route information is kept in sink only and is encapsulated in the packet. The nodes on its way use this information to take routing

U.K. Roy (✉)
Department of Information Technology, Jadavpur University, Kolkata, India
e-mail: u_roy@it.jusl.ac.in

© Springer Science+Business Media Singapore 2017   645
J.K. Mandal et al. (eds.), *Proceedings of the First International Conference on Intelligent Computing and Communication*, Advances in Intelligent Systems and Computing 458, DOI 10.1007/978-981-10-2035-3_65

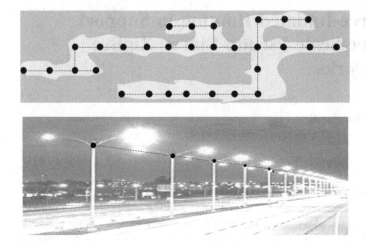

Fig. 1 Pseudo-linear networks in mine (*above*) and roadside (*below*)

decisions. The strength of the method is that intermediate nodes do not need and routing table. Consequently devices having limited memory and computing power can run also such algorithm. Experimental results show that this flexible mechanism exhibits excellent packet delivery performance. Since the scheme does not also involve complex calculation and logic, the text of this paper also looks very simple.

2 Related Work

We investigated [6] the ZigBee routing protocol and found that it is essentially an integration AODV and of tree routing adding many optimizations considering static network topologies.

Then we provided [2] a novel scheme based on mobile IP where we solved the address exhaustion drawback by borrowing address. It can easily used to expand the network of arbitrary diameter beyond 16 hops. A routing algorithm, was also suggested.

We extended [3] the ZigBee Tree routing for harsh and asymmetric networks. Then we suggested [4] a scheme for the dynamic network topologies, It was based on multi-channel routing scheme. We also and solved the link breakage issue by multi channeling. The proposed scheme had a very little overhead.

In [7, 8], we proposed a simple, variable-sized addressing mechanism together with a new resource-friendly, table-free routing scheme. It makes use of the features of prefix code, Devices, in this scheme, may have any number of children and does not also restrict network depth.

In [9], we provided another extended version of addressing mechanism suggested by ZigBee. It had similar features of actual addressing algorithm. So, we can still use table-free ZigBee routing scheme. Additionally, it can handle asymmetric tree topologies which the actual mechanism cannot handle.

There are very few works done in the area of 802.15.4/ZigBee. In [10, 11], its applications have been discussed thoroughly. This is one of the primitive studies on MAC sub-layer. In [12], the performance analysis of 802.15.4/ZigBee has been studied and described.

3 Proposed Algorithm

Since, IEEE 802.15.4 is very cheap technology; scientists want to use the same virtually everywhere such as in mines, in smart cities for monitoring environmental metrics and traffic controlling, in agriculture, in remote places for volcano monitoring what's not. These networks have often some common characteristics:

- Typically have a sink node (usually a full-fledge computer) and other nodes send sensed data (temperature, humidity, pictures etc.) to sink.
- Networks are often linear or pseudo-linear (Fig. 1) in nature.
- Nodes are stationary and are attached and detached the network infrequently once a network is formed.
- Sink, occasionally sends data to other nodes (e.g. traffic control network)
- A non-sink node hardly sends data to another non-sink node.

Keeping these requirements and the above network characteristics in mind, it is possible to device a simplified routing algorithm that can run on those devices. For example, we can use a tree network where no routing algorithm is needed to send data from sensor node to sink. Each sensor node (having exactly one parent) forwards data to its parent and this procedure continues until the data reach to the sink.

However, if sink wants to send data to a sensor node (for traffic control network, say) a routing algorithm is needed. The proposed scheme leverages the properties that a pseudo-linear network has limited branches. This means nodes in these kind of networks have a very few (0, 1, 2, 3 or 4) children. This property is cleverly used so that a packet can be routed to a destination based on small routing information encapsulated in the packet itself. Before, discussing the routing algorithm, let us understand how network is formed.

3.1 Network Formation

So, we form the tree networks in the following:

Each router of the tree network marks its all outgoing children (if there is any) using a unique (locally) binary string. The order of labeling is not important. One such tree is shown in Fig. 2. For example, node C has 3 children. So, it uses 2 bits to label its outgoing links. Similarly, node D4 has only two children. So, it uses

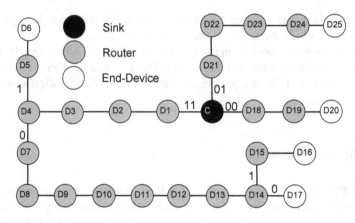

Fig. 2 Pseudo-linear network formation

only 1 bit to label its outgoing links. Nodes having only child (most of the nodes are of this kind) need not label the link.

Suppose a router R has C_R children. Then the number of bits necessary to mark each child:

$$N(C_R) = \begin{cases} C_R & \text{if } C_R = 0 \quad \text{or} \quad C_R = 1 \\ \lceil \lg(C_R) \rceil & \text{if } C_R > 1 \end{cases} \tag{1}$$

Since for a network line roadside-network, most nodes will be non-crossover nodes and C_R for such node is 1 and no bit is required for link labeling. There are a very few crossover nodes where C_R is limited and often ≤ 4 (a crossover of 5 roads) and $N(C_R) \leq 2$. These indicate that the proposed method has very small overhead.

3.2 Routing

To keep network protocol simple, proposed method uses MAC addresses as network addresses. However, any other addressing scheme may also be used. Once a pseudo-linear tree network is setup, sensor nodes send data to the sink. No routing algorithm is needed for this. Every node forwards packets to its parent and the packet eventually reaches to the sink. When a packet traverses from a sensor node to sink, intermediate nodes *append* link label to the packet. For example, the packet received by sink C from D16 will have following routing string:

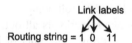

where first 1 is appended by node D14, next 0 is appended by D4 and last two bits 11 are appended by C itself. Sink keeps this information in a table. This process

Table 1 Routing string for all nodes stored in sink

Nodes	Routing string	Nodes	Routing string
D_1, D_2, D_3, D_4	11	$D_7, D_8, D_9, D_{10}, D_{11}, D_{12}, D_{13}, D_{14}$	011
D_5, D_6	111	D_{18}, D_{19}, D_{20}	00
D_{15}, D_{16}	1011	$D_{21}, D_{22}, D_{23}, D_{24}, D_{25}$	01
D_{17}	0011		

Table 2 Augmented table containing routing string for all nodes stored in sink

Nodes	Routing string	Nodes	Routing string	Nodes	Routing string
D_1–D_4	11	D_{15}–D_{16}	1011	D_{18}–D_{20}	00
D_5–D_6	111	D_{17}	0011	D_{21}–D_{25}	01
D_7–D_{14}	011				

continues and eventually sink will have routing string for all sensor nodes which looks as shown in the Table 1.

Since, sink is typically a full-fledge computer, it has enough memory to store this information. This table contains only total 20 bits of routing information and 25x (size of device address in bits) bits device addresses which can further be reduced by carefully assigning the device addresses. For example, the previous table can be restructured as shown in Table 2.

This table contains only 13 network address. Let us have an estimation of amount memory required in practice.

Consider a city with 10×10 km area with roads laid as a grid separated by 100 m apart. The transmission range of a ZigBee device is 20 m (say). So, there will be approximately $50 \times 100 \times 100 = 500000$ devices. If the network is formed carefully and the sink is placed in the centre of the city, average length of label string will have very few bits. So, sink needs a few MB of memory which even a very old computer possesses.

3.3 Examples

We shall consider the network in Fig. 2 to demonstrate the routing algorithm (Fig. 3). Consider the coordinator C sends data to the device D16.

The sink looks up the table and finds that the routing string for device D16 is 1011. Since, it has 3 children, it extracts least significant N(3) (=2) bits from 1011 which happens to be 11. It puts the remaining string 10 in the packet and forwards it through the link labeled with 11. The packet comes to device D1.

Since D1 exactly one child, it forwards the packet to D2. Devices D2, D3 do the same thing and the packet reaches to D4. Since, it has 2 children, it extracts least significant N(2) (=1) bits from 10 which happens to be 0 and puts the remaining string 1 in the packet and forwards it along the link marked as 0. It is then received

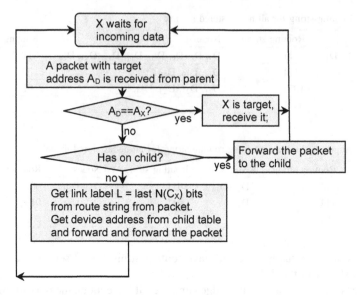

Fig. 3 Routing algorithm

by the device D7. Device D7–D13 do as D1 did and packet eventually reaches to D14. Since, it has 2 children, it extracts least significant N(2) (=1) bits from 1 which happens to be 1. It puts the remaining null string in the packet and forwards it along the link marked with 1. It is received by the device D15 which as one child. The device forwards the packet to its child and the packet is finally received by the taget device D16.

3.4 Restructuring

It may be noted that, that length of the link label might change if new roads are laid down. We shall consider the topology as shown in Fig. 4.

The number of children of router $R1$ (C_{R1}) is 2(two). So, the number of bits N (C_{R1}) necessary to mark each outgoing link of $R1$ is 1(one). Now, if a new road is

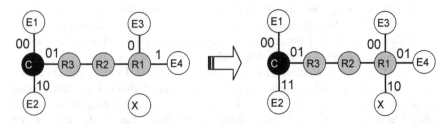

Fig. 4 "Restructuring" links (*left*) before (*right*) after

laid (i.e. *X* wants to join to *R*1), the $N(C_{R1})$ will have value 2. As a result, all links of *R*1 (including the new link) must be re-marked by 2 bits. It means that routing information for all descendants of *R1* must be updated to the sink. This procedure is called *restructuring*, which clearly imposes some overhead. Since, roads are laid down very in-frequently; thus overhead is almost negligible.

The restructuring may be done by sending a single update message to sink. The sink extracts routing string RS from this packet and updates the table by substituting all routing string of the form RSX by RS0X. The entry for X is added when sink receives any data from X. An example is shown in Fig. 5.

4 Experimental Results

We used Digi International's IEEE 802.15.4 complaint XBee and XBee-PRO OEM RF modules. It is a frequently used embedded solution suitable for low-cost and low-power scenarios. These modules are easy-to-use, share a common footprint and require minimal development time and risk.

Due to cost constraints, we have used only 6 devices (one sink, two routers and 3 end-devices) to form a pseudo-linear network as shown in the Fig. 5. All the devices except sink sent a randomly generated temperature value (from 20 to 25 °C) to the sink. Sink also sent data to the other devices. The experiment was carried out 1 h. The experiment is carried out 10 numbers of times and the packet delivery rate (average) was obtained as 100 % (Fig. 6).

R_1,R_2,R_3	E_1	E_2	E_3	E_4	=>	R_1,R_2,R_3	E_1	E_2	E_3	E_4	X
01	00	10	010	011		01	00	01	0100	0101	0110

Fig. 5 Route update by sink node for X (*left*) before (*right*) after

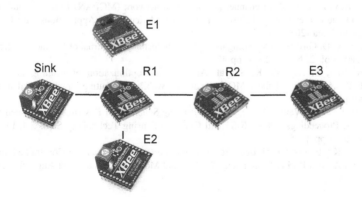

Fig. 6 Experimental test bed using MaxStream XBee OEM kit

Since, we couldn't arrange large number of devices physically; we also carried out a simulation using NS3 simulator.

The 1360 × 640 pixel area was taken for the simulation. The coordinator comes first and sets up a network containing itself. It then starts send beacons periodically. Other devices (routers and end devices) are then attached every after one second to the network area. The devices wais for some time and receives beacons from surrounding devices and sends a JOIN_REQUEST packet and attaches itself to the network. If the device is a router, it also starts sending beacons once it joins to the network. The entire specification was specified by the ZigBee consortium.

A simulation was also carried out for a large pseudo-linear network having 2000 nodes. The experiment was repeated over 100 times and the packet delivery rate was also observed as 100 %. These two experiments prove the correctness of the proposed scheme.

5 Summary

This paper proposes a table-free, light-weight routing algorithm based on source routing that supports arbitrary long networks. The proposed algorithm is especially suitable for networks that look almost linear. Theoretical and simulation results prove the correctness of the proposed scheme.

References

1. ZigBee Doc. 02130r7, Dft 0.90: Network Spec., July, 2004.
2. U. K. Roy, D. Giri: Address borrowing in wireless personal area network. Proceedings. of the Intl Adv. Computing Conf. (2009) pp 1074–1079.
3. U. K. Roy, D. Giri: Single-level address reorganization in wireless personal area network. Intl Conf. on Computers & Devices for Communication. (2009) pp 1–4.
4. U. K. Roy, D. Giri: Multi-channel personal area network (MCPAN) formation and routing. Proc. of the conference on Industrial Engineering Science and Applications, April 2-4 NIT Durgapur India (2014) pp 73–79.
5. U. K. Roy, D. Giri: WPAN routing using huffman technique. Journal of Engineering Sciences, February Vol. 2, No. 2 (2013) pp 49–61.
6. U. K. Roy, S. Ray, D. K. Sanyal: Analysis and optimization of Routing Protocols in IEEE802.15.4. Proc. of the AMOC Asian Intl Mobile Computing Conf. JU Calcutta India (2006) pp. 172–181.
7. Roy U. K: Light-Weight Addressing and Routing Schemes for Resource-Constrained WPAN Devices. Proceedings. Of 8th ICST Intl Conf. on Sensing Technology Sept. 2-4, Liverpool, UK. (2014) pp 46–51.
8. Roy U. K.: Extending ZigBee Tree Routing Protocol for Resource-Constrained Devices. APWiMob Asia Pacific Conference Wireless and Mobile, Bali Indonesia Aug. 28-30, (2014) pp 48–53.

9. Roy U. K., P. Bhaumik: Enhanced ZigBee Tree Addressing for Flexible Network Topologies. Proc. of the AIMoC, Intl Conf. on Applications and Innovations in Mobile Computing, February 12-14, Kolkata India (2015) pp. 6–11.
10. P. Gorday, L. Hester E. Callaway (ed): Home networking with IEEE 802.15.4: Developing Standard for Low-Rate wireless personal area networks. IEEE Comm. Magazine, August (2002).
11. M. J. Lee, J. Zheng: A comprehensive performance study of IEEE 802.15.4 (2004).
12. IEEE 802.15.4. WMAC (Wireless MAC) and PHY (PHysical Layer) Spec. for low-rate WPANs.

SSeS: A Self-configuring Selective Single-Path Routing Protocol for Cloud-Based Mobile Wireless Sensor Network

Sandip Roy, Rajesh Bose and Debabrata Sarddar

Abstract Mobile wireless sensor networks (MWSN) are most popular for new generation of sensor networks. Now a days' many applications such as health monitoring, environment monitoring or surveillance, the performance of MWSNs are much more versatile than WSNs. Some protocols from MANET such as AODV, DSR and GPSR are able to work on this environment. This manuscript has focused on two major issues. Firstly, formulating an energy-aware selective stable single-path routing protocol for MWSNs that is pertinent to the present challenges of the researchers. Secondly, presenting a germane architecture that can accumulate the data and synchronize with cloud database at periodic interval in order to predict about ominous situation. Coda of the manuscript presents the simulation result of our proposed algorithm and behavioural analysis of successful packet delivery from source to sink node with respect to node density and node mobility condition. Our proposed single-path selection methodology ameliorates the overall network lifetime in case of high network traffic.

Keywords Cloud computing · Internet of things (IoT) · MANET · MWSN

1 Introduction

MWSN is the promising wireless network comprises of mobile sensor nodes which are highly adaptable than WSNs. Internet of Things (IoT) and Cloud Computing technologies became boon in recent years, where WSN being as the basic infrastructure for data accumulation. Disposition of static sensor nodes usually degrade

Sandip Roy (✉) · Rajesh Bose · Debabrata Sarddar
University of Kalyani, Kalyani, Nadia 741235, West Bengal, India
e-mail: sandiproy86@gmail.com; sandip@klyuniv.ac.in

Rajesh Bose
e-mail: bose.raj00028@gmail.com

Debabrata Sarddar
e-mail: dsarddar1@gmail.com

© Springer Science+Business Media Singapore 2017
J.K. Mandal et al. (eds.), *Proceedings of the First International Conference on Intelligent Computing and Communication*, Advances in Intelligent Systems and Computing 458, DOI 10.1007/978-981-10-2035-3_66

performance of many applications such as monitoring environment conditions, health conditions of human beings or many surveillance monitoring services [1, 2]. Mobile sensors are capable of sensing data related to practical environment. Hence mobile sensor node works as stratagem in this context. Absence of steady routing topology in MWSNs summons to the transmission of the perceived sensor data from the target to destination cloud ends for analyzing time-to-time.

Due to the popularity of IoT, experts forecast that by the year 2020 fifty billion devices will be connected with the internet which is seven times more than world's population [3]. WSNs are the one of the key elements of the IoT paradigm and the main purpose of WSNs is to provide sensing data to the users periodically. In this manuscript we have articulated a self-configuring selective single-path routing protocol for MWSNs which intends to find out most steady route between source node to sink node. The remaining part of the paper is formulated as follows. Reviews of already alike problems are described in Sects. 2 and 3 demonstrates the problem formulation of our proposed methodology, Sect. 4 presents our proposed single-path routing methodology. Simulation result of proposed algorithm is detailed in Sects. 5 and 6 brings down curtain with conclusion and future scope.

2 State of the Art Review

Now a day's sensors can be deployed everywhere to sense any kind of intended data. Smart phones are comprising of several types of sensors in order to sense relevant data from the environment. Sensors in factories helps in controlling CO_2 emissions, in the forest sensors are useful for fast fire detection and many other applications. Due to this augmented performance in different application domain using sensors, more and more researches are upbringing advancement in wireless sensor network. The phrase "Internet of Things" (IoT) was revealed by K. Ashton, which is based on WSN and boon to the recent years [4]. It comprises of internet based things that have unique identities and are connected to the Internet. MWSN based IoT is congruous to the recent research field where mobility is corned and static sensors failed to achieve the expected performance. Finding a stable route for mobile sensor nodes are one of the critical and fundamental issue. Proactive route discovery algorithm like Dynamic Source Routing (DSR) also works in the context of MWSNs [5]. Considering the angle between two active nodes, one is sensor and another node is receiver, a new routing technique was formulated namely Angle-based Dynamic Source Routing by Kwangcheol et al. [6]. In the mobility-centric environments, the LEACH-Mobile protocol attains more successful data transfer rate with respect to the non-mobility centric LEACH protocol [7]. A novel selective, multi-path routing for effective load balancing was presented by Chakraborty et al. [8]. In their proposed work, trust value based node selection procedure also helped to improve the lifetime of the whole network.

3 Problem Formulation

The proposed routing scheme for MWSNs will help to optimize the network life and throughput through its energy efficient routing strategy.

3.1 Determining the Stable Path from Source to Sink Node

To determine the best-steady path among the all available paths between a given source node to the target destination node, we are interested in measuring a path score. We represent the set of available paths from a source to destination node as *Path*. Therefore, *Path* is defined as follows:

$Path = \{Path_i\}$, where, $1 \le i \le k$, k represents total number of paths.

3.2 Presenting the Number of Links in a Given Path

A path is formed by adjoining links between the nodes belong to the path. Now, to represent the number of links in a given path connecting one sensor node to another, we define L_{ij} as link between two sensor nodes; say N_i and N_j in a given path. Thus, L_{ij} is defined as (N_i, N_j).

3.3 The Two-State Markov Chain

The probability of forming link L_{ij} between two nodes N_i and N_j is defined by two-state Markov chain. Thus, availability of a link between two nodes depends on the energy level of corresponding nodes. If the node attains the defined energy level then the link can be formed, otherwise link is not formed. Thus, the recipient node could accept or reject the link establishment.

Figure 1 represents the state diagram of two-state Markov chain. The following transmission matrix can be represented as a two-state Markov chain as shown below:

Fig. 1 State transition diagram of two-state Markov Chain

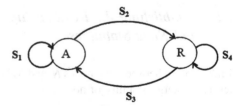

$$P = \begin{pmatrix} S_1 & S_2 \\ S_3 & S_4 \end{pmatrix} \tag{1}$$

where, S_1, S_2, S_3 and S_4 are defined as the probability which is explained below. Here two-state Markov chain delineates a state of acceptance as state A and the state of rejection to be state R.

S_1 represents the probability that the next state is A given that present state is A state,

S_2 represents the probability that the next state is R given that present state is A state,

S_3 represents probability that the next state is A given that present state is R state,

S_4 represents the probability that the next state is R given that present state is R state

The sum of each row of the transition matrix is 1, i.e. $S_1 + S_2 = 1$ and $S_3 + S_4 = 1$

3.4 Computing the Node Stability

To establish an optimum path from the target node to sink node, energy stability of each node N is computed by the sum of two factors; a) strength of residual energy level of a node that asserts link establishment with node N and b) chances of premature death of the node N. Energy stability of a given node N is computed as follows:

$$node\_stability(N) = P + Q \tag{2}$$

where, $P = w_1 \times probability(accept\_state) = w_1 \times Maximum(S_1, S_3)$ $Q = (-w_2) \times probability$ $(premature\_death)$ and $|w_1| + |w_2| = 1, w_1 \rangle w_2$.

The chance of premature death of sensor node can be calculated by Poisson Probability Distribution. Thus, the probability mass function is given as follows:

$$P(X = k) = \frac{e^{-\mu}\mu^k}{k!} \tag{3}$$

where $k = 0, 1, 2, 3 \ldots$ and μ is the expected value of X.

3.5 Establishing the Relationship Between Link Stability and Node Stabilities

Link stability between two nodes N_i and N_j is calculated by the following formula in terms of energy stability of nodes.

$$link\_stability(N_i, N_j) = \frac{node\_stability(N_i) + node\_stability(N_j)}{2} \quad (4)$$

3.6 Calculating Path Score

The *path score* is computed by summing three distinct terms; (a) stability of each path, (b) route expiration time and c) hop count of the given path.

Path score is to be evaluated maximizing the following objective function:

$$f(.) = c_1 \times \left(\frac{Path_i\_stability}{\max(Path\_stability)}\right) + c_2 \times \left(\frac{RET_i}{\max(RET_i)}\right) + c_3 \times \left(\frac{hop\_count_i}{\max(hop\_count)}\right)$$

$$(5)$$

where, $|c_1| + |c_2| + |c_3| = 1$ and $c_1 \rangle c_2 \rangle c_3$.

The path stability, route expiration time and hop count are explained below:
Path stability i.e. *path_stability* is defined by *PS (Path$_i$)*:

$$PS(Path_i) = \prod_{L_k \in (Path_j)} probability(L_k) \quad (6)$$

Link expiration time (LET) can be calculated by the prediction of future disconnected time between two neighbor nodes in motion. Let us consider two nodes n_1 and n_2 at (x_1, y_1) and (x_2, y_2) coordinates with uniform transmission radius r and initially they are within transmission range. Let nodes n_1 and n_2 have speed v_1 and v_2 along the directions Θ_1 and Θ_2 respectively. Then link expiration time (LET) between n_1 and n_2 is defined as follows [9]:

$$LET = \frac{-(ab+cd) + \sqrt{(a^2+c^2)r^2 - (ad-cb)^2}}{(a^2+c^2)} \quad (7)$$

where $a = v_1 \cos\theta_1 - v_2 \cos\theta_2$, $b = x_1 - x_2$, $c = v_1 \sin\theta_1 - v_2 \sin\theta_2$, and $d = y_1 - y_2$.

Route Expiration Time (RET) is equal to the minimum of the set of LETs for the feasible path.

Lastly, hop count represents the number of intermediate sensor nodes in a given path.

3.7 Determining the Highest Path Score to Select the Best Path

Therefore, to select the best path from source to sink node, *path_score* of the entire path are computed and is the maximum path score of the all paths from the source node to sink node.

$$Path\_score = \max_{\forall Path_i} f(Path_i) \tag{8}$$

4 Our Proposed Single-Path Routing Methodology

Our proposed system architecture is partitioned into the Sensing Layer, the Coordinating Layer and the Supervising Layer. Multiple sensor nodes collect data from external environment and send this data to the coordinating layer after some periodic intervals. After that coordinator node collects data and temporary stores this data into their buffer. Then coordinator node periodically sends the data to the supervising layer. At last in the supervising layer, that data is stored in a cloud for further examination. Figure 2 shows a schematic diagram of our proposed model. The sensor layer consists of different sensors (like DHT22, MICS-5525) with Raspberry Pi mini-computer which deployed different location in an area and senses the temperature, humidity, Carbon Monoxide (CO) or other relevant data from the external environment. After receiving the data sensor nodes transmit this data to the coordination layer using ZigBee protocol. The coordinator node then stores this data into its buffer and periodically sends this data to the supervising layer. Finally, in supervising layer sensed data is stored in cloud database and publish it on the internet. In this manuscript, we concern about mobile wireless sensor network where sensor nodes have mobile characteristics. The primary goal of the researcher is to search a steadiest route between source and destination node. Our proposed routing protocol consists of two major phases: Neighbor Discovery and Route Establishment which is discussed.

4.1 Neighbor Discovery

Each sensor node has to recognize its neighbors transmitting a HELLO packet to them. Receiving HELLO packets triggers an ACK packet to be send by the sensor node containing the stability of the recipient node which is computed using Eq. (2) assuming weight values as $w_1 = 0.7$ and $w_2 = -0.3$. For calculating node stability, at first we have calculated activeness of a certain node by their residual energy (RE). If the residual energy is more than 75 % of the initial energy, then P_1

Fig. 2 Layer based architecture for our proposed methodology

transition matrix is used. If the residual energy in the range of 25–74 % of the initial energy, then P_2 transition matrix is used otherwise when the residual energy below 25 % then P_3 transition matrix is used. P_1, P_2, and P_3 transition matrix is presented below in this context. After receiving ACK packet receiver sensor node calculated link stability between sender and receiver sensor nodes using Eq. (4).

$P_1 = \begin{pmatrix} 0.9 & 0.1 \\ 0.9 & 0.1 \end{pmatrix}$	$P_2 = \begin{pmatrix} 0.7 & 0.3 \\ 0.7 & 0.3 \end{pmatrix}$	$P_3 = \begin{pmatrix} 0.3 & 0.7 \\ 0.3 & 0.7 \end{pmatrix}$
RE > 75 %	25 % > RE > 75 %	RE < 25 %

4.2 Route Establishment

In order to deduce a stable route from source node to a sink node, a ROUTE_REQUEST packet is initiated by a source node towards the sink node. Prior to the sending of ROUTE_REQUEST packet, each sensor node verifies the value of link stability of its entire neighbors. The sensor node forwards the ROUTE_REQUEST packet to its only neighbor if the link stability value of that neighbor is greater than 0.5. The sender sensor node forwards packet to the highest and the second highest stable link node if the maximum value of link stability of its neighbor is in between 0.22 to 0.49. Otherwise the ROUTE_REQUEST packet is sent to all the neighbor sensor nodes. Consequently, link expiration time (LET) can be obtained by the selected neighbor sensor node(s) using Eq. (7). For preventing looping error receiving the ROUTE_REQUEST packet receiver node keep the corresponding sensor node's ID and insert its' own ID. A sensor node cannot forward the ROUTE_REQUEST where the sensor node's ID is already kept. After receiving the ROUTE_REQUEST packet, the sink node computed path score of the entire feasible paths between source sensor node and sink sensor node using Eq. (5) (assuming $c_1 = 0.5$, $c_2 = -0.3$ and $c_3 = -0.2$). Stable path is the maximum path score of all paths from source to sink node by Eq. (8). As soon as a stable path is determined, a ROUTE_REPLY packet is initiated from the sink node to the source node. Then the source sensor node affirms a path to the corresponding sink sensor node. Now data packet forwards towards the sink node along with the stable path. After receiving the data stores it in cloud database and publish it on the internet.

Let us considered a scenario where ten sensor nodes are moving with constant velocity of 10 m/s along in a given direction which is enlisted in Table 2. Residual energy of all nodes is enlisted in Table 1.

Table 1 Residual energy of each sensor node along with ID	Node ID	RE (%)	Node activeness
	1	90	0.9
	2	85	0.9
	3	78	0.9
	4	70	0.7
	5	68	0.7
	6	85	0.9
	7	50	0.7
	8	60	0.7
	9	78	0.9
	10	95	0.9

Table 2 Velocity and moving direction of each sensor node along with ID	Node ID	Constant velocity	Moving direction
	1	10 m/s	90°
	2		63°
	3		61°
	4		50°
	5		56°
	6		60°
	7		34°
	8		83°
	9		34°
	10		56°

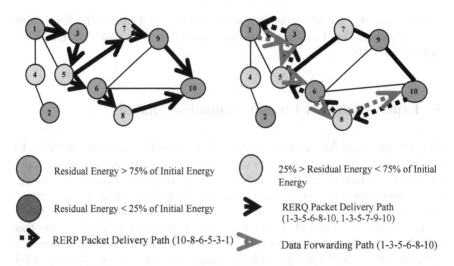

Residual Energy > 75% of Initial Energy

25% > Residual Energy < 75% of Initial Energy

Residual Energy < 25% of Initial Energy

RERQ Packet Delivery Path
(1-3-5-6-8-10, 1-3-5-7-9-10)

RERP Packet Delivery Path (10-8-6-5-3-1)

Data Forwarding Path (1-3-5-6-8-10)

Fig. 3 RERP, RERQ and data packet delivery using our proposed methodology

In the example shown in Fig. 3, we assume that node ID #1 wants to send data to a sink node ID #10. We assume that each node receives ACK packets from its corresponding neighbor sensor nodes. Every ACK packet which is transmitted contains a node stability value of the issuing node. As seen in the Fig. 3, node ID #1 determines that the maximum link stability is accorded by node ID #3 with a value of 0.44 which is computed by Eq. (4). Hence, node ID #1 sends a ROUTE_REQUEST packet to node ID #3 which is determined to be the most stable link. In a similar way, node ID #3 calculates link stabilities of its neighboring nodes. It finds node ID #5 to be most stable with a value of 0.51. Now, node ID #5 computes link stabilities of its neighbor nodes and find the highest link stability is 0.44. So node ID #5 sends the ROUTE_REQUEST packet to the node ID #6 and #7 both. This process goes on till the ROUTE_REQUEST packet reaches sink node ID #10. Upon receipt of this packet, the sink node computes the path score of all possible paths and able to pick out the most stable path after sorting the path score values in descending order with the highest score at the top. A ROUTE_REPLY packet is then transmitted back to the initiating node through the stable path which is resulting out our proposed process. The transmission is unicast and does not affect nodes which are not part of the path determined to be as the most stable for the transmission. The source node begins transferring data over this stable path to the sink node. As shown Fig. 3, two feasible paths which are mentioned below. Path I: (1, 3, 5, 6, 8, 10) and Path II: (1, 3, 5, 7, 9, 10) Path score of Path I and Path II is calculated below:

$$W_1 = 0.5 \times (0.0470/0.0470) - 0.3 \times (22.125/37.8) - 0.2 \times (5/5) = 0.1244$$
$$W_2 = 0.5 \times (0.0470/0.0470) - 0.3 \times (37.8/37.8) - 0.2 \times (5/5) = 0$$

Since $W_1 \rangle W_2$, according our proposed methodology Path I is the most stable route. Therefore, Path I is selected for data transmission from source node ID #1 to sink node ID #10.

5 Experimental Set-Up and Simulation Analysis

Our proposed algorithm has been successfully simulated using Matlab R2012b. To simulate the algorithm, we have considered here almost 300 mobile sensor nodes randomly distributed over 500 m × 500 m terrain area for our experimental set-up. Each sensor node is capable of communicating within the range of 100 meters and all the sensor nodes are considered to be able to harvest energy by solar panel. Neighbour nodes are considered to be capable of communicating with each other directly within their transmission range. It is assumed that each of sensor nodes is comprised of 5 J initial energy. MICA2 energy model is used in this context for receiving and transmitting signals [10]. Energy required for receiving and transmitting signals are 0.234 µJ/bit and 0.312 µJ/bit respectively. The size of the packet has taken as 96 bits. The simulation results are generated in account of two considerations. Firstly, we have considered the nature of packet density ratio (PDR) with respect to sensor node density from 100 to 280 sensor nodes. Secondly we have considered the same metric with respect to the node mobility from 5 m/s to 18 m/s and premature death of at most 5 % in a round is assumed to be 0.4. Our simulation results are generated in these context considering 10 rounds and 15 rounds respectively over the random network. Lastly we have measured the average network life time with respect to the node density.

5.1 Packet Delivery Ratio (PDR)

The percentage ratio of the total number of packets send by the source node and the total number of packets received by the destination node is defined as Packet delivery ratio (PDR).

$$PDR = (R/S) \times 100 \tag{9}$$

R   Total number of packets send by the source node
S   Total number of packets received by the destination node

Figures 4 and 5 shows that the packet delivery ratio is with changing node density and mobility speed of each node respectively. For the less node density, packet density ratio is almost 100 %, but the packet delivery ratio decreases with increasing the congestion of the network. However, the packet density ratio is quite stable with higher node density.

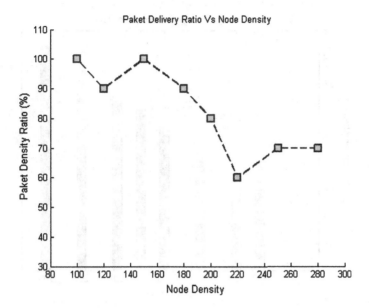

Fig. 4 PDR versus node density

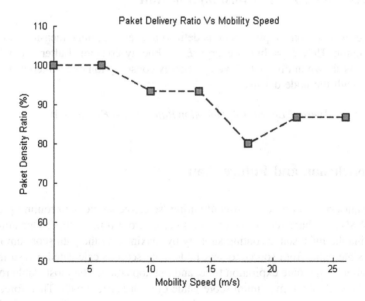

Fig. 5 PDR versus mobility speed

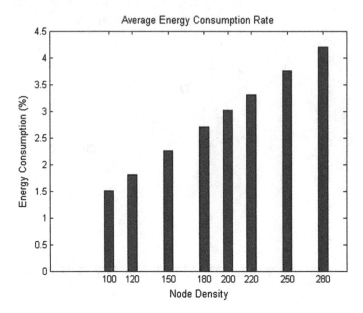

Fig. 6 Average energy consumption rate versus node density

5.2   *Average Energy Consumption Rate*

Average Energy Consumption Rate is defined as the percentage measure of energy consumption. Here E_I = Initial energy E_C = Energy consumed after n number of rounds. As shown in Fig. 6, the average energy consumption rate is increasing quite linearly with the node density.

$$Average\,Energy\,Consumption\,Rate = (E_I - E_C/E_I) \times 100 \qquad (10)$$

6   Conclusion and Future Plan

This manuscript proposed a self-configuring, selective single path routing protocol for MWSNs to share sensor data to the Internet from time to time. Our aim is to determine the most stable routing strategy by maximizing the path score among all the possible paths from source to sink node. Path score is calculated measuring the path stability, the route expiration time, and the hop count. The most stable route is selected based on the maximum score value of each feasible path. Therefore, more path score represents more reliable path from source to destination. Lastly Simulation results show that average 80 % successful packet delivery at different node density and several mobility conditions.

Our future research will focus on cloud-based smart forest fire detection and monitoring system using MWSNs.

References

1. Cheng, L., Wu, C. D., Zhang, Y. Z., Wu, H., Li, M.X., Maple, C.: A survey of localization in wireless sensor network. Int. J. Distrib. Sens. Netw. (2012) 1–12.
2. Dai, G., Miao, C., Li, Y., Mao, K., Chen, Q.: Study on Tree-Based Clustering MDS Algorithm for Nodes Localization in WSNs. Adv. in WSNs: The 8th China Conference, CWSN 2014. (2014) 176–186.
3. Swan, M.: Sensor Mania! The Internet of Things, Wearable Computing, Objective Metrics, and the Quantified Self 2.0. J. of Sens. and Actu. Netw. Vol. 1 (2012) 217–253.
4. Ashton, K.: That 'Internet of Things' Thing. In the real world, things matter more than ideas. RFID Journals. (2009).
5. Johnson, D., Maltz, D. A.: Dynamic Source Routing Protocol in Ad Hoc Networks. Kluwer Academic Publishers. (1996).
6. Kwangcheol, S., Kim, K., Kim, S.: ADSR: Angle-Based Multi-hop Routing Strategy for Mobile Wireless Sensor Networks. In Proc. of the IEEE Asia-Pacific Services Computing Conference (APSCC). (2011) 373–376.
7. Kim, D., Chung, Y.: Self-Organization Routing Protocol Supporting Mobile Nodes for Wireless Sensor Network. In proc. Of the 1st International multi-symposium on Computer and Computational Sciences (IMSCCS'06). (2006) 622–626.
8. Chakraborty, M., Chaki. N.: ETSeM: A Energy-Aware, Trust-Based, Selective Multi-Path Routing Protocol. Computer Information Systems and Industrial Management. Vol. 7564 (2012) 351–360.
9. Wang, N-C., Chen, J-C.: A Stable On-Demand Routing Protocol for Mobile Ad Hoc Networks with Weight-Based Strategy. In Proc. Of the 7th Int. Conf. on Par. and Dist. Comp., App. and Tech. (PDCAT'06) (2006).
10. Hussain, Md. Z., Singh, M. P., Singh, R. K.: Analysis of Lifetime of Wireless Sensor Network. IJAST. Vol. 53 (2013) 117–126.

Link Quality Modeling for Energy Harvesting Wireless Sensor Networks

Moumita Deb and Sarbani Roy

Abstract In energy harvesting wireless sensor networks (EH-WSNs) sensor nodes are capable of harvesting energy from environmental sources. Usually, knowledge about the link quality improves the performance of WSNs. During data routing, selection of good quality links are important to maintain stable communication. Thus helps to reduce the unnecessary energy wastage. Obtaining the link state information is more challenging for EH-WSN as different nodes has different energy profiles and state of the node depends on several environmental conditions. In this paper, we have studied different factors affecting the link quality and model it using finite state *Markov Model*. Energy availability of the harvesting devices through real time traces is considered for modeling the network. This model can significantly provide relevant information which is very effective to improve the routing decisions as the next hop decision will be more accurate. The usefulness and validity of the proposed approach is illustrated through simulations for specific examples.

Keywords Energy harvesting · Energy profile · Link quality · Markov chain

1 Introduction

In recent past years, the study of WSN has been well exercised. Especially energy harvesting options grab the attention most since it has immense application in present scenario. Often sensor networks fail to meet the design goal in terms of cost,

Moumita Deb (✉)
Department of Information Technology, RCCIIT, Kolkata, India
e-mail: moudeb@gmail.com

Sarbani Roy (✉)
Department of Computer Science and Engineering, Jadavpur University, Kolkata, India
e-mail: sarbani.roy@cse.jdvu.ac.in

© Springer Science+Business Media Singapore 2017   669
J.K. Mandal et al. (eds.), *Proceedings of the First International Conference on Intelligent Computing and Communication*, Advances in Intelligent Systems and Computing 458, DOI 10.1007/978-981-10-2035-3_67

lifetime, sensing transmission coverage and sensing reliability. One of the primary reasons behind it is sensor nodes are powered by battery. So the lifetime of a network depends solely on the lifetime of the battery installed in the sensor nodes [1]. To overcome this barrier, the concept of energy harvesting, i.e. harvesting energy from the environment or other sources and converting it to electrical energy comes into existence.

The researchers and engineers are seeking for strategies to remove batteries and wires so that unlimited and green energy sources can be used to develop autonomous WSNs with theoretically unlimited lifetimes. Moreover these new green sources are based on ambient energies and refer to any available harvesting technology, such as solar cell, thermoelectric, piezo-electric harvester, micro and wind turbine or other transducer [1]. Energy harvesting module in sensor nodes can exploit different sources of energy from environment, such as electromagnetic sources, mechanical vibrations, light, airflow, acoustic, temperature and heat variations etc. Though, energy output of a sensor node varies with time and immensely depends on environmental conditions. Potentially, sensors with harvesting subsystems enable WSN to last forever. One of the issues which significantly affect the network performance as well as lifetime of WSNs is link quality. Therefore, link quality estimation is a crucial problem, because its accuracy affects the design and the efficiency of networking protocols and applications. A large number of packets gets corrupted or lost because of insufficient knowledge of link quality and as a result energy consumption increases due to the retransmission of those packets. Furthermore, it reduces throughput and increases packet delivery latency. This anomaly becomes more severe in EH-WSN, since the high variability and unreliability of the medium imposes stricter constraints on the timing when estimation has to be carried out.

In this work, a simple theoretical approach for analyzing and optimizing various system parameters that determine the link quality of sensor nodes in EH-WSN is developed. We have simulated the EH-WSN environment and proposed methodology in Castalia simulator [2]. Nodes with mobility feature are also considered. Energy traces of nodes are collected through real life deployment and measurement studies reported in [3].

The paper mainly contributes the following:

1. Modeling and predicting link quality of sensor nodes in EH-WSN using Markov model.
2. Uses of real life trace information to model the energy source.

This paper is organized as follows. In the next section, related works are stated. The assumptions and system model are laid out, and the key equations that relate the link quality of EH-WSN to the Markov Model are presented in Sects. 3 and 4 describes simulation and analysis. Finally Sect. 5 concludes the paper.

2 Related Work

Several research works have been carried out [4–8] in the field of link quality estimation of WSN, which basically deals with the factors affecting the performance of link quality. In [4], the spatio-temporal characteristics of the link quality have been analyzed. They basically measured packet loss rate (PLR), link quality index (LQI) and received signal strength indicator (RSSI) from both indoor and outdoor experiments to judge the link quality and by using a sinusoidal function the temporal correlation of a link is modeled, the parameters of which depends on the quality of the link. The authors of [5] argue that estimating links quality require combining the network, link, and physical layer information. They propose a narrow, protocol-independent interface for the layers, which provide four bits of information: one bit from physical layer, one bit from link layer and two bits from the network layer. In [6], a comparative study of the link quality estimators for wireless sensor network has been performed. They have done the statistical analysis of the properties of these estimators individually and then explore their effect onto the routing protocol. In [7] a new kind of metric is proposed i.e. triangle metric which is actually geometric combinations of signal to noise ratio (SNR), LQI and packet reception ratio (PRR). In [8] a Fuzzy link quality metric (F-LQE) has been proposed which estimates link quality on the basis of link asymmetry, packet delivery, link stability and channel quality. Linguistic terms, the natural language of Fuzzy Logic is used to define of these properties. The overall quality of the link is measured by a set of fuzzy rules whose evaluation returns the membership of the good link from the fuzzy subset.

But none of these approaches are focused on link quality modeling. In [9], authors attempt to represent channel quality variation using a finite state birth-death Markov model using signal to noise ratio as a metric to characterize and predict the link quality. In [10] authors have modeled the link quality using oriented birth-death (OBD) model considering different mobility pattern. In [11], remaining connectivity forecasting in mobile environment is done by a genetic machine learning algorithm.

Unlike the existing approaches, in our approach, described in the next section, SNR and PRR are considered to characterize and predict the link quality of sensor nodes in EH-WSN environment. Moreover, we have considered the mobility feature and also the harvesting capabilities of nodes. The proposed formulation implicitly accounts for the energy harvested from the environment.

3 System Model

In this paper a methodology to model link quality in energy harvesting sensor network using a finite-state Markov chain Model with state space $S = \{s_1, s_2, \ldots, s_n\}$ is depicted where n is the possible number of states representing

Fig. 1 Birth death model

certain quality of link. Here, n is the parameter need to be fixed. The approach would start with any arbitrary choice for the transition probabilities. Then system is observed at each time interval to note the transitions. Thus the transition probabilities are empirically calculated on the basis of the observations.

It is assumed that the link may increase its quality with probability α or may decrease its quality with probability β or may remain in the same state with probability $1 - \alpha - \beta$, at the end of each time slot. Simple traffic model considered here. Thus, in each time slot, ρ is the probability that an event occurs.

Sensor nodes are capable of harvesting energy and they are having different energy profile. Rather than being connected to an energy source, a harvesting device read the energy availability traces collected from real life deployment. The future N i.e., the state N second later is predicted. States are predicted on the basis of the transition matrix and initial probability distribution.

The key feature of this model is that it takes into account the energy profile of different nodes since our concentration is on the energy harvesting sensor nodes. At the same time it does not use any location information to compute the link quality. Link quality is computed using some basic parameters like PRR and SNR. Unlike [9, 10], the Birth Death (BD) model is not used here, which is a variant of Markov chain model to predict the link quality. Since we have found that in a complex network it is not possible to have link states which gradually change over time like BD Model as depicted in Fig. 1. In BD model states can only progress to next higher state or it can degrade to previous lower state. But in reality based on several wireless channel factors the change of state of the link is not so smooth, they may be dynamic in nature. So instead of BD model here the generalized Markov Chain model is used as the next state can be any random state. The state transition diagram of Markov chain Model has been given in Fig. 2.

Markov chain modeling is a stochastic process, has an advantage: it is a memory less process. Thus the next sate depends only on the current state no matter how it reaches to the current state. Therefore, no need to keep historic data because history is already embedded in the successive transition probability.

Fig. 2 General markov model

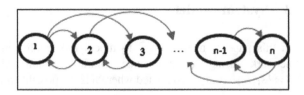

This model can use any number of states. But as it was mentioned in [9], at least three states are necessary to represent link quality variation. We can use as many states as wish but unnecessarily it will have states which represent a very small amount of change in link quality and the number of transition matrix would also be very huge. So, in the modeling there is a need to fix up the number of states, which is not a straight forward problem. From the variation of the trace value generated from the simulation we have to decide it, which can predict the link quality optimally.

A discrete Finite State Markov chain model can be defined by a triplet $<S, A, \lambda>$ where S corresponds to the state space, A is the transition probability matrix representing transitions from one state to another and the initial probability distribution of the states in S is represented by λ. The fundamental property of Markov model is the next step depends only on the previous state ignoring all other states. If vector $s(t)$ denotes the probability vector of all the states at time 't', then

$$s(t) = s(t-1) . A \qquad (1)$$

If markov chain has n states, then the transition probability matrix A will be of size n $\times$ n. Elements of matrix $A[s_i, s_j]$ can be calculated as follows

$$A_{i,j} = \frac{n(s_i, s_j)}{\sum n(s_i, s_j)} \qquad (2)$$

where $n(s_i, s_j)$ represents the number of times state s_i follows by state s_j. And

$$\lambda[s] = \frac{n(s_i)}{\sum n(s_n)} \qquad (3)$$

where $n(s_i)$ corresponds to the number of times state s_i was visited over all the states.

In our experiment we have gathered the link state information in the form of PRR from simulation and the state space S has been defined from that and any state can be taken as initial state A. Eqs. 2 and 3 is used to calculate transition matrix and initial probability. As we want to know the next state of the link for our experiment, which can be calculated using Eq. 1 as follows

$$s^n = \lambda * A^n \qquad (4)$$

Here s^n is the vector state n step ahead and A^n is the transition matrix powered by n. From Eq. 4 the state having the highest probability would be chosen as the next state.

To validate the model the accuracy has to be checked. The prediction accuracy is measured as the ratio between numbers of correct predictions (P_{corr}) over the total predictions (P_{tot}) done.

So, if *Acu* represent accuracy then,

$$Acu = P_{corr} / P_{tot} \tag{5}$$

Average of all the predictions is considered as the result.

So basically the proposed solution is cost effective, close to real life application and helpful for any routing decision to be made as because unnecessary transmission and retransmissions can be cut short.

4   Simulation and Analysis

The main goal of the simulation is to model the link quality and to measure the prediction accuracy. Because this accuracy represents the effectiveness of our approach.

Since we have considered a network where link quality is highly dynamic, predicting the next state knowing the current state is our aim. The model is not only going to predict the next link quality state just for next point of time but it is going to predict the state after n seconds ahead. But the accuracy will decrease as time increases. Here PRR and SNR are used as an estimator for link quality. The simulation is performed using Castalia simulator [2] which is an exclusive wireless sensor network simulator for early-phase algorithm/protocol testing introduced by Networks and Pervasive Computing program of National ICT Australia. It has some specialized feature which makes it a perfect simulator for realistic channel and radio modeling. It provides support for defining versatile physical processes. It also supports to enhance the modeling of the sensing devices and provide support for working on node clock. It is built on OMNET++ [12] framework. For simulating energy harvesting nodes Green Castalia [13] is used.

4.1   Simulation Scenario

As we want to simulate EH-WSN where sensor nodes has different energy profile, the energy related data has been gathered form National Renewable Energy Laboratory (NREL) [3], they aim to provide creative and dependable answers to today's energy challenges. Researchers of NREL are transforming the way the nation and the world use energy. In this simulation, 100 Nodes are deployed on the ground of $250 \times 250 \text{ m}^2$ area. 60 % of the nodes are static and 40 % of the nodes are mobile. 70 % of the nodes are energy harvesting nodes. Here two types of energy harvesting sources are taken care of solar cell and wind turbine. The Solar Cell module models a photo-voltaic panel, and converts raw irradiance data provided by the Energy

Table 1 Solar and wind data collected from NREL

NREL data snapshot				
Date	Time	Solar irradiance (W/m²)	Time	Wind speed (m/s)
7/4/2015	5:17	0.006347	4:00	1.012
7/4/2015	5:18	0.076469	4:01	1.3
7/4/2015	5:19	0.401693	4:02	0.862
7/4/2015	5:20	0.758958	4:03	0.925
7/4/2015	5:21	0.765305	4:04	0.763
7/4/2015	5:22	1.10988	4:05	0.925
7/4/2015	5:23	0.860705	4:06	2.4
7/4/2015	5:24	1.15418	4:07	0.712

Table 2 General parameter for simulation

Parameters	Value
Ground size	250 × 250 m²
No. of nodes	100
Packet rate	4 per sec
Simulation time	100 s
Collision model	0 (No collision)
MAC	Tunable MAC
Radio	CC2420
Speed	{10, 15, 20} m/s

Source module into harvested power. The wind Turbine cell also models a turbine which converts the wind energy. The data gathered from NREL, used to represent the amount of energy harvested over time. Solar irradiance has been measured in terms of watt/m² and wind speed as m/s. To model energy harvesting nodes, we need to feed the simulator with raw irradiance and speed traces collected over time. Table 1. Provides snapshot of data collected as provided by [3]. The general parameters for setting up the simulation environment have been presented in Table 2. First the dynamic nature of the link quality has been analyzed, as shown in Fig. 3. PRR has been taken as the estimator of link, as it can directly identify the link quality.

We have also tested the harvesting nodes varying the initial energy, if the node is out of energy then PRR becomes 0, otherwise it remains almost same. For prediction accuracy checking randomly we have taken 4 solar and 4 wind nodes which may be mobile in nature or static to show the diversion of the link nature. 10 simulations of 100 s each were performed. We have collected the traces of each simulation and consult them for generating the transition matrix. The number of states of the Markov chain depends on the variety of value that we get from the trace. Here, we have tested it for 3, 4 and 5 state.

Fig. 3 Link quality between
two nodes measured by PRR
with packet rate 4

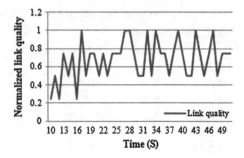

4.2 Results

The prediction accuracy of the proposed approach has been shown in Figs. 4, 5 and 6. The accuracy has been checked in terms of the number of steps ahead measured in seconds. Figure 4 presents the link quality estimation of a randomly selected wind static node. Link quality estimation in case of solar mobile node and wind mobile node scenarios are presented in Figs. 5 and 6.

Fig. 4 Link quality
estimation for a static, wind
energy harvesting node

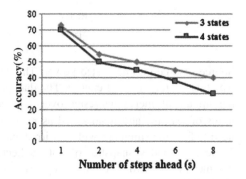

Fig. 5 Link quality
estimation for a mobile, solar
energy harvesting node

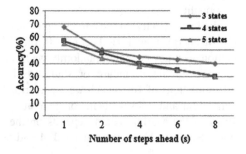

Fig. 6 Link quality estimation for mobile, wind energy harvesting node

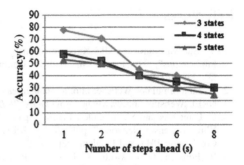

So for the given data set, the proposed model can predict the next state with 70 % accuracy for a randomly selected wind static node and for a mobile node it can also predict on an average with 70 % accuracy. The prediction accuracy after 8 s later is 30 %, the accuracy is not high because of the dynamic nature of the wireless medium. With increase in the number of state the accuracy decreases. The number of states is actually considered based on the range of values available from trace. We have taken here 3 and 4 states for static wind node and 3, 4 and 5 states for solar and wind mobile node.

Results can be analyzed from two perspectives (a) as the no. of steps increase and (b) as the no. of states increase. As the number of steps increase prediction accuracy decrease, since we are going to predict more future the link state information. And as the number of state increase the option of the next link state also increase thus decrease the accuracy.

5   Conclusion and Future Scope

Here energy harvesting sensor nodes are modeled and link quality has been measured in terms of PRR and SNR. Real life traces are concerned during design of the harvesting nodes. Dynamic nature of the wireless medium responsible for degradation in prediction with time but based on the prediction the routing decision can be improved. Accuracy can be enhanced if we can impose threshold on the traced values. Also need to consider several factors that are directly or indirectly affecting the link quality. Speed of the node is not taken into consideration. Some expensive GPS based solutions are there for connectivity prediction but our approach is less complex and cost saving. Markov chain modeling is used here for link quality prediction. Whether the next state is exactly same as the predicted state or not is taken into consideration in accuracy checking, this can be modified if we take tendency of the link rather than the exact value. Future work would be to refine this approach by (i) Taking real time energy harvesting sensor nodes and creating an environment which is more realistic. (ii) Considering speed into concern for modeling mobility along with the harvesting features.

References

1. S. Sudevalayam and P. Kulkarni, "Energy Harvesting Sensor Nodes: Survey and Implications", published in IEEE Communications Surveys and Tutorials 13 (3): 443–461 (2011).
2. Castalia Simulator Website: http://castalia.research.nicta.com.au.DOA. 1/4/2015
3. NREL Website: http://www.nrel.gov/.DOA. 7/4/2015
4. C. Umit Bas and Sinem Coleri Ergen, "Spatio-Temporal Characteristics of Link Quality in Wireless Sensor Networks", published in IEEE Wireless Communications and Networking Conference: PHY and Fundamentals, 2012.
5. R. Fonseca, O. Gnawali, K. Jamieson, P. Levis," Four-Bit Wireless Link Estimation", published in HotNets'07, 2007.
6. N. Baccour, A. Koubaa., H. Youssef. Ben Jamaa, M. and M. Alves. "A Comparative Simulation Study of Link Quality Estimators in Wireless Sensor Networks", MASCOTS '09. IEEE International Symposium on. (Sept. 2009).
7. Carlo Alberto Boano, Marco Antonio Zúñiga Zamalloa, Thiemo Voigt, Andreas Willig, Kay Römer, "The Triangle Metric: Fast Link Quality Estimation for Mobile Wireless Sensor Networks", ICCCN 2010, pages 1–7.
8. N. Baccour, A. Koub, H. Youssef, M. Ben Jamˆaa, Denis do Rosˊario, Mˊario Alves, and Leandro B. Becker," F-LQE: A Fuzzy Link Quality Estimator for Wireless Sensor Networks", 7th European conference on Wireless Sensor Networks, EWSN'10, Pages 240–255.
9. Ratul K. Guha, Saswati Sarkar, "Characterizing temporal SNR variation in 802.11 networks", in IEEE Wireless Communications and Networking Conference (WCNC-2008).
10. G. M. de Araújo, J. Kaiser, L. Buss Becker, "An optimized Markov model to predict link quality in mobile wireless sensor networks", IEEE Symposium on Computers and Communications (ISCC), 2012.
11. G. M.de Araújo, A. R. Pinto, J. Kaiser, L. Buss Becker, "An Evolutionary Approach to Improve Connectivity Prediction in Mobile Wireless Sensor Networks ", Procedia Computer Science 10 (2012), pages 1100–1105.
12. A. Varga et al., "The OMNeT++ discrete event simulation system," in Proceedings of the European Simulation Multiconference, 2001, pages. 319–324.
13. David Benedetti, Chiara Petrioli, Dora Spenza, "GreenCastalia: An Energy-Harvesting-Enabled Framework for the Castalia Simulator", ENSSys'13, November 13 2013.

GA Based Energy Efficient and Balanced Routing in k-Connected Wireless Sensor Networks

Suneet Kumar Gupta, Pratyay Kuila and Prasanta K. Jana

Abstract In the past few years network layer activities in wireless sensor networks gain enormous attention to improve network lifetime. Development of routing algorithms with energy efficacy is one of the most popular techniques to improve it. In this article, energy efficient and balanced route generation algorithm is proposed with considering both energy efficacy and energy balancing issues. Here, we consider the distance and residual energy of the nodes as energy efficiency parameters and energy is balanced by diverting the incoming traffic to other nodes having comparably lower incoming traffic. To develop the routing schedule, we have applied Genetic Algorithm which can quickly compute the routing schedule as per the current state of the network. It is observed that the performance of proposed algorithm is better than existing algorithm in terms of first node die and energy consumption in the network.

Keywords Energy efficient · Routing · Wireless sensor networks · Genetic algorithm

1 Introduction

In last several years, the field of wireless sensor networks (WSNs) has paid gigantic attention due to its various applications in real life, e.g. environment monitoring, military application and in industry etc. [1]. Sensor network is the network of small

S.K. Gupta (✉)
Department of Computer Science and Engineering, O.P. Jindal Institute of Technology,
Raigarh 496001, India
e-mail: suneet.banda@gmail.com

Pratyay Kuila
National Institute of Technology Sikkim, Ravangla 737139, Sikkim, India
e-mail: pratyay_kuila@yahoo.com

P.K. Jana
Indian School of Mines, Dhanbad 826004, India
e-mail: prasantajana@yahoo.com

© Springer Science+Business Media Singapore 2017   679
J.K. Mandal et al. (eds.), *Proceedings of the First International Conference
on Intelligent Computing and Communication*, Advances in Intelligent Systems
and Computing 458, DOI 10.1007/978-981-10-2035-3_68

tiny devices known as sensor, which are operated by the limited power battery. The most important responsibility of these sensors is to sense the region and forward the sensed data to BS directly or using intermediate nodes. In this kind of networks, energy is one of the most important constraint because once the sensors are deployed, it is very difficult to replace the battery due to least human intervention. To reduce the energy consumption clustering of sensors is one of the methods. In clustering [2] sensor nodes form the clusters and forward the data to respective cluster head, which is also known as CH. Normally CHs are selected amongst the sensor nodes, but due to less energy of these CHs, CHs may die quickly. In this regard, many researchers proposed the use of relay node as cluster head [3–5]. But only clustering of sensors is not sufficient because relay nodes are also battery operated and these CHs die quickly due to long haul communication. The lifetime of relay nodes is very crucial because failure of relay node(s) may partition the network, so some of relay node(s) may not transmit the data to BS. Therefore, energy efficient routing through relay nodes is extremely important to reduce the energy consumption in network [6]. Suppose there is a WSN with k relay nodes with p valid one hop relay nodes, then there are p^k possible routes. It means that the computational complexity is very high to discover the route for a large scale WSN using brute force approach. To solve such problem Genetic Algorithm (GA) [7] is one of the efficient tool.

In past years, many routing algorithms have been developed by the researchers for the WSN, which can be found in [8–10] and their references. Here we only discuss about those routing algorithms, which are relevant to our proposed work. In GAR [9], authors proposed GA based approach to minimize overall transmission in a round. During the mutation operation GAR finds out the critical node which contributes maximum distance and it is replaced by another the valid node. In MHRM [11], authors developed a routing algorithm which selects the route with minimum number of intermediate relay nodes. Bari et. al. [10] discussed about the routing algorithm amongst relay nodes using GA. In this algorithm authors developed fitness function based on the network lifetime and during mutation operation authors minimize the incoming traffic in nodes.

In this article, we propose a GA based approach to generate the routing schedule for transmitting the aggregated data from relay nodes to the BS directly or using other relay nodes in k-connected WSNs. In order to be energy efficient, the algorithm uses transmission distance, residual energy of the node and routing overhead. The algorithm can quickly computes the route for entire relay nodes to the BS based on the current network topology. For comparison, we also ran GAR [9] and MHRM [11]. The experimental results reveal that the proposed algorithm performs better than GAR and MHRM in terms of first node die and energy consumption.

The organization of the article is as follows. Proposed work and used terminologies are discussed in Sect. 2. The experimental results and conclusion are explained in Sects. 3 and 4 respectively.

2 Proposed Work

Here, we have assumed a WSN model in which few sensor nodes are placed randomly or manually and after deployment all of the sensor nodes are stationary. The relay nodes are placed with the help of an algorithm discussed in [3] so that the network is designed as k-connected. The main work of the senor nodes is to sense the region and collect the local data and finally send it to their corresponding CHs. After receiving the data, CHs first minimize the redundancy within their cluster and then they route the aggregated data to the BS. Therefore, each relay node acts as a source and the base station acts as a destination. It is obvious to note that for multi-hop communication, each relay node has to select a neighbor relay node in single-hop distance in order to forward the data to the BS. Our algorithm deals with the selection of the best neighbor relay node to find the optimal route with respect to energy consumption. The data gathering operation is divided into rounds. A single round is completed, when entire relay nodes forward the data to BS. We have used the same network and energy model, as discussed in [9, 12].

The terminologies used in proposed work is explained as follows:

1. $R = r_1, r_2, r_3, \ldots r_N$ represents the set of relay nodes and r_{N+1} represents the sink or base station.
2. C_R denotes the communication range of the relay nodes.
3. $Dist(r_i, r_j)$ represents the distance between relay node r_i and r_j.
4. $E_{rs}(r_j)$ denotes the residual energy of r_j.
5. $Next(r_i)$ denotes the set of relay nodes, those are within communication range of r_i. Therefore, it can be defined as follows:

$$Next(r_i) = \{ r_j \mid \forall r_j \in R \land Dist(r_i, r_j) \leq C_R \}$$

6. $RouteOhead(r_j)$ represents the number of relay nodes for which r_j acts as an intermediate forwarder.

In following subsection, we have presented routing algorithm for k-connected WSNs using GA. First, we have explained the representation of chromosome and generation of initial population. Then we have presented the other operations of GA, i.e., selection, crossover and mutation.

2.1 Chromosome Representation and Initial Population

We represent the chromosome as a string of relay node number. The length of each chromosome is kept same for a given network and it is equal to the number of relay nodes in that network. For a chromosome, ith gene value (say j) represents the next hop relay node of r_i is r_j. The value say j of the ith position gene is randomly selected such that $r_j \in Next(r_i)$. Initial population is a randomly generated set of chromosomes.

Fig. 1 **a** A sub graph of a WSN with 9 relay nodes and 2-connectivity. **b** Chromosome representation for the sub graph

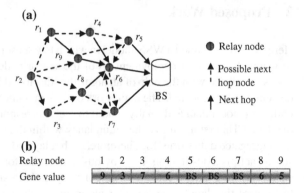

(b)

Relay node	1	2	3	4	5	6	7	8	9
Gene value	9	3	7	6	BS	BS	BS	6	5

A sub graph of WSN is depicted in Fig. 1a, where only 2 connectivity amongst relay nodes is shown. For example, relay node r_1 is connected with r_4 and r_9, similarly r_6 is connected with base station and r_7. A randomly generated chromosome from the Fig. 1a is presented in Fig. 1b. It can be observed that the gene value at the position 1, 4, 7 and 8 is 9, 6, BS and 6 respectively. It implies that relay node r_9 acts as a next hop node of r_1. Similarly, r_6 is the next hop of r_4, BS is the next hop of r_7 and r_6 is the next hop of r_8.

2.2 Fitness Function

The fitness value of a chromosome represents its qualification. This value can discriminate two chromosomes in terms of which one of these two can be a better solution for the given problem. It is well known that in a sensor network maximum energy is consumed in data transmission, which heavily depends on the transmission distance. Moreover, few relay nodes consume their energy to act as an intermediate node. Therefore, in proposed work we have designed the fitness function for selecting the next hop based on the transmission distance, routing overhead and residual energy.

To efficiently derive the fitness function, we have calculated the $cost(r_i, r_j)$ for each relay node r_i to transmit the data to the next hop node r_j. The $cost(r_i, r_j)$ is represented by the following Eq. 1.

$$cost(r_i, r_j) = \frac{E_{rs}(r_j)}{Dist(r_j, BS) \times Dist(r_i, r_j) \times RouteOhead(r_j)} \tag{1}$$

When $j = BS$, $cost(r_i, r_j) = E_{ie}$, i.e., the initial energy of all the relay nodes. It is clearly seen that the *cost* function is directly proportional to residual energy of the next hop node and inversely proportional to the distance between sender and receiver, the distance between next hop node and BS and routing overhead of the next hop node, (i.e., *RouteOhead*). Therefore, as the distance between sender and receiver or distance between next-hop node and BS or *RouteOhead* increases, the cost value decreases. The minimum cost value represents poor next hop selection. The *Fitness* function is represented as follow.

$$Fitness = \sum_{\forall r_i \in R} cost(r_i, r_j) \tag{2}$$

Higher is the *Fitness* value, better is the chromosome.

2.3 Selection and Crossover

Selection and crossover operations are the basic intermediate and iterative phases of GA. In proposed work, we use Roulette-Wheel selection method to select the 10 % best population. The selected population is used for 1-point crossover operation. It is a operation, which leads the population towards a local optimum. During the crossover operation two parent chromosomes share their information and produce two child chromosomes. If the fitness value of parent chromosomes is not better than the child chromosome then child chromosome(s) replace the parent(s). The crossover operation is depicted in Fig. 2.

2.4 Mutation

Mutation is the most important operation of GA. With the help of this operation we strengthen the chromosome by varying the gene value. Here we apply uniform mutation operation. During the mutation operation, we select that relay node which contributes minimum *cost* value and try to replace the next of that relay node in such a manner so that the *cost* is increased.

Fig. 2 Crossover operation

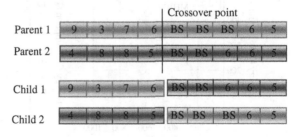

3 Experimental Results

For performing the experiments, we have considered two different network scenarios, WSN#1 and WSN#2. The size of WSN#1 is 150×150 square meter area and position of the base station is taken as (175, 175). The size of WSN#2 is 300×300 square meter area and position of the base station is taken as (350, 350). For the execution of GA based algorithm, an initial population of 70 chromosomes has been considered. For the crossover operation, best 10 % chromosomes has been selected with the help of Roulette Wheel selection model. The algorithm executed up to 50 epochs, but it returns the optimal results after 32 epochs. After completion of crossover operation, we have applied uniform mutation operation to strengthen the chromosome.

For the performance evaluation purpose, GAR [9] and MHRM [11] algorithms are also executed. For all the algorithms simulation parameters are same and as discussed in [12]. The algorithms are compared in terms of energy consumption and First Node Die (FND). Figure 3a and b represents the energy consumption amongst the algorithms in scenarios WSN#1 and WSN#2. From the Fig. 3a and b, it is marked that proposed algorithm consumes less energy than GAR and greater than MHRM. The reason for the same is that MHRM generates a route with minimum number of intermediate nodes and it is well known that less participation of intermediate nodes always saves the energy in WSN. The fitness function of GAR is designed in such a fashion that it tries to select the next hop with minimum distance, so it increases the number of intermediate nodes. Moreover, participation of intermediate nodes increases the energy consumption due to aggregation of data and forwarding it to the next hop. In the proposed algorithm, selection of next hop is based on the distance and routing overhead. The parameter distance tries to make the algorithm energy efficient and routing overhead tries to balance the energy consumption. Due to this energy consumption of proposed algorithm is lies between GAR and MHRM.

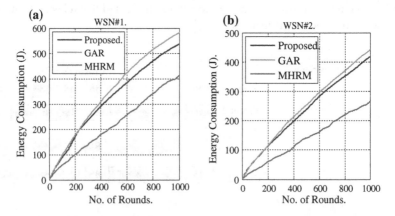

Fig. 3 Comparison in terms of energy consumption for **a** WSN#1 and **b** WSN#2

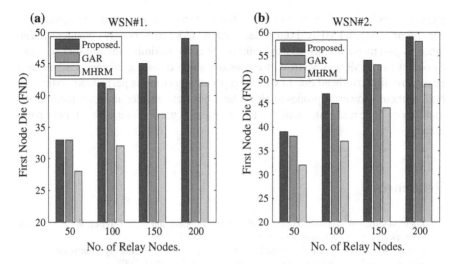

Fig. 4 Algorithms comparison in terms of first node die (FND) for **a** WSN#1 and **b** WSN#2

The comparsion of proposed algorithm with GAR and MHRM in terms of first node die is represented in Fig. 4a and b and it is clearly demonstrated that proposed algorithm performs better than existing algorithms in two different scenario namely WSN#1 and WSN#2. In MHRM node dies quickly because there is a long haul communication for minimizing the intermediate nodes. Moreover, it is well known that in WSN the energy consumption is increased as the distance between sender and receiver is increased, so in MHRM nodes quickly depletes the energy. In GAR, routing schedule is generated on the basis of distance covered in a round, but there is no mechanism for avoiding the bottleneck situation in the network, due to which some relay nodes are extra overloaded and die quickly. The proposed algorithm is performed better than GAR and MHRM because fitness function is designed with considering the distance as well as balancing the incoming traffic in each node.

4 Conclusion

In this research article, we have discussed a GA based routing scheme for k-connected wireless sensor networks. The algorithm has been provided with suitable representation of chromosomes, explanation for fitness function, crossover and mutation operations. In proposed work, the fitness function is based on distance between relay and its next hop node, the distance between next hop and the base station, residual energy and routing overhead of the next hop. Consideration of distance and residual energy always makes the algorithm energy efficient and other parameter, e.g., routing overhead tries to balance the energy consumption by distributing the incoming traffic of a relay node to other relay node, which has comparably less incoming traffic.

Experimentally, it is demonstrated that the performance of proposed algorithm is better than GAR and MHRM due to fact that in GAR, authors only consider the distance as parameter and in MHRM, authors tries to minimize the number of hops. In contrast to GAR and MHRM, we consider some other parameters, i.e., residual energy, routing overhead due to which proposed algorithm is performed better than both of them. However, proposed algorithm does not consider any QoS parameters, mobility of node and sink, so as our future research we will consider all these parameters.

References

1. Ian F Akyildiz, Weilian Su, Yogesh Sankarasubramaniam, and Erdal Cayirci. Wireless sensor networks: A survey. *Computer networks*, 38(4):393–422, 2002.
2. Ameer Ahmed Abbasi and Mohamed Younis. A survey on clustering algorithms for wireless sensor networks. *Computer communications*, 30(14):2826–2841, 2007.
3. Suneet K Gupta, Pratyay Kuila, and Prasanta K Jana. Genetic algorithm for k-connected relay node placement in wireless sensor networks. In *Proceedings of the Second International Conference on Computer and Communication Technologies*, pages 721–729. Springer, 2015.
4. Pratyay Kuila and Prasanta K Jana. Heap and parameter-based load balanced clustering algorithms for wireless sensor networks. *International Journal of Communication Networks and Distributed Systems*, 14(4):413–432, 2015.
5. Suneet Kumar Gupta, Pratyay Kuila, and Prasanta K Jana. Genetic algorithm approach for k-coverage and m-connected node placement in target based wireless sensor networks. *Computers & Electrical Engineering*, http://dx.doi.org/10.1016/j.compeleceng.2015.11.009, 2015.
6. Giuseppe Anastasi, Marco Conti, Mario Di Francesco, and Andrea Passarella. Energy conservation in wireless sensor networks: A survey. *Ad Hoc Networks*, 7(3):537–568, 2009.
7. Pratyay Kuila, Suneet K. Gupta, and Prasanta K. Jana. A novel evolutionary approach for load balanced clustering problem for wireless sensor networks. *Swarm and Evolutionary Computation*, 12:48–56, 2013.
8. Suneet K Gupta and Prasanta K Jana. Energy efficient clustering and routing algorithms for wireless sensor networks: GA based approach. *Wireless Personal Communications*, pages 1–21.
9. Suneet K. Gupta, Pratyay Kuila, and Prasanta K. Jana. GAR: An energy efficient GA-based routing for wireless sensor networks. In *International Conference on Distributed Computing and Internet Technology 2013, LNCS (Springer)*, volume 7753, pages 267–277. Springer, 2013.
10. Ataul Bari, Shamsul Wazed, Arunita Jaekel, and Subir Bandyopadhyay. A genetic algorithm based approach for energy efficient routing in two-tiered sensor networks. *Ad Hoc Networks*, 7(4):665–676, 2009.
11. Shao-Shan Chiang, Chih-Hung Huang, and Kuang-Chiung Chang. A minimum hop routing protocol for home security systems using wireless sensor networks. *Consumer Electronics, IEEE Transactions on*, 53(4):1483–1489, 2007.
12. Wendi B Heinzelman, Anantha P Chandrakasan, and Hari Balakrishnan. An application-specific protocol architecture for wireless microsensor networks. *Wireless Communications, IEEE Transactions on*, 1(4):660–670, 2002.

An Approach Towards Energy Efficient Channel Reduction in Cellular Networks

Spandan Chowdhury and Parag Kumar Guha Thakurta

Abstract An energy efficient model in cellular networks is proposed in this paper to reduce the number of long range channels as well as the energy consumption of the mobile devices. The devices having high energy in terms of battery level are used to establish communication between the base station and the requesting mobile devices. The overall energy consumed by all devices is estimated if one such high energy device receives data from the base station and subsequently propagates the same to other mobile devices. The energy consumption is assessed for increasing number of bridging devices. The proposed approach reduces the number of long range channels used and the overall energy consumption of devices. The experimental results show the effectiveness of the proposed approach.

Keywords Energy · Mobile device · Base station · Cellular networks · Unicast · Multicast

1 Introduction

The advancement of mobile applications has led to an explosive growth in mobile device (MD) usage. Thus, an enormous number of cellular subscribers is resulting into high rates of data downloads and ultimately, causing congestion in cellular networks. This in turn is affecting the end user quality of service [1].

For example, consider the simultaneous distribution of a common content such as news download or multimedia multicasting to a group of MDs over a network [2]. In order to accomplish the job, a server transmits the content to the MDs via a Base Station (BS). However, two different modes of interaction between the BS and

S. Chowdhury · P.K. Guha Thakurta (✉)
Department of CSE, NIT, Durgapur 713209, West Bengal, India
e-mail: parag.nitdgp@gmail.com

S. Chowdhury
e-mail: spandie990@gmail.com

© Springer Science+Business Media Singapore 2017   687
J.K. Mandal et al. (eds.), *Proceedings of the First International Conference on Intelligent Computing and Communication*, Advances in Intelligent Systems and Computing 458, DOI 10.1007/978-981-10-2035-3_69

MDs can be observed. The first approach provides dedicated one-to-one long range (LR) channels from the BS to every requesting MD. In another approach, the BS transfers the content to some of those requesting MDs and subsequently, these MDs further unicast or multicast the same content in a cooperative manner over short range (SR) channels. Thus an amount of energy is consumed by MDs in both approaches.

The measurement of such energy consumption [3] is associated with the average battery levels of MDs. However, the dependence of MDs on limited battery levels imposes a constraint on the earlier approaches. Moreover, the current age of green communication concerns the reduction of energy consumption in networks. This can be the key in designing an optimal framework that has a multi objective approach of minimizing the energy consumption of the MDs while simultaneously reducing the number of dedicated LR channels.

In this paper, an energy efficient model is proposed to reduce the number of LR channels as well as the energy consumption of the MDs for supporting the earlier mentioned two-pronged approach. The terms MD and node are interchangeably used in this work. Primarily, the nodes having high energy in terms of battery level are used to establish communication between the BS and the requesting MDs. Here, it is important to note that such high energy nodes are selected from the set of requesting MDs. Further, the overall energy consumed by all MDs is estimated if one such high energy MD receives the data from the BS and subsequently propagates the same to other MDs. Carrying on the same approach, the energy consumption is assessed if the number of such high energy bridging MDs is increased. Finally, the solution provides the reduction in the number of long range channels used and the overall energy consumption of devices. In addition, the experimental results of the proposed approach confirm the theoretical findings and highlights its effectiveness.

The rest of this paper is organized as follows: Sect. 2 discusses related works. Section 3 presents the system model, and the problem is formulated in Sect. 4. The proposed solution is introduced in Sect. 5. Next, the experimental results are shown in Sect. 6 followed by the conclusion and its future scope in Sect. 7.

2 Related Work

The recent popularization of cellular networks has caused the explosive growth of mobile users with ubiquitous internet access. Therefore, the demand for huge amount of data traffic has posed a big challenge to the cellular networks. Sometimes the traffic exceeds the capacity of cellular networks and hence deteriorates the network quality [1]. To handle such immense volume of data, MDs spend a considerable amount of energy. Under such scenario, in order to provide high throughput and low latency.

Requirements, protocols are developed by simultaneously accounting for the concerns on energy efficiency [4]. Due to limited transmission rate of MDs, the energy expenditure becomes very high. Thus reducing energy consumption in the wireless networks manifests into a major goal for supporting the emerging services.

The emergence of MDs with multiple wireless interfaces as discussed earlier is expected to address the issues of energy-efficiency by allowing cooperation among MDs over wireless interfaces [5]. Such cooperation takes into account the power limitation in the MDs. Bearing this fact in mind, cooperation among MDs have major advantages in terms of increasing the network throughput and decreasing the energy consumption of MDs [6].

In short, the previous literature studies introduce the concept of energy efficiency in perspective of cooperation among MDs. This cooperation based design methodology is further enhanced in this paper for reducing the required number of LR channels as well as for facilitating a more energy efficient approach by considering multicasting operation as introduced next.

3 System Model

Consider the scenario as introduced in the Fig. 1. Here, "k" number of MDs as $[N_1, N_2, N_3, \ldots N_k]$ can download the same content from the server (S) through BS. For the sake of simplicity, it is assumed that the MDs are uniformly distributed around the BS and at a one hop distance from each other. In this work, LR channels are used for the data transmission from BS to MDs, whereas SR channels are used for inter-MD communication.

Mainly the problem addressed in this work is associated with two major components. These are—(a) LR factor (LR_f): The ratio between the number of LR channels used and the total number of LR channels allowed for transferring the data

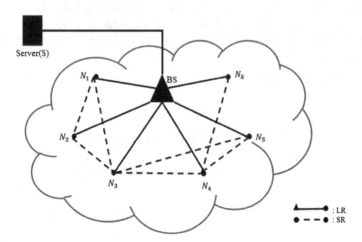

Fig. 1 Organization of the system

Table 1 Various rates of energy consumption in Joules/Mbit

Symbol	Description
R_{LR}	Energy consumed for receiving data on LR
R_{SRu}^{send}	Energy consumed for sending data over SR on unicast
R_{SRm}^{send}	Energy consumed for sending data over SR on multicast
R_{SRu}^{recv}	Energy consumed for receiving data over SR on unicast
R_{SRm}^{recv}	Energy consumed for receiving data over SR on multicast

to the various MDs, and (b) E factor (E_f): The ratio between the total energy used and the maximum amount of energy that is allowed to be used for facilitating the required communication. So, LR_f and E_f are represented as (LR_{used}/LR_{max}) and (E_{used}/E_{max}) respectively. In addition, E_{used} is estimated as follows:

$$E_{used} = Energy\ consumed\ by\ bridging\ MDs$$
$$+ Energy\ consumed\ by\ other\ requesting\ MDs \tag{1}$$

where, bridging MDs denote those devices that can receive data directly from BS and further transmit the same to other requesting MDs in a cooperative manner. Further, (1) can be discussed in more detail as in (2).

$$E_{used} = \sum LR_{recv} + \sum X + \sum Y \tag{2}$$

In (2), $\sum LR_{recv}$ and $\sum X$ refer to the summation of energy consumed for receiving data and the summation of energy consumed for either unicast or multicast transmission by all bridging MDs respectively. Again, the term $\sum Y$ used in (2) denotes the summation of energy consumed by all MDs to receive data by unicast or multicast over SR. In this work, the energy consumed by an MD for transmitting one unit (say 1 Mbit) of data is represented by Table 1.

The energy consumption for MDs in various scenarios with respect (2) are estimated by the followings. In this context, the data size, denoted as S_T, is also considered in our work.

- For each bridging MD:

$$LR_{recv} = S_T \times R_{LR} \tag{3}$$

$$X = S_T \times R_{SRu}^{send} | unicast\ communication \tag{4}$$

$$X = S_T \times R_{SRm}^{send} | multicast\ communication \tag{5}$$

- For each non-bridging MD,

$$Y = S_T \times R_{SRu}^{recv} | unicast\ communication \tag{6}$$

$$Y = S_T \times R_{SRm}^{recv} | multicast\ communication \tag{7}$$

4 Problem Formulation

Several high energy bridging MDs are involved in capturing the data from BS and subsequently forwarding the data to other requesting MDs in a cooperative manner. Therefore, a number of $(LR_f + E_f)$ values are obtained for varying number of bridging MDs. Hence, the energy-efficient solution under such scenario is obtained by selecting the minimum value of such $(LR_f + E_f)$. Thus the problem is formulated as follows:

$$minimum[(LR_f + E_f)_1, (LR_f + E_f)_2, \ldots, (LR_f + E_f)_{LR_{max}}] \tag{8}$$

5 Proposed Approach

The problem addressed in this paper deals only with the nodes that have the minimum energy to receive the data directly from BS over an LR. Along with each request, a node sends its energy level to the BS. It is to be noted that a limit for the usage of LRs (LR_{max}) can also be provided. By default, the limit is set to the total number of requesting nodes.

After the BS receives the request, it sorts the energy levels of all MDs in decreasing order. Then LR_{max} number of $(LR_f + E_f)$ values are calculated for two different scenarios as explained below:

(i) A few high energy bridging MDs receive the data on LR and unicasts the same to the other MDs;
(ii) A few high energy bridging MDs receive the data on LR and multicast the same to the other MDs

The calculations for each scenario are begun by considering only the highest energy MD as the bridging node and then calculating the energy consumption and LR requirements for the scenarios (i) and (ii) as mentioned earlier. Then the same approach is extrapolated further by considering the two highest MDs as bridging nodes and calculating the same parameters as mentioned earlier. This is continued until the total number of bridging MDs equal the number of requesting nodes or they equal the number of the maximum number of LR channels allowed. Finally the approach selects the lowest value of $(LR_f + E_f)$ as solution. This procedure is described by the following algorithm.

Algorithm:

1. $B \leftarrow$ *Sorted array of energy levels of requesting nodes*
2. $num \leftarrow$ *number of requesting nodes*
3. $minFinal \leftarrow Float.MAX\_VALUE$
4. *for* $i \leftarrow 0$ *to* $LR_{max} - 1$
5.   *unicast* $\leftarrow 0$
6.   *multicast* $\leftarrow 0$
7.   *for* $j \leftarrow 0$ *to* i
8. $B[j] \leftarrow B[j] - (S_T * R_{LR})$
9. $LR \leftarrow LR + 1$
10. $unicast \leftarrow unicast + \dfrac{B[j]}{R_{SRu}^{send} * S_T}$
11.   *if* $\dfrac{B[j]}{R_{SRm}^{send} * S_T} > 0$
12. $multicast \leftarrow multicast + 1$
13.   *end of j loop*
14.   *if unicast* $\geq$ *number of non bridging nodes*
15.   *calculate total energy consumption for unicasting*
16.   *calculate LR factor* + *E factor for unicasting (plan i: unicast)*
17.   *if multicast* > 0
18.   *calculate total energy consumption for multicasting*
19.   *calculate LR factor* + *E factor for multicasting (plan i: multicast)*
20. $min \leftarrow minimum(LR\ factor + E\ factor: unicasting, LR\ factor$
                 $+ E\ factor: multicasting)$
21.   *if min* $\leq$ *minFinal*
22. $minFinal \leftarrow min$
23. *end of i loop*
24. *return plan i whose LR factor* + *E factor* = *minFinal*

Remark The energy consumption is related with the testing of LR_{max} number of iterations. Further, every node in each iteration is checked for the compatibility of its energy level with the transmission or reception of data. Thus the algorithm in our proposed approach executes in $O(n^2)$ time, where "n" denotes the number of nodes considered in the network.

6 Experimental Results

In this section, the experimental results confirm the theoretical findings for the proposed approach. The analysis is carried out with Intel core i5 third generation processor, 4 GB RAM and Java programming language. The input values for the symbols used in Table 1 are considered here as 2 J/Mbit for R_{LR}, 0.4 J/Mbit for R_{SRu}^{send} and R_{SRm}^{send}, 0.8 J/Mbit for R_{SRu}^{recv} and R_{SRm}^{recv}. In addition, the data size (S_T) is taken into account as 2 Mbit.

In Fig. 2, the energy consumed in the network is estimated against number of requesting MDs. This figure shows that the energy consumption gradually increases proportionally with the increasing number of MDs for unicast communication. Similar characteristic is achieved for multicasting. However, the energy consumed in case of multicasting is significantly less than unicasting due to the lower requirements of the number of LR and SR channels.

Figure 3a shows the nature of requirement of LR channels against number of requesting MDs. In this figure, it is observed that the number of LR channels increase with increasing number of requesting MDs for unicasting. However, the

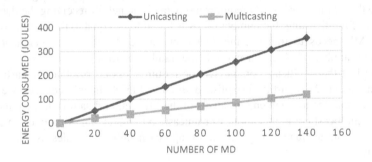

Fig. 2 Energy consumed in network versus number of requesting MDs

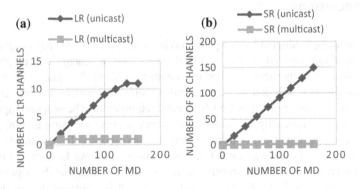

Fig. 3 a. Number of required LR channels against number of requesting MDs, **b.** Number of required SR channels against number of requesting MDs

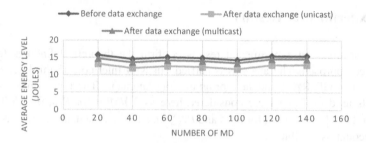

Fig. 4 Average energy level obtained against number of requesting MDs

rate of increase in percentage of LR channels slowly decreases after achieving stability with higher values of MDs. However, in case of multicasting, it is observed that the minimum energy consumption scenario arises when the number of LRs is reduced to one. Owing to the one hop nature of the network, a single MD is found to be capable of multicasting the data to all other requesting MDs, and hence reduces the LR requirement to 1. In Fig. 3b we have observed that the number of required SR channels complement the number of required LRs during unicasting, due to the fact that every MD requires a separate channel for receiving the data. But in case of multicasting, the number of SR is found to be equal to the number of LRs as an MD needs to transmit the data over multicast only once, and hence the required number of SR never exceeds the number of LRs.

The fact that multicast communication requires less energy than unicasting, as shown in Fig. 2, is further established through Fig. 4. In this context, we have observed that the average energy levels of the MDs, after data transfer, are more in case of multicast communication over unicast communication. Moreover, in all the scenarios, an average energy level of approximately 15 Joules is maintained initially.

7 Conclusions

The methodology adopted by this paper has been directed towards the simultaneous reduction of the required number of LR channels, and the energy consumption for communication in the network. The procedure to achieve this goal is involved in obtaining the best solution among several possible alternatives for reducing the energy consumption as well as the LR channel requirement. The proposed approach reveals that a significantly less amount of energy is spent by using a combination of LRs and SRs, rather than directly receiving data from the BS by all MDs. The implementation of a unicast model of communication shows how the number of dedicated channels for data transfer from the BS to MDs can be significantly reduced. Moreover, from the experimental results, it is observed that in order to achieve the best result, it is recommended to follow a multicast approach.

This approach is properly suited for a single hop network, and requires both minimum number of LR channels and the least amount of energy for successful communication. Further, the approach proposed in this paper can be extended by applying other powerful optimization techniques in order to obtain enhanced results.

References

1. Lina Al-Kanj, H. Vincent Poor, and Zaher Dawy, "Optimal Cellular Offloading via Device-to-Device Communication Networks With Fairness Constraints", IEEE Transactions On Wireless Communications, Vol. 13, No. 8, August 2014, pp. 4628–4643.
2. Lina Al-Kanj and Zaher Dawy, "Offloading Wireless Cellular Networks via Energy-Constrained Local Ad Hoc Networks", In proceedings of Globecom 2011, Texas, USA.
3. Xianfu Chen, Jinsong Wu, Yueming Cai, Honggang Zhang, and Tao Chen, "Energy-Efficiency Oriented Traffic Offloading in Wireless Networks: A Brief Survey and a Learning Approach for Heterogeneous Cellular Networks", IEEE Journal On Selected Areas In Communications, Vol. 33, No. 4, April 2015, pp. 627–640.
4. Lina Al-Kanj, Zaher Dawy, and Elias Yaacoub, "Energy-Aware Cooperative Content Distribution over Wireless Networks: Design Alternatives and Implementation Aspects", IEEE Communications Surveys & Tutorials, Vol. 15, No. 4, Fourth Quarter 2013, pp. 1736–1760.
5. M. Dohler, D.-E. Meddour, S.-M. Senouci and A. Saadani, "Cooperation in 4G - hype or ripe?," IEEE Technol. Society Mag., vol. 27, no. 1, pp. 13–17, Spring 2008.
6. L. Al-Kanj, W. Saad and Z. Dawy, "A game theoretic approach for content distribution over wirless networks with mobile-to-mobile cooperation," in Proc. 22nd IEEE Personal Indoor and Mobile Radio Communications (PIMRC), (Toronto, Canada), pp. 1567–1572, September 2011.

Military Robot Path Control Using RF Communication

Sandeep Bhat and Manjalagiri Meenakshi

Abstract This paper illustrates how to solve some risky tasks in military and rescue operations. The task like the continuous firing of bullets, provide weapons to soldiers in the war field, rescue operations in flood, earthquakes etc., a robot needs proper guideline to perform the task. In this work these tasks are solved by Radio Frequency (RF) based communication, because command is transmitted by a letter. Hence encryption of the data is not required. And also it presents the efficient way of utilizing the power and replaces the soldiers by many swam like robots. The communication on this system takes place by using RF signals. The robot path is controlled by remote controlled RF signal or using a mobile phone if the mobile carrier service is available. The real time result parameter are comparing with the vision based robot path planning system.

Keywords ASK · DTMF · Path planning · RF · Robot

1 Introduction

Radio Frequency communication plays a major role in our daily life. TV, Mobile, Internet MODEM, remote controller etc. are RF communication based gadgets. RF frequency is in the range of 3 kHz to 300 GHz [1]. It is well known fact that, most of the tasks in military applications are more dangerous than others. For example, walking through minefields, clearing out hostile buildings, are some of the most dangerous tasks a person is asked to perform in the line of duty [2]. Self-directed military hardware robotic system tasks are replaced direct human participation in the war field [3].

Sandeep Bhat (✉)
Department of E&CE, Srinivas Institute of Technology, Mangaluru, India
e-mail: sandeepsirsi@yahoo.co.in; sandusirsi@rediffmail.com

Manjalagiri Meenakshi
Department of Instrumentation Technology, Dr. AIT, Bengaluru, India
e-mail: meenakshi_mbhat@yahoo.com

© Springer Science+Business Media Singapore 2017
J.K. Mandal et al. (eds.), *Proceedings of the First International Conference on Intelligent Computing and Communication*, Advances in Intelligent Systems and Computing 458, DOI 10.1007/978-981-10-2035-3_70

The bullets can be fired by a military robot through its two barrel gun battlement. To a secure firing edge military robot have two cameras in synchronization with the turret which can turn around all the direction [4]. The terrorism problem can be solved by designing the RF based detective military robot which consists of the wireless camera. This robot can softly go through into rival locale and sends the information through wireless camera. Because of the colour sensor used in military robot senses the colour of the surface and according to that robot will change its colour. Hence the robot can be simply escaped by enemies [5].

The hand gesture recognition is developed for identifying the different auction signs in the MEMS based Gesture Controlled Robot. The gesture is based on the hand movement, hence increases the normal technique of intelligence [6]. The military robot includes low power Zigbee wireless sensor network to detect the intruders and the essential action will be taken involuntarily by the robot. The robotic system in the military is intended to save human life and thereby strengthen the country [7]. The existing military and business satellite-based RF communications structure property are both insufficient and costly; a system for global voice and data communications has been established recently to allow the Military Sealift Command (MSC) to achieve its task [8]. In a present multi-complex building utilizing RF tools and RF transmission path gain magnitude were made in classic narrow straight and wide straight interior corridors at 2400 MHz to solve vulnerabilities [9, 10].

As mentioned above, the most part of the work mentioned in the literature has its own particular disadvantages of RF based communication. Furthermore, they are confused, needs more scientific methodologies. Accordingly, to address the above issues, this paper introduces RF communication based robot path planning to fire the bullets automatically in the war field.

The organization of this paper is as follows: Sect. 2 gives the working principle. Next the concept of implementation is given in Sect. 3. Results and discussion are given in Sect. 4 followed by conclusion in Sect. 5.

2 Working Principle

The Block Diagram of RF Communication Module is as shown in the Fig. 1. In which the appropriate task is received by the sensor or by manual operation through the military officer. The received signal is later send to the microcontroller unit after

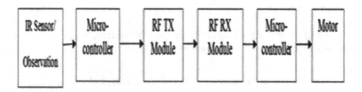

Fig. 1 Block diagram of RF communication module

Fig. 2 Block diagram of
mobile controlled robot

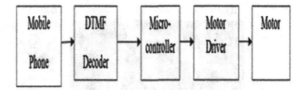

the digital conversion. The suitable command is transmitted by the RF Transmitter to RF Receiver according to a command received by the Microcontroller unit at the transmitter location. The microcontroller unit at the receiver site sends desired signal to the motor through the motor driver depending on the signal received by the RF Receiver Module.

The block diagram of Mobile Controlled Military Robot is as shown in the Fig. 2.

For a short distance robot path planning mobile phone based control is suitable. Depending on the obstacle position, the robot moving direction can be controlled by the key pressed from the mobile phone. The frequency tone generated by a mobile phone after pressing a key is converted into a binary signal by Dual Tone Multi Frequency (DTMF) decoder. The microcontroller receives the binary signal from DTMF decoder and sends appropriate signal to the motor driver. The motor gets proper direction and power to move from motor driver and move accordingly.

3 Implementation

The Fig. 3 shows the hardware implementation of RF transmitter and receiver modules. As shown in Fig. 3a is an RF transmitter on 433 MHz with an antenna and microcontroller unit. The input from the sensor is fed to the microcontroller after digital conversion of the analog signal. Microcontroller processed the digital signal and according to the information in the signal it sends the appropriate signal to the RF receiver module through RF transmitter module. The information is transmitted is being modulated by a RF carrier of frequency 433 MHz.

Figure 3b shows an RF receiver and microcontroller unit. The input from the RF receiver is fed to the microcontroller. Microcontroller sends appropriate signal to the robot for its movement to reach the destination.

Mobile Robot is as shown in the Fig. 4. It consists of DTMF decoder IC CS9370DGP, decoder to the Mobile Phone connector, microcontroller unit with IR sensor. From the mobile phone press '2' for forward, '8' backward, '4' for left, '6' for right and '0' for stop. The frequency tone is generated by a mobile phone after pressing a key is converted into a binary signal by Dual Tone Multi Frequency (DTMF) decoder CS9370DGP and send to microcontroller 89V51RD2. The motor gets the required data from the microcontroller through the motor driver L293D.

(a)   (b)

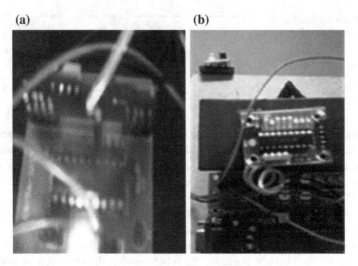

Fig. 3 RF communication for robot

Fig. 4 Mobile robot

4 Result and Discussion

The RF communication prototype is as shown in the Fig. 5.

Fig. 5 RF receiver
communication prototype

4.1 Test Point-1

The RF communication prototype consists of transmitter and receiver. The condition like war or rescue operation more weapons, firing equipment or food and medicines are required towards enemies/rescue premises. To carry these more carriers are required in the same direction. In the RF communication a command sent by a master control unit (command signal can be transmitted by a sensor or manually) can be received by a number of RF receivers mounted on robots.

The signal from the sensor is fed to the microcontroller after converting to digital signal by a comparator. The microcontroller 89V51RD2 processed digitally and generates the output to the RF transmitter unit. ASK modulator uses 433 MHz carriers and an information signal is the output of the microcontroller. The modulated signal is radiated by the helical antenna as shown in the Fig. 5. RF receiver demodulates the modulated signal, later the appropriate signal is applied to the motor of the robot via microcontroller. Hence, many robots which carry the weapons or to fire the bullets or food, medicines in a war field/rescue place are ready to fight.

Mobile controlled Robot Prototype is as shown in Fig. 6.

4.2 Test Point-2

Switch ON the power of microcontroller unit and robot units in the military surveillance area. As mentioned in the TEST POINT-1, to fight with the enemy in the war field for short distances. In which direction firing the bullets and weapons to carry out, is depending on the appropriate key pressed in the mobile phone.

Fig. 6 Mobile controlled
robot prototype

The DTMF converts frequency tone generated by the mobile key to the microcontroller compatible digital signal. Desired direction of the robot is achieved by the motor, by receiving the output signal of the microcontroller.

Thus, from the above two test points, RF communication in the military/rescue operation is implemented and tested for long and short distances. Instead of using many machines/rollers/robots for similar applications individually in the military/rescue operation, these approaches use only one master and many slaves to fight in the war field. Hence, utilizing the power and save the life of soldiers and citizens. The Vision based Robot Prototype is as shown in the Fig. 7.

Figure 8 shows the comparison of Embedded controlled RF Robot and Vision based Robot Path Planning in terms of efficiency and processing time. The accuracy of the Embedded System is 93 % (because identifying obstacle is not accurate) and for Vision based System is 96 % (it depends on camera resolution). The Processing Time of the Embedded System is 1030 ms and for Vision based System Processing Time is 1290 ms (it depends on image processing time).

Fig. 7 Vision based robot
prototype. (*Source* Ref. [2])

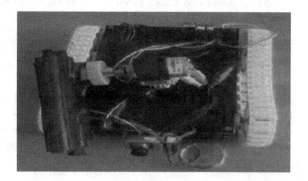

Fig. 8 Comparison of Embedded controlled RF robot and vision based robot

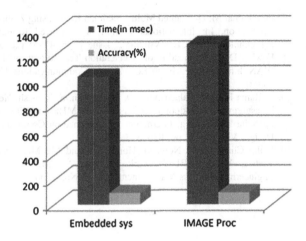

5 Conclusions

This paper illustrated the technique of solving risky task like the continuous firing of bullets; provide weapons to soldiers in the war field, rescue operations in flood, earthquakes etc. All these tasks are solved by Radio Frequency based communication. In this work long distance monitoring of robot path planning is achieved by a Radio Frequency based communication. And for the short distance, robot path is controlled by Mobile phone. The real time results are compared with the vision based robot path planning system.

Acknowledgments Authors highly acknowledge the "Skyfi Labs", Coimbatore, India for providing the lab and training to prepare these robots.

References

1. Alexander Wyglinski, Maziar Nekovee and Thomas Hou: Cognitive Radio Communications and Networks: Principles and Practice. Elsevier Publication (2009)
2. Sandeep Bhat, M. Meenakshi:Vision Based Robotic System for Military Applications – Design and Real Time Validation. In: IEEE Fifth International Conference on Signal and Image Processing (ICSIP), Jeju Island, pp. 20–25 (2014)
3. Paul J. Springer: Security Studies/Science Technology, and Security.: A Reference Handbook, Military Robots and Drones, ABC-CLIO, Volume 1, Pages 297(2013)
4. Naskar, S. Das S., Seth, A.K., Nath, A.: Application of Radio Frequency Controlled Intelligent Military Robot in Defense. In IEEE International Conference on Communication Systems and Network Technologies (CSNT), pp. 396–401(2011)
5. Kalyanee N. Kapadnis,:RF Based Spy Robot. In: Int. Journal of Engineering Research and Applications, Vol. 4, Issue 4, pp. 06–09 (2014)
6. N.V. MaruthiSagar, D.V.R. Sai Manikanta Kumar, N.Geethanjali: MEMS Based Gesture Controlled Robot Using Wireless Communication. In: International Journal of Engineering Trends and Technology (IJETT), Volume 14, Issue 4, pp. 185–188(2014)

7. Premkumar M: Unmanned Multi-functional Robot using Zigbee Adopter Network for Defense Application. In: International Journal of Advanced Research in Computer Engineering &Technology(IJARCET),Volume 2, Issue 1, pp. 47–55 (2013)
8. Fowler, D.G.: Application of Acceleration Technology to Military Sealift Command Afloat WAN Infrastructure. In: IEEE Military Communications Conference, MILCOM pp. 1–7 (2006)
9. T. Rama Rao, D. Balachander, D. Murugesan, S. Ramesh: Near Ground/Floor RF Path Gain Measurements in Indoor Corridors at 2400 MHz for Wireless Sensor Communications. In: Procedia Engineering, International Conference on Communication Technology and System Design, Volume 30, pp. 836–843 (2012)
10. Philip Chan, David Nowicki, Hong Man and Mo Mansouri: Managing Vulnerabilities of Tactical Wireless RF Network Systems: A Case Study. In: International Journal of Engineering Business Management, Vol. 3, No. 4, pp. 22–33(2011)

Author Index

© Springer Science+Business Media Singapore 2017

J.K. Mandal et al. (eds.), *Proceedings of the First International Conference on Intelligent Computing and Communication*, Advances in Intelligent Systems and Computing 458, DOI 10.1007/978-981-10-2035-3

Printed in the United States
By Bookmasters